# 精讲面向软件公司的低代码平台

## 以Oinone为例

陈鹏程 著

清华大学出版社
北京

## 内 容 简 介

本书是 Oinone 开源项目的配套图书，意在系统化地介绍如何基于 Oinone 开源项目，简单快速地开发出高质量的软件系统。全书共分为 7 章，第 1、2 章介绍设计 Oinone 的初衷；第 2～6 章重点面向研发人员，助力研发人员快速上手并做出业务系统；第 7 章面向非研发人员，讲解使用 Oinone 设计器完成对系统的适用性修改，并在可视化设计器不敷用时从低无一体中找到方式，寻求研发帮助。

本书是由创始人担任首席产品体验官潜心编写而成的培训实战课程。研发及非研发业务人员通过学习本书均能快速上手。书中每个细节均经由创始人验收，确保进入 Oinone 生态的每一位伙伴都能获得更好的体验。

版权所有，侵权必究。举报：010-62782989，beiqinquan@tup.tsinghua.edu.cn。

**图书在版编目 (CIP) 数据**

精讲面向软件公司的低代码平台：以 Oinone 为例 / 陈鹏程著．
北京：清华大学出版社，2025.3.-- ISBN 978-7-302-68297-4
Ⅰ．TP311.52
中国国家版本馆 CIP 数据核字第 2025QF3574 号

责任编辑：申美莹
封面设计：杨玉兰
版式设计：方加青
责任校对：徐俊伟
责任印制：刘 菲

出版发行：清华大学出版社
    网　　址：https://www.tup.com.cn，https://www.wqxuetang.com
    地　　址：北京清华大学学研大厦 A 座　　邮　　编：100084
    社 总 机：010-83470000　　邮　　购：010-62786544
    投稿与读者服务：010-62776969，c-service@tup.tsinghua.edu.cn
    质 量 反 馈：010-62772015，zhiliang@tup.tsinghua.edu.cn
印 装 者：河北鹏润印刷有限公司
经　　销：全国新华书店
开　　本：188mm×260mm　　印　　张：38.25　　字　　数：1105 千字
版　　次：2025 年 4 月第 1 版　　印　　次：2025 年 4 月第 1 次印刷
定　　价：169.00 元

产品编号：099406-01

# 序篇

## 庄卓然

从 2009 年加入阿里巴巴至今，经历了"三淘"时期、天猫时期、"双十一"，到最后的 all in 无线手淘时期，几乎赶上了淘系发展的所有历史性事件。在这个过程中，每一次业务的变革都催生了技术的变迁，倒逼着我们用技术的方式去解决业务问题：在存储、I/O、网络等环节满足不了淘系的业务规模时，开始去 IOE，最后演化成了阿里云；当业务的规模大到不能通过简单增加机器的方式去做调整，当开发的规模大到所有人在一起开发会互相影响的时候，我们开始做 SOA 改造，最后演化成了业务中台；在经历了几届"双十一"的巨大挑战后，我们开创了里程碑式的全链路压测；在手淘时代，为了解决动态发版问题，我们植入容器概念，搭建了可动态插拔的三层架构，一年实现了 500 多次的发版；为了同时满足写一套代码就解决多端开发和高并发的性能问题，我们做了 weex，最后还发给了开源社区……

每一次的业务需求推动技术进步，而技术的进步永远会超出我们的想象！

同为技术宅，我在 Oinone 身上能清晰地感受到技术演进的脉络，企业在数字化时代，需要一个能快速上手、全面设计、灵活适应且低成本的技术工具，时代的变迁推动了 Oinone 的诞生。Oinone 是一种全新的开发方式，在数字化时代，Oinone 在提升研发效率上做出的创新性"低无一体"的设计对传统软件代码开发或者无代码开发一定会产生巨大冲击，这种冲击会对软件市场格局造成什么样的变化，我拭目以待。

最后，愿我们这些追光人，在时代的洪流中，都能留下一抹印迹，不辜负时代，不辜负自己。

现任阿里巴巴副总裁，飞猪总裁
曾任阿里大文娱 CTO 兼优酷 COO、淘宝 CTO　庄卓然（南天）

## 陈浩

自 2017 年中国推进数字建设以来，数字经济规模持续增长，"十四五"规划和 2035 年远景目标纲要中明确强调企业和政府需大力推动数字化转型，中国正在迈进一个崭新的数字经济时代。

在这个过程中，软件已经从工具变成信息化的基础设施，如何有效应对该变化所带来的一系列新的核心技术挑战，是整个软件行业发展遇到的另一难题。我认为，开源创新是解决这些难题的有效手段之一，也是未来软件发展的重要方向。如果说，数字化转型是时代趋势，那么开源创新也已成为时代主流。"十四五"规划纲要首提开源，2021 年 11 月工信部印发《"十四五"软件和信息技术服务业发展规划》中提到开源重塑软件发展新生态，并将其作为"十四五"期间我国软件产业的四大发展形势之一进行重点阐述。支持国产化开源创新体系发展，建设自己的开源社区和开源平台，利用开源体系所具有的大众协同、开放共享、持续创新等特点，可有效推动各行业自主可控的数字化转型。

Oinone 所倡导的开源理念和生态共建，与国家开源战略不谋而合：将开源作为一种合作手段，通过完善社区实现开源治理，吸引更多的企业和个体参与其中。湖南大学作为首批国家示范性软件学院的双一流建设高校，一直致力于推进和引导国产化开源软件体系的建设，并为此开展多种形式的产学研研究和实践。基于 Oinone 微服务分布式的设计理念和面向生态的开源特性，湖南大学结合自身在大数据分布式存储、多元异构数据汇聚融合和大数据智能分析等方面的研究成果，与 Oinone 展开了深度的技术创新合作，并在多个大中型企业数字化应用和数字政府应用中取得了良好的效果。

随着 Oinone 的开源，相信能激发更多的开发者参与到国产软件建设中，通过开源模式实现更广泛参与方的共享、共创、共生、共赢，构建价值驱动的数字创新生态平台，为我国数字经济发展贡献科技力量。

湖南大学教授　陈浩

## 李强

我们常说"在今天所有的不确定性当中，数字化是最大的确定性"，数字化一定会全面改造所有的行业更是确定的。在菜鸟九年的探索中，我们最大的感受是"未来，任何一个物流企业都会是一个技术公司，真正拉开差距的是技术与实体产业的结合有多深"。菜鸟"简单极致，贴地疾飞"的技术文化也深刻体现了这一点——好的技术要能解决实际问题。数字化并不是简单地上线一个或几个系统，这是一个贴近业务持续迭代的过程，伴随着这个过程，我相信会诞生非常多的创新技术。

在本书中我看到了工程思维在推进技术创新过程中的缩影，把难的问题转化为简单的问题，用成熟实用的技术分而解之。高性能的微服务框架、CDM、元数据、低代码、无代码等，都是当下非常热门的技术课题，Oinone 把这一切都有机地结合起来，形成了一种具备先进理念的全新一代软件产品，每一个特性都贴合企业数字化遇到的实际问题。Oinone 的产品设计，把"大道至简，软件自造"贯穿始终，用最简单的方式，帮助企业驾驭数字化，相信会给企业带来不一样的体验。

正如本书作者所言，企业视角由内部管理转向业务在线、生态在线（协同）带来一系列新的诉求。这一大背景，以及云、端等新技术的发展，对研发人员的需求越来越大，同时要求越来越高，低代码平台是提升研发效率，降低研发成本的核心手段，低代码已经不是需不需要的问题，而是怎么选的问题。菜鸟网络自身也在推进自有低代码开发平台，我们有幸邀请本书作者陈鹏程来到菜鸟网络进行了分享交流，收获非常大。如您正在选型低代码开发平台，向您推荐这本书，低无一体的 Oinone 肯定会打动您。

菜鸟网络 CTO　李强（在宽）

## 梅丛银

认识陈鹏程及数式核心团队同学已经有一段时间了，在我们多次的交流讨论中时常会谈及：未来中国哪家软件企业能从互联网云原生时代走出来超越传统软件企业？史昂说这是他的梦想，也是他们团队这么多年坚持技术和产品研发与应用优先思考之路。史昂及数式核心团队面向企业应用市场，历经三年的潜心研发和实战交付，推出了 Oinone 产品及配套的低代码平台工具，相比国内外应用软件平台，在开放生态和云原生方面均有它的继承性和独特性，其最大亮点在于，将技术平台赋予各种业务领域属性，便于企业客户和开发伙伴二次开发并快速搭建各类企业核心应用场景。Oinone 的内在特点之一是参考了全球最大开源 ERP Odoo 的元数据模型设计，同时基于业务中台架构和云原生技术，形成了自己的一套国际化快速开发平台、建模规范和应用产品，通过自己进场落地很多品牌企业的应用中台化不断迭代升级,走出了一条具有显著特色的新应用软件之路。

  陈鹏程及数式核心团队行事低调沉稳、善于思考,善于与生态伙伴合作,这是他们能够走得更远更长的基础基因。祝愿 Oinone 能为中国企业在云时代的数字化实践做出更多的贡献,为软件产业构建强大的应用生态和开发社区,真正树立起 Oinone 自己的软件品牌形象。

资深 IT 咨询专家、浩鲸科技云智能专家学院院长 梅丛银

## 蒋江伟

  企业数字化转型经过多年演进,其趋势与价值已经毋庸置疑。近些年,随着流媒体平台的崛起,企业的营销方式、渠道建设方式甚至供应链都面临新的挑战,我们可以清晰地感觉到世界每时每刻都在发生变化。在未来的企业竞争中,谁数字化走在前沿,谁就更能掌握主动权。数字化是为了满足业务的持续创新,只有持续创新才能更好地迎接未知变化。而过去很多企业的技术路径是一个采购型的发展路径,买来的 ERP 和 CRM,升级都是各自管各自的,有一天推出一个新概念或者业务发生新需求,又去采购另外一家企业的 ERP 和 CRM,整个替换掉了,烟囱式的迭代演进。企业不怕重复建设,怕的是不断重复建设,企业不怕系统延期上线,怕的是错过业务发展的机会窗口。

  本书主要介绍了一种全新的数字化构建理念和技术落地方式——用低代码的方式一站式支撑企业的商业场景并能满足商业化持续创新,和其他低代码不同的是:既结合了中台架构,又兼顾了传统企业的 IT 发展水平,更符合企业数字化发展需求,对各行业做数字化选型有很大的帮助。

  很高兴看到阿里校友陈鹏程(本书作者)在这条路上发光发热,也把此书推荐给 IT 从业者、程序员以及爱好计算机应用软件的所有同学,希望对大家学习新型、更高效的系统构建方式有所启发。

阿里巴巴高级研究员 蒋江伟(小邪)

# 目录

开篇　致读者 ································································································· 1
书籍纲领 ······································································································ 1

## 第 1 章　揭开面纱，理解 Oinone ································································ 2
1.1　Oinone 的萌芽 ······················································································ 2
1.2　Oinone 的致敬 ······················································································ 4
　　1.2.1　数字化时代 Oinone 接棒 Odoo ························································ 4
　　1.2.2　Oinone 与 Odoo 的不同之处 ··························································· 4
1.3　Oinone 的生态思考 ················································································ 5
　　1.3.1　与中台的渊源 ············································································· 6
　　1.3.2　找解决方案 ················································································ 6
　　1.3.3　生态建设 ··················································································· 8
1.4　Oinone 与行业对比 ················································································ 9
　　1.4.1　整体视角对比 ············································································· 9
　　1.4.2　从技术角度对比 ·········································································· 9
　　1.4.3　从产品角度对比 ········································································· 10

## 第 2 章　Oinone 的技术独特性 ···································································· 12
2.1　数字化时代软件业的另一个本质变化 ······················································· 12
2.2　互联网架构作为最佳实践为何失效 ··························································· 13
2.3　Oinone 独特性之源，元数据与设计原则 ···················································· 15
2.4　Oinone 的三大独特性 ············································································ 17
　　2.4.1　Oinone 独特性之单体与分布式的灵活切换 ······································· 18
　　2.4.2　Oinone 独特性之每一个需求都可以是一个模块 ································· 20
　　2.4.3　Oinone 独特性之低无一体 ····························································· 23

## 第 3 章　Oinone 入门 ················································································· 28
3.1　环境搭建 ····························································································· 28
　　3.1.1　环境准备（macOS 版）································································ 29
　　3.1.2　环境准备（Windows 版）···························································· 33
3.2　Oinone 以模块为组织 ············································································ 41
　　3.2.1　构建第一个 Module ····································································· 41
　　3.2.2　启动前端工程 ············································································ 52
　　3.2.3　应用中心 ·················································································· 55

## 3.3 Oinone 以模型为驱动 ··············································· 57

- 3.3.1 构建第一个模型 ··············································· 57
- 3.3.2 模型的类型 ··············································· 61
- 3.3.3 模型的数据管理器 ··············································· 72
- 3.3.4 模型的继承 ··············································· 74
- 3.3.5 模型编码生成器 ··············································· 95
- 3.3.6 枚举与数据字典 ··············································· 98
- 3.3.7 字段之序列化方式 ··············································· 107
- 3.3.8 字段类型之基础与复合 ··············································· 110
- 3.3.9 字段类型之关系与引用 ··············································· 114

## 3.4 Oinone 以函数为内在 ··············································· 124

- 3.4.1 构建第一个 Function ··············································· 124
- 3.4.2 函数的开放级别与类型 ··············································· 130
- 3.4.3 函数的相关特性 ··············································· 131

## 3.5 Oinone 以交互为外在 ··············································· 141

- 3.5.1 构建第一个 Menu ··············································· 141
- 3.5.2 构建第一个 View ··············································· 144
- 3.5.3 Action 的类型 ··············································· 163
- 3.5.4 Ux 注解详解 ··············································· 173
- 3.5.5 设计器的结合 ··············································· 174
- 3.5.6 DSL 配置 ··············································· 179
- 3.5.7 前端组件自定义（初级篇） ··············································· 217

# 第 4 章 Oinone 的高级特性 ··············································· 261

## 4.1 后端高级特性 ··············································· 261

- 4.1.1 模块之 yml 文件结构详解 ··············································· 261
- 4.1.2 模块之启动指令 ··············································· 271
- 4.1.3 模块之生命周期 ··············································· 275
- 4.1.4 模块之元数据详解 ··············································· 278
- 4.1.5 模型之持久层配置 ··············································· 280
- 4.1.6 模型之元数据详解 ··············································· 284
- 4.1.7 函数之元数据详解 ··············································· 294
- 4.1.8 函数之事务管理 ··············································· 297
- 4.1.9 函数之元位指令 ··············································· 301
- 4.1.10 函数之触发与定时 ··············································· 304
- 4.1.11 函数之异步执行 ··············································· 314
- 4.1.12 函数之内置函数与表达式 ··············································· 325
- 4.1.13 Action 之校验 ··············································· 336
- 4.1.14 Search 之非存储字段条件 ··············································· 337
- 4.1.15 框架之网关协议 ··············································· 338
- 4.1.16 框架之网关协议——RSQL 及扩展 ··············································· 342

- 4.1.17 框架之网关协议——GraphQL 协议 · 344
- 4.1.18 框架之网关协议——Variables 变量 · 344
- 4.1.19 框架之网关协议——后端占位符 · 346
- 4.1.20 框架之 Session · 347
- 4.1.21 框架之分布式消息 · 350
- 4.1.22 框架之分布式缓存 · 361
- 4.1.23 框架之信息传递 · 362
- 4.1.24 框架之分库分表 · 364
- 4.1.25 框架之搜索引擎 · 370

## 4.2 前端高级特性 · 377
- 4.2.1 组件之生命周期 · 377
- 4.2.2 框架之 MessageHub · 380
- 4.2.3 框架之 SPI 机制 · 384
- 4.2.4 框架之网络请求——HttpClient · 385
- 4.2.5 框架之网络请求——Request · 389
- 4.2.6 框架之网络请求——拦截器 · 392
- 4.2.7 框架之翻译工具 · 399

## 4.3 Oinone 的分布式体验 · 402
## 4.4 Oinone 的分布式体验进阶 · 413
## 4.5 研发辅助 · 418
- 4.5.1 研发辅助之插件——结构性代码 · 418
- 4.5.2 研发辅助之 SQL 优化 · 421

# 第 5 章  Oinone 的 CDM · 423

## 5.1 CDM 的背景介绍 · 423
## 5.2 CDM 之工程模式 · 425
## 5.3 基础支撑之用户与客户域 · 427
## 5.4 基础支撑之商业关系域 · 434
## 5.5 基础支撑之结算域 · 437
## 5.6 商业支撑之商品域 · 439
## 5.7 商业支撑之库存域 · 442
## 5.8 商业支撑之执行域 · 443

# 第 6 章  Oinone 的通用能力 · 446

## 6.1 文件与导入 / 导出 · 446
## 6.2 集成平台 · 458
## 6.3 数据审计 · 491
## 6.4 国际化之多语言 · 494
## 6.5 权限体系 · 498
## 6.6 消息 · 504

# 第 7 章　Oinone 的设计器 …… 511

## 7.1　设计器总览 …… 511
## 7.2　实战训练（积分发放） …… 517
## 7.3　实战训练（全员营销为例） …… 573
### 7.3.1　去除资源上传大小限制 …… 573
### 7.3.2　原业务加审批流程 …… 582
## 7.4　Oinone 的低无一体 …… 596

# 附录 A　下载说明 …… 599

# 开篇
# 致读者

欢迎来到 Oinone 生态，我们为您提供了一站式低代码商业支撑平台，数式科技已经用其服务了如中烟、得力、上海电气、中航金网、雾芯科技等多个知名企业，我们的技术实力在商业场景得到了很好的验证。我们希望通过开源 Oinone 项目，为中国软件行业带来变革，提升整体工程化水平，与广大软件工程师一起为客户创造价值！《精讲面向软件公司的低代码平台——以 Oinone 为例》一书是 Oinone 开源项目的配套书籍，系统化地介绍了如何基于 Oinone 开源项目，快速开发高质量的软件系统。此外，如果您开发的项目是用于商业化而非企业自用，我们还为您提供了一种可选的商业化变现途径：申请成为 Oinone 的合作伙伴，并提交相关产品，详情请访问 www.oinone.top 网站。

## 书籍纲领

本书的章节安排如下：

第 1 章至第 2 章："揭开面纱，理解 Oinone"和"Oinone 的技术独特性"。这两个章节可以帮助您更好地理解我们设计 Oinone 的初衷以及特性的由来。

第 3 章：面向研发人员的"Oinone 的基础入门"。如果您是专业的研发人员，本章可以帮助您快速上手并做出业务系统。只要按着里面的 case 一步步操作下来就可以。

第 4 章至第 6 章：面向研发人员的"Oinone 的高级特性""Oinone 的 CDM""Oinone 的通用能力"。这三篇章节重点介绍了 Oinone 的高级技术特性、提供的通用数据模型和通用基础能力。它们能够帮助我们更快地进行业务开发，从容应对业务的特殊场景要求，比较适合进阶的研发人员。

第 7 章：面向非研发人员的"Oinone 的设计器们"和"Oinone 的低无一体"。如果您并不是专业的研发人员，本章可以帮助您通过使用 Oinone 的无代码可视化设计器轻松自主解决业务需求，并且当可视化设计器满足不了的时候，您还可以在"Oinone 的低无一体"中找到方式，并寻求研发帮助。

# 第1章　揭开面纱，理解 Oinone

本章从以下几个维度逐步揭开 Oinone 的面纱，介绍 Oinone 的初心与愿景，以及它是如何站在软件领域的巨人肩膀上，结合企业数字化转型的趋势，形成全新的理念，帮助企业完成数字化转型。

具体来说，本章会从以下四个方面逐一展开：
1. 中国软件行业的发展和自身职业发展经历，以此谈论 Oinone 的初心和愿景。
2. Oinone 向西方软件业新贵 Odoo 致敬。
3. 从企业数字化转型的困境出发，阐述 Oinone 的全新思路。
4. 通过行业对比，让读者从不同的视角来理解 Oinone。

## 1.1　Oinone 的萌芽

在信息化时代，中国并没有涌现出一家世界知名的软件公司。这是因为像 SAP、Oracle、IBM、Salesforce、NetSuite、Odoo 等西方巨头所拥有的最佳实践在业务、技术和模式方面，给予了它们在企业信息化建设中高额利润的优势。中国软件业在这个时代的角色是学习和追随者，而最优秀的追随者是金蝶和用友，它们能在国家推行会计电算化的机遇中占据领先地位。但是，追随者始终只是追随者，没有真正的突破。

我自己进入软件行业的经历可以追溯到 2015 年，当时资本市场非常热门，大家都在创业。我认为这是一个时代的机会，就像国家改革开放一样，于是，我和很多同事一起开始了创业之旅。在数式之前，我加入并创办了三家公司：500mi、数列和端点，这个过程给了我宝贵的经验和启示，帮助我找到了最终想要的方向。

在 500mi 公司时，我从技术岗位转型为业务经营，起步并不顺利。然而，我从这份经历中学到了重要的一课：做自己擅长的事情，有助于渡过创业启动期最艰难的阶段。同时，市场调研为我提供了一个信号：传统企业对于 IT 的需求正逐渐向互联网靠拢。这个信号像注入了一剂强心剂，激励我继续前行。

2016 年，我和三个曾在阿里工作的同事一起创办了一家新公司——数列，我们决定专注于我们最擅长的领域，即软件服务商。在没有任何商务资源的情况下，我们第一年就签订了 1000 多万的合同，这相较之前是一个非常成功的开端。然而，对于公司未来的发展方向，我们花费了长达大半年的时间进行思考：应该坚持做底层的 PaaS 还是专注于企业可见的上层应用和业务产品？我倾向于后者。尽管我们持续存在分歧，但凭借着多年的革命友情，最终我们友好地分道扬镳。数列此前的成功让我更加坚信：在数字化时代，软件需求将会有井喷式的增长，数字化软件服务将是未来 5～10 年的重要方向。而在这个领域，专业的技能将是应对未来不确定性的真正力量。

提到数字化，就不得不提阿里巴巴的中台理念。中台理念在 2008 年被阿里巴巴提出，当时引起了广泛的关注和讨论。企业之所以认同中台理念，是因为他们的核心需求已经从内部转向外部：从关注管理、流程、效率的提升，转向关注外部协同、运营、创新。他们已经不再只担心企业的效率和成本，而是担心自己是否有能力跟上时代的快速变化。现今做生意的渠道已经不再是单一的线下渠道，而是包括淘宝、天猫、京东、拼多多、抖音、快手等的线上渠道，以及海外市场，这种变化速度非常快。而中台的核心理念是敏捷响应、低成本快速创新，正好解决了企业主的核心焦虑。

企业的视角正在从内部管理向业务在线和生态在线（协同）转变，这种转变带来了一系列新的需求（如图 1-1 所示）。这种转变不仅是为了支持现有业务的发展，也为企业未来的业务发展和创新提供了支持，并将变化实时反映到上下游合作伙伴中。

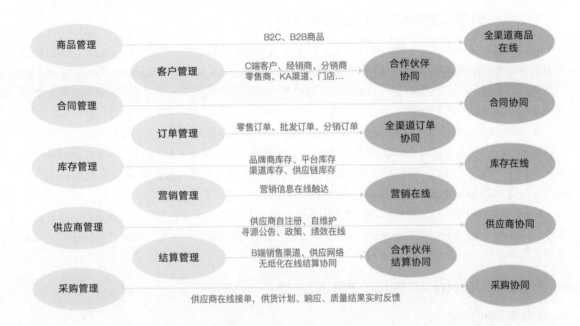

图 1-1　企业视角转变带来一系列新的诉求

在 2017 年下半年，阿里云收购了端点科技，打算重启阿里软件。那个时候，市场上涌现出一批中台厂商，整个行业也比较混乱，很多人对互联网架构本身的理解不够深入，快速学习拿到阿里云认证后就开始做定制化的中台架构开发，但最终的效果无法达到预期。因此，阿里云和端点科技的联姻是为了弥补阿里云没有向外输出上层应用产品能力的缺陷。多年来，软件市场一直被国外厂商掌控，中国一直缺乏一个强大的本土软件公司。阿里收购端点，承载着无数中国人的软件梦想，在这种背景下，我回到了阿里体系，加入了端点科技。后来，我参与了许多中台项目，深刻地认识到通过搭建中台技术架构和一些基础能力，上层应用场景落地并不难。但是，当客户接手扩展中台能力和新的上层应用场景时，效果往往不尽如人意，这并不是中台架构理念的问题，而是因为传统企业客户的 IT 能力大多较弱，这是一个硬伤。许多文章都在讲述中台战略，长篇大论地描述组织中台、技术中台、业务中台、数据中台，我们不去评论这些方法论的对错，从技术角度回到初衷，我们只关注一个问题：技术是为商业服务的，中台如何快速满足企业业务多变的需求？

我们经历了多个行业的中台建设，每次都向客户强调第一阶段是打好基础，因此需要较长的周期，并且每个项目都需要顶级架构师来把控整体项目。如何找到互联网架构与传统软件良好结合点，降低对组织的要求，实现中台架构的标准化输出？这是我回归阿里后致力于解决的问题。然而，随着阿里云对端点战略发展思路的变化，阿里不再提供 SaaS 服务，而只愿意做平台，被其他企业集成。因此，我离开了端点，并决定把自己的技术思考转化为现实，于是数式科技诞生了。

在数字化时代，无论是业务、技术还是商业模式的最佳实践，都源自中国。中国已经从追随者转变为互联网领域的全面引领者。我们有理由相信，中国一定会崛起一家世界级的软件公司，而 Oinone 将始终以此为愿景。

## 1.2 Oinone 的致敬

站在巨人的肩膀上，天地孤影任我行。

### 1.2.1 数字化时代 Oinone 接棒 Odoo

在数字化时代，中国在互联网化的应用、技术的领先毋庸置疑，但在软件的工程化、产品化输出方面仍有许多改进的空间。这时，我了解到了 Odoo——一个国外非常优秀的开源 ERP 厂商，全球 ERP 用户数量排名第一，服务全球客户。Odoo 的工程化能力和商业模式深深吸引了我，它是软件行业典型的产品制胜和长期主义者的胜利之一。

在 2019 年，也就是数式刚成立的时候，我们跟很多投资人聊起公司的对标，数式不是要成为数字化时代的 SAP，而是要成为 Odoo。然而，当时大部分国内投资人并不了解 Odoo，尽管它已经是全球最大的 ERP 厂商之一，因为当时 Odoo 还没有明确的估值。直到 2021 年 7 月份获得 Summit Partners 的 2.15 亿美元投资后，Odoo 才正式成为 IT 独角兽企业。

Odoo 对我们提供了极大的启示，因此我们致敬 Odoo，同样选择开源，每年对产品进行升级发布。如今，Odoo15 已经发布，而 Oinone 也已推出第三版，恰好相隔 12 年，这是一个时代的接棒，从信息化升迁至数字化。

### 1.2.2 Oinone 与 Odoo 的不同之处

**1. 技术方面的不同**

在技术上，Oinone 和 Odoo 有相同之处，也有不同之处。它们都是基于元数据驱动的软件系统，但是它们在如何让元数据运作的机制上存在巨大差异。Odoo 是企业管理场景的单体应用，而 Oinone 则致力于企业商业场景的云原生应用。因此，它们在技术栈的选择、前后端协议设计、架构设计等方面存在差异。

**2. 场景方面的不同**

在场景上，Oinone 和 Odoo 呈现许多差异。相对于 SAP 这些老牌 ERP 厂商，Odoo 算是西方在企业级软件领域的后起之秀，其软件构建方式、开源模式和管理理念在国外取得了非凡的成就。然而，在国内，Odoo 并没有那么成功或者并没有那么知名。国内做 Odoo 的伙伴普遍认为，Odoo 与中国用户的交互风格不符，收费模式、设计以及外汇管制使商业活动受到限制，本地化服务不到位，国内生态没有形成合力，伙伴们交流合作都非常少。另外，Odoo 在场景方面主要围绕内部流程管理，与国内老牌 ERP 如用友、金蝶重叠，市场竞争激烈。相比之下，Oinone 看准了企业视角由内部管理转向业务在线、生态在线（协同）带来的新变化，聚焦新场景，利用云、端等新技术的发展，从企业内外部协同入手，以业务在线驱动企业管理流程升级。它先立足于国内，做好国内生态服务，再着眼未来的国际化。

**3. 无代码设计器的定位**

Odoo 的无代码设计器是一个非常轻量的辅助工具，因为在 ERP 场景下，一个企业实施完以后，基本几年不会变，流程稳定度非常高。与之相反，Oinone 为了适应企业业务在线化后，所有的业务变化与创新都需要通过系统来触达上下游，从而实现敏捷响应和快速创新的需求，重点打造了五大设计器（如图 1-2 所示）。

在数字化时代中国软件将接棒世界，而 Oinone 也要接棒 Odoo，把数字化业务与技术的最佳实践赋能给企业，帮助企业数字化转型不走弯路！

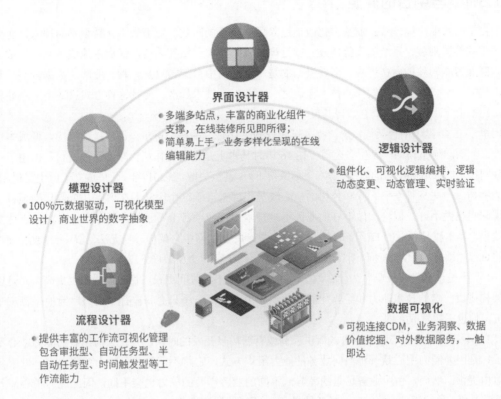

图 1-2　Oinone 五大设计器

## 1.3　Oinone 的生态思考

Oinone 致力于以"企业级软件生态"的方式去帮助企业建立"一站式的商业智能软件"。

通过观察从信息化到数字化的软件行业发展历程（如图 1-3 所示），我们可以发现，企业真正需要的是一站式的软件产品。然而，一站式的软件产品往往都是从单个领域的需求满足开始，在信息化时代和数字化时代都是如此。在信息化时代，以 ERP 为终点的一站式趋势逐渐形成；而在数字化时代，中台概念的提出则标志着一站式的趋势重新开始。本文将从企业数字化转型所临的困境出发，探讨 Oinone 的生态思考。

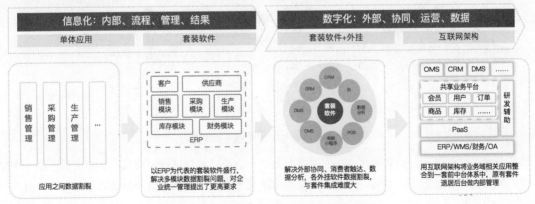

图 1-3　软件行业从信息化到数字化发展历程

### 1.3.1 与中台的渊源

中台概念的提出标志着企业数字化改造进入了一个新的时代。随着数字化转型不断深入，企业面临着严重的数据割裂、系统隔离等问题。在这样的背景下，"敏捷响应，低成本地快速创新"成为了一站式商业智能软件的内在诉求。需要澄清的是，互联网中台架构只是一种企业解决数据割裂、系统隔离，建立一站式商业智能软件的技术概念之一，并不是技术标准。而且这种方式只适用于企业自建模式，在多供应商环境下，则会适得其反，导致建立更复杂的烟囱系统。

阿里于15年提出中台架构概念，抓住了企业数字化转型的核心诉求，即"敏捷响应，低成本快速创新"。然而，阿里作为一家生态公司，在16年时基本上是带着合作伙伴来给企业交付，但由于伙伴对互联网技术的理解和能力的限制，基本上都做得不好，甚至失败。在2017年，阿里成立了原生交付团队，希望能够树立一些标杆案例。我和公司的核心成员也都来自于这个团队。在做完几个客户的项目后，我发现阿里也做不好，但这次做不好的原因不是技术不行或项目上不了线，而是上线以后没有达到预期的效果，其本质是企业的IT组织能力无法驾驭复杂的互联网中台架构。当无法驾驭中台架构的时候，所谓的目标"敏捷响应，快速创新"就无从说起了。结果客户会反馈以下三类问题：

（1）不是说敏捷响应吗？为什么改个需求这么慢，不但时间更长，付出的成本也更高了？是因为中台架构需要一定的技术能力和经验才能有效地应用，就像一个只会骑自行车的人，给他一辆汽车或者飞机，他也不能驾驭它们。

（2）不是说能力中心吗？当引入新供应商或有新场景开发的时候，为什么前期做的能力中心不能支撑了？是因为能力中心是一种面向业务的能力组织方式，它将不同的业务能力抽象出来，以服务的形式对内提供。然而，由于业务场景的差异，不同的业务需要的能力也会不同，因此能力中心需要不断迭代和升级。对于新引入的供应商或新场景开发，需要根据实际情况对能力中心进行定制化和扩展化，但谁来负责呢？新项目的供应商还是客户自己？

（3）不是说性能好吗？为什么我投入的物理资源更多了？是因为中台架构采用微服务来解决单点瓶颈问题，提高系统性能和可用性，但是在初始阶段，投入的资源可能会更多。每个模块至少需要两个实例来保障高可用性，因此物理资源的投入量可能会比以前更多。

### 1.3.2 找解决方案

在考虑解决方案之前，我们需要思考企业数字化软件的最终状态将是什么样子。目前有两种主要的方案（如图1-4所示）。

图1-4　企业数字化的桎梏和图圄

第一种是以自建研发团队为核心。中国的大型企业已经开始尝试这种模式，看起来似乎是一个时下比较流行的可行性方案。然而，绝大多数企业由于成本、人才团队等原因而难以坚持下去，只能与供应商合作开发。

第二种是以供应商为核心。由于大多数企业无法选择第一种路径，他们必须接受目前分散的情况，并通过系统集成尽可能拉通各个系统。尽管如此，在数字化时代中，真正意义上的一站式商业智能软件供应商还未出现。

对企业来说，这两种方案都非常艰难，但在大规模数字化历程中又不得不做出选择。此外，我们还能清晰看到以下几点：

（1）"敏捷响应，低成本地快速创新"成为企业推行一站式商业智能软件的内在诉求。

（2）目前没有一家软件供应商能满足企业所有外围商业场景，也不可能有这样的供应商。

（3）绝大部分企业需要软件供应商，而不是自建软件。

如何突破这种局面也成为中国软件行业发展的一个机遇。强化第一种路线只能服务于极少数头部企业，因此我们更希望优化第二种路线并形成全新的模式。因此，我的思考是：

（1）我们的目标不是让企业学会复杂的互联网架构，而是降低互联网架构的门槛，让更多企业真正拥有"敏捷响应，低成本快速创新"的能力。

（2）我们的目标不是输出中台方法论，而是提供中台建设的技术平台。

（3）我们的目标不是只服务大企业，而是真正赋能不同 IT 组织能力的企业，让它们都具备持续创新的能力。

今天，许多中台软件公司告诉企业："中台是持续演进和快速迭代的过程，因此企业需要组建中台架构团队来实现项目落地，而他们则通过中台项目落地将中台建设方法论传授给企业。"这句话的前半部分是正确的，因为我们之前提到，企业需要具备敏捷响应业务的能力，即应变能力，因为应变是不断变化的。然而，后半部分是不正确的，因为今天的企业已经有能力组建团队，那么这些中台软件公司到底有什么用呢？企业真的缺少方法论吗？在 2019 年，我就提出了自己的看法：没有低代码能力的中台公司都在收取智商税，都在欺诈，因为很多企业根本找不到足够懂互联网架构的人才。明白流氓在哪里了吗？这些流氓公司赚了很多钱，最后责怪企业无法招到人才，这是企业的责任。因此，我仍然认为"最好的赋能是降低门槛，而不是让客户提高技术水平"。

最终，我们得出了一个服务模式的想法：构建企业级的软件生态。企业级软件生态的确切定义是：通过开放的方式，让企业本身以及不同的软件供应商共同参与，遵循相同的技术和数据规范，打造一体化、无须集成的各类企业级软件。如果要打造企业级软件生态，我们列出了六个要点（如图1-5所示）。

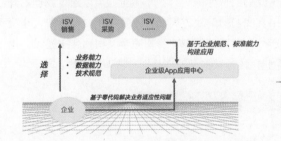

图 1-5　打造企业级软件生态需要具备的六大能力

我很幸运地有机会通过"企业级软件生态"的方式，为企业建立"一站式的商业支持平台"提供帮助。我们的 Oinone 平台结合了低代码开发、通用数据模型和业务产品的优势（如图1-6所示）。

我们对 Oinone 一站式低代码商业支撑平台展开介绍，它大致分为 4 部分：

（1）以低代码开发平台为基础，输出具备互联网架构的软件快速开发标准。这可以帮助企业快速构建符合互联网架构标准的应用程序，从而实现快速响应和低成本创新。

（2）以通用数据模型为基础，满足不同软件基于同一套数据标准的扩展能力。这可以确保不同软件系统之间的数据兼容性和互操作性，避免数据孤岛和信息隔离。

（3）在业务产品层面上，企业和伙伴基于相同的技术标准和数据标准共同提供解决方案。这可以帮助企业和伙伴共同开发出符合标准的商业支撑平台，以提高业务效率和创新能力。

（4）最后是无代码设计器，用于满足项目开展中，超出业务标品范围之外的需求，或者针对标品的临时需求。这可以帮助业务人员在没有专业软件支持的情况下，自主解决业务需求，并支持部门间的协同工作。

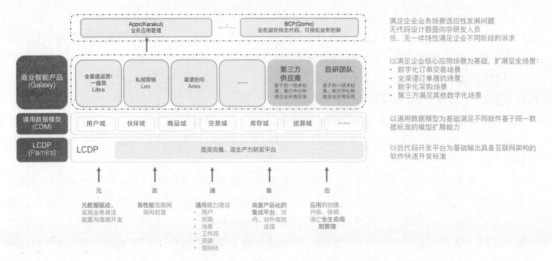

图 1-6　Oinone= 低代码开发平台＋通用数据模型＋业务产品

## 1.3.3　生态建设

Oinone 致力于打造全球最大的无须集成的商业应用程序及其生态系统，通过开源内核、汇集数千名开发人员和业务专家，为企业提供成本效益、一体化、模块化的解决方案，解决所有商业需求，让不同技术之间的合作变得简单易行，摆脱令人烦恼的集成问题。

在客户和场景领域，我们严格限定了自身的专注领域，如图 1-7 所示。针对超大型头部企业，我们专注于树立标杆；而对于大、中、小型企业，则交由我们的伙伴来支持；小微企业可以通过我们的开源社区版获得覆盖。在企业数字化转型的核心领域中，我们的解决方案涵盖了数字化交易场景、全渠道订单履约场景、数字化采购场景、数字化营销等产品。在其他领域，我们完全交由伙伴来建设。由于我们自身在企业协同商务领域拥有深厚的背景，因此在该领域提供的产品拥有特别的优势。

图 1-7　企业数字化转型核心领域

## 1.4 Oinone 与行业对比

随着企业数字化转型的推进，软件公司获得了许多机会。尽管竞争日趋激烈，但由于需求旺盛，各种模式仍在不断涌现。因此，当前市场上存在各种各样的数字化转型解决方案，围绕企业的各个方面展开，每种解决方案都有其优点和缺点。本文将从定位、技术和产品等方面对 Oinone 和其他数字化转型解决方案进行简单比较，帮助您从不同的视角了解 Oinone 的差异。

### 1.4.1 整体视角对比

**1. 与对标公司 Odoo 的对比**

Oinone 与对标公司 Odoo 的对比如表 1-1 所示。

表 1-1　Oinone 与对标公司 Odoo 的对比

|  | Odoo | Oinone |
| --- | --- | --- |
| 定位 | 一站式全业务链管理平台，赋能企业信息化升级 | 一站式低代码商业支撑平台，赋能企业数字化升级 |
| 需求变化 | 关注单一企业的管理、流程、效率的提升 | 关注企业价值链的网络竞争，围绕外部协同、运营、数据、商业展开 |
| 技术更替 | 关注稳定、安全、功能丰富度 | 除了稳定、安全、功能丰富度以外，更强调需求响应速度、用户体验、系统承载极限与弹性扩展、智能化 |

**2. 与国内低代码或无代码公司对比**

Oinone 与国内低代码 / 无代码公司对比如表 1-2 所示。

表 1-2　Oinone 与国内低代码 / 无代码公司对比

|  | 低代码 / 无代码公司 | Oinone |
| --- | --- | --- |
| 定位 | 低代码开发工具，提供各类系统模板，基于模板快速搭建和个性化配置，但系统模板无法再升级 | 平台型 SaaS，提供各类系统产品，产品安装后客户可以根据需求进行个性化调整，同时产品永远在线可升级 |
| 场景差异 | 只能支持企业内部人员使用，以完成部门级边缘系统为主，一般没有专业软件厂商支撑，具有强临时性特性 | 从内外部协同的商业场景出发，关注企业核心业务场景，适应"企业业务在线化后，所有的业务变化与创新都需要通过系统来触达上下游"的时代背景，以敏捷响应业务的变化与创新为目标 |
| 技术代差 | 单表支撑 100 万数据已是业内天花板 | 支撑单模型数据过亿，无单点瓶颈，封装互联网架构并且做到单体与分布式的灵活部署，为不同大小公司提供不同技术支撑 |

### 1.4.2 从技术角度对比

我们不会与其他无代码平台进行比较，因为它们不能解决业务复杂性的问题。相反，我们将重点介绍三种不同的低代码平台模式（如图 1-8 所示）。

第一种模式是最基础的低代码平台，也被称为代码生成器。它通过预定义应用程序模板和必要的配置生成代码，简化了工程搭建并提供了一些基础逻辑。虽然在信息化时代，内部流程标准化方面较为适合，但在数字化时代外部协同业务在线的情况下就不那么合适了。因为这种模式不能减少研发难度和提高效率，也无法体现敏捷迭代快速创新的优势。

第二种模式是经典的低代码平台，以元数据为基础，以模型为驱动。当无法满足需要时，通过特定方式将代码以插件的形式注入平台，作为低代码平台的内置逻辑，供设计器使用。它的优点在于降低了研发门槛，当无法满足需求时才需要编写代码。它可以实现企业内部的复杂流程和复杂逻辑，但其性能和工程管理存在局限性。性能问题使其不适合处理互联网化的在线业务，而工程管理问题则使其不适合处理快速变化的业务。这也是许多研发人员反对低代码的核心原因之一，因为研发人员变成了辅助角色，而软件工程是一门需要技术能力的学科，让没有技术能力的人主导是违反常理的。对于软件产品公司来说，产品需要迭代规划，需要多人协作，需要工程化管理。

第三种模式是 Oinone 提出的基于互联网架构的低代码平台，它采用低无一体的设计。首先，Oinone 屏蔽了互联网架构带来的复杂性。其次，同样以元数据为基础，以模型为驱动，但是元数据的生成方式有两种：一种是使用无代码设计器（与经典低代码相同），另一种是通过代码来描述元数据。通过使用代码来描述元数据，可以无缝地与代码衔接，并在不改变研发习惯的情况下降低门槛、提高效率，并进行工程化管理。

最后总结来说：低无一体不仅仅是指两种模式的结合，还包括两种模式的融合应用方式。具体来说，这种融合应用方式可以分为两种情况：

（1）当开发核心产品时，主要采用低代码开发，无代码设计器作为辅助。这种方式可以提高开发效率和代码质量，同时保证产品的快速迭代和升级。

（2）当需要满足个性化或非产品支持的需求时，主要采用无代码设计器，低代码作为辅助。这种方式可以快速地满足客户需求，并且避免对产品的核心代码产生影响。

简单来说，低代码模式适用于产品的迭代升级，而无代码设计器则适用于满足个性化和非产品支撑的额外需求。低代码和无代码模式在整个软件生命周期中都有各自的价值，在不同场景下可以相互融合，发挥最大的优势。

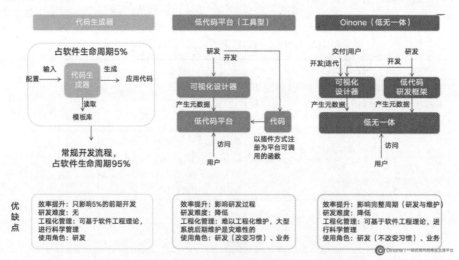

图1-8　代码生成器、低代码平台与 Oinone 的优缺点对比

### 1.4.3　从产品角度对比

产品上的对比，从客户、满足度、销售三个方面来做简易的对比。

**1. Oinone vs 数字化软件服务商**

Oinone 与数字化软件服务商对比如表1-3所示。

表 1-3　Oinone 与数字化软件服务商对比

| | 客户 | 满足度 | 销售 |
|---|---|---|---|
| Oinone | 一站式商业智能软件，更高性价比<br>客户范围：5000 万～5 亿、5 亿～100 亿；<br>标杆：100 亿～1000 亿、1000 亿以上 | 满足企业核心业务需求，并联合伙伴一起满足企业所有需求，无须集成提供统一工作台、数据接口、底层协议，无论基于 Oinone 的开源框架还是增加其他应用都有很好的扩展性 | 支持 OP 和 SaaS 两种模式，收费方式不同：OP 按买断方式进行，SaaS 按效果付费跟账号数无关，新的模块可进行二次销售 |
| 数字化软件服务商 | 针对成熟的大型企业，需投入巨大资源和成本<br>客户范围：100 亿～1000 亿、1000 亿以上 | 满足企业部分需求，无法输出技术标准，无法解决多供应商一起开发的问题，只能通过集成实现对接 | OP 模式进行销售，通过设置权限来进行二次销售或无法进行二次销售 |

**2. Oinone vs 低代码或无代码行业**

Oinone 与低代码或无代码行业对比如表 1-4 所示。

表 1-4　Oinone 与低代码或无代码行业对比

| | 客户 | 满足度 | 销售 |
|---|---|---|---|
| Oinone | 一站式商业智能软件<br>客户范围：5000 万～5 亿、5 亿～100 亿；<br>标杆：100 亿～1000 亿、1000 亿以上 | 从外部商业场景出发，强业务场景驱动，符合企业从信息化管理到业务创新的数字化转变的趋势。提供统一工作台、数据接口、底层协议，无论基于 Oinone 的开源框架还是增加其他应用都有很好的扩展性 | 支持 OP 和 SaaS 两种模式，收费方式不同：OP 按买断方式进行，SaaS 按效果付费跟账号数无关，新的模块可进行二次销售 |
| 低代码或无代码公司 | 针对小微企业内部信息化管理诉求，以表单流程为主<br>客户范围：5 亿以下 | 满足企业部门级信息化的适应性需求，无法满足企业核心业务管理与业务创新诉求 | 按应用模块进行收费，新的模块可进行二次销售 |

**3. Oinone vs 国外对标公司 Odoo**

Oinone 与国外对标公司 Odoo 对比如表 1-5 所示。

表 1-5　Oinone 与国外对标公司 Odoo 对比

| | 客户 | 满足度 | 销售 |
|---|---|---|---|
| Oinone | 一站式商业智能软件<br>客户范围：5000 万～5 亿、5 亿～100 亿；<br>标杆：100 亿～1000 亿、1000 亿以上 | 从外部商业场景出发，强业务场景驱动，符合企业从信息化管理到业务创新的数字化转变的趋势。基线产品覆盖：采购、营销、服务、销售、交易等企业商业领域。主要涉及行业：零售品牌。其他领域或行业靠合作伙伴共建方式进行 | 支持 OP 和 SaaS 两种模式，收费方式不同：OP 按买断方式进行，SaaS 按效果付费跟账号数无关，新的模块可进行二次销售 |
| Odoo | 一站式企业管理软件<br>客户范围：5000 万～5 亿、5 亿～100 亿；<br>标杆：100 亿～1000 亿、1000 亿以上 | 从企业内部管理需求出发，逐渐拥有互联网相关应用组件，但还是属于强内部管理、弱外部场景。<br>基线产品覆盖：业务财务一体化、人财务、进销存。主要涉及行业：建造业。其他领域或行业靠合作伙伴共建方式进行 | 支持 OP 和 SaaS 两种模式，收费方式相同：<br>按用户数+应用模块进行收费<br>新的模块可进行二次销售 |

# 第 2 章　Oinone 的技术独特性

本章的主要目的是通过分析企业商业支撑软件的项目特性和关注点，找到企业软件发展的另一个本质变化——新技术流派的产生。在对"互联网架构作为最佳实践为何失效"的思考基础上，我们分析互联网中台架构的发展历史以及企业实际现状，找出其水土不服的原因。进而引出 Oinone 的低代码开发平台如何结合互联网架构并完成创新，以满足企业数字化转型的需求。

具体而言，本章包括以下内容：

1. 企业软件发展的另一个本质变化：新技术流派的产生；
2. 最佳实践为何失效？Oinone 如何打造具有企业特色的互联网架构；
3. Oinone 独特性之源：元数据与设计原则；
4. Oinone 独特性之单体与分布式的灵活切换；
5. Oinone 独特性之每一个需求都是一个模块；
6. Oinone 独特性之低无一体。

## 2.1　数字化时代软件业的另一个本质变化

随着企业从信息化向数字化转变，软件公司提供的产品也由传统的企业管理软件向企业商业支撑软件发展，如图 2-1 所示。这一变化带来了许多技术上的挑战和机遇。在之前的章节中，我们提到企业的视角已经从内部管理转向业务在线和生态在线协同，这也带来了一系列新的需求。但是，我们常常会忽视这一变化所带来的对系统要求的变化。在本章中，我们将探讨这些技术上的变化，以及这些变化所带来的机遇和挑战。

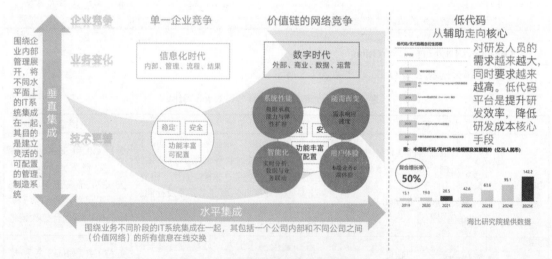

图 2-1　从信息化到数字化软件本质变化

在信息化时代，企业的业务围绕着内部管理效率展开，借鉴国外优秀的管理经验，企业将其管理流程固化下来，典型的例子是 ERP 项目。这类项目上线后往往长期稳定，不轻易更改，因此信息化时代软件的技术流派侧重于通过模型对业务进行全面支持。例如，SAP 具有丰富的配置能力，将已有企业管理思想抽象到极致，其功能基本上可以通过配置来实现，因此其模型设计特别复杂。但是，我们也应该清楚地了解到，配置是面向已知问题的。在数字化时代，创新和业务迭代速度非常快，这种方法可能就不太适合了，因为，模型抽象是在设计时具有前瞻性的，一旦不适合，修改起来就会异常困难。

随着数字化时代的到来，企业主的关注点已经从单一企业内部管理转变为了围绕企业上下游价值链的协同展开。这种变化给企业信息化系统提出了更高的要求，例如业务需求的响应速度、系统性能和用户体验等方面。现在，企业对软件的需求不仅是管理需求的承载，更是业务在线化的承载。传统的重模型设计软件模式已经不再适用，因为业务本身不断创新和变化。因此，数字化时代需要新的软件技术流派，这种流派必须是轻模型加上低代码技术的结合体。通过模型抽象实现 80% 的通用场景，剩余的 20% 个性化需求可以通过技术手段来完成。这样的设计可以让每家企业的研发人员轻松理解模型，而不像 ERP 模型那样异常复杂，无法进行修改。此外，配合低代码技术可以快速研发和上线模型。如果说配置化是面向已知问题的，那么低代码就是面向未知问题设计的。虽然低代码的概念可以追溯到 20 世纪 80 年代，当时是为了满足企业内部部门之间有协同需求，但又没有专业软件支撑，定制化开发又不划算的辅助场景。但现在它的核心原因是企业数字化的核心场景不稳定，变化很快，每家企业都有强烈的个性化需求。数字化时代的低代码需要具备处理复杂场景的能力，而不仅仅是围绕着内部管理展开。

企业在数字化转型的过程中需要考虑到的不仅有成熟的全链路业务解决方案，还要应对数字化场景的快速变化和持续创新的需求。为此，Oinone 打造了一站式低代码商业支撑平台，从业务与技术两个维度来帮助企业建立开放、链接、安全的数字化平台。这将在水平和垂直两个维度上全面推动企业数字化转型。

另外，低代码的另一个好处是完成了软件本身的数字化建设。通过基于元数据进行设计，元数据成为软件中逻辑和交互的数据。想象一下，AI 了解软件的元数据后可以自我运作，人在极少情况下才需要参与，人机交互也会发生大的改变，使软件和 AI 结合，创造更多可能性。未来的软件交互不再需要研发提前预设，而是能够实现用户所需即所呈现的效果。

## 2.2 互联网架构作为最佳实践为何失效

随着业务和生态的发展，企业对效率、性能、体验和智能化等方面的要求越来越高，但很多企业的系统面临着严重的系统架构落后和系统间割裂等问题，这些问题导致原有系统在业务发展下面临着效率和性能的双重挑战。与此同时，互联网平台的技术水平远远领先于传统企业系统，但是是否可以直接将互联网架构照搬到企业数字化转型中呢？显然，这是不合适的，因为互联网架构在企业数字化转型中面临着许多水土不服的问题。本节将结合互联网中台架构的发展，分析这些问题的原因。

● 借鉴互联网中台理念

我们要先看互联网架构，是如何一步步发展到今天提的中台架构概念的，每一步又解决了什么具体问题，我们以阿里架构变迁史为例（如图 2-2 所示）。

在 2009 年，淘宝上线了五彩石项目，这标志着淘宝从单体应用向服务化应用迈出了一步。那么，淘宝为什么要开发五彩石项目呢？因为当时淘宝面临两个非常严峻的问题：一个是性能问题，数据库连接不足，数据库成为了瓶颈；另一个是效率问题，当时淘宝有百余个研发人员，但核心系统只有一套供测试、预发、线上环境使用，导致研发需求排队等待。在开始五彩石项目之前，淘宝还做了千岛湖项目，用来验证服务化架构的可行性，将用户中心独立出来。随后，淘宝开启了五彩石项目，目标是通过增加人力来提升效率，通过增加机器来提升性能。

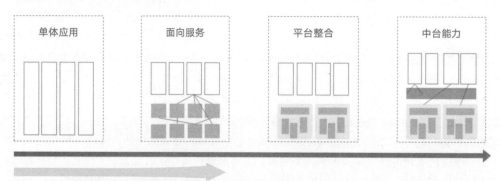

图 2-2 阿里架构变迁史

随着淘宝的业务发展,他们又面临了一个问题:各个服务之间有很多重复的建设,效率低下。为了解决这个问题,淘宝开始从服务化转向平台化,并创立了"共享业务事业部",将重复建设的公共业务分配给这个事业部,以避免成本浪费。这些公共业务包括商品平台、交易平台和结算平台等。平台化的目标是规避服务化没有规划导致的重复建设问题。

但是随着业务的快速发展,淘宝变成了一个拥有几十个事业部的巨型企业,而这带来了新的问题:效率问题。例如,如果需要在一个业务线上做出改动,需要与十几个平台进行沟通,这是非常低效的。同时,对于一个平台来说,需要面对来自不同事业部的需求,这需要平台研发人员具备理解和抽象所有业务线需求的能力,这让平台研发人员感觉回到了单体应用时代,所有的需求都要排队,即使增加人力也无法提高效率。这个问题主要表现在交易平台上。

为了解决这个问题,淘宝提出了中台的概念,中台是在一套规范下建立的,让具有专业技能的团队自主决策业务系统发展的平台。中台的目标是弱化平台的业务特性,提供通用能力。简而言之,就是将"共享业务"中的"业务"两个字去掉,只提供通用能力。

我们将每个阶段的核心目标总结为一句话:

(1)从单体到服务化:通过增加人员和机器来提高效率和性能。

(2)从服务化到平台化:解决服务化阶段因缺乏规划而导致的重复建设问题。

(3)平台化到中台化:在一套规范下,让各业务团队自行决定业务系统发展,适用于多个业务线或多个场景应用的独立发展。

类似地,在企业数字化转型过程中,也面临着类似的问题:

(1)企业业务在线化,对系统性能和稳定性提出了更高的要求,但由于内部系统之间的割裂,导致很多重复建设。因此,我们需要对企业进行服务化和平台化;

(2)没有一个供应商能够解决企业所有的商业场景问题,所以需要多个供应商共同参与。我们可以将供应商类比为业务线,在一套规范下让供应商或业务线自行决定业务系统的发展。

然而,阿里的中台架构方案并不能直接照搬到企业中,因为阿里的中台架构采用了平台共建模式,即让业务线基于平台设计的规范共同开发。这本质上还是平台主导模式,对企业来说历史包袱较大。在企业中,让不同背景的研发一起共建交易或商品平台是非常复杂的事情。平台化已经足够复杂,再加上共建会导致企业架构的负载过重，这对企业来说就不再是赋能，而是"内耗"。

● **互联网中台架构在企业实践中遇到的问题**

在 1.3 节"Oinone 的生态思考"中，"与中台的渊源"部分提到，在阿里云为企业提供数字化项目时，客户经常会对以下三个问题提出质疑，这些问题非常突出：

（1）我们听说你们具备敏捷响应能力，但为什么改动需求如此缓慢？不仅所需时间更长，而且成本更高？

（2）我们听说你们有能力中心，但为什么当我们引入新供应商或开发新场景时，前期建立的能力中心无法支持我们？

（3）我们听说你们的性能很好，但为什么我们需要投入更多的物理资源来支持项目？

在探讨互联网架构的适用性时，我想提出以下两个问题：

（1）企业应用程序的性能问题是否与互联网平台公司遇到的性能问题相同？

（2）企业应用程序的开发效率问题是否与互联网平台公司遇到的效率问题相同？

综合这些问题，通过比较企业和互联网之间的差异，我们可以了解水土不服的核心原因，如表 2-1 所示。

表 2-1 从企业与互联网的对比，看水土不服的核心原因

| 企业 | 互联网 |
| --- | --- |
| 企业 IT 组织能力无法与数字化转型的速度匹配，缺乏足够的人才支持。为提高开发效率，企业需要寻找工具和技术来降低开发难度，同时提高个人开发效率 | 互联网企业拥有众多优秀的人才，需要解决团队协作和知识共享的问题，即协同开发的效率 |
| 企业无法制定并主导技术规范，这导致了能力复用的不足。为了提高效率和减少开发成本，企业需要建立统一的技术规范和标准，以便能力复用和组织协同 | 互联网企业可以自定义技术规范，因此能力复用更易于保障 |
| 企业往往当前业务量相对小，期望数字化建设能打动业务发展，对业务发展的预期比较高，所以企业的诉求是既满足当下成本效应又能兼顾未来对发展预期 | 互联网企业起步时的系统目标负载较高，通常会忽略资源起步门槛的问题，当然也可以通过自动扩容、云计算等方式来解决初期的负载问题 |

我们可以看到企业和互联网架构在很多方面存在着不同的需求和问题。因此，在提供数字化服务时，Oinone 需要注意与企业的组织能力进行匹配，并根据企业自身的特性来提供在线化的服务能力。

## 2.3 Oinone 独特性之源，元数据与设计原则

让我们来揭开 Oinone 元数据的神秘面纱，了解它的核心组成、获取方式、面向对象特性以及带来的好处。您或许会想，这些特性能否解决企业数字化转型中互联网架构遇到的挑战呢？

元数据是本文多次提到的重要概念。作为低代码开发平台的基础，元数据支持企业所有研发范式，它数字化地描述了软件本身，包括数据、行为和视图等方面。在描述数据时，元数据本身就是数据的数据；在描述行为时，它就是行为的数据；在描述视图时，它就是视图的数据。只有深入理解元数据，才能全面了解 Oinone 的其他特性。

本节将介绍元数据的整体概览（如图 2-3 所示），带领您了解其核心组成、面向对象特性以及组织方式。请注意，本节将不会详细展开元数据的细节，这些细节将在后续的相关章节中深入介绍。

**1. 元数据的核心组成**

① 模块（Module）：将程序划分成若干个子功能，每个模块完成了一个子功能，再把这些模块整合起来组成一个整体。它是按业务领域划分和管理的最小单元，是一组功能、界面的集合。

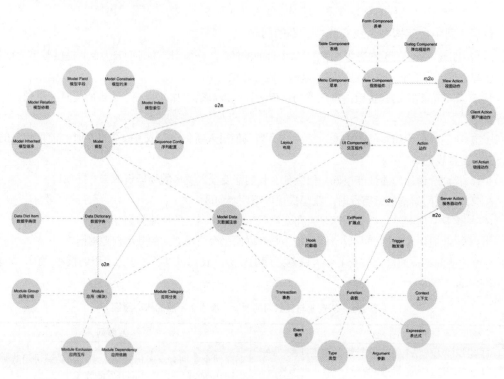

图 2-3　元数据整体视图

② 模型（Model）：Oinone 一切从模型出发，是数据及行为的载体。它是对所需要描述的实体进行必要的简化，并用适当的变现形式或规则把它的主要特征描述出来所得到的系统模仿品。它包括元信息、字段、数据管理器和自定义函数。同时遵循面向对象设计原则，拥有包括封装、继承和多态等特性。

③ 交互组件（UI Component）：它用菜单、视图和 Action 来勾绘出模块的前端交互拓扑，并且用组件化的方式统一管理、布局和视图。它用 Action 来描述所有可操作行为。

④ 函数（Function）：它是 Oinone 可执行逻辑单元，跟模型绑定则对应模型的方法。它的描述满足数学领域函数定义，含有三个要素：定义域 A、值域 C{f（x），x 属于 A} 和对应法则 f。其中核心是对应法则 f，它是函数关系的本质特征。它满足面向对象原则，可以设置不同的开放级别，可以在本地与远程智能间切换。

⑤ 元数据注册表：它是以模块为单位的安装记录，在模块安装时，相关的元数据都会在元数据注册表中记录。

**2. 元数据的产生方式**

既可以通过代码注解扫描获取，也可以通过可视化编辑器直接添加。

（1）从代码注解中扫描获取，代码如下。

```
1   @Model.model(ResourceBank.MODEL_MODEL)
2   @Model(displayName = "银行",labelFields = "name")
3 ▾ public class ResourceBank extends IdModel {
4
5       public static final String MODEL_MODEL = "resource.ResourceBank";
6
7       @Field.String
8       @Field(required = true, displayName = "名称")
9       private String name;
10
11      @Field.String
```

```
12        @Field(required = true, displayName = "银行识别号码", summary = "Bank Identifie
    r Code, BIC 或者 Swift")
13        private String bicCode;
14        ......
15     }
```

（2）通过可视化的编辑器添加，具体介绍详见 7.1 节"设计器总览"。

**3. Oinone 的原则**

Oinone 是一种通用低代码开发平台，其元数据设计满足应用开发所需的所有元素，并支持所有研发范式。

它基于元数据的具体实现秉承以下原则：

（1）部署与研发无关。

（2）以模型驱动，符合面向对象设计原则。

（3）代码与数据相互融合，编辑器产生的元数据以面向对象的方式继承、扩展标准产品的元数据。

这些原则的集合使整个平台能够实现以下功能特点：

（1）开发分布式应用与单体应用一样简单，部署方式由后期决定。如果要部署为分布式应用，则需要在 boot 工程中引入 Oinone 的 rpc 包。详见 4.3 节"Oinone 的分布式体验"。

（2）面向对象的特性使得每个需求都可以是独立模块，独立安装与卸载，让设计系统像搭建乐高积木一样。

（3）支持两种元数据产生方式，融合的原则确保标准与个性化，真正做到低无一体。

这些特性刚好也解决了 2.2 节"互联网架构作为最佳实践为何失效"中客户挑战的三个刺眼问题。如表 2-2 所示。

表 2-2　互联网架构落地企业数字化转型面临的问题及 Oinone 应对策略

| 互联网架构落地企业数字化转型面临的问题 | Oinone 应对的策略 |
| --- | --- |
| 不是说敏捷响应吗？为什么改个需求这么慢，不但时间更长，付出的成本也更高了？ | 特点（1）、特点（2）、特点（3） |
| 不是说有能力中心吗？当引入新供应商或有新场景开发的时候，为什么前期做的能力中心不能支撑了？ | 特点（2）、特点（3） |
| 不是说性能好吗？为什么我投入的物理资源更多了？ | 特点（1） |

## 2.4　Oinone 的三大独特性

Oinone 在技术方面通过整合互联网架构和低代码技术，实现了三个独特的关键创新点（如图 2-4 所示）：

（1）独立模块化的个性化定制：每个需求都可以被视为一个独立的模块，从而实现个性化定制，提高软件生产效率。此外，这些独立模块也不会影响产品的迭代和升级，为客户带来无忧的体验。

（2）灵活的部署方式：单体部署和分布式部署的灵活切换，为企业业务的发展提供了便利，同时适用于不同规模的公司，有助于有效地节约企业成本，提升创新效率，并让互联网技术更加亲民。

（3）低代码和无代码的结合：低无一体为不同的 IT 组织和业务用户提供了有效的协同工作方式，能够快速部署安全、可扩展的应用程序和解决方案，帮助企业/组织更好地管理业务流程并不断优化。

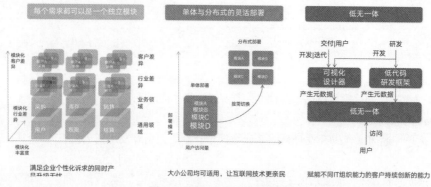

图 2-4 Oinone 的三大独特性

## 2.4.1 Oinone 独特性之单体与分布式的灵活切换

企业数字化转型需要处理分布式带来的复杂性和成本问题。尽管这些问题令人望而却步，但分布式架构对于大部分企业仍然是必须的选择。如果一个低代码平台缺乏分布式能力，那么它的性能就无法满足客户的要求。相比之下，Oinone 平台通过对部署的创新（如图 2-5 所示），成功实现了分布式架构的支持，而且能够按照客户的业务发展需求，灵活选择不同的部署模式，同时节约企业成本，提升创新效率。这一创新是 Oinone 平台与其他低代码平台的重要区别，能够满足客户预期发展并兼顾成本效益。

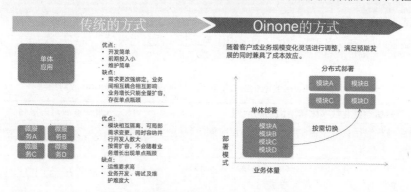

图 2-5 传统部署方式 vs Oinone 部署方式

**实现原理**

要实现灵活部署的特性，必须满足两个基本要求：

（1）开发过程中不需要过多关注分布式技术，就像开发单体应用一样简单。代码在运行时应该能够根据模块是否在运行容器中，来决定路由通过本地还是远程。这样可以大大减少研发人员的工作量和技术复杂度。

（2）研发与部署要分离，即"开发单体应用一样开发分布式应用，而部署形式由后期决定"。为此，我们的工程结构支持多种启动模式，图 2-6 逐一介绍了针对不同场景的工程结构类型。这样可以让客户在后期根据业务发展情况和需求，选择最适合的部署模式，从而达到灵活部署的目的。

在整个工程结构上，我们秉承了 Spring Boot 的规范，不会改变大家的工程习惯。而 Oinone 的部署能力则可以让我们更灵活地应对各种情况。现在，我们来逐一介绍几种常规的工程结构以及它们适用的场景：

（1）单模块工程结构（常规操作）。

这是非常标准的 Spring Boot 工程，适用于简单的应用场景开发以及入门学习。

（2）多模块工程结构（常规操作）。

这是非常标准的多 Spring Boot 工程，可以实现分布式独立启动，适用于常规的分布式应用场景开发。

（3）多模块工程结构（独立 boot 工程模式）。

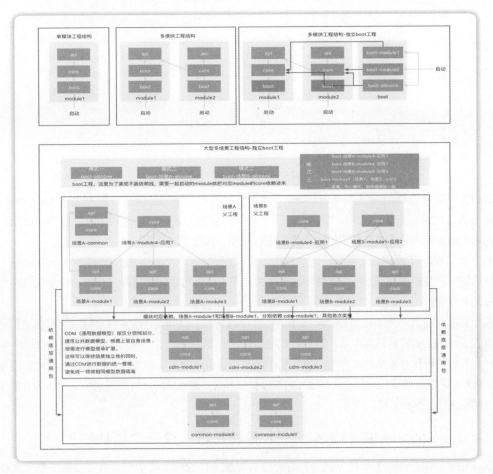

图 2-6　Oinone 工程结构梳理

这种工程结构在多模块工程的基础上，通过独立的 boot 工程来支撑多部署方式。适用于中大型分布式应用场景开发。

然而，随着工程越来越多，我们也会面临一些问题：

● 研发：环境准备非常困难，每个模块都要单独启动，研发调试跟踪困难。

● 部署：分布式的高可靠性保证需要每个模块至少有两个部署节点，但在模块较多的情况下，起步成本非常高。同时，企业初期业务不稳定且规模较小，使用多模块工程的第二种模式会增加问题排查难度和成本。

此时，Oinone 的多模块工程下的独立 boot 工程模式部署就可以发挥其灵活性，让研发和业务起步阶段可以选择 all-in-one 模式，等到业务发展到一定规模的时候，只需要把线上部署模式切换成模块独立部署，而研发还可以保留 all-in-one 模式的优势。

值得注意的是，分分合合的部署模式在传统互联网架构和低代码或无代码平台上都是有代价的，但是 Oinone 却可以灵活适配，只需要在 boot 工程的 yml 文件中写入需要加载的模块就可以解决。此处我们仅介绍多模块加载配置，选择性忽略其他无关配置，具体配置如图 2-7 所示。

（4）大型多场景工程结构（独立 boot 工程模式）。

多模块工程结构基础上的加强版，增加 CDM 层设计，让不同场景既保持数据统一，又保持逻辑独立。这种工程结构特别适用于大型企业软件开发，其中涉及多个场景的情况，例如 B 端和 C 端的应用，或者跨不同业务线的应用，这种工程结构能够保证数据的一致性，同时也能够保持逻辑独立，避免不同场景间的代码冲突。

```yaml
pamirs:
  boot:
    init: true
    sync: true
    modules:
      - base
      - resource
      - sequence
      - user
      - auth
      - web
    tenants:
      - pamirs
```

图 2-7　Oinone yml 配置图

这种工程结构是 Oinone 支撑"企业级软件生态"的核心，我们可以把场景 A 当作我们官方应用，场景 B 当作其他第三方伙伴应用。在这个工程结构下，我们的客户可以定制化开发自己的应用，同时我们也可以通过这种模式来支持我们的伙伴们进行开发，实现多方共赢。

基于独立 boot 工程模式，我们使用多种部署模式应对不同情况，并统一管理所有伙伴应用。这种工程结构的优点是扩展性好，可以支持不同规模的应用，并且可以根据需要进行快速扩展或缩小规模，具有很高的灵活性。

基于标准产品的二开工程结构，是指基于标准产品进行二次开发，满足客户特定需求的工程结构。这种模式下，Oinone 提供标准产品，客户可以根据自己的需求进行二次开发，实现定制化需求，同时可以利用我们的模块化开发特性，将每一个需求作为一个模块进行开发和管理。这种工程结构的优点是能够快速满足客户特定需求，同时也具有很好的可维护性和可扩展性，因为每个需求都是一个独立的模块，可以方便地进行维护和扩展。

### 2.4.2　Oinone 独特性之每一个需求都可以是一个模块

Oinone 平台采用模型驱动的方式，并符合面向对象设计原则，每个需求都可以是一个独立模块，可以独立安装、升级和卸载。这让搭建系统真正像搭建乐高积木一样，具有高度的灵活性和可维护性。

大部分低代码或无代码平台的应用市场上的应用往往是模板式的，也就是说，个性化只能通过在应用上直接修改实现，而且一旦修改就不能升级。这对于软件公司和客户来说都非常痛苦。客户无法享受到软件公司产品的升级功能，而软件公司在服务大量客户时，也会面临不同版本的维护问题，成本也非常高。而我们的 Oinone 平台完全避免了这些问题，让客户和软件公司都可以从中受益（如图 2-8、图 2-9 所示）。

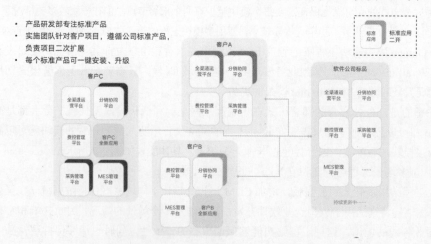

图 2-8　软件公司与客户项目的关系——让标准与个性化共存

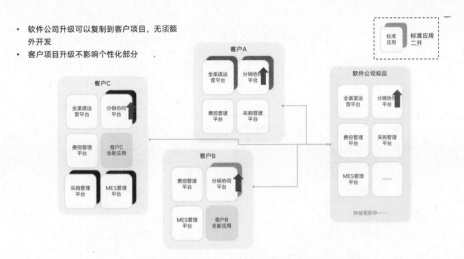

图 2-9　软件公司与客户项目的关系——让升级无忧

**实现原理**

在满足客户个性化定制需求时,传统的方法通常是直接修改标准产品源码,但这样做会带来一个问题:标准产品无法持续升级。相反,无论是在 OP 模式还是 SaaS 模式下,Oinone 都采用全新的模块为客户进行个性化开发,保持标准产品和个性化模块的独立维护和升级。这是因为在元数据设计时,Oinone 采用了面向对象的设计原则,实现了元数据设计与面向对象设计思想的完美融合。

面向对象设计的核心特征包括封装、继承、多态,而 Oinone 的元数据设计完全融入了这些思想。下面是几个例子,说明 Oinone 的元数据设计如何体现面向对象设计的核心特征,并带来了什么好处:

(1)继承:在继承原有模型的字段、逻辑、展示的情况下,增加一段代码来扩展模型的字段、逻辑、展示。

(2)多态:在继承原有模型的字段、逻辑、展示的情况下,增加一段代码来覆盖模型的原有字段、逻辑、展示。

(3)封装:外部无需关心模型内部如何实现,只需按照不同场景调用模型对应开放级别的字段、逻辑、展示。

这些特征和优势使得 Oinone 在满足客户个性化需求时更加灵活和可持续,同时使得标准产品的维护和升级变得更加容易和高效。

在 Java 语言设计中,万物皆对象,一切都以对象为基础。而 Oinone 的元数据设计则是以模型为出发点,作为数据和行为的承载体。图 2-10 清晰地描述了 Java 面向对象编程中封装、继承、多态与 Oinone 元数据的对应关系。Oinone 元数据描述了 B 对象继承 A 对象并拥有其所有属性和方法,并覆盖了 A 对象的属性 1 和方法 1,同时新增了属性 3 和方法 3。

此外,Oinone 的面向对象特性是用元数据来描述的。一方面,我们基于 Java 编码规范收集相关元数据,以保持 Java 编程习惯。另一方面,方法和对象的挂载是松耦合的,只要按照元数据规范进行挂载,就能轻松地将其附加到模型上。在不改变原有 A 对象的情况下,我们可以直接增加方法和属性(如图 2-11 所示)。

Oinone 函数不仅支持面向对象的继承和多态特性,还提供了面向切面的拦截器和 SPI 机制的扩展点,以应对方法逻辑的覆盖和扩展,以及系统层面的逻辑扩展(如图 2-12 所示)。这些扩展功能可以独立地在模块中维护。

其中,拦截器可以在不侵入函数逻辑的情况下,根据优先级为满足条件的函数添加执行前和执行后的逻辑。

扩展点是一种类似于 SPI 机制的逻辑扩展机制,用于扩展函数的逻辑。通过这一机制,可以对函数逻辑进行灵活的扩展,以满足不同的业务需求。

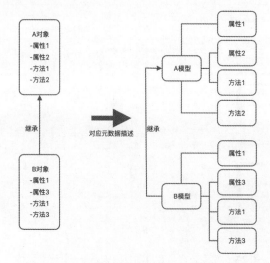

图 2-10　Java 面向对象在 Oinone 元数据中对应

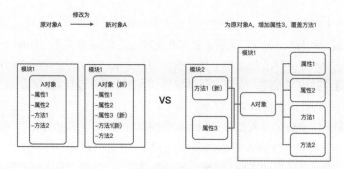

图 2-11　Java 对象的修改对比 Oinone 元数据模型的修改

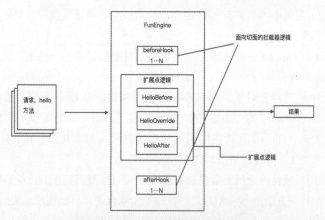

图 2-12　Oinone 函数拦截与扩展机制

　　不管是对象、属性还是方法，都可以以独立的模块方式来扩展，这就使得每一个需求都可以成为一个独立的模块，方便我们在研发标准产品时进行模块化的划分，同时也让我们在以低代码模式为客户进行二次开发时，能够更好地支持"标准产品迭代与个性化保持独立"的需求。在 2.4.3 节"Oinone 独特性之低无一体"中，我们也提到了这个特性，但那是在低无一体的情况下，通过元数据融合来实现的。让我们看看基于低代码开发模式下，典型的 Oinone 二次开发工程结构（如图 2-13 所示，图中简称为二开），就可以更好地理解这个特性了！

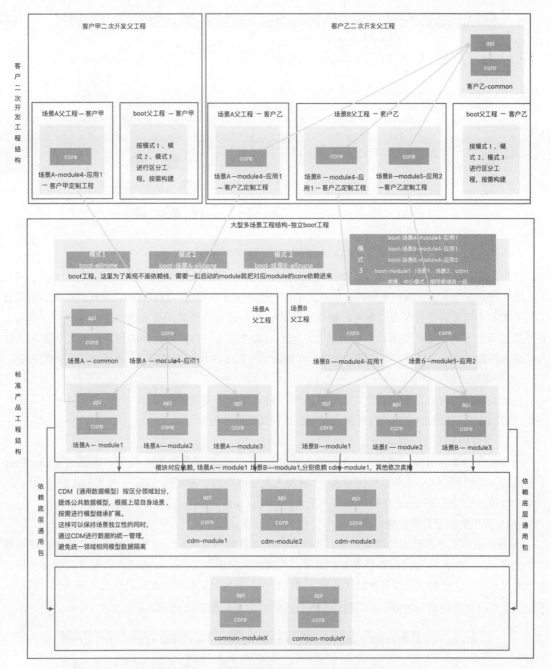

图 2-13　Oinone 典型的二次开发工程结构

## 2.4.3　Oinone 独特性之低无一体

现在的软件开发行业，低代码和无代码已经成为热门话题。它们的优势很明显：加速软件开发周期、减少代码开发时间、降低开发成本、易于维护等。而 Oinone 作为一个低无一体的开发平台，更是在这些优势上做出了巨大的创新。

**1. 技术亮点**

低代码可以在不改变研发习惯的前提下，提升效率，降低难度（如图 2-14 所示）。

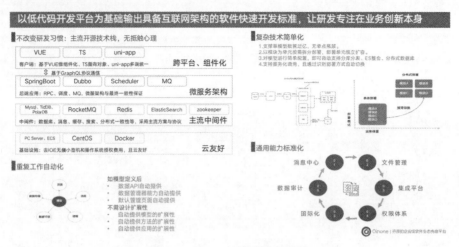

图 2-14　Oinone 低代码特性介绍

（1）提高专业开发人员效率。

低代码开发模式大大减少了烦琐、重复的工作，模型定义完成后，数据 API、数据管理器、基础管理的界面都不需要再进行开发。同时，低代码模式让分布式微服务架构的系统开发变得简单，研发人员不需要考虑分布式部署能力和大数据能力，也不需要去关心一些业务无关的通用能力，如权限、导入导出、国际化翻译、消息、审计等。这样，开发人员可以专注于业务研发，从而大幅提高开发效率。

（2）提升系统扩展性。

在研发标品的时候，低代码模式让开发人员不再需要关心系统的扩展性。与传统模式不同，低代码模式更加注重元数据的管理，这样就可以更好地保障系统扩展性。

（3）保留研发人员习惯。

Oinone 平台非常开放，满足开发人员的各种习惯，比如保留原有的 IDE 环境、熟悉的 Spring Boot 工程结构等。而且在 Oinone 的低代码模式下，研发人员还可以通过无代码方式，在线可视化地修改应用。这样，即使在使用低代码模式的情况下，开发人员也可以保留原有的习惯，提升开发效率。

（4）提供更加开放的解决方案。

Oinone 提供了非常开放的解决方案，让开发人员可以自由定制和组合各种功能。当行业出现特殊的功能需求时，开发人员可以整合成平台组件，并集成到应用中。Oinone 的低代码模式具有高度的开放性和灵活性，这使得它在与其他低代码平台的比较中具有明显的优势。相比其他低代码平台，Oinone 不会在无法满足特定需求的情况下限制开发人员的创造力（如图 2-15 所示）。

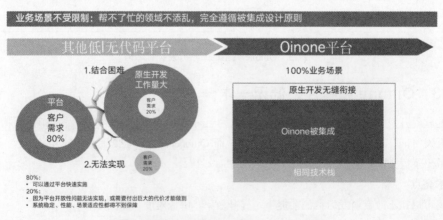

图 2-15　Oinone 低代码的被集成特性示意图

## 2. 无代码：五大设计器覆盖研发方方面面，让业务、实施也能参与

如表 2-3 的五大设计器是低代码平台（Low-Code Develpment Platform，LCDP）的产品化呈现，是露在外面大家看得到的冰山，核心还是在 LCDP 本身。这部分实时在演进迭代，如您想体验最新版本，可以在 Oinone 官网注册，可扫左侧二维码获取官网的网址。

Oinone 官网

表 2-3 Oinone 无代码 - 五大设计器简述

| 设计器 | 说明 | 产品展示 |
| --- | --- | --- |
| 模型设计器 | 以模型为驱动，有模型、数据字典、数据编码等设计功能，我们就可以完整地定义产品数据模型，模型设计器默认整体呈现区别于普通 ER 图，以当前模型为核心视角展开，可以单击关联模型切换主视角；<br>多种模式可切换：专家与经典模式切换，图与表模式的切换 |  |
| 界面设计器 | 界面设计器旨在帮助用户快速搭建页面；<br>所见即所得和根据不同视图类型设计契合的搭建交互就变得尤为重要；<br>拥有多端页面设计能力 |  |
| 流程设计器 | 为业务流程和审批流程提供可自动执行的流程模型，通过定义流转过程中的各个动作、规则，以此实现流程自动化；<br>流程可以跨应用设计，不同应用的模型之间可以执行同一流程 |  |
| 逻辑设计器 | 组件化、可视化逻辑编排，逻辑动态变更、动态管理，实施验证 |  |
| 数据可视化 | 从内部系统模型获取数据内容后，根据业务需求自定义图表，目的是为企业提供更高效的数据分析工具；<br>可以智取业务系统模型，系统自动解析选择的模型、接口、表格中的字段后进行数据分析；<br>降低对数据分析人员的研发能力要求，提升数据分析的效率 |  |

## 3. 真正的低无一体，体现在一体化的融合能力上

在做核心产品的时候以低代码开发为主，以无代码为辅助，见低代码开发的基础入门篇中 3.5.5 节"设计器的结合"。

在实现实施或临时性需求时则是以无代码为主，低代码为辅助，见本教材的最后一篇"Oinone 的

低无一体",这种模式比较特殊,只在 SaaS 模式下提供。当客户个性化需求部分无法通过无代码设计器完成,则可以通过 SaaS 提供的"低无一体"模块提供的"反向生成 API 代码"的功能,生成对应的扩展工程和 API 依赖包,再由专业研发人员基于扩展工程,利用 API 包进行开发并上传至平台。具体的场景适配如表 2-4 所示。

表 2-4  不同场景适配方式说明

| 场景 | 融合形式 | 具体操作 |
|---|---|---|
| 标准产品以低代码开发为主,以无代码为辅助 | 标品开发时,结合无代码设计器来完成页面开发,可以把设计后的页面元数据装载为标准产品的一部分。详细教程见 3.5.5 节"设计器的结合" | ①向Oinone技术支持获取专属设计器Docker镜像 ②在指定jar路径下放自有模块的jar包如应的依赖包比如:pamirs-demo-api-1.0.0-SNAPSHOT.jar和pamirs-demo-core-1.0.0-SNAPSHOT.jar ③启动模块,并进入 ④通过接口导出设计器产生的元数据 ⑤把元数据文件放到应用的指定文件路径里 |
| 项目交付以无代码为主,以低代码为辅助 | 当设计器无法支持特殊需求时,则可以通过低无一体应用的代码模式来完成,支持两种使用模式:上传 jar 包模式、源码托管模式。详细教程见 7.4 节"Oinone 的低无一体" | |

**4. 解读低无一体**

我们将从以下三个方面来解读 Oinone 的低无一体。

(1) 低无一体的设计原则及好处。

真正的低无一体平台应该确保标准产品迭代与个性化保持独立,让软件企业具备为客户提供在线化的快速响应、个性化定制、持续更新等服务的能力,让企业客户能够真正自主做到敏捷响应和快速创新。所以 Oinone 的元数据融合方案跟其他平台有所区别(如图 2-16 所示)。

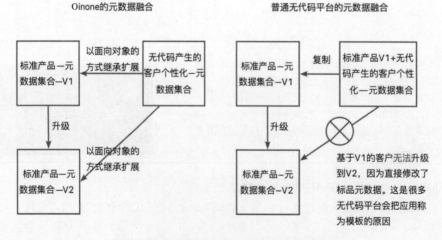

图 2-16  Oinone 与其他平台的元数据融合对比图

(2）低无一体中低与无的关系。

无代码是低代码平台的图形化呈现,是低代码的一个子集,它将无限接近低代码的能力,同时也将成为低代码平台的必备特征,是通过低代码开发的标准产品的二次开发配套工具。

(3）低无一体中低与无的定位。

通过表2-5可以看出,低代码和无代码在Oinone的体系中相互融合,共同构成了一个完整的低无一体模式,提供更加开放、灵活和可扩展的解决方案,让用户能够更加轻松地完成开发和实施。

表2-5 Oinone低代码开发平台的两种开发模式对比

|  | 低代码模式 | 无代码模式 |
| --- | --- | --- |
| 用户群体 | 专业研发 | 产品经理、需求分析师、直接业务人员 |
| 支撑场景 | 企业全场景软件以及二次开发 | 企业全场景软件以及二次开发,专业化场景比较高的则需低代码支持 |
| 核心能力 | 不改变研发习惯,提升研发效率 | 可视化编程无须专业编程语言知识 |
| 核心定位 | 开发标准模块 | 标准模块的二次开发;<br>无标品支撑场景的新模块开发 |

# 第 3 章　Oinone 入门

本章从环境准备、构建第一个 Oinone Module、完成一个个小功能等方面引领读者进入 Oinone 的世界，这将是一个美妙的开始。

1. 环境搭建：Windows 以及 Mac 版环境准备
2. 全面了解 Oinone 之 Oinone 以模块为组织
3. 全面了解 Oinone 之 Oinone 以模型为驱动
4. 全面了解 Oinone 之 Oinone 以函数为内在
5. 全面了解 Oinone 之 Oinone 以交互为外在

## 3.1　环境搭建

**1. 基础环境说明**

基础环境说明见表 3-1。

表 3-1　基础环境说明

| 内容 | 是否必须 | 说明 |
| --- | --- | --- |
| 后端基础环境 | | |
| JDK1.8 | 必须 | Java 的基础运行环境 |
| MySQL | 必须 | 8.0.26 版本以上<br>需要注意点：<br>修改：my.cnf（macOS）/ my.ini（windows）<br>时区、大小写敏感设置<br>lower_case_table_names = 2<br>default-time-zone = '+08:00' |
| IDEA | 必须 | 需要注意点：<br>（1）禁用 Lombok 插件<br>（2）Java Compiler 增加 -parameters（不然 Java 反射获取方法入参名会变成 arg*）指令<br>（3）安装 Oinone 插件 |
| DB GUI | 非必须 | Datagrip、MySQLWorkbench、DBEaver 选其一 |
| Insomnia | 非必须 | GraphQL 测试工具 |
| Git | 必须 | 2.2.0 以上 |
| Maven | 必须 | 3.8.1 以上<br>需要注意点：<br>（1）配置 mvn 的 settings 文件下载地址见 Oinone 开源社区群公告，也可以联系 Oinone 合作伙伴或服务人员<br>（2）把 settings.xml 拷贝一份到 maven 安装目录 conf 目录下 |
| RocketMQ | 必须 | 4.7.1 以上 |
| Redis | 必须 | 5.0.2 以上 |
| ZooKeeper | 必须 | 3.5.8 以上 |
| 前端基础环境 | | |
| nvm | 非必须 | 方便 node 的版本管理 |
| Node.js | 必须 | 版本要求为 12.12.0<br>注意事项：<br>（1）npm 的源配置为 http://nexus.shushi.pro/repository/kunlun/<br>（2）源的用户名、密码见 Oinone 开源社区群公告，也可以联系 Oinone 合作伙伴或服务人员 |
| vue-cli | 必须 | vue 脚手架工具 |

其他：canal 和 Es 的环境搭建见具体学习章节。

**2. 基础知识准备**

基础知识准备见表 3-2。

表 3-2 基础知识准备

| 前端必备知识 | vue3、typescript、graphql |
|---|---|
| 后端必备知识 | SpringBoot、MybatisPlus |

**3. 学习安装**

Mac 见"环境准备（Mac 版）"一节。

Windows 见"环境准备（Windows 版）"一节。

### 3.1.1 环境准备（macOS 版）

为了降低环境准备难度，基础环境全程用安装包无脑模式配置，安装下载参见附录（提供 Mac 版本安装包，其他操作系统请自行网上下载与安装）。

**1. 后端相关**

1）基础环境准备

（1）安装 jdk 1.8（下载地址见附录）。

（2）安装 MySQL 8.0.26（下载地址见附录）。

① 安装 MySQL，并配置环境变量，参见步骤（8）。

② 如果 MySQL 启动失败，则在命令行执行以下命令。

```
1    mysql --initalize-insecure
2    sudo chmod -R a+rwx /user/local/mysql/data/
```

（3）安装 IDEA 社区版（官方下载链接见附录）。

① 根据不同版本下载不同的 IDEA 插件，并安装。

● 下载后去除后缀 .txt，请根据各自 IDEA 版本下载对应插件，如图 3-1 所示（具体参见附录）。

| idea版本 | 对应插件 |
|---|---|
| 2020.2.4 | pamirs-intellij-lombok-plugin-2.2.0-SNAPSHOT-2020.2.4.zip.txt (4.9 MB) |
| 2021.1 | pamirs-intellij-plugin-3.0.1-2021.1.zip.txt (5 MB) |
| 2021.2 | pamirs-intellij-plugin-3.0.1-2021.2.zip.txt (4.9 MB) |
| 2021.3 | pamirs-intellij-plugin-3.0.1-2021.3.zip.txt (4.9 MB) |
| 2022.1 | pamirs-intellij-plugin-3.0.1-2022.1.zip (4.9 MB) |
| 2022.2 | pamirs-intellij-plugin-3.0.1-2022.2.zip.txt (5 MB) |
| 2022.3 | pamirs-intellij-plugin-3.0.1-2022.3.zip.txt (4.9 MB) |
| 2023.1 | pamirs-intellij-plugin-3.0.9-2023.1.zip (5 MB) |
| 2023.2 | pamirs-intellij-plugin-4.0.1-2023.2.zip (5 MB) |
| 2023.3 | pamirs-intellij-plugin-4.0.1-2023.3.zip (5 MB) |
| 2024.1 | pamirs-intellij-plugin-4.7.8-2024.1.zip (5 MB) |

图 3-1 不同版本下载不同的 IDEA 插件

● 单击 Preferences 菜单（快捷键 comand+，）。

● 选择 Plugins，进入插件管理页面，接下来按图 3-2 所示操作即可。

图 3-2 插件管理页面操作示意

● 选择 .zip 文件，不需要解压。

② 如果安装了 Lombok，请禁用。

③ 选择 IDEA 的 Java Complier，不然 Java 反射获取方法入参名会变成 arg*，导致元数据默认取值出错。或者 pom 中加入 Complier 插件，此方法为正解，不然上线也会有问题，我们学习的工程都会选用 mvn 插件方式，如图 3-3 所示。

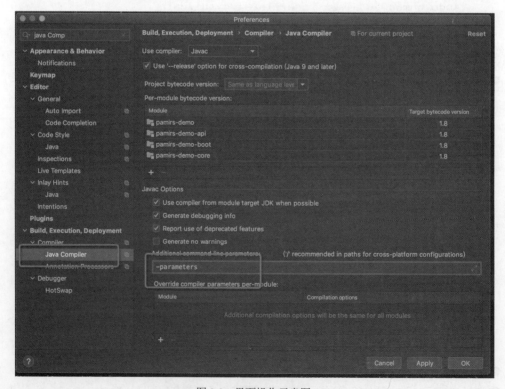

图 3-3 界面操作示意图

完整代码如下:

```xml
<plugin>
    <groupId>org.apache.maven.plugins</groupId>
    <artifactId>maven-compiler-plugin</artifactId>
    <configuration>
        <compilerArgument>-parameters</compilerArgument>
        <source>${maven.compiler.source}</source>
        <target>${maven.compiler.source}</target>
        <encoding>${project.build.sourceEncoding}</encoding>
    </configuration>
</plugin>
```

(4)安装 dataGrip 最新版本。

当前版本过期时可以删除"~/Library/Application\ Support/JetBrains/DataGrip202xxxx"相关的目录,就可以获得无限期试用,或者安装其他 MySQL GUI 工具。

(5)安装 git 2.2.0(下载地址见附录)。

(6)安装 GraphQL 客户端工具 Insomnia。

第一次使用看 demo "Module 启动后如何用该工具验证后端启动成功",更多使用技巧自行百度,下载文件见附录,下载后需修改文件名去除 .txt 后缀。

(7)安装 maven,并配置环境变量(下载地址见附录)。

① 配置 mvn 的 settings,下载附件 settings-open.xml,并重命名为 settings.xml,建议直接放在~/.m2/下面。下载地址见 Oinone 开源社区群公告,也可以联系 Oinone 合作伙伴或服务人员。

② 把 settings.xml 拷贝一份到 maven 安装目录 conf 目录下。

(8)环境变量设置 vi ~/.bash_profile,并执行 source ~/.bash_profile

```bash
##按实际情况设置
export PATH=$PATH:/usr/local/mysql/bin
export PATH=$PATH:/usr/local/mysql/support-files
export JAVA_HOME=/Library/Java/JavaVirtualMachines/jkd1.8.0_0221.jdk/Contents/Home
##替换掉${mavenHome},为你的实际maven的安装路径
export M2_HOME=${mavenHome}
export PATH=$PATH:$M2_HOME/bin
```

① 查看主机名,代码如下。

```bash
#查看主机名
echo $HOSTNAME
```

② 根据主机名配置 /etc/hosts 文件。此步如果没有配置,可能导致 Mac 机器在启动模块时出现 dubbo 超时,从而导致系统启动非常慢,记得把 oinonedeMacBook-Pro.local 换成自己的主机名,代码如下。

```
#oinonedeMacBook-Pro.local #需要换成自己对应的主机名,自己的主机名用 echo $HOSMNAME
127.0.0.1 oinonedeMacBook-Pro.local
::1 oinonedeMacBook-Pro.local
```

2)必备中间件安装脚本(rocketmq、zk、redis)

(1)zk(下载地址见附录),解压并:

① vi ~/.bash_profile,追加以下两行,并执行 source ~/.bash_profile,代码如下。

```bash
#### 替换掉${basePath}为你的实际安装路径
export ZOOKEEPER_HOME=${basePath}/apache-zookeeper-3.5.8-bin
export PATH=$PATH:$ZOOKEEPER_HOME/bin
```

② 启动 zk,代码如下。

```
1  ##启动
2  zkServer.sh start
3  ##停止
4  zkServer.sh stop
```

(2) rocketmq(下载地址见附录)。

① vi ~ /.bash_profile,追加以下两行,并执行 source ~ /.bash_profile,代码如下。

```
1  #### 替换掉${basePath},为你的实际安装路径
2  export ROECET_MQ_HOME=${basePath}/rocketmq-all-4.7.1-bin-release
3  export PATH=$PATH:$ROECET_MQ_HOME:$ROECET_MQ_HOME/bin
```

② 到 bin 目录下修改配置文件 runserver.sh 和 runbroker.sh,代码如下。

```
1  ##注释掉下面一行
2  ##choose_gc_log_directory
3  ##修改java启动所需内存,按自己实际情况改,1g或者512m
4  JAVA_OPT = "${JAVA_OPT} -server -Xms1g -Xmx1g -Xmn1g -XX:MetaspaceSize=128m -XX:MaxMetaspaceSize=320m
```

③ 启停 rocketmq,代码如下。

```
1  ##启动 nameserver
2  nohup mqnamesrv &
3  ##启动 broker
4  nohup mqbroker -n localhost:9876 &
5
6  ##停止
7  mqshutdown broker
8  mqshutdown namesrv
```

(3) redis(下载地址见附录)。

① 安装 redis,代码如下。

```
1  ## 替换掉${redisHome},为你的实际安装路径
2  cd ${redisHome}
3  make
4  make install PREFIX=${redisHome}
```

② vi ~ /.brah_profile,追加以下两行,并执行 source ~ /.brah_profile,代码如下。

```
1  ##替换掉${redisHome},为你的实际安装路径
2  export REDIS_HOME=${redisHome}
3  export PATH=$PATH:$REDIS_HOME/bin
```

③ 启停 redis,代码如下。

```
1  ##启动
2  nohup redis-server &
3  ##停止
4  redis-cli shutdown
```

3) 晋级中间件安装脚本(canel,es)

canal、es 相关的安装与使用,我们放到后续教程中再介绍,它们不是必须的,只有用到异步事件、增强模型等高级特性时,才需要安装。

**2. 前端环境准备**

(1) 安装 nvm,便于 node 的版本管理:

```
1  curl -o- https://pamirs.oss-cn-hangzhou.aliyuncs.com/pamirs/software/install.sh |
   bash
```

(2) vi ~/.bash_profile,追加以下两行,并执行 source ~/.bash_profile,代码如下。

```
1  export NVM_DIR="$([ -z "${XDG_CONFIG_HOME-}" ] && printf %s "${HOME}/.nvm" || prin
   tf %s "${XDG_CONFIG_HOME}/nvm")"
2  [ -s "$NVM_DIR/nvm.sh" ] && \. "$NVM_DIR/nvm.sh" # This loads nvm
```

(3) 用 nvm 安装 Node.js 版本 12.12.0,代码如下。

```
1  #看nvm是否安装成功
2  nvm -v
3  nvm install 12.12.0
```

(4) vi ~/.bash_profile,追加以下一行,并执行 source ~/.bash_profile,代码如下。

```
1  nvm use 12.12.0
```

(5) 安装 vue-cli,代码如下。

```
1  #@vue/cli需要固定,安装成功后可vue -V查看
2  sudo npm install @vue/cli@4.5.17 -g
```

(6) 配置 npm 源,代码如下。

```
1  #通过config命令,其他方式自行百度
2  npm config set registry http://nexus.shushi.pro/repository/kunlun/
```

(7) 登录 npm 账号,代码如下。

```
npm login --registry "http://nexus.shushi.pro/repository/kunlun/"
npm info underscore   #如果上面配置正确这个命令会有字符串response
```

username、password、email 请见 Oinone 开源社区群公告,也可以联系 Oinone 合作伙伴或服务人员。

(8) 安装 cnpm,代码如下。

```
1  #参加https://www.npmjs.com/package/cnpm
2  sudo npm install cnpm -g --registry=https://registry.nlark.com
```

至此所有环境都准备好了,可以开始学习如何使用 Oinone 进行业务开发了。

**3. 其他说明**

解决一些新版 Mac 系统"每次都需 source ~/.bash_profile"的问题,vi ~/.zshrc 新建 .zshrc 文件,添加内容为 source .bash_profile,保存退出后执行 source ~/.zshrc 命令。

### 3.1.2 环境准备(Windows 版)

**1. 后端基础环境准备**

1) 安装 JDK 1.8(下载地址见附录)

(1) 打开 Windows 环境变量配置页面。

此电脑 → 右键属性 → 系统高级设置 → 环境变量。

（2）配置环境变量。

在用户环境变量中新建变量名为 JAVA_HOME 的项，值为 JDK 安装地址，如图 3-4 所示。

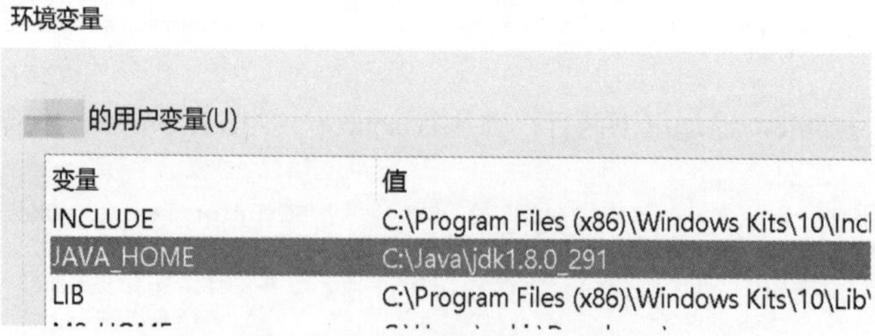

图 3-4　新建变量 JAVA_HOME

编辑变量为 Path 的项添加一个值 %JAVA_HOME%\bin，如图 3-5 所示。

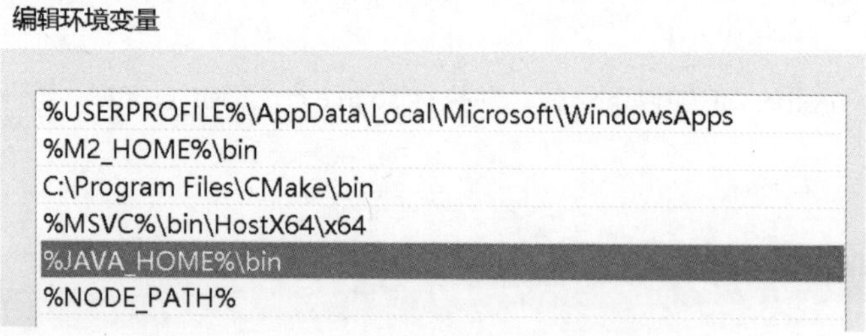

图 3-5　添加一个值 %JAVA_HOME%\bin

（3）在 PowerShell 或者 CMD 中验证，如图 3-6 所示。

图 3-6　验证

输出与图 3-6 所示类似信息为安装配置成功。

2）安装 Apache Maven 3.8+（下载地址见附录）

（1）删除 Maven 安装目录下的 conf/settings.xml。

（2）下载到 C:\Users\ 你的用户名 \.m2 目录中并重命名为 settings.xml。

（3）配置环境变量。

在用户环境变量中新建变量名为 M2_HOME 的项，值为 Maven 安装路径，如图 3-7 所示。

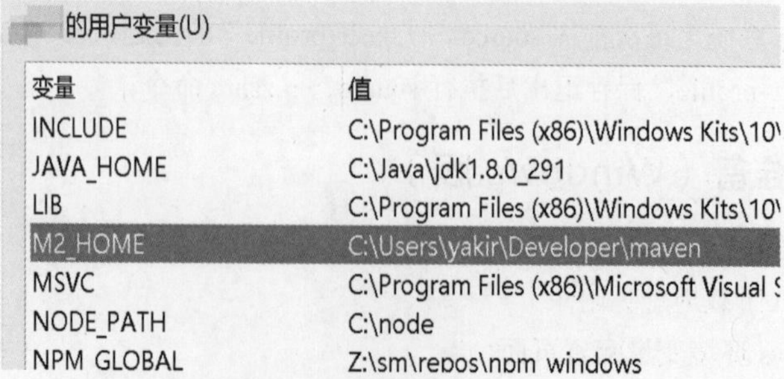

图 3-7　新建变量 M2_HOME

编辑变量为 Path 的项添加一个值 %M2_HOME%\bin，如图 3-8 所示。

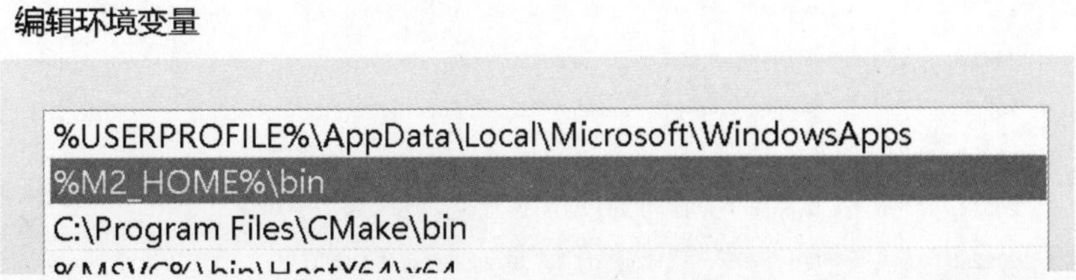

图 3-8 添加一个值 %M2_HOME%\bin

（4）验证，如图 3-9 所示。

图 3-9 验证

3）安装 Jetbrains IDEA 2020.2.4（下载地址见附录）

（1）插件安装（详见附录）。

（2）如果 IDEA 安装了 Lombok 插件，请禁用 Lombok 插件。

（3）单击菜单项 File → Settings → Plugins，如图 3-10 所示。

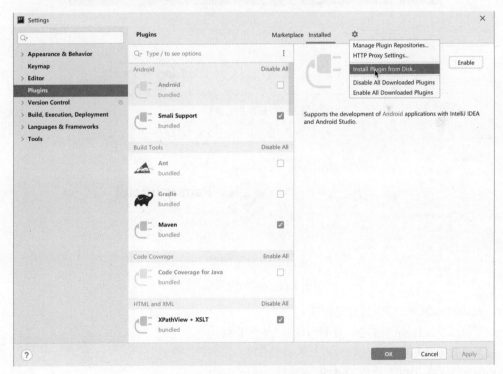

图 3-10 单击菜单项 File → Settings → Plugins

选择下载的插件包，下载后去除后缀 .txt，根据各自 IDEA 版本下载对应插件（具体参见附录），如图 3-11、图 3-12、图 3-13 所示。

| idea 版本 | 对应插件 |
|---|---|
| 2020.2.4 | pamirs-intellij-lombok-plugin-2.2.0-SNAPSHOT-2020.2.4.zip.txt (4.9 MB) |
| 2021.1 | pamirs-intellij-plugin-3.0.1-2021.1.zip.txt (5 MB) |
| 2021.2 | pamirs-intellij-plugin-3.0.1-2021.2.zip.txt (4.9 MB) |
| 2021.3 | pamirs-intellij-plugin-3.0.1-2021.3.zip.txt (4.9 MB) |
| 2022.1 | pamirs-intellij-plugin-3.0.1-2022.1.zip (4.9 MB) |
| 2022.2 | pamirs-intellij-plugin-3.0.1-2022.2.zip.txt (5 MB) |
| 2022.3 | pamirs-intellij-plugin-3.0.1-2022.3.zip.txt (4.9 MB) |
| 2023.1 | pamirs-intellij-plugin-3.0.9-2023.1.zip (5 MB) |
| 2023.2 | pamirs-intellij-plugin-4.0.1-2023.2.zip (5 MB) |
| 2023.3 | pamirs-intellij-plugin-4.0.1-2023.3.zip (5 MB) |
| 2024.1 | pamirs-intellij-plugin-4.7.8-2024.1.zip (5 MB) |

图 3-11 不同 IDEA 版本对应插件

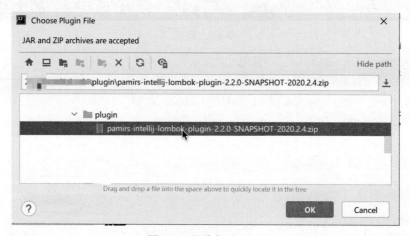

图 3-12 操作指引（1）

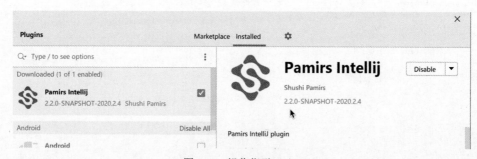

图 3-13 操作指引（2）

4）安装 MySQL 8（下载地址见附录）

（1）解压下载的 zip 安装包，复制到自定义安装目录。

（2）设置环境变量。

MYSQL_BASE_DIR 替换成 MySQL 安装目录路径。

（3）把 %MYSQL_BASE_DIR%\bin 加入到系统环境变量中。

在 PowerShell 中可以使用 Get-Command mysqld 命令验证环境变量是否配置成功，执行成功输出 mysqld 所在的路径。

（4）初始化。

在命令行中执行 mysqld --initialize-insecure --user=mysql。

（5）安装。

mysqld -install

（6）启动 MySQL 服务。

mysqld -install

（7）设置 root 用户密码。

```
alter user 'root'@'localhost' identified with mysql_native_password by 'Oinone';
flush privileges;
```

5）安装 DB GUI 工具

Datagrip、MySQLWorkbench、DBEaver 选其一。

6）安装 Git

下载地址见附录。

7）安装 GraphQL 测试工具 Insomnia

下载地址见附录。

8）安装 RocketMQ（下载地址见附录）

（1）修改安装目录下 bin 中的文件默认配置。

① 修改 runserver.cmd 文件内容 -Xms2g -Xmx2g 为 -Xms1g -Xmx1g。

② 修改 runbroker.com 文件内容 -Xms2g -Xmx2g 为 -Xms1g -Xmx1g，以及 -XX:G1HeapRegionSize=1m。

（2）启动 RocketMQ NameServer 命令。

```
RocketMQ 安装目录 \bin\mqnamesrv.cmd start
```

（3）启动 RocketMQ Broker 命令。

```
RocketMQ 安装目录 \bin\mqbroker.cmd -n localhost:9876
```

（4）停止 RocketMQ 命令。

```
mqshutdown.cmd broker
mqshutdown.cmd namesrv
```

9）安装 ElasticSearch 版本 8.4.1（下载地址见附录）

ES 运行时需要 JDK18 及以上版本 JDK 运行环境，ES 安装包中包含了一个 JDK18 版本。

```
set JAVA_HOME=ES 安装路径 \jdk
```

（1）启动。

ES 安装路径 \bin\elasticsearch.bat。

（2）停止。

Ctrl+C 或者关闭 cmd、PowerShell 的窗口。

10）安装 Redis（下载地址见附录）

（1）解压安装包到安装目录；

（2）在 PowerShell 进入到 Redis 安装目录；

（3）在 PowerShell 中执行 .\redis-server.exe，输出与图 3-14 所示类似信息。

图 3-14　在 PowerShell 中执行 .\redis-server.exe

（4）新开 PowerShell 窗口，进入到 Redis 安装目录，执行 .\redis-cli.exe 按 Enter 键，输入 ping，输出 PONG 即表示 Redis 安装成功，如图 3-15 所示。

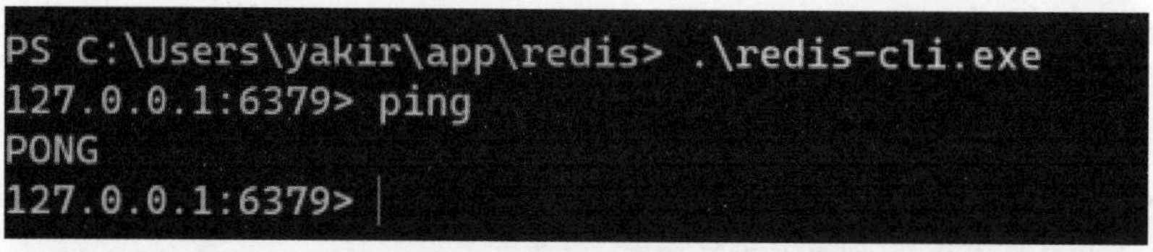

图 3-15　执行 .\redis-cli.exe 按 Enter 键

11）ZooKeeper 安装（下载地址见附录）

（1）解压安装，在 PowerShell 中执行 "tar zxvf 安装包路径（tar.gz 包）-C Zk 安装目录"。

（2）进入 Zk 安装目录：\conf\ 复制 zoo_sample.cfg 文件为 zoo.cfg。

（3）修改 zoo.cfg 文件的内容（其中 dataDir 需要自己设定），代码如下。

```
1  tickTime=2000
2  initLimit=10
3  syncLimit=5
4  clientPort=2181
5  dataDir=需要设置路径用来保存zk数据
6  maxClientCnxns=120
7  autopurge.snapRetainCount=3
8  autopurge.purgeInterval=1
9  admin.enableServer=false
```

（4）进入 Zk 安装目录 \bin\ 执行 zkServer.cmd。

（5）新开 PowerShell 窗口，进入到 Zk 安装目录，执行 zkCli.cmd，连接成功后输入 ls /，输出与图 3-16 所示类似信息即安装成功。

图 3-16　操作演示

**2. 前端环境准备**

1）安装 Node.js 版本 12.12.0（下载地址见附录）

（1）下载 zip 包之后解压到安装目录。

（2）配置环境变量。

① 添加 NODE_HOME 环境变量，如图 3-17 所示。

图 3-17　添加 NODE_HOME 环境变量

② 编辑 PATH 环境变量，如图 3-18 所示。

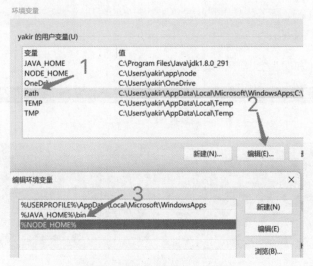

图 3-18　编辑 PATH 环境变量

（3）打开 PowerShell 输入，代码如下。

```
1  node --version
2  npm --version
```

输出与图 3-19 所示类似信息，即为成功安装 node 与 npm。

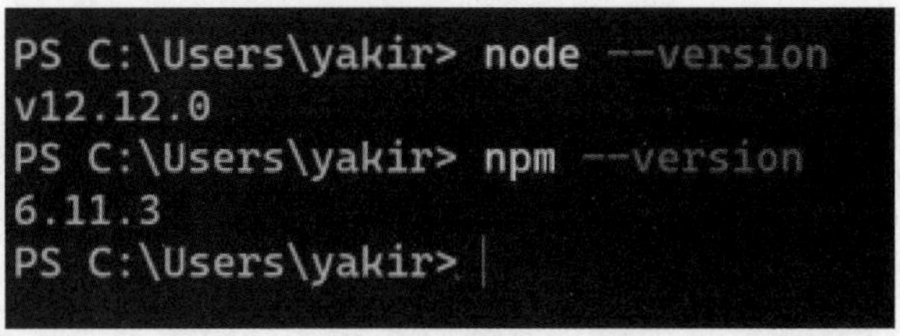

图 3-19　成功安装 node 与 npm

2）安装 vue-cli

安装 vue-cli，代码如下。

```
1  npm install @vue/cli@4.5.17 -g
```

安装完成之后执行 vue --version，如图 3-20 所示。

```
PS C:\Users\yakir> vue.cmd --version
@vue/cli 4.5.17
```

图 3-20　执行 vue --version

输出与图 3-20 所示类似信息，即为成功安装 vue/cli。

3）配置 NPM 源

配置 NPM 源，代码如下。

```
1  npm config set registry http://nexus.shushi.pro/repository/kunlun/
```

4）登录 NPM 源账号

登录 NPM 源账号指令，代码如下。

```
1  npm login --registry "http://nexus.shushi.pro/repository/kunlun/"
2  # username、password、email 请见oinone开源社区群公告，也可以联系oinone合作伙伴或服务人员
3  npm info underscore
```

输出信息如图 3-21 和图 3-22 所示，即为登录成功。

```
Logged in as deploy on http://nexus.shushi.pro/repository/kunlun/.
```

图 3-21　登录 NPM 源账号（1）

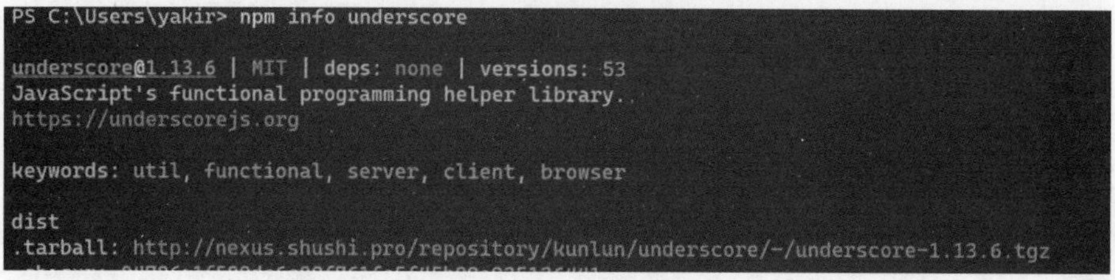

```
PS C:\Users\yakir> npm info underscore

underscore@1.13.6 | MIT | deps: none | versions: 53
JavaScript's functional programming helper library..
https://underscorejs.org

keywords: util, functional, server, client, browser

dist
.tarball: http://nexus.shushi.pro/repository/kunlun/underscore/-/underscore-1.13.6.tgz
```

图 3-22　登录 NPM 源账号（2）

5）安装 cnpm（参见 https://www.npmjs.com/package/cnpm）

安装 cnpm 指令，代码如下。

```
1  npm install cnpm -g --registry=https://registry.nlark.com
2  cnpm.cmd --version
```

输出信息如图 3-23 所示，即为安装成功。

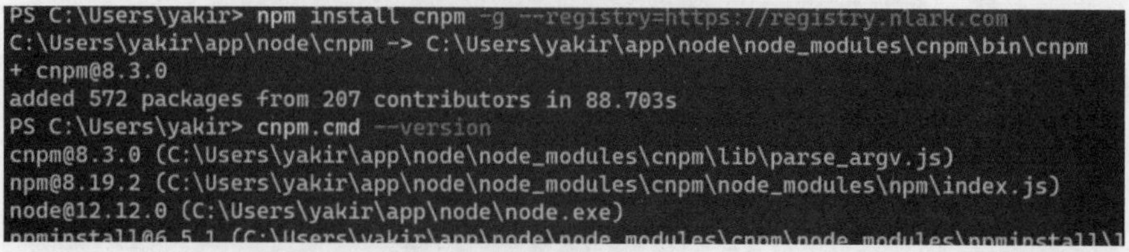

```
PS C:\Users\yakir> npm install cnpm -g --registry=https://registry.nlark.com
C:\Users\yakir\app\node\cnpm -> C:\Users\yakir\app\node\node_modules\cnpm\bin\cnpm
+ cnpm@8.3.0
added 572 packages from 207 contributors in 88.703s
PS C:\Users\yakir> cnpm.cmd --version
cnpm@8.3.0 (C:\Users\yakir\app\node\node_modules\cnpm\lib\parse_argv.js)
npm@8.19.2 (C:\Users\yakir\app\node\node_modules\cnpm\node_modules\npm\index.js)
node@12.12.0 (C:\Users\yakir\app\node\node.exe)
npminstall@6.5.1 (C:\Users\yakir\app\node\node_modules\cnpm\node_modules\npminstall\l
```

图 3-23　安装 cnpm

## 3.2 Oinone 以模块为组织

模块（Module）是按业务领域划分和管理的最小单元，是一组功能、界面的集合。

以下分 3 个小节介绍模块的构建与启动方式，以及 Oinone 模块所包含的内容。

### 3.2.1 构建第一个 Module

现在所有环境准备就绪，就让我们踏上 Oinone 的奇妙之旅吧。先做一个 demo 模块（展示名为"Oinone 的 Demo 工程"，名称为"demoCore"，编码为"demo_core"）试试看，本节学习目的就是能把它启动起来，有个大概的认知。

**1. 后端工程脚手架**

使用如下命令利用项目脚手架生成启动工程。

（1）新建 archetype-project-generate.sh 脚本，或者直接下载。

```
1   #!/bin/bash
2
3   # 项目生成脚手架，用于新项目的构建
4   # 脚手架使用目录
5   # 本地 local，本地脚手架信息存储路径 ~/.m2/repository/archetype-catalog.xml
6   archetypeCatalog=local
7
8   # 以下参数以pamirs-demo为例
9   # 新项目的groupId
10  groupId=pro.shushi.pamirs.demo
11  # 新项目的artifactId
12  artifactId=pamirs-demo
13  # 新项目的version
14  version=1.0.0-SNAPSHOT
15  # Java包名前缀
16  packagePrefix=pro.shushi
17  # Java包名后缀
18  packageSuffix=pamirs.demo
19  # 新项目的pamirs platform Version
20  pamirsVersion=3.0.1-SNAPSHOT
21  # Java类名称前缀
22  javaClassNamePrefix=Demo
23  # 项目名称 module.displayName
24  projectName=oinoneDemo工程
25  # 模块 MODULE_MODULE 常量
26  moduleModule=demo_core
27  # 模块 MODULE_NAME 常量
28  moduleName=DemoCore
29  # spring.application.name
30  applicationName=pamirs-demo
31  # tomcat server address
32  serverAddress=0.0.0.0
33  # tomcat server port
34  serverPort=8090
35  # redis host
36  redisHost=127.0.0.1
37  # redis port
38  redisPort=6379
```

```
39    # 数据库名
40    db=demo
41    # zookeeper connect string
42    zkConnectString=127.0.0.1:2181
43    # zookeeper rootPath
44    zkRootPath=/demo
45
46    mvn archetype:generate \
47      -DinteractiveMode=false \
48      -DarchetypeCatalog=${archetypeCatalog} \
49      -DarchetypeGroupId=pro.shushi.pamirs.archetype \
50      -DarchetypeArtifactId=pamirs-project-archetype \
51      -DarchetypeVersion=3.0.1-SNAPSHOT \
52      -DgroupId=${groupId} \
53      -DartifactId=${artifactId} \
54      -Dversion=${version} \
55      -DpamirsVersion=${pamirsVersion} \
56      -Dpackage=${packagePrefix}.${packageSuffix} \
57      -DpackagePrefix=${packagePrefix} \
58      -DpackageSuffix=${packageSuffix} \
59      -DjavaClassNamePrefix=${javaClassNamePrefix} \
60      -DprojectName="${projectName}" \
61      -DmoduleModule=${moduleModule} \
62      -DmoduleName=${moduleName} \
63      -DapplicationName=${applicationName} \
64      -DserverAddress=${serverAddress} \
65      -DserverPort=${serverPort} \
66      -DredisHost=${redisHost} \
67      -DredisPort=${redisPort} \
68      -Ddb=${db} \
69      -DzkConnectString=${zkConnectString} \
70      -DzkRootPath=${zkRootPath}
```

（2）Linux/Unix/Mac 需要执行以下命令添加执行权限，代码如下。

```
1    chmod +x archetype-project-generate.sh
```

（3）根据脚本中的注释修改项目变量（demo 工程无须编辑）。

（4）执行脚本，代码如下。

```
1    ./archetype-project-generate.sh
```

**2. 后端工程结构介绍**

通过脚手架生成的 demo 工程是我们在 2.4.1 节中介绍的单模块工程结构，属于入门级的一种，"麻雀虽小，五脏俱全"，特别适合新手学习。

（1）结构示意图

后端工程结构示意图如图 3-24 所示。

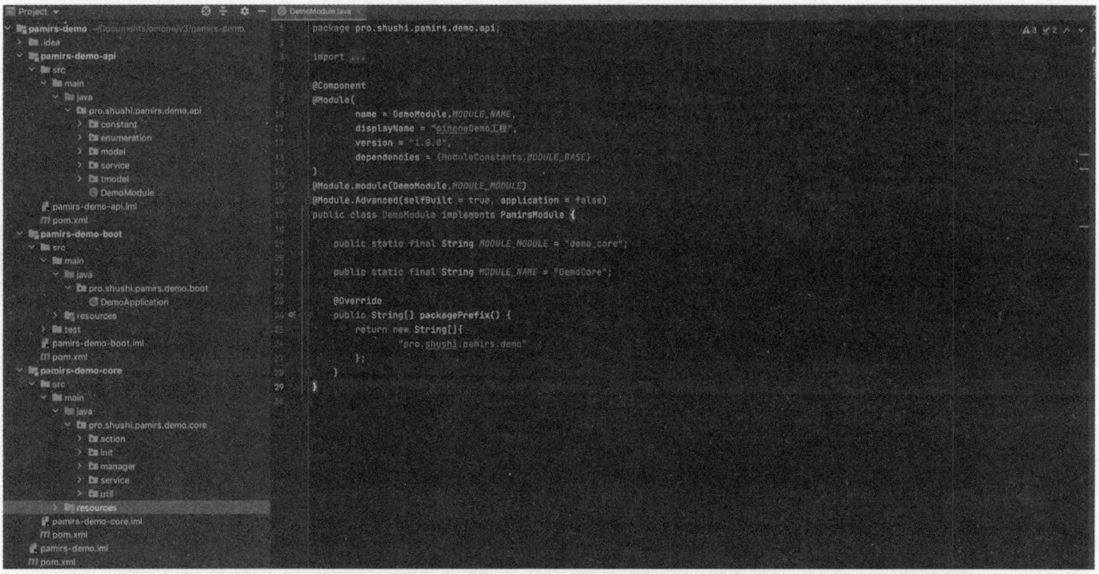

图 3-24　结构示意图

（2）工程结构说明

工程结构说明见表 3-3。

表 3-3　工程结构说明

| 工程名 | 包名 | 说明 |
| --- | --- | --- |
| pamirs-demo-api | | 对外 api 包，如果有其他模块需要依赖 demo 模块，则可以在其 pom 中引入 pamirs-demo-api 包 |
| | constant | 常量的包路径 |
| | enumeration | 枚举类的包路径 |
| | Model | 该领域核心模型的包路径 |
| | service | 该领域对外暴露接口 api 的包路径 |
| | tModel | 存放该领域的非存储模型，如：用于传输的临时模型 |
| | DemoModule | 该类是 demo 模块的定义 |
| pamirs-demo-boot | | demo 模块的启动类 |
| | boot | 启动类的包路径 |
| | DemoApplication | demo 模块的应用启动类，遵循 Spring Boot 规范 |
| | resources/config/application-dev.yml | 研发环境的 yml 配置文件，遵循 Spring Boot 规范 |
| | resources/bootstrap.yml | 启动的 yml 配置文件，遵循 spring boot 规范 |
| pamirs-demo-core | | |
| | action | 模型对外交互的行为的包路径 |
| | init | 模块初始化工作的包路径 |
| | manager | manager 是 service 的一些公共逻辑，不会定义为独立的 function 的类 |
| | service | service 是对应 api 工程中 service 接口的实现类，是模型的 function |

**3. pom.xml**

1）父 pom 的依赖管理

（1）启动包依赖。

① pamirs-boot 里有多种启动模式依赖包。这里只介绍标准启动模式。

② pamirs-base-standard。

（2）为了方便介绍下面的 demo-boot 工程和匹配 Oinone 的前端应用，这里我们直接引入 pamirs-core，它提供了一些基础业务功能，比如国际化、消息、用户、权限、商业关系等，这些基础功能模块为前端工程提供了一些基础能力，后续会介绍。如果不用 Oinone 提供的前端工程，可以不使用依赖，如图 3-25 所示。

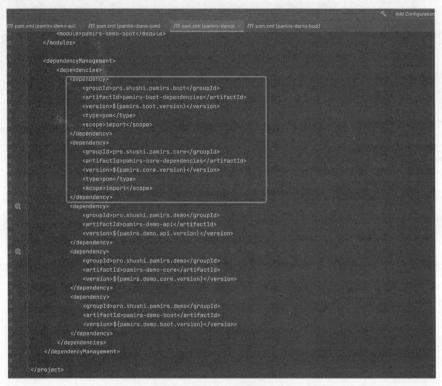

图 3-25　父 pom 的依赖管理

2）pamirs-domo-api

pamirs-domo-api 是一个标准 Java 工程，可以看出只是依赖了 pamirs-base-standard，如图 3-26 所示。

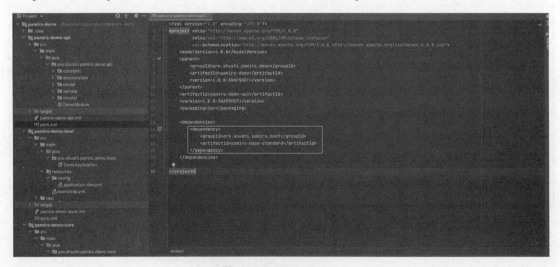

图 3-26　pamirs-domo-api

3）pamirs-demo-core

pamirs-demo-core 是一个标准 Java 工程，可以看出只是依赖了 pamirs-demo-api，如图 3-27 所示。

图 3-27　pamirs-demo-core

4）pamirs-demo-boot

把需要启动的 Module（模块）对应的 jar 进行依赖引入，这样就可以在 yml 文件中启动 module 的配置。

（1）所需 module（模块）对应的 jar 如图 3-28 所示。

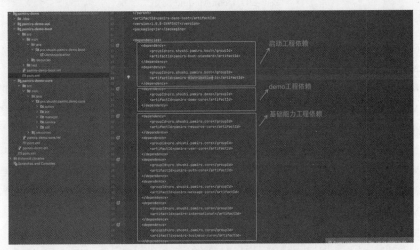

图 3-28　所需 module（模块）对应的 jar

（2）boot 工程的 yml 配置详情参见 4.1.1 节，如图 3-29 所示。

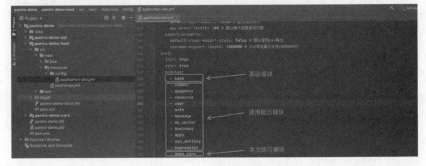

图 3-29　yml 文件基础配置

**4. DemoModule 的定义**

到此 Oinone 中一个 Module 对应的工程就介绍完了。到目前为止跟一般的 spring boot 的工程没有太多区别，只是多做了一个 DemoModule 的定义，Oinone 的所有 Module 都继承自 PamirsModule，如图 3-30 所示。

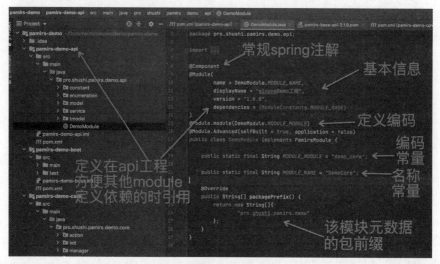

图 3-30　DemoModule 的定义

1）配置注解

通过 @Module 的 name 属性配置模块技术名称，前端与后端交互协议使用模块技术名称来定位模块。

通过 @Module 的 displayName 属性配置模块展示名称，在产品视觉交互层展现。

通过 @Module 的 version 属性配置模块版本，系统会比较版本号来决定模块是否需要进行升级。

通过 @Module 的 priority 属性配置模块优先级（数字越小，优先级越高），系统会根据优先级取优先级最高的应用设置的首页来作为整个平台的首页。

通过 @Module 的 dependencies 属性和 exclusions 属性来配置模块间的依赖互斥关系，值为模块编码数组。如果模块继承了另一模块的模型或者与另一模块的模型建立了关联关系，则需要为该模块的依赖模块列表配置另一模块的模块编码。

通过 @Module.module 配置模块编码，模块编码是模块在系统中的唯一标识。

通过 @Module.Advanced 的 selfBuilt 属性配置模块是否为平台自建模块。

通过 @Module.Advanced 的 application 属性配置模块是否为应用（具有视觉交互页面的模块）。模块切换组件只能查看到应用。

通过 @UxHomepage 注解配置首页。@UxHomepage 注解的 Model 属性指定跳转页面的模型编码，name 属性指定跳转页面的视图动作，默认为列表页。如果配置首页为列表页且未定义 ViewAction，系统会根据首页配置自动生成。

更多 Module 的详细元数据描述参见 4.1.4 节。

2）配置扫描路径

设置扫描模型配置的包路径：

● 使用 packagePrefix 方法配置模块需要扫描模型配置的包路径。

● 使用 dependentPackagePrefix 方法配置依赖模块的元数据所在包路径；如果不配置，默认根据依赖模块的自有配置的包路径扫描元数据。

● 不同模块配置的包路径，如果相互包含，会导致加载模块时元数据出问题。

3）命名规范

命名规范见表 3-4。

表 3-4 命名规范

| 属性 | 默认取值规则 | 命名规范 |
| --- | --- | --- |
| Module | 无默认值<br>开发人员定义规范示例：<br>{项目名称}_{模块功能示意名称} | 使用下画线命名法<br>仅支持数字、大写或小写字母、下画线<br>必须以字母开头<br>不能以下画线结尾<br>长度必须小于或等于 128 个字符 |
| name | 无默认值 | 使用大驼峰命名法<br>仅支持数字、字母<br>必须以字母开头<br>长度必须小于或等于 128 个字符 |

**5. DemoModule 的启动**

（1）修改 yml 文件中数据源配置，数据库地址与密码修改为自己的。

（2）需要为启动或依赖模块配置扫描路径：pmairs 是 Oinone 的基础包（必选）；himalaya、tanggula 是业务模型基础包（可选）；pro.shushi.pamirs.demo 是测试项目包路径（必选，只不过它也在 pamirs 路径下，不填也可以）。

数据源配置如图 3-31 所示。

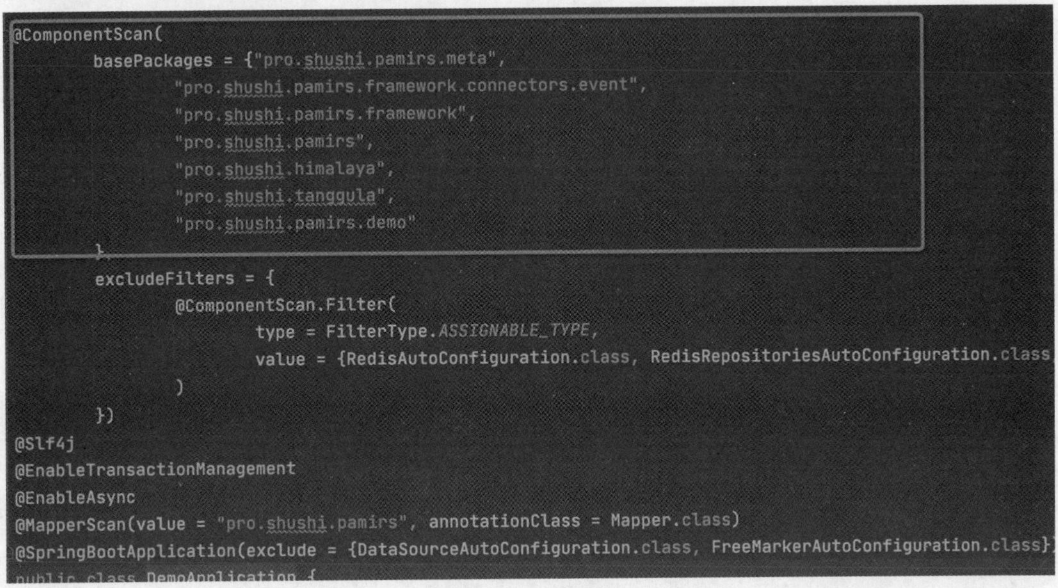

```
@ComponentScan(
        basePackages = {"pro.shushi.pamirs.meta",
                "pro.shushi.pamirs.framework.connectors.event",
                "pro.shushi.pamirs.framework",
                "pro.shushi.pamirs",
                "pro.shushi.himalaya",
                "pro.shushi.tanggula",
                "pro.shushi.pamirs.demo"
        },
        excludeFilters = {
                @ComponentScan.Filter(
                        type = FilterType.ASSIGNABLE_TYPE,
                        value = {RedisAutoConfiguration.class, RedisRepositoriesAutoConfiguration.class}
                )
        })
@Slf4j
@EnableTransactionManagement
@EnableAsync
@MapperScan(value = "pro.shushi.pamirs", annotationClass = Mapper.class)
@SpringBootApplication(exclude = {DataSourceAutoConfiguration.class, FreeMarkerAutoConfiguration.class})
public class DemoApplication {
```

图 3-31 数据源配置

（3）启动应用，我们会发现报错，如图 3-32 所示。这个错误主要因为 Oinone 默认的是 RELOAD（重启）模式，这种模式下只会从数据库中读取已安装模块并与 yml 文件中配置需要加载的模块对比，如图 3-33 所示。如果数据库中没有则会报没有安装过 ** 模块，而我们是第一次启动，前面必然没有安装过模块，所以我们要把模式变为 INSTALL，启动参数项见 4.1.2 节。如果整个平台第一次启动则会报 "'base_sequence_config' doesn't exist"。

图 3-32　启动应用会发现报错

图 3-33　因为首次启动没有安装相关模块

（4）把模式变为 INSTALL 再次启动，如图 3-34 所示。

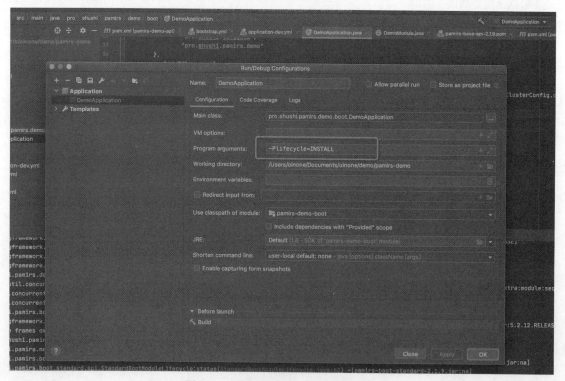

图 3-34　把模式变为 INSTALL 再次启动

（5）查看日志，看到对应启动字样才代表启动完成，而非 Spring Boot 的启动完成日志，如图 3-35 所示。Oinone 是在 Spring Boot 启动完以后异步启动的。

图 3-35　查看日志

（6）打开 GraphQL 客户端 Insomnia 看对应服务是否暴露。

① 创建一个 demo 项目，如图 3-36、图 3-37 所示。

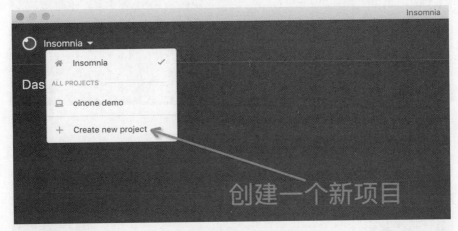

图 3-36　创建一个新项目

图 3-37　命名为 demo

② 为 demo 项目创建一个请求集合，如图 3-38、图 3-39、图 3-40 所示。

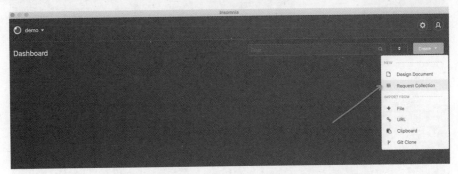

图 3-38　为 demo 项目创建一个请求集合（第一步）

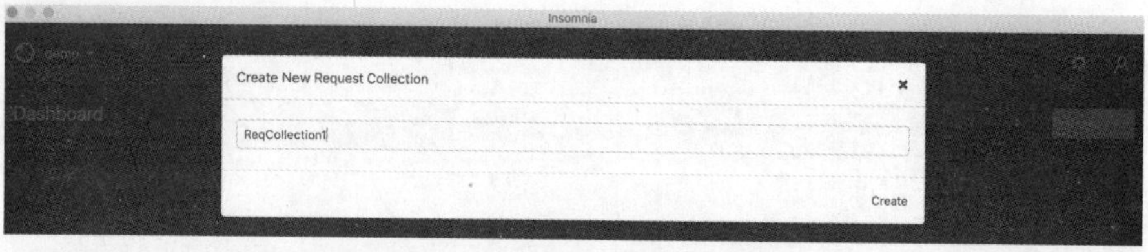

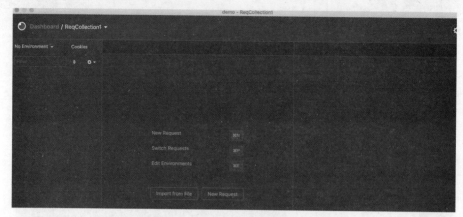

图 3-39　为 demo 项目创建一个请求集合（第二步）

图 3-40　为 demo 项目创建一个请求集合（第三步）

③ 创建一个请求 command+N，如图 3-41～图 3-44 所示。

图 3-41　创建一个请求 command+N

图 3-42　选择 POST

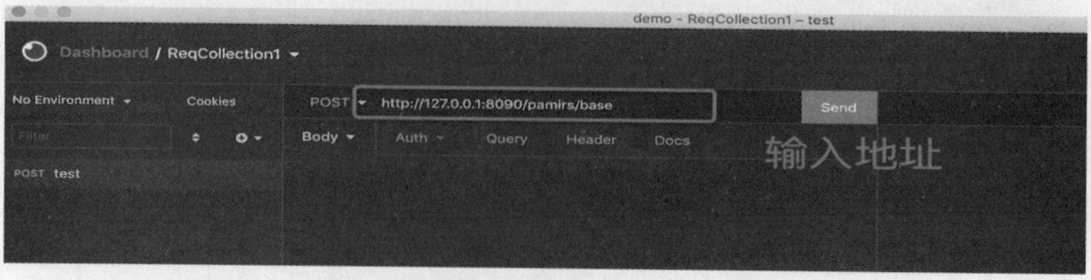

图 3-43　输入地址

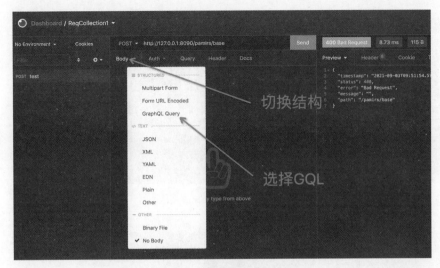

图 3-44 切换结构并选择 GQL

④ 检验结果，对应后端提供的服务已经有提示了，如图 3-45 所示。

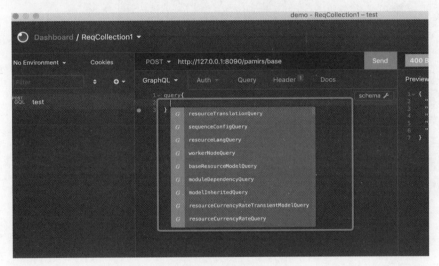

图 3-45 检验结果

⑤ 查看 schema 文档，如图 3-46～图 3-48 所示。

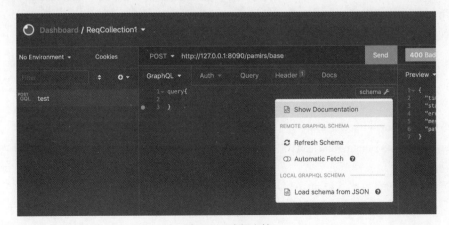

图 3-46 选择文档

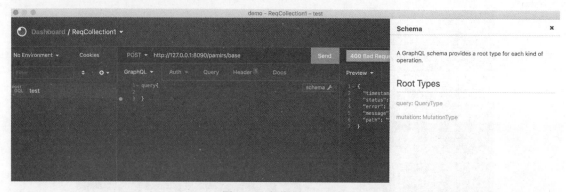

图 3-47　查看 schema 文档

图 3-48　查看细节说明

恭喜第一个 Oinone 模块已经启动完成，下面将通过页面来看看 Oinone 的系统是什么样的。

### 3.2.2　启动前端工程

让我们直观地感受一下上面构建的 demo 模块，并搭建前端环境为后续学习打下基础。

**1. 使用 vue-cli 构建工程**

代码如下：

```
##demo-front是项目名，可以替换成自己的
vue create --preset http://ss.gitlab.pamirs.top/:qilian/pamirs-archetype-front3 --c
```

如果启动报错，清除 node_modules 后重新执行 npm i。

mac 清除命令：npm run cleanOs。

windows 清除命令：npm run clean。

**2. 启动前端工程**

找到 README.MD 文件，根据文件一步一步操作。

（1）找到 vue.config.js 文件，修改 devServer.proxy.pamirs.target 为后端服务的地址和端口。

```js
1  const WidgetLoaderPlugin = require('@kunlun/widget-loader/dist/plugin.js').default
2  const Dotenv = require('dotenv-webpack');
3
4  module.exports = {
5    lintOnSave: false,
6    runtimeCompiler: true,
7    configureWebpack: {
8      module: {
9        rules: [
10         {
11           test: /\.widget$/,
12           loader: '@kunlun/widget-loader'
13         }
14       ]
15     },
16     plugins: [new WidgetLoaderPlugin(), new Dotenv()],
17     resolveLoader: {
18       alias: {
19         '@kunlun/widget-loader': require.resolve('@kunlun/widget-loader')
20       }
21     }
22   },
23   devServer: {
24     port: 8080,
25     disableHostCheck: true,
26     progress: false,
27     proxy: {
28       pamirs: {
29         // 支持跨域
30         changeOrigin: true,
31         target: 'http://127.0.0.1:8090'
32       }
33     }
34   }
35 };
```

注：要用 localhost 域名访问，.env 文件这里也要改成 localhost。如果开发中一定要使前后端域名不一致，老版本 Chrome 会报错，修改可以参考 https://www.cnblogs.com/willingtolove/p/12350429.html，或者下载新版本 Chrome。

（2）进入前端工程 demo-front 文件目录下，执行 npm run dev，最后出现如下所示界面就代表启动成功。

```
App running at:
- Local:   http://localhost:8081/
- Network: http://192.168.125.4:8081/

Note that the development build is not optimized.
To create a production build, run npm run build.
```

（3）使用 http://127.0.0.1:8081/login 进行访问，并用 admin 账号登录，默认密码为 admin，如图 3-49 所示。

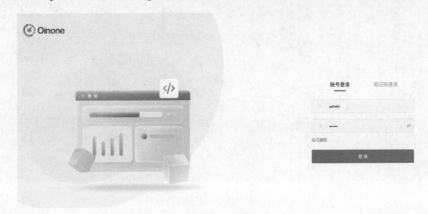

图 3-49　登录界面

（4）单击左上角进行应用切换，会进入 App Finder 页面，可以看到所有已经安装的应用，可以对照 boot 工作的 yml 配置文件查看。细心的小伙伴应该已经注意到了，在 App Finder 页面出现的应用跟我们启动工程 yml 配置文件中加载的启动模块数不是一一对应的，同时也没有看到我们的 demo 模块，如图 3-50 所示。

图 3-50　已安装应用界面

表 3-5　boot 工作的 yml 文件中加载模块及 App Finder 应用说明

| boot 工作的 yml 文件中加载模块 | App Finder 的应用 | 说明 |
| --- | --- | --- |
| - base<br>- common<br>- sequence<br>- international | 无 | 模块的 application = false |
| - resource<br>- user<br>- auth<br>- business<br>- message<br>- apps | 有 | 模块的 application = true |
| - demo_core | 无 | 刚建的 OinoneDemo 工程，默认为 false |
| 设计器：无 | 设计器：无 | 因为 boot 中没有加载设计器模块，所以 App Finder 中的设计器 tab 选项卡下没有应用 |

（5）只需要修改 OinoneDemo 工程的模块定义，如图 3-51 所示，就可以在 App Finder 页面看见"OinoneDemo 工程"，如图 3-52 所示。

图 3-51　修改 OinoneDemo 工程的模块定义

图 3-52　在 App Finder 页面看见"OinoneDemo 工程"

目前 Oinone 的 demo 模块还是一个全空的，所以我们还单击不了。在后续的学习过程中我们会不断完善该模块。

至此恭喜您，前端工程已经启动完成。

**3. 前端工程结构介绍**

```
├── public 发布用的目录，index.html 入口文件将在这里
│
├── src 源代码
│   ├── actions 扩展动作的目录
│   ├── assets css/image 等静态资源目录
│   ├── field 模型字段对应控件目录
│   ├── layout 布局控件目录
│   ├── middleware 中间件
│   │   └── network-interceptor 网络请求拦截处理
│   ├── view 自定义视图目录
│   └── main.ts 应用入口文件，这里会注册 providers/application.ts
├── .env.example 启动的环境变量，后端 api 的请求地址在这里
├── package.json 包描述文件
├── tsconfig.json ts 配置文件，可配置语法校验
└── vue.config.js vue 的配置文件，里面可以配置 webpack 参数和开发模式的后端 api 请求地址
```

### 3.2.3　应用中心

在 App Finder 中单击"应用中心"可以进入 Oinone 的"应用中心"，可以看到 Oinone 平台所有应用列表、应用大屏以及技术可视化内容。

**1. 应用列表**

标准版本不支持在线安装，只能通过 boot 工程的 yml 文件来配置安装模块。在 www.oinone.top 官方 SaaS 平台，客户可以在线管理应用生命周期，如安装、升级、卸载。同时针对已安装应用可以进行无代码设计（前提安装了设计器），针对应用类的模块则可进行收藏，收藏后会在 App Finder 的"我收藏的应用"中出现。在应用列表中可以看到我们已经安装的应用以及模块，Oinone demo 工程也在其中，如图 3-53、图 3-54 所示。

图 3-53　在线管理应用生命周期

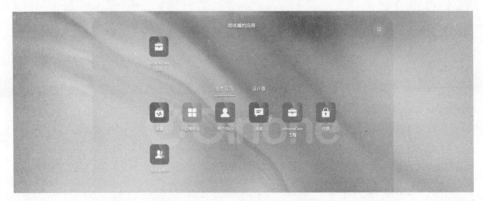

图 3-54　收藏后会在 App Finder 的"我收藏的应用"中出现

## 2. 应用大屏

但我们的测试应用没有设置应用类目,则无法在应用大屏中呈现,如图 3-55 所示。

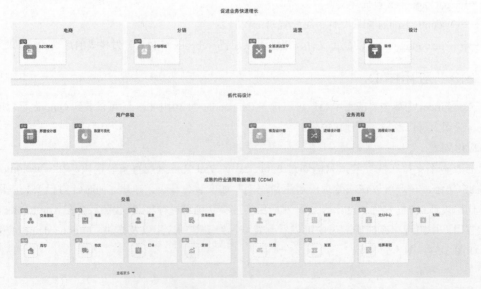

图 3-55　未设置应用类目则无法在应用大屏中呈现

**3. 技术可视化**

在技术可视化页面，展示已经安装模块的元数据，并进行分类呈现，如图 3-56 所示。

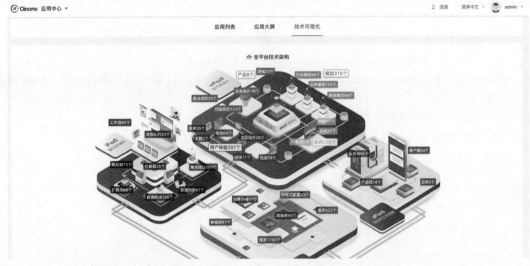

图 3-56  技术可视化分类呈现

## 3.3  Oinone 以模型为驱动

模型（Model）：Oinone 一切从模型出发，模型是数据及行为的载体。

模型是对所需要描述的实体进行必要的简化，并用适当的形式或规则把它的主要特征描述出来所得到的系统模仿品。模型由元信息、字段、数据管理器和自定义函数构成。模型符合面向对象设计原则，包括、封装、继承、多态。

### 3.3.1  构建第一个模型

定义模型，并配上相应的菜单或配置模块的 homepage，模块就具备了可访问该模型对应的列表页、新增页、修改页、删除记录和导入/导出功能。

都说 Oinone 是以模型为驱动，下面通过一个实例说明。

**1. 构建宠物店铺模型**

构建宠物店模型，代码如下。

```java
 9  @Model.model(PetShop.MODEL_MODEL)
10  @Model(displayName = "宠物店铺",summary="宠物店铺")
11  public class PetShop extends IdModel {
12      public static final String MODEL_MODEL="demo.PetShop";
13
14      @Field(displayName = "店铺名称",required = true)
15      private String shopName;
16
17      @Field(displayName = "开店时间",required = true)
18      private Time openTime;
19
20      @Field(displayName = "闭店时间",required = true)
21      private Time closeTime;
22  }
```

**2. 配置注解**

（1）必须使用 @Model 注解来标识当前类为模型类。

（2）可以使用 @Model.Model、@Fun 注解模型的编码（也表示命名空间），先取 @Model.Model 注解值，若为空则取 @Fun 注解值，若皆为空则取全限定类名。

（3）使用 @Model.Model 注解配置模型编码，模型编码唯一标识一个模型。

（4）请勿使用 Query 和 Mutation 作为模型编码和技术名称的结尾。

上方示例使用模型注解和 Field 注解来定义一个实体模型。displayName 属性最终会作为 label 展现在前端界面上。

模型的详细元数据描述参见 4.1.6 节。

**1. 模型命名规范**

模型命名规范见表 3-6。

表 3-6　模型命名规范

| 模型属性 | 默认取值规范 | 命名规则规范 |
| --- | --- | --- |
| name | 默认取 Model.Model 的点分割最后一位 | 仅支持数字、字母<br>必须以字母开头<br>长度必须小于或等于 128 个字符 |
| module | 无默认值<br>开发人员定义规范示例：<br>{ 项目名称 }_{ 模块功能示意名称 } | 仅支持数字、大写或小写字母、下画线<br>必须以字母开头<br>不能以下画线结尾<br>长度必须小于或等于 128 个字符 |
| Model | 默认使用全类名，取 lname 的值<br>开发人员定义规范示例：<br>{ 项目名称 }.{ 模块功能示意名称 }.{ 简单类名 } | 仅支持数字、字母、点<br>必须以字母开头<br>不能以点结尾<br>长度必须小于或等于 128 个字符 |
| display_name | 空字符串 | 长度必须小于或等于 128 个字符 |
| lname | 符合 Java 命名规范，真实的 Java 全类名，无法指定，要符合 Model 的约束，即为"包名+类名" | lname 是不能定义的，为全类名：包名+类名，和 Model 一样的校验规则：包名和类名的校验 |
| summary | 默认使用 displayName 属性 | 不能使用分号<br>长度必须小于或等于 128 个字符 |
| description | NULL，注解无法定义 | 长度必须小于或等于 65535 个字符 |
| table | 默认使用 name 字段生成表名时，table 字段的命名规则约束同样生效（大小驼峰命名转为下画线分割的表名称） | 仅支持数字、字母、下画线<br>长度必须小于或等于 128 个字符（此限制为系统存储约束，与数据库本身无关） |
| type | Java 属性类型与数据库存储类型可执行转换即可 | ModelTypeEnum 枚举值 |

**2. 字段命名规范**

字段命名规范见表 3-7。

表 3-7　字段命名规范

| 字段属性 | 默认取值规范 | 命名规则规范 |
| --- | --- | --- |
| name | 默认使用 Java 属性名 | 仅支持数字、字母<br>必须以小写字母开头<br>长度必须小于或等于 128 个字符 |
| field | 默认使用 Java 属性名 | 与 name 使用相同命名规则约束 |
| display_name | 默认使用 name 属性 | 长度必须小于或等于 128 个字符 |

续表

| 字段属性 | 默认取值规范 | 命名规则规范 |
| --- | --- | --- |
| lname | 使用 Java 属性名，符合 Java 命名规范，真实的属性名称，无法指定 | 与 name 使用相同命名规则约束 |
| column | 列名为属性名的小驼峰转下画线格式 | 仅支持数字、字母、下画线<br>长度必须小于或等于 128 个字符（此限制为系统存储约束，与数据库本身无关） |
| summary | 默认使用 displayName 属性 | 不能使用分号<br>长度必须小于或等于 500 个字符 |

我们重启 Demo 应用以后，打开 Insomnia 刷新 GraphQL 的 schema，就可以看到 PetShop 默认对应的读写服务了，如图 3-57、图 3-58 所示。

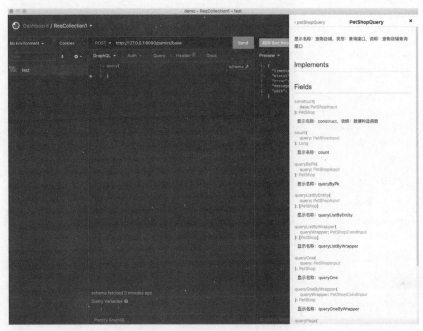

图 3-57　PetShopQuery 默认读写服务

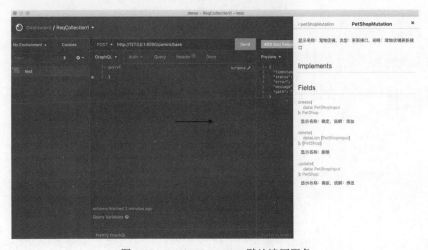

图 3-58　PetShopMutatiom 默认读写服务

**3. 配置模块的主页为宠物商店的列表页**

为了方便大家对模型有个更加直观的了解，接下来我们通过前端交互来感受一下。

在 3.2.2 节，在模块下拉列表中"OinoneDemo 工程"还是不能单击。是因为该模块没有配置主页，我们现在把主页设置为宠物商店的列表页，只需要在 DemoModule 这个类上增加一个注解 @UxHomepage[@UxRoute（PetShop.MODEL_MODEL）]。

```
11  @Component
12  @Module(
13          name = DemoModule.MODULE_NAME,
14          displayName = "oinoneDemo工程",
15          version = "1.0.0",
16          dependencies = {ModuleConstants.MODULE_BASE}
17  )
18  @Module.module(DemoModule.MODULE_MODULE)
19  @Module.Advanced(selfBuilt = true, application = true)
20  @UxHomepage(@UxRoute(PetShop.MODEL_MODEL))
21  public class DemoModule implements PamirsModule {
22      public static final String MODULE_MODULE = "demo_core";
23      public static final String MODULE_NAME = "DemoCore";
24
25      @Override
26      public String[] packagePrefix() {
27          return new String[]{"pro.shushi.pamirs.demo"};
28      }
29  }
```

重启 demo 应用，打开前端页面登录后，通过 App Finder 切换至"OinoneDemo 工程"，我们跟着以下操作一步一步来体验。

（1）在 App Finder 的应用列表选中 OinoneDemo 工程，我们会看到系统进入了宠物商店列表，如图 3-59 所示。

图 3-59　进入了宠物商店列表

（2）单击宠物商店列表页面中新增按钮进入宠物商店的新增页面，如图 3-60 所示。

图 3-60　进入宠物商店的新增页面

（3）单击宠物商店的新增页面的"确定"按钮，回到宠物商店列表，并能看到刚刚新增的记录，如图 3-61 所示。

图 3-61　新增记录产生

（4）对宠物商店列表的数据记录可以进行编辑、详情操作。如单击"编辑"则进入数据记录的宠物商店数据记录的编辑页，如图 3-62 所示。

图 3-62　宠物商店数据记录的编辑页

**4. 查看数据库**

大家可能会问：我们的数据存储在哪里呢？

我们这个 demo 应用一共生成了两个库，即 demo 和 demo_base，分别对应 yml 文件两个数据源，即 pamirs 和 base：base 库存放系统元数据；pamirs 库存放 demo 模块以及其他安装模块的数据，如图 3-63 所示。

图 3-63　pamirs 库存放了 demo 模块以及其他安装模块的数据

我们来找到 demo 库，就可以看到有一个 demo_core_pet_shop 表以及我们新增的数据。

库、表的规则是 boot 工程里的 yml 文件配置的，如何配置可以参考 4.1.1 节。

定义模型，并配上相应的菜单或配置模块的 homepage 为该模型，模块就具备了可访问的该模型对应的列表页、新增页、修改页、删除记录和导入 / 导出功能。

注：本节中的所有页面都是系统默认的，更多有关前端交互包括菜单、页面、行为知识可以在 3.5 节中详细学习。

### 3.3.2　模型的类型

模型分为元模型和业务模型：元模型是指用于描述内核元数据的一套模式集合；业务模型是指用于描述业务应用元数据的一套模式集合。

元模型分为模块域、模型域和函数域三个域。域的划分规则是根据元模型定义数据关联关系的离散性来判断的，离散程度越小越聚集到一个域。4.1.4 节中 ModuleDefinition 就是元模型，而我们在开发中涉及的就是业务模型。

**1. 模型类型**

● 抽象模型：往往是提供公共能力和字段的模型，它本身不会直接用于构建协议和基础设施（如表结构等）。

● 传输模型：用于表现层和应用层之间的数据交互，本身不会存储数据，没有默认的数据管理器，只有数据构造器。

● 存储模型：存储模型用于定义数据表结构和数据的增删改查（数据管理器）功能，是直接与连接器进行交互的数据容器。

● 代理模型：用于代理存储模型的数据管理器能力的同时，扩展出非存储数据信息的交互功能的模型。

**2. 模型定义种类**

模型定义就是模型描述，不同定义的类型，代表计算描述模型的元数据规则不同。

● 静态模型定义：模型元数据不持久化，不进行模型定义的计算（默认值、主键、继承、关联关系）。

● 静态计算模型定义：模型元数据不持久化但初始化时进行模型定义计算获得最终的模型定义。

● 动态模型定义：模型元数据持久化且初始化时进行模型定义计算获得最终的模型定义。

静态模型定义需要使用 @Model.Static 进行注解；静态计算模型定义使用 @Model.Static(compute=true) 进行注解；动态模型定义不注解。

**3. 安装与更新**

使用 @Model.Model 来配置模型的不可变更编码。模型一旦安装，无法再对该模型编码值进行修改，之后的模型配置更新会依据该编码进行；如果仍然修改该模型的编码值，则系统会将该模型识别为新模型，存储模型会创建新的数据库表，而原表将会重命名为废弃表。

如果模型配置了 @Base 注解，表明在"Oinone 的设计器"中该模型配置不可变更；如果字段配置了 @Base 注解，表明在"Oinone 的设计器"中字段配置不可变更。

**4. 基础配置**

1）模型基类

所有的模型都需要继承以下模型中的一种，来表明模型的类型，同时继承以下模型的默认数据管理器（详见 3.3.3 节）。

● 继承 BaseModel，构建存储模型，默认无 id 属性。

● 继承 BaseRelation，构建多对多关系模型，默认无 id 属性。

● 继承 TransientModel，构建临时模型（传输模型），临时模型没有数据管理器，也没有 id 属性。

● 继承 EnhanceModel，构建数据源为 ElasticSearch 的增强模型。

2）快捷继承

● 继承 IdModel，构建主键为 id 的模型。继承 IdModel 的模型数据管理器会增加 queryById 方法（根据 id 查询单条记录）。

● 继承 CodeModel，构建带有唯一编码 code 的主键为 id 的模型。可以使用 @Model.Code 注解配置编码生成规则，也可以直接覆盖 CodeModel 的 generateCode 方法或者自定义新增的前置扩展点自定义编码生成逻辑。继承 CodeModel 的模型数据管理器会增加 queryByCode 方法（根据唯一编码查询单条记录）。

● 继承 VersionModel，构建带有乐观锁，唯一编码 code 且主键为 id 的模型。

● 继承 IdRelation，构建主键为 id 的多对多关系模型。

3）模型继承关系

模型继承关系如图 3-64 所示。

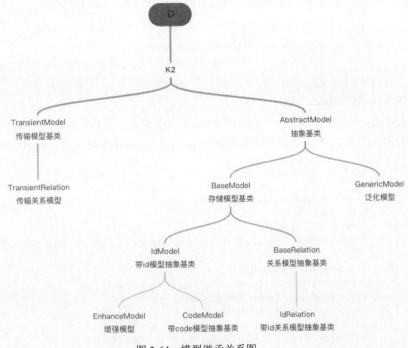

图 3-64　模型继承关系图

● AbstractModel 抽象基类是包含 createDate 创建时间、writeDate 更新时间、createUid 创建用户 id、writeUid 更新用户 id、aggs 聚合结果和 activePks 批量主键列表等基础字段的抽象模型。

● TransientModel 传输模型基类是所有传输模型的基类，传输模型不存储，没有数据管理器。

● TransientRelation 传输关系模型是所有传输关系模型的基类，传输关系模型不存储，用于承载多对多关系，没有数据管理器。

● BaseModel 存储模型基类提供数据管理器功能，数据模型主键可以不是 id。

● IdModel 带 id 模型抽象基类，在 BaseModel 数据管理器基础之上提供根据 id 查询、更新、删除数据的功能。

● BaseRelation 关系模型抽象基类用于承载多对多关系，是多对多关系的中间模型，数据模型主键可以不是 id。

● IdRelation 带 id 关系模型抽象基类，在 BaseRelation 数据管理器基础之上提供根据 id 查询、更新、删除数据的功能。

● CodeModel 带 code 模型抽象基类，提供按配置生成业务唯一编码功能，根据 code 查询、更新、删除数据的功能。

● EnhanceModel 增强模型，提供全文检索能力。此模型会在 4.1.25 节中展开介绍。

**5. 抽象模型（举例）**

抽象模型本身不会直接用于构建协议和基础设施（如表结构等），而是通过继承的机制供子模型复用其字段和函数。子模型可以是所有类型的模型。

比如 demo 模块要管理的一些公共模型字段，我们可以建一个 AbstractDemoIdModel 和 AbstractDemoCodeModel，demo 模块中的实体模型就可以继承它们。我们来为 demo 模块的模型统一增加一个数据状态字段，用作数据的生效与失效管理。

Step1. 引入 DataStatusEnum 类。

使用 pamirs-demo-api 的 pom.xml 包增加依赖，便于引入 DataStatusEnum 类，当然也可以自己建，Oinone 提供了统一的数据记录状态的枚举，以及相应的通用方法，可以直接引入。

```
1  <dependency>
2      <groupId>pro.shushi.pamirs.core</groupId>
3      <artifactId>pamirs-core-common</artifactId>
4  </dependency>
```

Step2. 修改 DemoModule。

DataStatusEnum 枚举类本身也会作为数据字典，以元数据的方式被管理起来。当一个模块依赖另一个模块的元数据相关对象，则需要改模块的依赖定义，为 DemoModule 增加 CommonModule 的依赖注解。

```
12  @Component
13  @Module(
14          name = DemoModule.MODULE_NAME,
15          displayName = "oinoneDemo工程",
16          version = "1.0.0",
17          dependencies = {ModuleConstants.MODULE_BASE, CommonModule.MODULE_MODULE}
18  )
19  @Module.module(DemoModule.MODULE_MODULE)
20  @Module.Advanced(selfBuilt = true, application = true)
21  @UxHomepage(@UxRoute(PetShop.MODEL_MODEL))
22  public class DemoModule implements PamirsModule {
23      public static final String MODULE_MODULE = "demo_core";
24      public static final String MODULE_NAME = "DemoCore";
25
26      @Override
27      public String[] packagePrefix() {
28          return new String[]{ "pro.shushi.pamirs.demo"};
29      }
30  }
```

Step3. 新建 AbstractDemoCodeModel 和 AbstractDemoIdModel。

新增 AbstractDemoIdModel 和 AbstractDemoCodeModel 分别继承 IdModel 和 CodeModel，实现 IDataStatus 接口不是必需的，DataStatus 有配套的通用逻辑，可以先加进去，具体使用会在下文的"代理模型"介绍。

新建 AbstractDemoIdModel，代码如下。

```
11  @Base
12  @Model.model(AbstractDemoCodeModel.MODEL_MODEL)
13  @Model.Advanced(type = ModelTypeEnum.ABSTRACT)
14  @Model(displayName = "AbstractDemoIdModel")
15  public abstract class AbstractDemoCodeModel extends CodeModel implements IDataSta
    tus {
16
17      public static final String MODEL_MODEL="demo.AbstractDemoCodeModel";
18
19      @Base
20      @Field.Enum
21      @Field(displayName = "数据状态",defaultValue = "DISABLED",required = true,summa
    ry = "作为基类给每一个继承模型增加一个数据状态字段")
22      private DataStatusEnum dataStatus;
23  }
```

新建 AbstractDemoIdModel，代码如下。

```
11    @Base
12    @Model.model(AbstractDemoIdModel.MODEL_MODEL)
13    @Model.Advanced(type = ModelTypeEnum.ABSTRACT)
14    @Model(displayName = "AbstractDemoIdModel")
15    public abstract class AbstractDemoIdModel extends IdModel implements IDataStatus
      {
16
17        public static final String MODEL_MODEL="demo.AbstractDemoIdModel";
18
19        @Base
20        @Field.Enum
21        @Field(displayName = "数据状态",defaultValue = "DISABLED",required = true,summa
      ry = "作为基类给每一个继承模型增加一个数据状态字段")
22        private DataStatusEnum dataStatus;
23    }
```

**Step4.** 修改 PetShop 的父类。

修改 PetShop 父类从 IdMode 变更为 AbstractDemoIdModel，代码如下。

```
 8    @Model.model(PetShop.MODEL_MODEL)
 9    @Model(displayName = "宠物店铺",summary="宠物店铺")
10    public class PetShop extends AbstractDemoIdModel {
11        public static final String MODEL_MODEL="demo.PetShop";
12
13        @Field(displayName = "店铺名称",required = true)
14        private String shopName;
15
16        @Field(displayName = "开店时间",required = true)
17        private Time openTime;
18
19        @Field(displayName = "闭店时间",required = true)
20        private Time closeTime;
21
22    }
23
```

**Step5.** 重启看效果。

宠物商店的列表页面和修改页面都增加数据状态字段，如图 3-65、图 3-66 所示。

图 3-65　宠物商店列表页面增加数据状态字段（1）

图 3-66　宠物商店修改页面增加数据状态字段（2）

### 6. 存储模型

存储模型用于定义数据表结构和数据的增删改查（数据管理器）功能，是直接与连接器进行交互的数据容器。

PetShop 就是一个存储模型，这里不过多举例子介绍。

### 7. 代理模型（举例）

代理模型是用于代理存储模型数据的管理器，同时又可以扩展出非存储数据信息的交互功能。

如果 PetShop 模型要展示创建人昵称，那么就可以建一个 PetShopProxy 类来完成，代理模型增加的字段都是非存储字段，只用于交互，包括展示或提交。

Step1. 引入 PamirsUser 类。

使用 pamirs-demo-api 的 pom.xml 包增加依赖，便于引入 PamirsUser 类，代码如下。

```xml
1  <dependency>
2      <groupId>pro.shushi.pamirs.core</groupId>
3      <artifactId>pamirs-user-api</artifactId>
4  </dependency>
```

Step2. 新建 PetShopProxy 模型。

新建一个 PetShopProxy 类继承 PetShop，并声明类型为 PROXY。同时增加一个 creater 字段，代码如下。

```java
9   @Model.model(PetShopProxy.MODEL_MODEL)
10  @Model.Advanced(type = ModelTypeEnum.PROXY)
11  @Model(displayName = "宠物店铺代理模型",summary="宠物店铺代理模型")
12  public class PetShopProxy extends PetShop {
13
14      public static final String MODEL_MODEL="demo.PetShopProxy";
15
16      @Field.many2one
17      @Field(displayName = "创建者",required = true)
18      @Field.Relation(relationFields = {"createUid"},referenceFields = {"id"})
19      private PamirsUser creater;
20  }
```

Step3. 修改 DemoModule。

PamirsUser 模型隶属于 UserModule。为 DemoModule 增加 UserModule 的依赖注解，同时修改 DemoModule 的 homepage 注解，默认进入 PetShopProxy 的管理页面，代码如下。

```java
13  @Component
14  @Module(
15          name = DemoModule.MODULE_NAME,
16          displayName = "oinoneDemo工程",
17          version = "1.0.0",
18          dependencies = {ModuleConstants.MODULE_BASE, CommonModule.MODULE_MODULE, UserModule.MODULE_MODULE}
19  )
20  @Module.module(DemoModule.MODULE_MODULE)
21  @Module.Advanced(selfBuilt = true, application = true)
22  @UxHomepage(@UxRoute(PetShopProxy.MODEL_MODEL))
23  public class DemoModule implements PamirsModule {
24
25      public static final String MODULE_MODULE = "demo_core";
26
27      public static final String MODULE_NAME = "DemoCore";
28
29      @Override
30      public String[] packagePrefix() {
31          return new String[]{
32                  "pro.shushi.pamirs.demo"
33          };
34      }
35  }
```

**Step4.** 新增 PetShopProxyAction 类。

为了展示效果覆盖，PetShopProxy 默认从 PetShop 继承数据管理器方法，但对于 Action 和 Function 这些这里不展开介绍，具体会在 3.4 和 3.5 两节中介绍。

（1）function 都挂在 PetShopProxy 这个模型载体上；

（2）PetShopProxyAction 继承 DataStatusBehavior 类配套 dataStatus 使用，定义一个"启用" serverAction，而启用逻辑则复用父类 DataStatusBehavior 的 dataStatusEnable 方法；

（3）覆盖"queryPage"Function，该方法名为前后约定协议中列表查询的默认方法，代码如下。

```
15  @Model.model(PetShopProxy.MODEL_MODEL)
16  @Component
17  public class PetShopProxyAction extends DataStatusBehavior<PetShopProxy> {
18
19      @Override
20      protected PetShopProxy fetchData(PetShopProxy data) {
21          return data.queryById();
22      }
23      @Action(displayName = "启用")
24      public PetShopProxy dataStatusEnable(PetShopProxy data){
25          data = super.dataStatusEnable(data);
26          data.updateById();
27          return data;
28      }
29
30      @Function.Advanced(type= FunctionTypeEnum.QUERY)
31      @Function.fun(FunctionConstants.queryPage)
32      @Function(openLevel = {FunctionOpenEnum.API})
33      public Pagination<PetShopProxy> queryPage(Pagination<PetShopProxy> page, IWrapper<PetShopProxy> queryWrapper){
34
35          Pagination<PetShopProxy> result = new PetShopProxy().queryPage(page,queryWrapper);
36          new PetShopProxy().listFieldQuery(result.getContent(),PetShopProxy::getCreater);
37          return result;
38      }
39
40  }
```

**Step5.** 重启看效果。

宠物商店的列表页面和修改页面都增加数据状态字段，如图 3-67 所示。

图 3-67  宠物商店的列表页面和修改页面都增加数据状态字段

多次单击启用会报对应的错误，如图 3-68 所示。当然如果场景不需要，则没有必要复用 DataStatusBehavior 逻辑，也就不需要实现 IDataStatus 接口了，启动按钮只在禁用状态展示，在后续的章节中会讲解到。

图 3-68　多次单击启用会报对应的错误

**8. 传输模型（举例）**

由于传输模型没有默认的数据管理器，只有数据构造器，所以在不自定义动作的情况下，传输模型可以打开详情页、新增表单以及修改表单和列表页，但是所有的动作全部为窗口动作。传输模型本身不会存储，如果是关联关系字段关联传输模型，可以将传输模型序列化存储在模型的关联关系字段上。因为没有数据管理器，所以传输模型的列表页没有分页能力。

场景举例：如果我们想批量修改 PetShop 的数据状态，需要在列表页选中数据记录，单击"批量修改"跳转到一个批量修改页面，选择要修改的数据状态，并单击"确认"提交。那么传输模型就可以承载批量修改页面、数据以及操作的载体模型。

Step1. 新建 PetShopBatchUpdate 模型。

新建一个 PetShopBatchUpdate 类继承 TransientModel，同时增加 petShopList 和 dataStatus 字段用于接收页面传入的值，代码如下。

```
@Model.model(PetShopBatchUpdate.MODEL_MODEL)
@Model(displayName = "批量修改宠物商店数据状态",summary = "批量修改宠物商店数据状态")
public class PetShopBatchUpdate extends TransientModel {

    public static final String MODEL_MODEL="demo.PetShopBatchUpdate";

    @Field(displayName = "数据状态",required = true)
    private DataStatusEnum dataStatus;

    @Field(displayName = "宠物商店列表",required = true)
    @Field.many2many
    private List<PetShopProxy> petShopList;


}
```

Step2. 新增 PetShopBatchUpdateAction 类。

传输模型没有默认的数据管理器，只有数据构造器（construct）。

（1）新建一个 PetShopBatchUpdateAction 类，Model.Model 设置为 PetShopBatchUpdate，也就是把所有 Action 和 Function 都挂在 PetShopBatchUpdateAction 这个模型载体上。

（2）覆盖数据构造器（construct），接收从宠物商店列表多选带过来的数据参数，非 PetShopBatchUpdate 模型参数不能放第一个，用"List<PetShopProxy> petShopList"来接收，进行数据组装逻辑处理，对应数据也是由 PetShopBatchUpdate 来承载返回给 PetShopBatchUpdate 的 Form 编辑页。

（3）定义一个"确定"按钮 serverAction，绑定 PetShopBatchUpdate 的 Form 编辑页。单击"确定"则批量修改 Form 页面中的宠物商店列表为指定的数据状态，代码如下。

```
16    @Model.model(PetShopBatchUpdate.MODEL_MODEL)
17    @Component
18    public class PetShopBatchUpdateAction {
19
20        @Function(openLevel = FunctionOpenEnum.API)
21        @Function.Advanced(type= FunctionTypeEnum.QUERY)
22        public PetShopBatchUpdate construct(PetShopBatchUpdate petShopBatchUpdate, List<PetShopProxy> petShopList){
23            PetShopBatchUpdate result = new PetShopBatchUpdate();
24            result.setPetShopList(petShopList);
25            return result;
26        }
27
28        @Action(displayName = "确定",bindingType = ViewTypeEnum.FORM,contextType = ActionContextTypeEnum.SINGLE)
29        public PetShopBatchUpdate conform(PetShopBatchUpdate data){
30            List<PetShopProxy> proxyList = data.getPetShopList();
31            for(PetShopProxy petShopProxy:proxyList){
32                petShopProxy.setDataStatus(data.getDataStatus());
33            }
34            new PetShopProxy().updateBatch(proxyList);
35            return data;
36        }
37
38    }
```

Step3. 初始化 ViewAction 窗口动作。

配置从宠物商店列表到 PetShopBatchUpdate 的窗口动作。

（1）新建 DemoModuleMetaDataEditor，MetaDataEditor 是 Oinone 在计算元数据时提供主动动态生成元数据的方式，它在自动扫描元数据注解之后进行。我们只要实现 MetaDataEditor 接口就可以做一些元数据初始化，常用场景就是初始化一些交互相关的元数据，如窗口动作。更多有关模块生命周期扩展的详情参见 4.1.3 节。

（2）通过指定工具（InitializationUtil），生成一个窗口动作，动作名为"批量更新数据状态"，指定在宠物商店列表页，当单条或多条记录被选中时出现。单击该窗口动作跳转到 PetShopBatchUpdate 的默认 Form 页，跳转方式为 DIALOG 弹出框，代码如下。

```
17    @Component
18    public class DemoModuleMetaDataEditor implements MetaDataEditor {
19        @Override
20        public void edit(AppLifecycleCommand command, Map<String, Meta> metaMap) {
21            InitializationUtil util = InitializationUtil.get(metaMap, DemoModule.MODULE_MODULE,DemoModule.MODULE_NAME);
22            if(util==null){
23                return;
24            }
25
26            //初始化自定义前端行为
27            viewActionInit(util);
28        }
29        private void viewActionInit(InitializationUtil util){
30            util.createViewAction("demo_petShop_batch_update","批量更新数据状态", PetShopProxy.MODEL_MODEL,
31                    InitializationUtil.getOptions(ViewTypeEnum.TABLE), PetShopBatchUpdate.MODEL_MODEL,ViewTypeEnum.FORM, ActionContextTypeEnum.SINGLE_AND_BATCH
32                    , ActionTargetEnum.DIALOG,null,null);
33        }
34    }
```

Step4. 重启看效果。

重启 demo 应用看效果，宠物商店的列表页面在选中记录时就多出了"批量更新数据状态"按钮，单击进入批量修改数据状态页面。

（1）宠物商店的列表页面勾选数据记录后出现"批量更新数据状态"按钮，因为我们初始化 ViewAction 的时候配置了 ActionContextTypeEnum.SINGLE_AND_BATCH，详细见 3.5.3 一节，如图 3-69 所示。

图 3-69　批量更新数据状态

（2）单击"批量更新数据状态"，进入"批量更新数据状态"页面，如图 3-70 所示。

图 3-70　进入"批量更新数据状态"页面

（3）"数据状态"选择"未启用"，如图 3-71 所示。

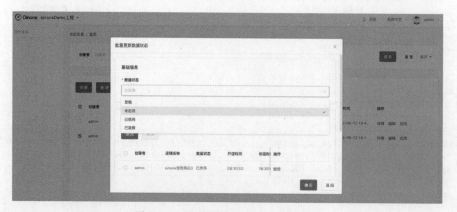

图 3-71　"数据状态"选择"未启用"

(4)单击"添加"按钮,可以继续追加宠物商店记录,如图 3-72 所示。

图 3-72 继续追加宠物商店记录

(5)在"批量更新数据状态"页面单击"确认",回到宠物商店页面查看数据状态已经变更为"未启用",如图 3-73 所示。

图 3-73 宠物商店页面查看数据状态已变更为"未启用"

Step5. 注解式初始化 ViewAction 窗口动作。

在 PetShopProxyAction 增加类注解,并注释掉 DemoModuleMetaDataEditor 的 viewActionInit 方法体代码。用 UxRouteButton 注解来申明一个 ViewAction。

(1)@Model.Model (PetShopProxy.MODEL_MODEL)代表 UxRouteButton 申明 viewAction 所在模型。

(2)@UxRoute.Model 代表 viewAction 的 resModel 路由的目标模型。

①在 PetShopProxyAction 增加类注解,代码如下。

```
1  @Model.model(PetShopProxy.MODEL_MODEL)
2  @UxRouteButton(action = @UxAction(name = "demo_petShop_batch_update", label = "批量更新数据状态",contextType = ActionContextTypeEnum.SINGLE_AND_BATCH), value = @UxRoute(model = PetShopBatchUpdate.MODEL_MODEL, viewType = ViewTypeEnum.FORM,openType = ActionTargetEnum.DIALOG))
3  @Component
4  public class PetShopProxyAction extends DataStatusBehavior<PetShopProxy> {
5  }
```

②注解式初始化 ViewAction 窗口动作,代码如下。

```
1  private void viewActionInit(InitializationUtil util){
2  //       util.createViewAction("demo_petShop_batch_update","批量更新数据状态", PetShopProxy.MODEL_MODEL,
3  //              InitializationUtil.getOptions(ViewTypeEnum.TABLE), PetShopBatchUpdate.MODEL_MODEL,ViewTypeEnum.FORM, ActionContextTypeEnum.SINGLE_AND_BATCH
4  //              , ActionTargetEnum.DIALOG,null,null);
5  }
```

重启系统看到的效果跟前面是一致的，更多 UX 系列注解见"UX 注解详解"一节。

### 3.3.3 模型的数据管理器

数据管理器和数据构造器是 Oinone 为模型自动赋予的 Function 内在数据管理能力。数据管理器针对存储模型，用于在编程模式下利用 Function 快速进行数据操作，数据构造器则主要用于模型初始化时字段默认值计算和页面交互。

**1. 数据管理器**

只有存储模型才有数据管理器。如果 @Model.Advanced 注解设置 dataManager 属性为 false，则表示在 UI 层不开放默认数据管理器。开放级别为 API 则表示 UI 层可以通过 HTTP 请求利用"Pamirs 标准网关协议"进行数据交互。

**2. 模型默认数据读管理器**

模型默认数据读管理器见表 3-8。

表 3-8　模型默认数据读管理器

| 函数编码 | 描述 | 开放级别 |
| --- | --- | --- |
| queryByPk | 根据主键查询单条记录，会进行主键值检查 | Local、Remote |
| queryByEntity | 根据实体查询单条记录 | Local、Remote、Api |
| queryByWrapper | 根据查询类查询单条记录 | Local、Remote |
| queryListByEntity | 根据实体查询返回记录列表 | Local、Remote |
| queryListByWrapper | 根据查询类查询记录列表 | Local、Remote |
| queryListByPage | 根据实体分页查询返回记录列表 | Local、Remote |
| queryListByPageAndWrapper | 根据查询类分页查询记录列表 | Local、Remote |
| queryPage | 分页查询返回分页对象，分页对象中包含记录列表 | Local、Remote、Api |
| countByEntity | 按实体条件获取记录数量 | Local、Remote |
| countByWrapper | 按查询类条件获取记录数量 | Local、Remote |

**3. 模型默认数据写管理器**

模型默认数据写管理器见表 3-9。

表 3-9　模型默认数据写管理器

| 函数编码 | 描述 | 开放级别 |
| --- | --- | --- |
| createOne | 提交新增单条记录 | Local、Remote |
| createOrUpdate | 新增或更新，需要为模型设置唯一索引，如果数据库检测到索引冲突，会更新数据，若未冲突则新增数据 | Local、Remote |
| updateByPk | 根据主键更新单条记录，会进行主键值检查 | Local、Remote |
| updateByUniqueField | 条件更新，条件中必须包含唯一索引字段 | Local、Remote |
| updateByEntity | 按实体条件更新记录 | Local、Remote、Api |
| updateByWrapper | 按查询类条件更新记录 | Local、Remote |
| createBatch | 批量新增记录 | Local、Remote |
| createOrUpdateBatch | 批量新增或更新记录 | Local、Remote |
| updateBatch | 根据主键批量更新记录，会进行主键值检查 | Local、Remote |
| deleteByPk | 根据主键删除单条记录，会进行主键值检查 | Local、Remote |
| deleteByPks | 根据主键批量删除，会进行主键值检查 | Local、Remote |
| deleteByUniqueField | 按条件删除记录，条件中必须包含唯一索引字段 | Local、Remote |

续表

| 函数编码 | 描述 | 开放级别 |
|---|---|---|
| deleteByEntity | 根据实体条件删除 | Local、Remote、Api |
| deleteByWrapper | 根据查询类条件删除 | Local、Remote |
| createWithField | 新增实体记录并更新实体字段记录 | Local、Remote、Api |
| updateWithField | 更新实体记录并更新实体字段记录 | Local、Remote、Api |
| deleteWithFieldBatch | 批量删除实体记录并删除关联关系 | Local、Remote、Api |

如果模型继承 IdModel，会自动设置主键为 id，并继承 queryById、updateById 和 deleteById 函数。
- queryById（详情，根据 id 查询单条记录，开放级别为 Remote）。
- updateById（提交更新单条记录，根据 id 更新单条记录，开放级别为 Remote）。
- deleteById（提交删除单条记录，根据 id 删除单条记录，开放级别为 Remote）。

如果模型继承 CodeModel，也会继承 IdModel 的数据管理器，编码字段 code 为唯一索引字段。在新增数据时会根据编码生成规则自动设置编码字段 code 的值，继承 queryByCode、updateByCode 和 deleteByCode 函数。
- queryByCode（详情，根据 code 查询单条记录，开放级别为 Remote）。
- updateByCode（提交更新单条记录，根据 code 更新单条记录，开放级别为 Remote）。
- deleteByCode（提交删除单条记录，根据 code 删除单条记录，开放级别为 Remote）。

没有主键或唯一索引的模型，在 UI 层不会开放默认数据写管理器。

**4. 使用场景**

使用场景如图 3-74 所示。

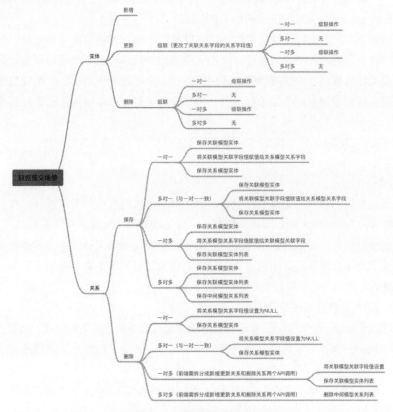

图 3-74　数据管理器使用场景

**5. 数据构造器**

模型数据构造器 construct 供前端新开页面构造默认数据。所有模型都拥有 construct 构造器，默认会将字段上配置的默认值返回给前端，另外可以在子类中覆盖 construct 方法。数据构造器 construct 函数的开放级别为 API，函数类型为 QUERY 查询函数，系统将识别模型中的以 construct 命名的函数，将其强制设置为 API 开放级别和 QUERY 查询类型。

可以使用 @Field 的 defaultValue 属性配置字段的默认值。注意，枚举的默认值为枚举的 name。

### 3.3.4 模型的继承

在我们的很多项目中，客户都是有个性化需求的，就像我们不能找到两片一模一样的树叶，何况是企业的经营与管理思路，多少都会有差异。常规的方式只能去修改标准产品的逻辑来适配客户的需求，导致后续标品维护非常困难。通过学习模型的继承，将会更加清晰认知到"Oinone 独特性之每一个需求都可以是一个模块"带来的好处。

**1. 继承方式**

继承方式可以分为以下五种：

● 抽象基类 ABSTRACT，只保存不希望为每个子模型重复键入信息的模型，抽象基类模型不生成数据表存储数据，只供其他模型的继承模型的可继承域使用，抽象基类可以继承抽象基类。

● 扩展继承 EXTENDS，子模型与父模型的数据表相同，子模型继承父模型的字段与函数。存储模型之间的继承默认为扩展继承。

● 多表继承 MULTI_TABLE，父模型不变，子模型获得父模型的可继承域生成新的模型；子父模型不同表，子模型会建立与父模型的一对一关联关系字段（而不是交叉表），使用主键关联，同时子模型会通过一对一关联关系引用父模型的所有字段。多表继承父模型需要使用 @Model.MultiTable 来标识，子模型需要使用 @Model.MultiTableInherited 来标识。

● 代理继承 PROXY，为原始模型创建代理，可以增删改查代理模型的实体数据，就像使用原始（非代理）模型一样。不同之处在于代理继承并不关注更改字段，可以更改代理中的元信息、函数和动作，而无须更改原始内容。一个代理模型必须仅能继承一个非抽象模型类。一个代理模型可以继承任意数量的没有定义任何模型字段的抽象模型类。一个代理模型也可以继承任意数量继承相同父类的代理模型。

● 临时继承 TRANSIENT，将父模型作为传输模型使用，并可以添加传输字段。

**2. 继承约束**

（1）通用约束。

对于扩展继承，查询的时候，父模型只能查询到父模型字段的数据，子模型可以查询到父模型及子模型的字段数据（因为派生关系所以子模型复刻了一份父模型的字段到子模型中）。

系统不会为抽象基类创建实际的数据库表，它们也没有默认的数据管理器，不能被实例化也无法直接保存，它们就是用来被继承的。抽象基类完全就是用来保存子模型共有的内容部分，达到重用的目的。当它们被继承时，它们的字段会全部复制到子模型中。

系统不支持非 jar 包依赖模型的继承。

多表继承具有阻断效应，子模型无法继承多表继承父模型的存储父模型的字段，需要使用 @Model.Advanced 注解的 inherited 属性显示声明继承父模型的父模型。但是可以继承多表继承父模型的抽象父模型的字段。

可以使用 @Model.Advanced 的 unInheritedFields 和 unInheritedFunctions 属性设置不从父类继承的字段和函数。

（2）跨模块继承约束。

如果模型间的继承是跨模块继承，应该与模型所属模块建立依赖关系；如果模块间有互斥关系，则不允许建立模块依赖关系，同理模型间也不允许存在继承关系。

跨模块代理继承，对代理模型的非 inJvm 函数调用将使用远程调用方式；跨模块扩展（同表）继承将使用本地调用方式，如果是数据管理器函数，将直连数据源。

（3）模型类型与继承约束。

● 抽象模型可继承抽象模型（Abstract）。

● 临时模型可继承抽象模型（Abstract）、传输模型（Transient）。

● 存储模型可继承抽象模型（Abstract）、存储模型（Store）、多表存储模型（Multi-table Store），不可继承多个 Store 或 Multi-table Store。

● 多表存储模型（父）可继承同扩展继承。

● 多表存储模型（子）在继承单个 Multi-table Store 后可继承抽象模型（Abstract）、存储模型（Store），不可继承多个 Store。

● 代理模型可继承：

> 抽象模型（Abstract），须搭配继承 Store、Multi-table Store 或 Proxy；

> 存储模型（Store），不可继承多个 Store 或 Multi-table Store；

> 存储模型（多表，Multi-table Store），不可继承多个 Store 或 Multi-table Store；

> 代理模型（Proxy），可继承多个 Proxy，但多个父 Proxy 须继承自同一个 Store 或 Multi-table Store，且不能再继承其他 Store 或 Multi-table Store。

同名字段以模型自身字段为有效配置，若模型自身不存在该字段，继承字段以第一个加载的字段为有效配置，所以在多重继承的情况下，为避免继承同名父模型字段的不确定性，在自身模型配置同名字段来确定生效配置。

**3. 继承的使用场景**

模型可以继承父模型的元信息、字段、数据管理器和函数。

● 抽象基类：解决公用字段问题。

● 扩展继承：解决开放封闭原则、跨模块扩展等问题。

● 多表继承：解决多型派生类字段差异问题和前端多存储模型组合外观问题。

● 代理继承：解决同一模型在不同场景下的多态问题（一表多态）。

● 临时继承：解决使用现有模型进行数据传输问题。

例如，前端多存储模型组合外观问题可通过多表继承的子模型，一对一关联到关联模型，同时使用排除继承字段去掉不需要继承的字段。子模型通过默认模型管理器提供查询功能给前端，默认查询会查询子模型数据列表并在列表行内根据一对一关系查出关联模型数据合并，关联模型数据展现形态在行内是平铺还是折叠，在详情是分组还是选项卡可以自定义 view 进行配置。

扩展继承父子同表，模型在所有场景都有一致化的表现，意味着原模型被扩展成了新模型，父子模型的表名一致，模型编码不同，可覆盖父模型的模型管理器、数据排序规则、函数。

多表继承父子多表，父子间有隐式一对一关系，即父子模型都增加了一对一关联关系字段，同时父模型的字段被引用到子模型，且引用字段为只读字段，意味着子模型不可以直接更改父模型的字段值，子模型不继承父模型的模型管理器、数据排序规则、函数，子模型拥有自己的默认模型管理器、数据排序规则、函数。多表继承具有阻断效应，子模型无法自动多表继承父模型的存储父模型，需要显式声明多表继承父模型的存储父模型。

代理继承，可覆盖父模型的模型管理器、数据排序规则、函数，同时可以使用排除继承字段和函数来达到不同场景不同视觉交互的效果。

继承的使用场景如图 3-75 所示。

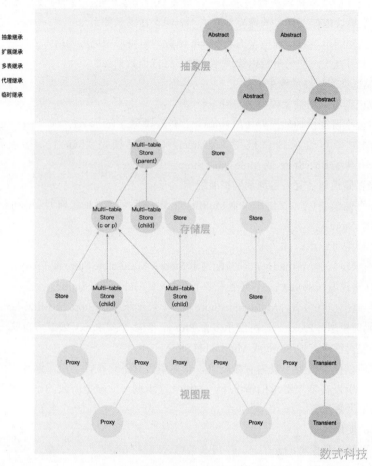

图 3-75　继承的使用场景

**4. 抽象基类（举例）**

参考前文中 3.3.2 节中"抽象模型"的介绍。

**5. 多表继承（举例）**

场景设计如图 3-76 所示。

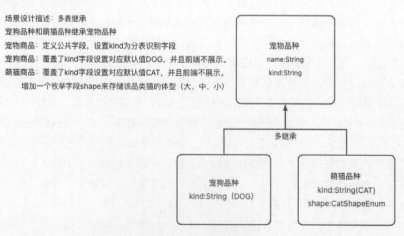

图 3-76　多表继承场景设计

Step1. 新建宠物品种、宠狗品种和萌猫品种模型。

（1）新建宠物品种模型，用 @Model.MultiTable (typeField = "kind")，声明为可多表继承父类，typeField 指定为 kind 字段，代码如下。

```
 7  @Model.MultiTable(typeField = "kind")
 8  @Model.model(PetType.MODEL_MODEL)
 9  @Model(displayName="品种",labelFields = {"name"})
10  public class PetType extends IdModel {
11
12      public static final String MODEL_MODEL="demo.PetType";
13
14      @Field(displayName = "品种名")
15      private String name;
16
17      @Field(displayName = "宠物分类")
18      private String kind;
19  }
```

（2）新建宠狗品种模型，用 @Model.MultiTableInherited (type = PetDogType.KIND_DOG)，声明以多表继承模式继承 PetType，覆盖 kind 字段（用 defaultValue 设置默认值，用 invisible = true 设置为前端不展示），更多模块元数据以及模型字段元数据配置请见"模型之元数据详解"一节，代码如下。

```
 6  @Model.MultiTableInherited(type = PetDogType.KIND_DOG)
 7  @Model.model(PetDogType.MODEL_MODEL)
 8  @Model(displayName="宠狗品种",labelFields = {"name"})
 9  public class PetDogType extends PetType {
10
11      public static final String MODEL_MODEL="demo.PetDogType";
12      public static final String KIND_DOG="DOG";
13
14      @Field(displayName = "宠物分类",defaultValue = PetDogType.KIND_DOG,invisible = true)
15      private String kind;
16  }
```

（3）新建萌猫品种模型，用 @Model.MultiTableInherited (type = PetCatType.KIND_CAT)，声明以多表继承模式继承 PetType，覆盖 kind 字段（用 defaultValue 设置默认值，用 invisible = true 设置为前端不展示），代码如下。

```
 6  @Dict(dictionary = CatShapeEnum.DICTIONARY,displayName = "萌猫体型")
 7  public class CatShapeEnum extends BaseEnum<CatShapeEnum,Integer> {
 8
 9      public static final String DICTIONARY ="demo.CatShapeEnum";
10
11      public final static CatShapeEnum BIG =create("BIG",1,"大","大");
12      public final static CatShapeEnum SMALL =create("SMALL",2,"小","小");
13  }
```

新增一个 CatShapeEnum 枚举类型的字段 shape，代码如下。

```
 7  @Model.MultiTableInherited(type = PetCatType.KIND_CAT)
 8  @Model.model(PetCatType.MODEL_MODEL)
 9  @Model(displayName="萌猫品种",labelFields = {"name"})
10  public class PetCatType extends PetType {
11
12      public static final String MODEL_MODEL="demo.PetCatType";
13      public static final String KIND_CAT="CAT";
14
15      @Field(displayName = "宠物分类",defaultValue = PetCatType.KIND_CAT,invisible = true)
16      private String kind;
17
18      @Field.Enum
19      @Field(displayName = "宠物体型")
20      private CatShapeEnum shape;
21  }
```

Step2. 配置菜单后，重启查看效果。

（1）为了帮助大家更好地理解，这里先配置对应的菜单，从页面来看看效果。下面以注解方式进行菜单配置，更多详情请见 3.5.1 节。代码如下。

```
10    @UxMenus
11  ▸ public class DemoMenus implements ViewActionConstants {
12        @UxMenu("宠物品种")@UxRoute(PetType.MODEL_MODEL) class PetTypeMenu{}
13        @UxMenu("萌猫品种")@UxRoute(PetCatType.MODEL_MODEL) class CatTypeMenu{}
14        @UxMenu("宠狗品种")@UxRoute(PetDogType.MODEL_MODEL) class DogTypeMenu{}
15    }
```

（2）在宠狗品种中新增、修改、列表都隐藏了品种类型字段，并附上了默认值 DOG，如图 3-77～图 3-79 所示。

图 3-77　宠狗品种新增页

图 3-78　宠狗品种列表页

图 3-79　宠狗品种数据表

（3）在萌猫品种中新增、修改、列表都增加宠物体型枚举字段，并且隐藏了品种类型字段，并附上了默认值 CAT，如图 3-80～图 3-82 所示。

图 3-80　萌猫品种新增页

图 3-81 萌猫品种列表页

图 3-82 萌猫品种数据表

**Step3.** 为宠狗和萌猫品种模型实现默认读写扩展点，保证父子表数据同步。

多表继承默认父子表是不同步的，不过 Oinone 为每一个模型的"默认数据管理器"都提供默认读写扩展点，我们可以主动利用这个特性来进行手动同步，更多扩展点知识详见 3.4.3 节中函数相关特性部分的"扩展点"部分，其中会介绍除了利用系统默认提供的扩展点，还有如何自定义扩展。

（1）PetCatType 和 PetDogType 都需要实现扩展点来完成与父模型的同步，所以这里对相同逻辑进行了提取，得到 AbstractPetTypeExtPoint，代码如下。

```java
public abstract class AbstractPetTypeExtPoint <T extends PetType> extends Default
ReadWriteExtPoint<T> {

    public T createBefore(T data) {
        data.construct();
        return data;
    }

    public T updateBefore(T data) {
        return data;
    }

    public T createAfter(T data) {
        CopyHelper.simpleReplace(data, new PetType()).create();
        return data;
    }

    public T updateAfter(T data) {
        CopyHelper.simpleReplace(data, new PetType()).updateById();
        return data;
    }

    public List<T> deleteBefore(List<T> data) {
        List<Long> petTypeIdList = new ArrayList();
        for(T item:data){
            petTypeIdList.add(item.getId());
        }
        Models.data().deleteByWrapper(Pops.<PetType>lambdaQuery().from(PetType.MODEL_MODEL).in(PetType::getId,petTypeIdList));
        return data;
    }
}
```

（2）PetCatTypeExtPoint 继承 AbstractPetTypeExtPoint<PetCatType> 类，并通过 @Ext (PetCatType.class) 声明扩展点扩展函数所在类为 PetCatType，并 @ExtPoint.Implement 声明对应扩展点的实现方法，代码如下：

```java
 9   @Ext(PetCatType.class)
10   public class PetCatTypeExtPoint extends AbstractPetTypeExtPoint<PetCatType>{
11
12       @Override
13       @ExtPoint.Implement
14       public PetCatType createBefore(PetCatType data) {
15           return super.createBefore(data);
16       }
17       @Override
18       @ExtPoint.Implement
19       public PetCatType updateBefore(PetCatType data) {
20           return super.updateBefore(data);
21       }
22       @Override
23       @ExtPoint.Implement
24       public PetCatType createAfter(PetCatType data) {
25           return super.createAfter(data);
26       }
27       @Override
28       @ExtPoint.Implement
29       public PetCatType updateAfter(PetCatType data) {
30           return super.updateAfter(data);
31       }
32       @Override
33       @ExtPoint.Implement
34       public List<PetCatType> deleteBefore(List<PetCatType> data) {
35           return super.deleteBefore(data);
36       }
37
38   }
```

（3）PetDogTypeExtPoint 继承 AbstractPetTypeExtPoint<PetDogType> 类，并通过 @Ext(PetDogType.class) 声明扩展点扩展函数所在类为 PetDogType，并 @ExtPoint.Implement 声明对应扩展点的实现方法实例，代码如下：

```java
 9   @Ext(PetDogType.class)
10   public class PetDogTypeExtPoint extends AbstractPetTypeExtPoint<PetDogType>{
11
12       @Override
13       @ExtPoint.Implement
14       public PetDogType createBefore(PetDogType data) {
15           return super.createBefore(data);
16       }
17       @Override
18       @ExtPoint.Implement
19       public PetDogType updateBefore(PetDogType data) {
20           return super.updateBefore(data);
21       }
22       @Override
23       @ExtPoint.Implement
24       public PetDogType createAfter(PetDogType data) {
25           return super.createAfter(data);
26       }
27       @Override
28       @ExtPoint.Implement
29       public PetDogType updateAfter(PetDogType data) {
30           return super.updateAfter(data);
31       }
32       @Override
33       @ExtPoint.Implement
34       public List<PetDogType> deleteBefore(List<PetDogType> data) {
35           return super.deleteBefore(data);
36       }
37
38   }
```

**Step4.** 重启查看效果：在萌猫品种和宠狗品种下新增、修改、删除，则会在宠物品种下看到对应记录变化。

重启看效果如图 3-83、图 3-84 所示。

图 3-83　父模型宠物品种列表页

图 3-84　父模型宠物品种数据表

**6. 扩展继承（举例）**

场景设计如图 3-85 所示。

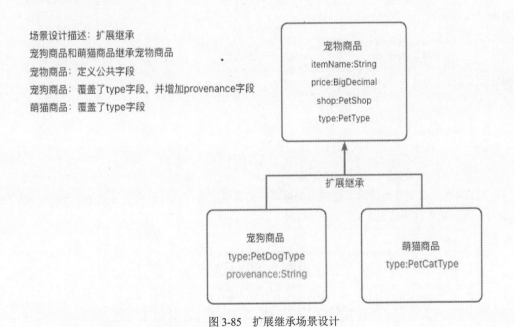

图 3-85　扩展继承场景设计

**Step1.** 新建宠物商品、宠狗商品和萌猫商品模型，并配置菜单。

（1）新建宠物商品模型为普通存储模型，用扩展继承父类设置公共字段，如店铺、品种等，注意 type 字段类型为 PetType，代码如下。

```java
 8      @Model.model(PetItem.MODEL_MODEL)
 9      @Model(displayName = "宠物商品",summary="宠物商品")
10      public class PetItem extends AbstractDemoCodeModel{
11
12          public static final String MODEL_MODEL="demo.PetItem";
13          @Field(displayName = "商品名称",required = true)
14          private String itemName;
15
16          @Field(displayName = "商品价格",required = true)
17          private BigDecimal price;
18
19          @Field(displayName = "店铺",required = true)
20          private PetShop shop;
21
22          @Field(displayName = "品种")
23          @Field.many2one
24          private PetType type;
25
26      }
```

（2）新建宠狗商品模型继承宠物商品模型，注意这里把 type 字段类型覆盖为 PetType 的子类 PetDogType，并新增原产地字段 provenance，代码如下。

```java
 6      @Model.model(PetDogItem.MODEL_MODEL)
 7      @Model(displayName = "宠狗商品",summary="宠狗商品")
 8      public class PetDogItem extends PetItem{
 9
10          public static final String MODEL_MODEL="demo.PetDogItem";
11
12          @Field(displayName = "品种")
13          @Field.many2one
14          private PetDogType type;
15
16          @Field(displayName = "原产地")
17          private String provenance;
18      }
```

（3）新建萌猫商品模型继承宠物商品模型，注意这里把 type 字段类型覆盖为 PetType 的子类 PetCatType，代码如下。

```java
 6      @Model.model(PetCatItem.MODEL_MODEL)
 7      @Model(displayName = "萌猫商品",summary="萌猫商品")
 8      public class PetCatItem extends PetItem{
 9
10          public static final String MODEL_MODEL="demo.PetCatItem";
11
12          @Field(displayName = "品种")
13          @Field.many2one
14          private PetCatType type;
15
16      }
```

（4）为了帮助大家更好地理解，这里先配置对应的菜单配置，从页面来看看效果。下面以注解方式进行菜单配置，更多详情请见 3.5.1 节。代码如下。

```java
10      @UxMenus
11      public class DemoMenus implements ViewActionConstants {
12          @UxMenu("商品管理")@UxRoute(PetItem.MODEL_MODEL) class ItemMenu{}
13          @UxMenu("宠狗商品")@UxRoute(PetDogItem.MODEL_MODEL) class DogItemMenu{}
14          @UxMenu("萌猫商品")@UxRoute(PetCatItem.MODEL_MODEL) class CatItemMenu{}
15          @UxMenu("宠物品种")@UxRoute(PetType.MODEL_MODEL) class PetTypeMenu{}
16          @UxMenu("萌猫品种")@UxRoute(PetCatType.MODEL_MODEL) class CatTypeMenu{}
17          @UxMenu("宠狗品种")@UxRoute(PetDogType.MODEL_MODEL) class DogTypeMenu{}
18      }
```

至此模型就建完了，还是很简单的吧。

Step2. 分别新建三条宠物商品、宠狗商品和萌猫商品记录，查看效果。

（1）父模型宠物商品的商品管理可以看到品类中可选柴犬、加菲猫，数据来源是品种的父模型

PetType，店铺下拉框展示为"-"，因为店铺模型没有配置 label 字段，可以在店铺模型的注解上增加"@Model (displayName = " 宠物店铺 ", summary=" 宠物店铺 ", labelFields = {"shopName"})"，这里先选择第一个往下执行，如图 3-86 所示。

图 3-86　宠物商品管理

（2）父模型宠狗商品的商品管理可以看到品类中可选柴犬，数据来源是品种的子模型 PetDogType，而且新增源产地字段，如图 3-87 所示。

图 3-87　父模型宠狗商品管理

（3）父模型萌猫商品的商品管理可以看到品类中可选加菲猫，数据来源是品种的子模型 PetCatType，如图 3-88 所示。

图 3-88　父模型萌猫商品管理

Step3. 分别观察宠物商品、宠狗商品和萌猫商品的列表页，我们会发现数据记录都是三条，但是展示字段会随着模型不同而有差异。

（1）父页面在品种一列中可以显示所有品种名称，如图 3-89 所示。

图 3-89　父页面在品种一列中可以显示所有品种名称

（2）宠狗页面在品种一列中只能显示 PetDogType 的品种名称，但多了原产地字段，如图 3-90 所示。

图 3-90　宠狗页面品种列展示

（3）萌猫页面在品种一列中只能显示 PetCatType 的品种名称，如图 3-91 所示。

图 3-91　萌猫页面在品种一列中只显示 PetCatType 的品种名称

（4）对应的数据库表：数据都在一张记录表中，新增了原产地字段，如图 3-92 所示。

图 3-92　对应的数据库表

Step4. 顺带优化下宠物店铺的展示 Labels。

（1）给 PetShop 增加 @Model (labelFields ={"shopName"} ) 注解，labelFields 为模型的"数据标题，用于前端展示"，其默认值为 name，但 PetShop 没有 name 字段，所以列表上展示不出来。更多元数据见 4.1.6 节，代码如下。

```
 8  @Model.model(PetShop.MODEL_MODEL)
 9  @Model(displayName = "宠物店铺",summary="宠物店铺",labelFields ={"shopName"} )
10  public class PetShop extends AbstractDemoIdModel {
11      public static final String MODEL_MODEL="demo.PetShop";
12
13      @Field(displayName = "店铺名称",required = true)
14      private String shopName;
15
16      @Field(displayName = "开店时间",required = true)
17      private Time openTime;
18
19      @Field(displayName = "闭店时间",required = true)
20      private Time closeTime;
21
22  }
```

（2）重启查看效果，店铺字段可以展示出来，如图 3-93 所示。

图 3-93　店铺字段可展示

Step5. 思考在扩展继承模式下数据隔离问题。

多表继承需要自己搞定父、子模型的数据同步问题，同表继承则需要自己搞定数据隔离问题。我们可以重写 queryPage 这个 Function，如 3.3.2 节中介绍"代理模型类型"时的 PetShopProxyAction 就覆盖了 queryPage。这里举例 PetCatItem 的 queryPage 覆盖，PetDogItem 留给大家自行练习。

（1）给 PetItem 显示增加一个 typeId 字段，方便在 PetCatItemAction 中用于过滤条件。为什么说显示增加呢？因为 type 字段是一个 many2one 的字段，在没有配置 @Field.Relation 的情况下，Oinone 会为 PetItem 模型自动推断出一个 typeId 字段去关联 PetType 模型的 id。@Field.Relation 的 relationFields 为本模型的关联字段，referenceFields 为目标模型关联字段。如符合推断规范可以不配置 @Field.Relation，就如这个场景配不配效果是一样的。代码如下。

```java
8   @Model.model(PetItem.MODEL_MODEL)
9   @Model(displayName = "宠物商品",summary="宠物商品")
10  public class PetItem extends AbstractDemoCodeModel{
11
12      public static final String MODEL_MODEL="demo.PetItem";
13      @Field(displayName = "商品名称",required = true)
14      private String itemName;
15
16      @Field(displayName = "商品价格",required = true)
17      private BigDecimal price;
18
19      @Field(displayName = "店铺",required = true)
20      private PetShop shop;
21
22      @Field(displayName = "品种")
23      @Field.many2one
24      @Field.Relation(relationFields = {"typeId"},referenceFields = {"id"})
25      private PetType type;
26
27      @Field(displayName = "品种类型",invisible = true)
28      private Long typeId;
29
30  }
```

（2）覆盖 PetCatItem 的 Function "queryPage"，增加 typeId 的过滤条件为 PetCatType 表对应的 id 列表，代码如下。

```java
21  @Model.model(PetCatItem.MODEL_MODEL)
22  @Component
23  public class PetCatItemAction extends DataStatusBehavior<PetCatItem> {
24
25      @Override
26      protected PetCatItem fetchData(PetCatItem data) {
27          return data.queryById();
28      }
29      @Action(displayName = "启用")
30      public PetCatItem dataStatusEnable(PetCatItem data){
31          data = super.dataStatusEnable(data);
32          data.updateById();
33          return data;
34      }
35
36      @Function.Advanced(type= FunctionTypeEnum.QUERY)
37      @Function.fun(FunctionConstants.queryPage)
38      @Function(openLevel = {FunctionOpenEnum.API})
39      public Pagination<PetCatItem> queryPage(Pagination<PetCatItem> page, IWrapper<PetCatItem> queryWrapper){
40
41          List<PetCatType> typeList = new PetCatType().queryList();
42          if(!CollectionUtils.isEmpty(typeList)) {
43              List<Long> typeIds = typeList.stream().map(PetCatType::getId).collect(Collectors.toList());
44              queryWrapper = Pops.<PetCatItem>f(queryWrapper).from(PetCatItem.MODEL_MODEL).in(PetCatItem::getTypeId, typeIds).get();
45          }
46          return new PetCatItem().queryPage(page,queryWrapper);
47      }
48
49  }
```

（3）重启查看效果，萌猫商品只能看到 PetCatType 对应的数据记录。赶紧自己动手试试 PetDogItem 的改造吧，如图 3-94 所示。

图 3-94　重启查看效果

**7. 代理继承（举例）**

我们在 3.3.2 节中有介绍过，代理模型看名字就知道其本身是通过继承方式代理另一个存储模型，这里不过多介绍。我们来尝试一下它继承的特殊性"一个代理模型也可以继承任意数量的同源代理模型，即继承相同父类的代理模型"。

场景设计如图 3-95 所示。

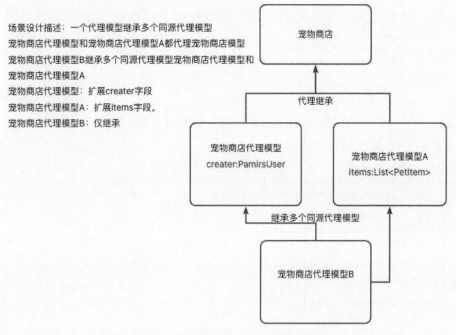

图 3-95　代理继承场景设计

Step1. 新建宠物店铺代理模型 A 和宠物店铺代理模型 B，同时修改 PetItem。

（1）新建宠物店铺代理模型 A 声明为代理模型，新增一个 one2many 字段 items，用 @Field. Relation 声明关联字段，代码如下。

```
12  @Model.model(PetShopProxyA.MODEL_MODEL)
13  @Model.Advanced(type = ModelTypeEnum.PROXY)
14  @Model(displayName = "宠物店铺代理模型A",summary="宠物店铺代理模型A")
15  public class PetShopProxyA extends PetShop {
16
17      public static final String MODEL_MODEL="demo.PetShopProxyA";
18
19      @Field.one2many
20      @Field(displayName = "商品列表")
21      @Field.Relation(relationFields = {"id"},referenceFields = {"shopId"})
22      private List<PetItem> items;
23
24  }
```

（2）新建宠物店铺代理模型 B 声明为代理模型，用 @Model.Advanced (inherited ={PetShopProxy.MODEL_MODEL,PetShopProxyA.MODEL_MODEL}) 声明继承多个同源代理模型，并且新增一个 one2many 字段 catItems 用 @Field.Relation 声明关联字段，代码如下。

```
11  @Model.model(PetShopProxyB.MODEL_MODEL)
12  @Model.Advanced(type = ModelTypeEnum.PROXY,inherited ={PetShopProxy.MODEL_MODEL,P
    etShopProxyA.MODEL_MODEL} )
13  @Model(displayName = "宠物店铺代理模型B",summary="宠物店铺代理模型B")
14  public class PetShopProxyB extends PetShop {
15
16      public static final String MODEL_MODEL="demo.PetShopProxyB";
17
18      @Field.one2many
19      @Field(displayName = "萌猫商品列表")
20      @Field.Relation(relationFields = {"id"},referenceFields = {"shopId"})
21      private List<PetCatItem> catItems;
22
23  }
```

（3）修改 PetItem 增加一个 labelFields={"itemName"} 注解，给 PetItem 显示增加一个 shopId 字段，方便在 Service 中用于过滤条件使用，代码如下。

```
8   @Model.model(PetItem.MODEL_MODEL)
9   @Model(displayName = "宠物商品",summary="宠物商品",labelFields={"itemName"})
10  public class PetItem  extends AbstractDemoCodeModel{
11
12      public static final String MODEL_MODEL="demo.PetItem";
13      @Field(displayName = "商品名称",required = true)
14      private String itemName;
15
16      @Field(displayName = "商品价格",required = true)
17      private BigDecimal price;
18
19      @Field(displayName = "店铺",required = true)
20      private PetShop shop;
21
22      @Field(displayName="店铺id",invisible = true)
23      private Long shopId;
24
25      @Field(displayName = "品种")
26      @Field.many2one
27      @Field.Relation(relationFields = {"typeId"},referenceFields = {"id"})
28      private PetType type;
29
30      @Field(displayName = "品种类型",invisible = true)
31      private Long typeId;
32  }
```

Step2. 覆盖 PetShopProxyB 的 Function "queryPage" 和 "queryOne"。

PetShopProxyB 的 queryPage 和 queryOne 都查 catItems\creater\items 三个字段，代码如下。

```java
19    @Model.model(PetShopProxyB.MODEL_MODEL)
20    @Component
21    public class PetShopProxyBAction  {
22
23        @Function.Advanced(type= FunctionTypeEnum.QUERY)
24        @Function.fun(FunctionConstants.queryPage)
25        @Function(openLevel = {FunctionOpenEnum.API})
26        public Pagination<PetShopProxyB> queryPage(Pagination<PetShopProxyB> page, IWrapper<PetShopProxyB> queryWrapper){
27            Pagination<PetShopProxyB> result = new PetShopProxy().queryPage(page,queryWrapper);
28            if(!CollectionUtils.isEmpty(result.getContent())) {
29                //自身继承"宠物店铺"后扩展的字段catItems
30                new PetShopProxyB().listFieldQuery(result.getContent(),PetShopProxyB::getCatItems);
31                //继承"宠物店铺代理模型"而来的字段creater, java语言限制"宠物店铺代理模型B"没有getCreate()方法
32                new PetShopProxyB().listFieldQuery(result.getContent(),"creater");
33                //继承"宠物店铺代理模型A"而来的字段items, java语言限制"宠物店铺代理模型B"没有getItems()方法
34                new PetShopProxyB().listFieldQuery(result.getContent(),"items");
35            }
36            return result;
37        }
38        @Function.Advanced(type= FunctionTypeEnum.QUERY)
39        @Function.fun(FunctionConstants.queryByEntity)
40        @Function(openLevel = {FunctionOpenEnum.API})
41        public PetShopProxyB queryOne(PetShopProxyB query){
42            query = query.queryById();
43            //自身继承"宠物店铺"后扩展的字段catItems
44            query.fieldQuery(PetShopProxyB::getCatItems);
45            //继承"宠物店铺代理模型"而来的字段creater, java语言限制"宠物店铺代理模型B"没有getCreate()方法
46            query.fieldQuery("creater");
47            //继承"宠物店铺代理模型A"而来的字段items, java语言限制"宠物店铺代理模型B"没有getItems()方法
48            query.fieldQuery("items");
49            return query;
50        }
51    }
```

Step3. 配置菜单，并重启看效果

（1）配置菜单：请参考本节前面介绍自行增加三个菜单商店管理、商店管理 A、商店管理 B 分别对应宠物店铺代理模型、宠物店铺代理模型 A、宠物店铺代理模型 B 的管理入口。

（2）宠物店铺代理模型覆盖了 queryPage 方法但没有覆盖 queryOne 方法，所以列表页有显示新增字段创建者，但详情页没有。具体参见 3.3.2 节中的"代理模型"部分。

"宠物店铺代理模型"列表页如图 3-96 所示。

图 3-96　"宠物店铺代理模型"列表页

"宠物店铺代理模型"详情页如图 3-97 所示。

图 3-97 "宠物店铺代理模型"详情页

（3）宠物店铺代理模型 A 只增加了一个商品列表字段，没有覆盖对应 query 的相关方法，所以列表页有这个字段但没有值，如图 3-98 所示。

图 3-98 宠物店铺代理模型 A 列表页

（4）宠物店铺代理模型 B 因为同源多表继承了宠物店铺代理模型和宠物店铺代理模型 A，所以拥有商品列表、萌猫商品列表、创建者三个字段，并且同时覆盖 queryPage 和 queryOne 对的 Function，所以列表和详情都有值，如图 3-99、图 3-100 所示。

图 3-99 宠物店铺代理模型 B 列表页

图 3-100　宠物店铺代理模型 B 详情页

Step4. 思考 PetShopProxyB 中的 PetCatItem 数据没有过滤，是不是与前面说到的扩展继承 PetCatItem 过滤是一回事。

提取 PetCatItem 的过滤逻辑，将其维护到独立的 Service 中去，并把 Service 的方法注解为 Function，如果 Function 的 namespace 跟模型的编码一致则代表挂载在模型上。当前例子的 namespace 不一致，用于提供内部服务。Function 的更多介绍请参见 3.4.1 节。

（1）新建 PetCatItemQueryService 接口类，声明三个方法，并在接口方法上增加 @Function 注解，这样依赖了接口但没有依赖实现类的工程，可以自动识别并远程调用，代码如下。

```
12  @Fun(PetCatItemQueryService.FUN_NAMESPACE)
13  public interface PetCatItemQueryService {
14
15      public static final String FUN_NAMESPACE = "demo.PetCatItem.PetCatItemQueryService";
16      @Function
17      Pagination<PetCatItem> queryPage(Pagination<PetCatItem> page, IWrapper<PetCatItem> queryWrapper);
18      @Function
19      List<PetCatItem> queryByShopId(Long shopId);
20      @Function
21      List<PetCatItem> queryByShopList(List<Long> shopIds);
22  }
```

（2）新建 PetCatItemQueryServiceImpl 实现 PetCatItemQueryService 接口，加上 @Fun 注解，并在方法上加 @Function 注解，默认开放级别是 Remote，代码如下。

```
18  @Fun(PetCatItemQueryService.FUN_NAMESPACE)
19      @Component
20      public class PetCatItemQueryServiceImpl implements PetCatItemQueryService {
21
22          @Override
23          @Function
24          public Pagination<PetCatItem> queryPage(Pagination<PetCatItem> page, IWrapper<PetCatItem> queryWrapper) {
25              List<PetCatType> typeList = new PetCatType().queryList();
26              if(!CollectionUtils.isEmpty(typeList)) {
```

```
27            List<Long> typeIds = typeList.stream().map(PetCatType::getId).col
    lect(Collectors.toList());
28            queryWrapper = Pops.<PetCatItem>f(queryWrapper).from(PetCatItem.M
    ODEL_MODEL).in(PetCatItem::getTypeId, typeIds).get();
29        }
30        return new PetCatItem().queryPage(page,queryWrapper);
31    }
32
33    @Override
34    @Function
35    public List<PetCatItem> queryByShopId(Long shopId) {
36        List<PetCatType> typeList = new PetCatType().queryList();
37        if(shopId ==null|| CollectionUtils.isEmpty(typeList)){
38            return Collections.EMPTY_LIST;
39        }
40        List<Long> typeIds = typeList.stream().map(PetCatType::getId).collect
    (Collectors.toList());
41        IWrapper<PetCatItem> queryWrapper =Pops.<PetCatItem>lambdaQuery().fro
    m(PetCatItem.MODEL_MODEL).eq(PetCatItem::getShopId, shopId).in(PetCatItem::getTyp
    eId, typeIds);
42        return new PetCatItem().queryList(queryWrapper);
43    }
44    @Override
45    @Function
46    public List<PetCatItem> queryByShopList(List<Long> shopIds) {
47
48        List<PetCatType> typeList = new PetCatType().queryList();
49        if(CollectionUtils.isEmpty(shopIds) || CollectionUtils.isEmpty(typeLi
    st)){
50            return Collections.EMPTY_LIST;
51        }
52        List<Long> typeIds = typeList.stream().map(PetCatType::getId).collect
    (Collectors.toList());
53        IWrapper<PetCatItem> queryWrapper = Pops.<PetCatItem>lambdaQuery().fr
    om(PetCatItem.MODEL_MODEL).in(PetCatItem::getShopId, shopIds).in(PetCatItem::getT
    ypeId, typeIds);
54        return new PetCatItem().queryList(queryWrapper);
55    }
56 }
```

（3）修改 PetShopProxyBAction 和 PetCatItemAction 的相关代码，把填充 PetCatItem 的逻辑做替换，代码如下。

```
20 @Model.model(PetCatItem.MODEL_MODEL)
21 @Component
22 public class PetCatItemAction extends DataStatusBehavior<PetCatItem> {
23
24    @Autowired
25    private PetCatItemQueryService petCatItemQueryService;
26
27    @Override
28    protected PetCatItem fetchData(PetCatItem data) {
29        return data.queryById();
30    }
31    @Action(displayName = "启用")
32    public PetCatItem dataStatusEnable(PetCatItem data){
33        data = super.dataStatusEnable(data);
34        data.updateById();
35        return data;
36    }
```

```
37
38          @Function.Advanced(type= FunctionTypeEnum.QUERY)
39          @Function.fun(FunctionConstants.queryPage)
40          @Function(openLevel = {FunctionOpenEnum.API})
41          public Pagination<PetCatItem> queryPage(Pagination<PetCatItem> page, IWrapper<PetCatItem> queryWrapper){
42
43  //          List<PetCatType> typeList = new PetCatType().queryList();
44  //          if(!CollectionUtils.isEmpty(typeList)) {
45  //              List<Long> typeIds = typeList.stream().map(PetCatType::getId).collect(Collectors.toList());
46  //              queryWrapper = Pops.<PetCatItem>f(queryWrapper).from(PetCatItem.MODEL_MODEL).in(PetCatItem::getTypeId, typeIds).get();
47  //          }
48  //          return new PetCatItem().queryPage(page,queryWrapper);
49  //
50              return petCatItemQueryService.queryPage(page,queryWrapper);
51          }
52      }
```

修改 PetCatItemAction 的相关代码。

```
--
24  @Model.model(PetShopProxyB.MODEL_MODEL)
25  @Component
26  public class PetShopProxyBAction {
27      @Autowired
28      private PetCatItemQueryService petCatItemQueryService;
29
30      @Function.Advanced(type= FunctionTypeEnum.QUERY)
31      @Function.fun(FunctionConstants.queryPage)
32      @Function(openLevel = {FunctionOpenEnum.API})
33      public Pagination<PetShopProxyB> queryPage(Pagination<PetShopProxyB> page, IWrapper<PetShopProxyB> queryWrapper){
34          Pagination<PetShopProxyB> result = new PetShopProxy().queryPage(page,queryWrapper);
35          if(!CollectionUtils.isEmpty(result.getContent())) {
36              //自身继承"宠物店铺"后扩展的字段catItems
37              List<Long> shopIds = result.getContent().stream().map(PetShopProxyB::getId).collect(Collectors.toList());
38              List<PetCatItem> catItems = petCatItemQueryService.queryByShopList(shopIds);
39              fillCatItems(result.getContent(),catItems);
40              //          new PetShopProxyB().listFieldQuery(result.getContent(), PetShopProxyB::getCatItems);
41              //继承"宠物店铺代理模型"而来的字段creater，java语言限制"宠物店铺代理模型B"没有getCreate()方法
42              new PetShopProxyB().listFieldQuery(result.getContent(),"creater");
43              //继承"宠物店铺代理模型A"而来的字段items，java语言限制"宠物店铺代理模型B"没有getItems()方法
44              new PetShopProxyB().listFieldQuery(result.getContent(),"items");
45          }
46          return result;
47      }
48      @Function.Advanced(type= FunctionTypeEnum.QUERY)
49      @Function.fun(FunctionConstants.queryByEntity)
50      @Function(openLevel = {FunctionOpenEnum.API})
51      public PetShopProxyB queryOne(PetShopProxyB query){
```

```
52          query = query.queryById();
53          //自身继承"宠物店铺"后扩展的字段catItems
54          query.setCatItems(petCatItemQueryService.queryByShopId(query.getId()));
55          //继承"宠物店铺代理模型"而来的字段creater, java语言限制"宠物店铺代理模型B"没有get
    Create()方法
56          query.fieldQuery("creater");
57          //继承"宠物店铺代理模型A"而来的字段items, java语言限制"宠物店铺代理模型B"没有getI
    tems()方法
58          query.fieldQuery("items");
59          return query;
60      }
61
62      private void fillCatItems(List<PetShopProxyB> shops,List<PetCatItem> catItem
    s){
63          for(PetShopProxyB shop:shops){
64              for(PetCatItem item:catItems){
65                  if(item.getShopId().equals(shop.getId())){
66                      if(shop.getCatItems()==null){
67                          List<PetCatItem> shopCatItems =new ArrayList<>();
68                          shopCatItems.add(item);
69                          shop.setCatItems(shopCatItems);
70                      }else{
71                          shop.getCatItems().add(item);
72                      }
73                  }
74              }
75          }
76      }
77  }
```

（4）重启看效果，我们发现商店管理 B 下面的列表页和详解页的萌猫商品列表已经过滤了相关数据，如图 3-101、图 3-102 所示。

图 3-101　商店管理 B 列表页萌猫商品列表已过滤相关数据（1）

图 3-102　商店管理 B 详解页萌猫商品列表已过滤相关数据（2）

**8. 临时继承（举例）**

参考前文"传输模型"的介绍，差别在于一个继承抽象基类，一个继承其他模型类型，不管什么类型都只是复用父类的字段，自己还是传输模型。

### 3.3.5　模型编码生成器

在日常开发中经常要对一些单据生成指定格式的编码，而且现在分布式环境下要考虑的事情会特别多。Oinone 提供了简易的编码生成能力。

**1. 编码生成器**

可以在模型或者字段上配置编码自动生成规则。在进行数据存储时，如果配置编码自动生成规则的字段值为空，则系统将根据规则自动生成编码。编码自动生成功能是通过序列生成器来支持的。可以在序列生成器生成的序列编码基础上，再进行组合配置的功能编码生成最终的编码。序列生成器可以配置初始序列、步长、日期格式、长度。

1）模型序列生成器（举例）

使用模型编码生成器，需要继承 CodeModel 或者有 Code 字段，那么使用 Model.Code 注解即可。

Step1. 为 PetShop 增加一个 @Model.Code 注解，并增加一个店铺编码（Code）字段，代码如下。

```
 8  @Model.model(PetShop.MODEL_MODEL)
 9  @Model(displayName = "宠物店铺",summary="宠物店铺",labelFields = {"shopName"})
10  @Model.Code(sequence = "DATE_ORDERLY_SEQ",prefix = "P",size=6,step=1,initial = 10
    000,format = "yyyyMMdd")
11  public class PetShop extends AbstractDemoIdModel {
12      public static final String MODEL_MODEL="demo.PetShop";
13
14      @Field(displayName = "店铺编码")
15      private String code;
16
17      @Field(displayName = "店铺名称",required = true)
18      private String shopName;
19
20      @Field(displayName = "开店时间",required = true)
21      private Time openTime;
22
23      @Field(displayName = "闭店时间",required = true)
24      private Time closeTime;
25
26  }
```

Step2. 重启查看效果。

（1）进入店铺新增页面，新增一个 Oinone 宠物店铺 003，如图 3-103 所示。

图 3-103　进入店铺新增页面，新增一个 Oinone 宠物店铺 003

（2）查看店铺列表页面，新增的记录中店铺编码一列，已经按 Model.Code 注解要求生成了，如图 3-104 所示。

图 3-104　店铺列表页面已按 Model.Code 注解要求生成

2）字段序列生成器

字段编码生成器，在对应的字段上增加，并使用 Field.Sequence 注解即可。

Step1. 为 PetShop 增加一个字段 codeTwo 并增加 @Field.Sequence 注解，代码如下。

```
 8  @Model.model(PetShop.MODEL_MODEL)
 9  @Model(displayName = "宠物店铺",summary="宠物店铺",labelFields = {"shopName"})
10  @Model.Code(sequence = "DATE_ORDERLY_SEQ",prefix = "P",size=6,step=1,initial = 10
    000,format = "yyyyMMdd")
11  public class PetShop extends AbstractDemoIdModel {
12      public static final String MODEL_MODEL="demo.PetShop";
13
14      @Field(displayName = "店铺编码")
15      private String code;
16
17      @Field(displayName = "店铺编码2")
18      @Field.Sequence(sequence = "DATE_ORDERLY_SEQ",prefix = "C",size=6,step=1,init
    ial = 10000,format = "yyyyMMdd")
19      private String codeTwo;
20
21      @Field(displayName = "店铺名称",required = true)
22      private String shopName;
23
```

```
24      @Field(displayName = "开店时间",required = true)
25      private Time openTime;
26
27      @Field(displayName = "闭店时间",required = true)
28      private Time closeTime;
29
30  }
```

Step2. 重启查看效果。

（1）进入店铺新增页面新增一个 Oinone 宠物店铺 004，如图 3-105 所示。

图 3-105　新增 Oinone 宠物店铺 004

（2）查看店铺列表页面，新增的记录中店铺编码 2 一列，已经按 Field.Sequence 注解要求生成了，如图 3-106 所示。

图 3-106　店铺列表页已按 Field.Sequence 注解生成

**2. 编码注解说明**

1）模型编码注解说明

（1）模型编码生成器规定仅针对 code 属性生效。

（2）Model.Code#sequence：序列生成函数。

● SEQ——自增流水号（不连续）。

● ORDERLY_SEQ——自增强有序流水号（连续）。

● DATE_SEQ——日期 + 自增流水号（不连续）。

● DATE_ORDERLY_SEQ——日期 + 强有序流水号（连续）。

● DATE——日期。

● UUID——随机 32 位字符串，包含数字和小写英文字母。

（3）Model.Code#prefix：前缀。

（4）Model.Code#suffix：后缀。

（5）Model.Code#size：长度。

（6）Model.Code#step：步长（包含流水号有效）。

（7）Model.Code#initial：起始值（包含流水号有效）。

（8）Model.Code#format：格式化（包含日期有效）。

（9）Model.Code#separator：分隔符。

2）字段编码注解说明

（1）字段编码生成器规定在字段上增加 Field.Sequence 注解即可。

（2）Field.Sequence#sequence：序列生成函数。

● SEQ——自增流水号（不连续）。

● ORDERLY——自增强有序流水号（连续）。

● DATE_SEQ——日期 + 自增流水号（不连续）。

● DATE_ORDERLY_SEQ——日期 + 强有序流水号（连续）。

● DATE——日期。

● UUID——随机 32 位字符串，包含数字和小写英文字母。

（3）Field.Sequence#prefix：前缀。

（4）Field.Sequence#suffix：后缀。

（5）Field.Sequence#size：长度。

（6）Field.Sequence#step：步长（包含流水号有效）。

（7）Field.Sequence#initial：起始值（包含流水号有效）。

（8）Field.Sequence#format：格式化（包含日期有效）。

（9）Field.Sequence#separator：分隔符。

### 3.3.6　枚举与数据字典

枚举是系统开发中经常用的一种类型，Oinone 不仅支持枚举类型，还做了相应的加强。

**1. 枚举系统与数据字典**

枚举是列举出一个有穷序列集的所有成员的程序。在元数据中，我们使用数据字典进行描述。

1）协议约定

枚举需要实现 IEnum 接口和使用 @Dict 注解进行配置，通过配置 @Dict 注解的 dictionary 属性来设置数据字典的唯一编码。前端使用枚举的 displayName 来展示，使用枚举的 name 进行交互；后端使用枚举的 value 进行交互（包括默认值设置也使用枚举的 value）。

枚举会存储在元数据的数据字典表中。枚举分为两类：异常类和业务类。异常类枚举用于定义程序中的错误提示，业务类枚举用于定义业务中某个字段值的有穷有序集。

2）编程式用法

编程式用法如图 3-107 所示。

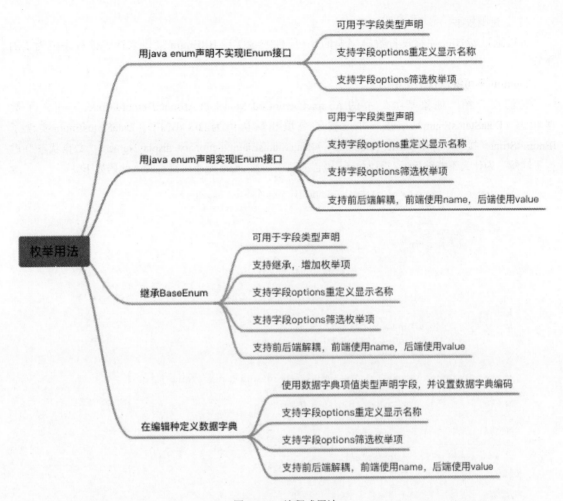

图 3-107　编程式用法

如果一个字段的类型被定义为枚举，则该字段就可以使用该枚举来进行可选项约束（options）。该字段的可选项为枚举所对应数据字典的子集。

3）可继承枚举

继承 BaseEnum 可以实现 Java 不支持的继承枚举，同时可继承枚举也可以用编程式动态创建枚举项。

可继承枚举也可以兼容无代码枚举，如图 3-108 所示。

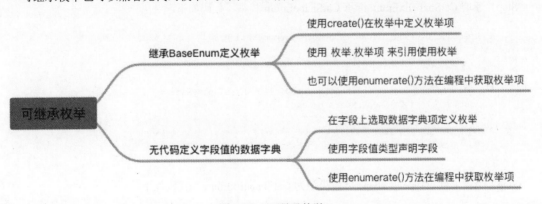

图 3-108　可继承枚举

4）二进制枚举

可以通过 @Dict 注解设置数据字典的 bit 属性或者实现 BitEnum 接口来标识该枚举值为 2 的次幂。

**2. enum 不可继承枚举（举例）**

我们在介绍"抽象基类"中的 AbstractDemoCodeModel 和 AbstractDemoIdModel 时引入了数据状态（DataStatusEnum）字段，并设置了必填和默认值为 DISABLED。DataStatusEnum 实现了 IEnum<String> 接口，并用 @Dict (dictionary = DataStatusEnum.dictionary, displayName = " 数据状态 ") 进行了注解。为什么不能继承呢？因为 Java 语言的限制导致 enum 是不可继承的。代码如下：

```java
 6   @Dict(dictionary = DataStatusEnum.dictionary, displayName = "数据状态")
 7   public enum DataStatusEnum implements IEnum<String> {
 8
 9       DRAFT("DRAFT", "草稿", "草稿"),
10       NOT_ENABLED("NOT_ENABLED", "未启用", "未启用"),
11       ENABLED("ENABLED", "已启用", "已启用"),
12       DISABLED("DISABLED", "已禁用", "已禁用");
13
14       public static final String dictionary = "partner.DataStatusEnum";
15
16       private String value;
17       private String displayName;
18       private String help;
19
20       DataStatusEnum(String value, String displayName, String help) {
21           this.value = value;
22           this.displayName = displayName;
23           this.help = help;
24       }
25
26       public String getValue() {
27           return value;
28       }
29
30       public String getDisplayName() {
31           return displayName;
32       }
33
34       public String getHelp() {
35           return help;
36       }
37   }
```

**3. BaseEnum 可继承枚举（举例）**

Step1. 新增 CatShapeExEnum 继承 CatShapeEnum 枚举，代码如下。

```java
 1   package pro.shushi.pamirs.demo.api.enumeration;
 2
 3   import pro.shushi.pamirs.meta.annotation.Dict;
 4
 5   @Dict(dictionary = CatShapeExEnum.DICTIONARY,displayName = "萌猫体型Ex")
 6   public class CatShapeExEnum extends CatShapeEnum {
 7
 8       public static final String DICTIONARY ="demo.CatShapeExEnum";
 9       public final static CatShapeExEnum MID =create("MID",3,"中","中");
10   }
```

Step2. 修改 PetCatType 的 shape 字段类型为 CatShapeExEnum，代码如下。

```
 7      @Model.MultiTableInherited(type = PetCatType.KIND_CAT)
 8      @Model.model(PetCatType.MODEL_MODEL)
 9      @Model(displayName="萌猫品种",labelFields = {"name"})
10    public class PetCatType extends PetType {
11
12          public static final String MODEL_MODEL="demo.PetCatType";
13          public static final String KIND_CAT="CAT";
14
15          @Field(displayName = "宠物分类",defaultValue = PetCatType.KIND_CAT,invisible = true)
16          private String kind;
17
18          @Field.Enum
19          @Field(displayName = "宠物体型")
20          private CatShapeExEnum shape;
21      }
```

Step3. 重启系统，查看效果，如图 3-109 所示。

图 3-109　重启系统，查看效果

另：可继承枚举的 Switch API。

继承 BaseEnum 可以实现 Java 不支持的可变枚举，可变枚举可以在运行时增加非 Java 代码定义的枚举项，同时可变枚举支持枚举继承。由于可变枚举不是 Java 规范中的枚举，所以无法使用 switch…case…语句，但是 K2 提供稍作变化的 switches（无须返回值）与 switchGet（需要返回值）方式实现相同功能与逻辑。

枚举的 switches 用法，代码如下。

```
1   BaseEnum.switches(比较变量，比较方式/*系统默认提供两种方式：caseName()和caseValue()*/,
2                    cases(枚举列表1).to(() -> {/*逻辑处理*/}),
3                    cases(枚举列表2).to(() -> {/*逻辑处理*/}),
4                    ...
5                    cases(枚举列表N).to(() -> {/*逻辑处理*/}),
6                    defaults(() -> {/*默认逻辑处理*/})
7   );
```

枚举的 switchGet 用法，代码如下。

```
1   BaseEnum.<比较变量类型，返回值类型>switchGet(比较变量，
2                    比较方式/*系统默认提供两种方式：caseName()和caseValue()*/,
3                    cases(枚举列表1).to(() -> {/*return 逻辑处理的结果*/}),
4                    cases(枚举列表2).to(() -> {/*return 逻辑处理的结果*/}),
5                    ...
6                    cases(枚举列表N).to(() -> {/*return 逻辑处理的结果*/}),
7                    defaults(() -> {/*return 逻辑处理的结果*/})
8   );
```

caseName () 使用枚举项的 name 与比较变量进行匹配比较；caseValue () 使用枚举项的 value 值与比较变量进行匹配比较。

例如，以下逻辑表示当 ttype 的值为 O2O、O2M、M2O 或 M2M 枚举值时返回 true，否则返回 false：

```
1  return BaseEnum.<String, Boolean>switchGet(ttype, caseValue(),
2                  cases(O2O, O2M, M2O, M2M).to(() -> true),
3                  defaults(() -> false)
4  );
```

**4. 二进制枚举（举例）**

可以通过 @Dict 注解设置数据字典的 bit 属性或者实现 BitEnum 接口来标识该枚举值为 2 的次幂。二进制枚举最大的区别在于值的序列化和反序列化方式是不一样的。更多有关序列化知识请见 3.3.5 节。

Step1. 新建店铺选项枚举，并添加为 PetShop 的一个字段。

（1）PetShopOptionEnum 继承 BaseEnum<PetShopOptionEnum,Long> 并实现 BitEnum 接口，增加三个枚举，值分别是 2 的 0 次幂，2 的 1 次幂，2 的 2 次幂。多选枚举 3 位枚举都选中，字段值为 7，代码如下。

```
5  @Errors(displayName = "demo模块错误枚举")
6  public enum  DemoExpEnumerate implements ExpBaseEnum {
7
8      SYSTEM_ERROR(ERROR_TYPE.SYSTEM_ERROR,90000000,"系统异常"),
9      PET_SHOP_BATCH_UPDATE_SHOPLIST_IS_NULL(ERROR_TYPE.BIZ_ERROR,90000001,"店铺列表
   不能为空");
10
11     private ERROR_TYPE type;
12     private int code;
13     private String msg;
14
15     DemoExpEnumerate(ERROR_TYPE type, int code, String msg) {
16         this.type= type;
17         this.code=code;
18         this.msg=msg;
19     }
20 }
```

（2）修改 PetShop，增加一个多选枚举字段 options，枚举类型为 PetShopOptionEnum，代码如下。

```
10 @Model.model(PetShop.MODEL_MODEL)
11 @Model(displayName = "宠物店铺",summary="宠物店铺",labelFields = {"shopName"})
12 @Model.Code(sequence = "DATE_ORDERLY_SEQ",prefix = "P",size=6,step=1,initial = 10
   000,format = "yyyyMMdd")
13 public class PetShop extends AbstractDemoIdModel {
14     public static final String MODEL_MODEL="demo.PetShop";
15
16     @Field(displayName = "店铺编码")
17     private String code;
18
19     @Field(displayName = "店铺编码2")
20     @Field.Sequence(sequence = "DATE_ORDERLY_SEQ",prefix = "C",size=6,step=1,init
   ial = 10000,format = "yyyyMMdd")
21     private String codeTwo;
22
23     @Field(displayName = "店铺名称",required = true)
24     private String shopName;
25
26     @Field(displayName = "开店时间",required = true)
27     private Time openTime;
```

```
28
29      @Field(displayName = "闭店时间",required = true)
30      private Time closeTime;
31
32      @Field(displayName = "店铺标志")
33      private List<PetShopOptionEnum> options;
34
35  }
36
```

Step2. 重启查看效果。

（1）模型宠狗商店的编辑页面可以看到店铺标志字段可多选"十年老店""七天无理由退货""正品认证"，如图 3-110 所示。

图 3-110　模型宠狗商店编辑页

（2）模型宠狗商店的列表页面可以看到店铺标志字段为"十年老店、七天无理由退货、正品认证"如图 3-111 所示。

图 3-111　模型宠狗商店列表页

（3）查看数据库对应的 options 字段值为 7，如图 3-112 所示。

图 3-112　查看数据库对应的 options 字段值为 7

### 5. 异常枚举（举例）

作为 Oinone 管理异常的规范，一般枚举都是用 @Dict 声明为数据字典，但是异常枚举会用 @Error 来注解，因为异常跟业务枚举有很大区别，异常往往数量非常多，如果用 @Dict 数据字典方式来管理，那么数据字典的量会非常大。

Step1. 新建一个异常枚举类 DemoExpEnumerate，实现 ExpBaseEnum 接口并加上 @Errors (displayName = "demo 模块错误枚举") 注解，增加对应错误枚举，代码如下。

```java
@Errors(displayName = "demo模块错误枚举")
public enum DemoExpEnumerate implements ExpBaseEnum {

    SYSTEM_ERROR(ERROR_TYPE.SYSTEM_ERROR,90000000,"系统异常"),
    PET_SHOP_BATCH_UPDATE_SHOPLIST_IS_NULL(ERROR_TYPE.BIZ_ERROR,90000001,"店铺列表不能为空");

    private ERROR_TYPE type;
    private int code;
    private String msg;

    DemoExpEnumerate(ERROR_TYPE type, int code, String msg) {
        this.type= type;
        this.code=code;
        this.msg=msg;
    }
}
```

Step2. 修改宠物商店批量更新数据状态逻辑。

增加一个 PetShopList 必选判断，如果没选则抛出异常并指定异常枚举为 PET_SHOP_BATCH_UPDATE_SHOPLIST_IS_NULL，代码如下。

```java
@Model.model(PetShopBatchUpdate.MODEL_MODEL)
@Component
public class PetShopBatchUpdateAction {

    @Function(openLevel = FunctionOpenEnum.API)
    @Function.Advanced(type= FunctionTypeEnum.QUERY)
    public PetShopBatchUpdate construct(PetShopBatchUpdate petShopBatchUpdate, List<PetShopProxy> petShopList){
        PetShopBatchUpdate result = new PetShopBatchUpdate();
        result.setPetShopList(petShopList);
        return result;
    }

    @Action(displayName = "确定",bindingType = ViewTypeEnum.FORM,contextType = ActionContextTypeEnum.SINGLE)
    public PetShopBatchUpdate conform(PetShopBatchUpdate data){
        if(data.getPetShopList() == null || data.getPetShopList().size()==0){
            throw PamirsException.construct(DemoExpEnumerate.PET_SHOP_BATCH_UPDATE_SHOPLIST_IS_NULL).errThrow();
        }
        List<PetShopProxy> proxyList = data.getPetShopList();
        for(PetShopProxy petShopProxy:proxyList){
            petShopProxy.setDataStatus(data.getDataStatus());
        }
        new PetShopProxy().updateBatch(proxyList);
        return data;
    }
}
```

Step3. 重启系统（如图 3-113 所示）。

图 3-113　重启系统看效果

平台的异常枚举如下。

**1. pamirs-framework 异常枚举**

每一个模块都可以包含一个或多个异常枚举类，枚举项定义了应用中异常的错误编码与描述。在应用需要抛出异常的位置，可在抛出异常的时候附带对应的错误枚举。我们使用 @Errors 注解来定义错误枚举类。

pamirs-framework 异常枚举见表 3-10。

表 3-10　pamirs-framework 异常枚举

| 工程名 | 定义位置 | 编码起始值 |
| --- | --- | --- |
| pamirs-meta-Model | MetaExpEnumerate | 10010000 |
| pamirs-meta-dsl | DslExpEnumerate | 10020000 |
| pamirs-framework-common | FwExpEnumerate | 10050000 |
| pamirs-framework-configure-annotation | AnnotationExpEnumerate | 10060000 |
| pamirs-framework-configure-db | MetadExpEnumerate | 10070000 |
| pamirs-framework-compute | ComputeExpEnumerate | 10080000 |
| pamirs-framework-compare | CompareExpEnumerate | 10090000 |
| pamirs-framework-faas | FaasExpEnumerate | 10100000 |
| pamirs-framework-orm | OrmExpEnumerate | 10110000 |
| pamirs-connectors-data | DataExpEnumerate | 10150000 |
| pamirs-connectors-data-dialect | DialectExpEnumerate | 10160000 |
| pamirs-connectors-data-sql | SqlExpEnumerate | 10170000 |
| pamirs-connectors-data-ddl | DdlExpEnumerate | 10180000 |
| pamirs-connectors-data-infrastructure | InfExpEnumerate | 10190000 |
| pamirs-connectors-data-tx | TxExpEnumerate | 10200000 |
| pamirs-gateways-rsql | RsqlExpEnumerate | 10500000 |
| pamirs-gateways-graph-Java | GqlExpEnumerate | 10510000 |
| pamirs-boot-api | BootExpEnumerate | 11000000 |
| pamirs-boot-uxd | BootUxdExpEnumerate | 11040000 |
| pamirs-boot-standard | BootStandardExpEnumerate | 11050000 |

续表

| 工程名 | 定义位置 | 编码起始值 |
|---|---|---|
| pamirs-base-api | BaseExpEnumerate | 11500000 |
| pamirs-sid | SidExpEnumerate | 11510000 |

**2. 通用异常码**

通用异常码见表3-11。

表3-11 通用异常码

| 错误 | 错误描述 | 定义位置 | 编码 |
|---|---|---|---|
| BASE_USER_NOT_LOGIN_ERROR | 用户未登录 | BaseExpEnumerate | 11500001 |
| BASE_CHECK_DATA_ERROR | 校验失败,数据错误 | FwExpEnumerate | 10050009 |

**3. pamirs-core 异常枚举(20010000-20290000)**

pamirs-core 异常枚举见表3-12。

表3-12 pamirs-core 异常枚举

| 工程名 | 编码起始值 | 数据字典名 |
|---|---|---|
| pamirs-core-common | 20010000 | error.core.common.exceptions |
| pamirs-sequence(原 pamirs-bid) | 20020000 | error.core.sequence.exceptions |
| pamirs-data-audit | 20030000 | error.core.data.audit.exceptions |
| pamirs-channel | 20040000 | error.core.channel.exceptions |
| pamirs-resource | 20050000 | error.core.resource.exceptions |
| pamirs-user | 20060000 | error.core.user.exceptions |
| pamirs-auth | 20070000 | error.core.auth.exceptions |
| pamirs-message | 20080000 | error.core.message.exceptions |
| pamirs-international | 20090000 | error.core.international.exceptions(未正确定义) |
| pamirs-translate | 20100000 | error.core.translate.exceptions(未正确定义) |
| pamirs-scheduler(已作废)<br>pamirs-data-audit | 20110000 | error.core.schedule.exceptions(已作废)<br>error.core.data.audit.exceptions |
| pamirs-trigger | 20120000 | error.core.trigger.exceptions(未正确定义) |
| pamirs-file2(原 pamirs-file) | 20130000 | error.core.file.exceptions(未正确定义) |
| pamirs-eip2(原 pamirs-eip2) | 20140000 | error.core.eip.exceptions(未正确定义) |
| pamirs-third-party-communication | 20150000 | error.core.third-party-communication.exceptions(未定义) |
| pamirs-third-party-map | 20160000 | error.core.third-party-map.exceptions(未定义) |
| pamirs-business | 20170000 | error.core.business.exceptions(未定义) |
| pamirs-web | 20180000 | error.core.web.exceptions |
| pamirs-studio(已作废) | 20190000 | error.core.studio.exceptions(未正确定义) |
| pamirs-workflow | 20200000 | error.core.workflow.exceptions |
| pamirs-apps | 20210000 | AppsExpEnumerate |
| pamirs-paas | 20220000 | PaasExpEnumerate |

**4. pamirs-Model-designer 异常枚举(20300000)**

pamirs-Model-designer 异常枚举见表3-13。

表 3-13　pamirs-Model-designer 异常枚举

| 工程名 | 编码起始值 | 数据字典名 |
|---|---|---|
| pamirs-Model-designer | 20300000 | ModelDesignerExp |

**5. pamirs-workflow 异常枚举（20310000-20320000）**

pamirs-workflow 异常枚举见表 3-14。

表 3-14　pamirs-workflow 异常枚举

| 工程名 | 编码起始值 | 数据字典名 |
|---|---|---|
| pamirs-workflow | 20310000 | WorkflowExpEnumerate |
| pamirs-workflow-designer | 20320000 | WorkflowDesignerExpEnumerate |

### 3.3.7　字段之序列化方式

本节核心是带大家全面了解 Oinone 的序列方式，包括支持的序列化类型、注意点、如何新增客户化序列化方式以及字段默认值的反序列化。

**1. 数据存储的序列化（举例）**

使用 @Field 注解的 serialize 属性来配置非字符串类型属性的序列化与反序列化方式，最终会以序列化后的字符串持久化到存储中。

Step1. 新建 PetItemDetail 模型，并为 PetItem 添加两个字段。

（1）PetItemDetail 继承 TransientModel，增加两个字段，分别为备注和备注人，代码如下。

```
 8    @Model.model(PetItemDetail.MODEL_MODEL)
 9    @Model(displayName = "商品详情",summary = "商品详情",labelFields = {"remark"})
10    public class PetItemDetail extends TransientModel {
11        public static final String MODEL_MODEL="demo.PetItemDetail";
12    
13        @Field.String(min = "2",max = "20")
14        @Field(displayName = "备注",required = true)
15        private String remark;
16    
17        @Field(displayName = "备注人",required = true)
18        private PamirsUser user;
19    
20    }
```

（2）修改 PetItem，增加两个字段 PetItemDetails 类型为 List<PetItemDetail> 和 tags 类型为 List<String>，并设置为不同的序列化方式，PetItemDetails 为 JSON（默认就是 JSON，可不配），tags 为 COMMA。同时设置 @Field.Advanced (columnDefinition = "varchar(1024)")，防止序列化后存储过长，代码如下。

```
11    @Model.model(PetItem.MODEL_MODEL)
12    @Model(displayName = "宠物商品",summary="宠物商品",labelFields = {"itemName"})
13    public class PetItem  extends AbstractDemoCodeModel{
14    
15        public static final String MODEL_MODEL="demo.PetItem";
16    
17        @Field(displayName = "商品名称",required = true)
18        private String itemName;
19    
20        @Field(displayName = "商品价格",required = true)
21        private BigDecimal price;
22    
23        @Field(displayName = "店铺",required = true)
24        @Field.Relation(relationFields = {"shopId"},referenceFields = {"id"})
25        private PetShop shop;
26    
27        @Field(displayName = "店铺id",invisible = true)
28        private Long shopId;
29    
30        @Field(displayName = "品种")
31        @Field.many2one
```

```
32      @Field.Relation(relationFields = {"typeId"},referenceFields = {"id"})
33      private PetType type;
34
35      @Field(displayName = "品种类型",invisible = true)
36      private Long typeId;
37
38      @Field(displayName = "详情", serialize = Field.serialize.JSON, store = Nullabl
     eBoolEnum.TRUE)
39      @Field.Advanced(columnDefinition = "varchar(1024)")
40      private List<PetItemDetail> petItemDetails;
41
42      @Field(displayName = "商品标签",serialize = Field.serialize.COMMA,store = Null
     ableBoolEnum.TRUE,multi = true)
43      @Field.Advanced(columnDefinition = "varchar(1024)")
44      private List<String> tags;
45
46  }
```

Step2. 重启系统看效果。

（1）模型宠狗商品的编辑页面可以看到详情字段和商品标签字段，用新增按钮添加详情记录，直接在"商品标签"的 input 框输入，按 Enter 键可以输入多值，如图 3-114 所示。

图 3-114　模型宠狗商品编辑页面

（2）模型宠狗商品的列表页面可以看到"详情"字段和"商品标签"字段，如图 3-115 所示。

图 3-115　模型宠狗商品的列表页面可看到"详情"字段和"商品标签"字段

（3）查看商品数据表，我们可以看到"详情"字段和"商品标签"字段按指定序列化方式进行存储，如图 3-116 所示。

图 3-116　商品数据表可看到"详情"字段和"商品标签"字段

Step3. 字段序列化注意点。

（1）必须使用 Field#store 属性将字段存储设置为 NullableBoolEnum.TRUE。

（2）使用 Field#serialize 属性指定序列化方式，默认为 JSON。

（3）如把 PetItemDetail 设置为存储模型，须在 PetItem 的 PetItemDetails 字段上使用 Field.Relation#store 属性将关联关系存储设置为 False。不然会同时存储 PetItemDetails 字段和对应的 PetItemDetail 表记录。

Step4. 字段序列化方式说明见表 3-15。

表 3-15　字段序列化方式说明

| 序列化方式 | 说明 | 备注 |
| --- | --- | --- |
| JSON | JSON 序列化 | 主要用于模型相关类型字段的序列化，是 @Field.serialize 默认选项 |
| DOT | 点拼接集合元素 | — |
| COMMA | 逗号拼接集合元素 | — |
| BIT | 按位与，2 次幂数求和 | 非 @Field.serialize 可选项列表，用于二进制枚举序列化不需要配置，由 Oinone 自动推断 |

**2. 注册自己的序列化器（举例）**

注册自己的序列化器（实现 pro.shushi.pamirs.meta.api.core.orm.serialize.Serializer 接口），如 Oinone 的 DOT 的序列化方式，用 type () 方法返回值做匹配，serialize 和 deserialize 分别对应序列化和反序列化方法，代码如下。

```
/**
 * 点表达式序列生成处理器实现
 *
 * @author d@shushi.pro
 * @version 1.0.0
 * date 2020/3/4 2:48 上午
 */
@SuppressWarnings("rawtypes")
@Slf4j
@Component
public class DotSerializeProcessor implements Serializer<Object, String> {

    @Override
    public String serialize(String ltype, Object value) {
        if (null == value) {
            return null;
        }
        if (List.class.isAssignableFrom(value.getClass())) {
            return StringUtils.join((List) value, CharacterConstants.SEPARATOR_DOT);
        } else {
            return StringUtils.join(Collections.singletonList(value), CharacterConstants.SEPARATOR_DOT);
```

```
36          }
37      }
38
39      @SuppressWarnings("unchecked")
40      @Override
41      public Object deserialize(String ltype, String ltypeT, String value, String format) {
42          if (null == value) {
43              return null;
44          }
45          String[] dots = value.split(CharacterConstants.SEPARATOR_ESCAPE_DOT);
46          List list = new ArrayList();
47          for (String dot : dots) {
48              Object object = TypeUtils.valueOfPrimary(ltypeT, dot, null);
49              list.add(object);
50          }
51          return list;
52      }
53
54      @Override
55      public String type() {
56          return SerializeEnum.DOT.value();
57      }
58  }
```

**3. 字段默认值的反序列化**

用 @Field.defaultValue 注解在字段上配置 defaultValue 属性时，将根据字段的 Ttype 类型及字段的 Ltype 等类型属性，自动进行反序列化。包括但不限于以下几种情况：

- OBJ、STRING、TEXT、HTML——保持不变。
- BINARY、INTEGER——转换为整数。
- FLOAT、MONEY——转换为浮点数。
- DATETIME、DATE、TIME、YEAR——根据 Field.Date#format 属性决定反序列化日期格式。
- BOOLEAN——仅允许 null、true、false。
- ENUM——使用 value 进行匹配。

### 3.3.8 字段类型之基础与复合

使用 @Field 注解来描述模型的字段。如果未配置字段类型，系统会根据 Java 代码的字段声明类型自动获取业务类型。建议配置 displayName 属性来描述字段在前端的显示名称。可以使用 defaultValue 配置字段的默认值。

**1. 安装与更新**

使用 @Field.field 来配置字段的不可变更编码。字段一旦安装，无法再对该字段编码值进行修改，之后的字段配置更新会依据该编码进行查找并更新；如果仍然修改该注解的配置值，则系统会将该字段识别为新字段，存储模型会创建新的数据库表字段，而原字段将会 rename 为废弃字段。

**2. 基础配置**

参考 3.3.2 节中的基础配置。

**3. 字段类型**

类型系统由基本类型、复合（组件）类型、引用类型和关系类型四种类型系统构成。通过类型系统描述应用程序、数据库和前端视觉视图如何进行交互，数据及数据间关系如何处理的协议。其中引用类型和关系类型介绍详见 3.3.9 节，字段命名规范参见 3.3.1 一节，这里不再赘述。

1）基本类型

字段基本类型见表 3-16。

表 3-16 字段基本类型

| 业务类型 | Java 类型 | 数据库类型 | 规则说明 |
| --- | --- | --- | --- |
| BINARY | Byte<br>Byte[] | TINYINT<br>BLOB | 二进制类型，不推荐使用 |

续表

| 业务类型 | Java 类型 | 数据库类型 | 规则说明 |
|---|---|---|---|
| INTEGER | Short<br>Integer<br>Long<br>BigInteger | smallint<br>int<br>bigint<br>decimal (size,0) | 整数，包括整数（10～11 位有效数字）、长整数（19～20 位有效数字）和大整数（超过 19 位）<br>数据库规则：默认使用 int。如果 size 少于 6 位数字则使用 smallint；如果 size 超过 6 位数字则使用 int；如果 size 超过 10 位数字，即大于或等于 11（包含符号位），则使用长整数 bigint；如果 size 超过 19 位数字，即大于或等于20（包含符号位），则使用大数 decimal。若未配置 size，则按 Java 类型推测<br>前端交互规则：整数使用 Number 类型，长整数和大整数前后端协议使用字符串类型 |
| FLOAT | Float<br>Double<br>BigDecimal | float (M,D)<br>double (M,D)<br>decimal (M,D) | 浮点数，包括单精度浮点数（7～8 位有效数字）、双精度浮点数（15～16 位有效数字）和大数（超过 15 位）<br>数据库规则：默认使用单精度浮点数 float。如果 size 超过 7 位数字，即大于或等于 8，则使用双精度浮点数 double；如果 size 超过 15 位数字，即大于或等于 16，则使用大数 decimal。若未配置 size，则按 Java 类型推测<br>前端交互规则：单精度浮点数 float 和双精度浮点数 double 使用 Number 类型（因为都使用 IEEE754 协议 64 位进行存储），大数前后端协议使用字符串类型 |
| BOOLEAN | Boolean | tinyint (1) | 布尔类型，值为 1，true（真）或 0，false（假） |
| ENUM | Enum | 与数据字典指定基本类型一致 | 前端交互规则：可选项从 ModelField 的 options 字段获取，options 字段值为字段指定数据字典子集的 JSON 序列化字符串。前后端传递的是可选项的 name，数据库存储使用可选项的 value。multi 属性为 true 则使用多选控件；multi 属性为 false 则使用单元控件 |
| STRING | String | varchar (size) | 字符串，size 为长度限制默认值参考，前端可以 view 中覆盖该配置 |
| TEXT | String | text | 多行文本，编辑态组件为多行文本框，长度限制为配置项 size 值 |
| HTML | String | text | 富文本编辑器 |
| DATETIME | Java.util.Date<br>Java.sql.Timestamp | datetime (fraction)<br>timestamp (fraction) | 日期时间类型<br>数据库规则：日期和时间的组合。<br>时间格式为 YYYY-MM-DD HH:MM:SS[.fraction]，默认精确到秒，在默认的秒精确度上，可以带小数，最多带 6 位小数，即可以精确到 microseconds（6 digits）precision。可以通过设置 fraction 来设置精确小数位数，最终存储在字段的 decimal 属性上<br>前端交互规则：前端默认使用日期时间控件，根据日期时间类型格式化的格式（format）来格式化日期 |
| YEAR | Java.util.Date | year | 年份类型<br>日期类型<br>数据库规则：默认"YYYY"格式表示的日期值<br>前端交互规则：前端默认使用年份控件，根据日期类型格式化的格式（format）来格式化日期 |
| DATE | Java.util.Date<br>Java.sql.Date | date<br>date | 日期类型<br>数据库规则：默认"YYYY-MM-DD"格式表示的日期值<br>前端交互规则：前端默认使用日期控件，根据日期类型格式化的格式（format）来格式化日期 |
| TIME | Java.util.Date<br>Java.sql.Time | time (fraction)<br>time (fraction) | 时间类型<br>数据库规则：默认"HH:MM:SS"格式表示的时间值<br>前端交互规则：前端默认使用时间控件，根据日期类型格式化的格式（format）来格式化日期 |

2）复合类型

字段复合类型见表3-17。

表3-17 字段复合类型

| 业务类型 | Java 类型 | 数据库类型 | 规则说明 |
| --- | --- | --- | --- |
| MONEY | BigDecimal | decimal (M,D) | 金额，前端使用金额控件，可以使用 currency 设置币种字段 |

3）不可变更字段

使用 immutable 属性来描述该字段前后端都无法进行更新操作，系统会忽略不可变更字段的更新操作。

4）自动生成编码的字段

详见 3.3.5 节。

5）字段的序列化与反序列化

使用 @Field 注解的 serialize 属性来配置非字符串类型属性的序列化与反序列化方式，最终会以序列化后的字符串持久化到存储中。

详见 3.3.7 节。

6）前端默认配置

可以使用 @Field 注解中的以下属性来配置前端的默认视觉与交互规则，也可以在前端设置覆盖以下配置：

- @Field (required)，是否必填；
- @Field (invisible)，是否不可见；
- @Field (priority)，字段优先级，列表的列使用该属性进行排序；
- 更多前端默认视图配置请见 3.5.4 节，如 readonly 是否只读等。

7）举例

（1）回顾我们前面学习的例子，现有 PetShop 代码如下。

```
11  @Model.model(PetShop.MODEL_MODEL)
12  @Model(displayName = "宠物店铺",summary="宠物店铺",labelFields ={"shopName"} )
13  @Model.Code(sequence = "DATE_ORDERLY_SEQ",prefix = "P",size=6,step=1,initial = 10
    000,format = "yyyyMMdd")
14  public class PetShop extends AbstractDemoIdModel {
15      public static final String MODEL_MODEL="demo.PetShop";
16
17      @Field(displayName = "店铺编码")
18      private String code;
19
20      @Field(displayName = "店铺编码2")
21      @Field.Sequence(sequence = "DATE_ORDERLY_SEQ",prefix = "C",size=6,step=1,init
    ial = 10000,format = "yyyyMMdd")
22      private String codeTwo;
23
24      @Field(displayName = "店铺名称",required = true)
25      private String shopName;
26
27      @Field(displayName = "开店时间",required = true)
28      private Time openTime;
29
30      @Field(displayName = "闭店时间",required = true)
31      private Time closeTime;
32
33      @Field(displayName = "店铺标志")
34      private List<PetShopOptionEnum> options;
35
36  }
```

（2）字段默认推断。

我们从 PetShop 代码中发现：

① 很多字段类型是由 Oinone 根据模型中的 Java 类型自动推断的；

② 大部分字段只是简单加了 @Field (displayName) 注解。

（3）新增未涉及基础字段以及其他注解配置。

① 为 PetShop 新增 FLOAT、BOOLEAN、TEXT、HTML、YEAR、DATE、MONEY 等字段。

② 设置 shopName 字段为 immutable=true。

③ 给 MONEY 类型字段，income 配置 @Field (priority = 1)。

基础字段以及其他注解配置代码如下。

```java
@Model.model(PetShop.MODEL_MODEL)
@Model(displayName = "宠物店铺",summary="宠物店铺",labelFields ={"shopName"} )
@Model.Code(sequence = "DATE_ORDERLY_SEQ",prefix = "P",size=6,step=1,initial = 10
000,format = "yyyyMMdd")
public class PetShop extends AbstractDemoIdModel {
    public static final String MODEL_MODEL="demo.PetShop";

    @Field(displayName = "店铺编码")
    private String code;

    @Field(displayName = "店铺编码2")
    @Field.Sequence(sequence = "DATE_ORDERLY_SEQ",prefix = "C",size=6,step=1,init
ial = 10000,format = "yyyyMMdd")
    private String codeTwo;

    @Field(displayName = "店铺名称",required = true,immutable=true)
    private String shopName;

    @Field(displayName = "开店时间",required = true)
    private Time openTime;

    @Field(displayName = "闭店时间",required = true)
    private Time closeTime;

    @Field.Enum
    @Field(displayName = "数据状态",defaultValue = "DRAFT",required = true,summary
= "枚举可选项举例")
    private DataStatusEnum dataStatus;

    @Field(displayName = "店铺标志")
    private List<PetShopOptionEnum> options;

    @Field(displayName = "一年内新店")
    private Boolean oneYear;

    @Field.Float
    @Field(displayName = "店内员工平均年龄")
    private BigDecimal averageAge;

    @Field.Text
    @Field(displayName = "描述")
    private String description;

    @Field.Html
    @Field(displayName = "html描述")
    private String descHtml;

    @Field.Date(type = DateTypeEnum.DATE,format = DateFormatEnum.DATE)
    @Field(displayName = "店庆")
    private Date anniversary;

    @Field.Date(type = DateTypeEnum.YEAR,format = DateFormatEnum.YEAR)
    @Field(displayName = "开店年份")
    private Date publishYear;

    @Field.Money
    @Field(displayName = "收入",priority = 1)
    private BigDecimal income;

}
```

④ 重启应用看效果。

在商店管理页面，单击数据记录的"修改"操作进入编辑页面（如图 3-117 所示），发现以下几点：
- "收入"字段排序靠前，但不是第一，这是因为默认 FORM 视图其他信息 GROUP 在前；
- 开店时间变为只读；
- 其他新增基础类型字段，前端提供默认组件展示；
- html 描述控件的图片上传会报错，需要引入 File 模块，请参见 6.1 节。

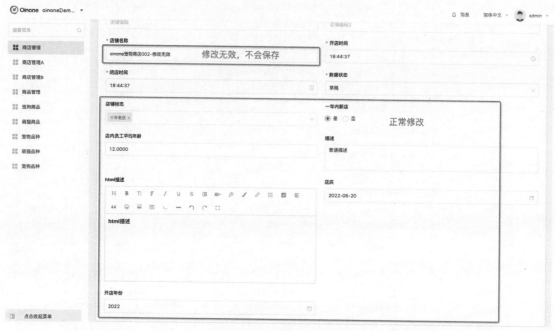

图 3-117　商店管理——编辑页

返回列表页查看数据，发现只有店铺名称的修改被忽略了，其他字段都赋值成功（如图 3-118 所示）。

图 3-118　商店管理——列表页

### 3.3.9　字段类型之关系与引用

有关系与引用类型才让 Oinone 具备完整的描述模型与模型之间关系的能力。

在 PetShop 以及其代理模型中已经用到了 O2M、M2O 字段，分别如 petItems (PetItem) 和 create (PamrisUser) 字段，但是没有过多的讲解。本节重点举例 RELATED、M2M（多对多）、O2M（一对多）、O2O（一对一），至于 M2O（多对一）留给大家自行尝试。

## 1. 引用类型（举例）

字段引用类型见表 3-18。

表 3-18  字段引用类型

| 业务类型 | Java 类型 | 数据库类型 | 规则说明 |
| --- | --- | --- | --- |
| RELATED | 基本类型或关系类型 | 不存储或 varchar、text | 引用字段<br>数据库规则：点表达式最后一级对应的字段类型；数据库字段值默认为 Java 字段的序列化值，默认使用 JSON 序列化<br>前端交互规则：点表达式最后一级对应的字段控件类型 |

Step1. 修改 PetShopProxy 类。

（1）为 PetShopProxy 类新增一个引用字段 relatedShopName，并加上 @Field.Related ("shopName") 注解。

（2）为 PetShopProxy 类新增一个引用字段 createrId，并加上 @Field.Related ({"creater","id"}) 注解，代码如下。

```
 9  @Model.model(PetShopProxy.MODEL_MODEL)
10  @Model.Advanced(type = ModelTypeEnum.PROXY)
11  @Model(displayName = "宠物店铺代理模型",summary="宠物店铺代理模型")
12  public class PetShopProxy extends PetShop {
13
14      public static final String MODEL_MODEL="demo.PetShopProxy";
15
16      @Field.many2one
17      @Field(displayName = "创建者",required = true)
18      @Field.Relation(relationFields = {"createUid"},referenceFields = {"id"})
19      private PamirsUser creater;
20
21      @Field.Related("shopName")
22      @Field(displayName = "引用字段shopName")
23      private String relatedShopName;
24
25      @Field.Related({"creater","id"})
26      @Field(displayName = "引用创建者Id")
27      private String createrId;
28
29  }
```

Step2. 重启系统查看效果。

我们发现商店管理—列表页面多出了两个有值字段：引用字段 shopName 和引用创建者 Id，如图 3-119 所示。

图 3-119  商店管理—列表页面新增两个有值字段

## 2. 关系类型

字段关系类型见表3-19。

表3-19 字段关系类型

| 业务类型 | Java类型 | 数据库类型 | 规则说明 |
|---|---|---|---|
| O2O | 模型/DataMap | 不存储或varchar、text | 一对一关系 |
| M2O | 模型/DataMap | 不存储或varchar、text | 多对一关系 |
| O2M | List<模型/DataMap> | 不存储或varchar、text | 一对多关系 |
| M2M | List<模型/DataMap> | 不存储或varchar、text | 多对多关系 |

多值字段或者关系字段需要存储，默认使用JSON序列化。多值字段数据库字段类型默认为varchar (1024)；关系字段数据库字段类型默认为text。

## 1. 关系字段

关联关系用于描述模型之间的关联方式（如图3-120所示）：
- 多对一关系（M2O），主要用于明确从属关系。
- 一对多关系（O2M），主要用于明确从属关系。
- 多对多关系（M2M），主要用于弱依赖关系的处理，提供中间模型进行关联关系的操作。
- 一对一关系（O2O），主要用于多表继承和行内合并数据。

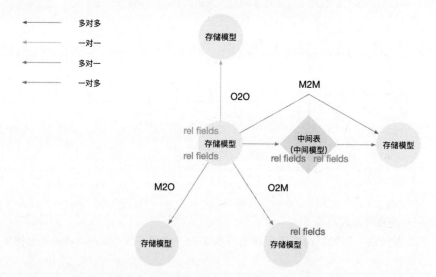

rel fields表示的是存储关联关系的字段所在模型
多对多关系的红字rel fields为throughRelationFields，蓝字rel fields为throughReferenceFields

图3-120 字段关联关系

## 2. 名词解释

关联关系比较重要的名词解释如下：
- 关联关系：使用relation表示，模型间的关联方式的一种描述，包括关联关系类型、关联关系双边的模型和关联关系的读写。
- 关联关系字段：业务类型为O2O、O2M、M2O或M2M的字段。
- 关联模型：使用references表示，自身模型关联的模型。
- 关联字段：使用referenceFields表示，关联模型的字段，表示关联模型的哪些字段与自身模型的哪些字段建立关系。

● 关系模型：自身模型。

● 关系字段：使用 relationFields 表示，自身模型的字段，表示自身模型的哪些字段与关联模型的哪些字段建立关系。

● 中间模型，使用 through 表示，只有多对多存在中间模型，模型的 relationship=true。

**3. M2M 关系类型（举例）**

多对多关系，主要用于弱依赖关系的处理，提供中间模型进行关联关系的操作。这也是在业务开发中很常见的，用于描述单据间关系。本例将列举两种方式描述多对多关系中间表，一是中间表没有在系统显示定义模型，二是中间表显示定义模型。第一种往往仅是维护多对多关系，第二种往往用于多对多关系，中间表自身也有业务含义，中间表模型还经常额外增加其他字段。

对于中间表没有在系统显示定义模型，如果出现跨模块的场景，在分布式环境下两个模块独立启动，有可能会导致系统关系表被删除的情况发生，因为没有显示定义中间表模型，中间表的模型所属模块会根据两边模型的名称计算，如果刚好被计算到非关系字段所属模型的模块，那么单独启动非关系字段所属模型的模块，则会导致删除关系表。

为什么不直接把中间表的模型所属模块设置为关系字段所属模型的模块？因为如果这样做，当模型两边都定义了多对多关系字段，则会导致 M2M 关系表的所属模块出现混乱。

所以这里建议大家都选用第二种中间表显示定义模型，不论扩展性还是适应性都会好很多。请用 through=XXXRelationModel.model_model 或者 throughClass=XXXRelationModel.class。

Step1. 新建宠物达人模型，并分别为宠物商品和宠物商店增加 many2many 到宠物达人模型的字段。

（1）新建宠物达人模型 PetTalent，代码如下。

```java
package pro.shushi.pamirs.demo.api.model;

import pro.shushi.pamirs.meta.annotation.Field;
import pro.shushi.pamirs.meta.annotation.Model;

@Model.model(PetTalent.MODEL_MODEL)
@Model(displayName = "宠物达人",summary="宠物达人",labelFields ={"name"})
public class PetTalent extends AbstractDemoIdModel{
    public static final String MODEL_MODEL="demo.PetTalent";

    @Field(displayName = "达人")
    private String name;

}
```

（2）修改宠物商品模型，新增 many2many 字段 PetTalents，类型为 List<PetTalent>，并加上注解 @Field.many2many (relationFields = {"petItemId"},referenceFields = {"petTalentId"},through = "PetItemRelPetTalent")，through 为指定关联中间表，代码如下。

```java
@Model.model(PetItem.MODEL_MODEL)
@Model(displayName = "宠物商品",summary="宠物商品",labelFields = {"itemName"})
public class PetItem  extends AbstractDemoCodeModel{

    public static final String MODEL_MODEL="demo.PetItem";

    @Field(displayName = "商品名称",required = true)
    private String itemName;

    @Field(displayName = "商品价格",required = true)
    private BigDecimal price;

```

```java
23        @Field(displayName = "店铺",required = true)
24        @Field.Relation(relationFields = {"shopId"},referenceFields = {"id"})
25        private PetShop shop;
26
27        @Field(displayName = "店铺Id",invisible = true)
28        private Long shopId;
29
30        @Field(displayName = "品种")
31        @Field.many2one
32        @Field.Relation(relationFields = {"typeId"},referenceFields = {"id"})
33        private PetType type;
34
35        @Field(displayName = "品种类型",invisible = true)
36        private Long typeId;
37
38        @Field(displayName = "详情", serialize = Field.serialize.JSON, store = NullableBoolEnum.TRUE)
39        @Field.Advanced(columnDefinition = "varchar(1024)")
40        private List<PetItemDetail> petItemDetails;
41
42        @Field(displayName = "商品标签",serialize = Field.serialize.COMMA,store = NullableBoolEnum.TRUE,multi = true)
43        @Field.Advanced(columnDefinition = "varchar(1024)")
44        private List<String> tags;
45
46        @Field.many2many(relationFields = {"petItemId"},referenceFields = {"petTalentId"},through = "PetItemRelPetTalent")
47        @Field(displayName = "推荐达人",summary = "推荐该商品的达人们")
48        private List<PetTalent> petTalents;
49
50    }
```

(3) 修改宠物商店模型。

① 新增 M2M 关联模型 PetShopRelPetTalent 并继承 BaseRelation，BaseRelation 为关系模型抽象基类，用于承载多对多关系，是多对多关系的中间模型，数据模型主键可以不是 ID。更多类似抽象基类请见 3.3.2 节。代码如下。

```java
7     @Model.model(PetShopRelPetTalent.MODEL_MODEL)
8     @Model(displayName = "宠物店铺与达人关联表",summary="宠物店铺与达人关联表")
9     public class PetShopRelPetTalent extends BaseRelation {
10
11        public static final String MODEL_MODEL="demo.PetShopRelPetTalent";
12        @Field(displayName = "店铺Id")
13        private Long petShopId;
14
15        @Field(displayName = "达人Id")
16        private Long petTalentId;
17
18    }
```

② 修改宠物商店模型，新增 many2many 字段 petTalents，类型为 List<PetTalent>，并加上注解 @Field.many2many (relationFields = {"petShopId"},referenceFields = {"petTalentId"},throughClass =PetShopRelPetTalent.class)，throughClass 为指定关联模型类，代码如下。

```java
15    @Model.model(PetShop.MODEL_MODEL)
16    @Model(displayName = "宠物店铺",summary="宠物店铺",labelFields ={"shopName"} )
17    @Model.Code(sequence = "DATE_ORDERLY_SEQ",prefix = "P",size=6,step=1,initial = 10000,format = "yyyyMMdd")
18    public class PetShop extends AbstractDemoIdModel {
19        public static final String MODEL_MODEL="demo.PetShop";
20
21        @Field(displayName = "店铺编码")
```

```java
22      private String code;
23
24      @Field(displayName = "店铺名称",required = true,immutable=true)
25      private String shopName;
26
27      @Field(displayName = "开店时间",required = true)
28      @Field.Advanced(readonly = true)
29      private Time openTime;
30
31      @Field(displayName = "闭店时间",required = true)
32      private Time closeTime;
33
34      @Field.Enum
35      @Field(displayName = "数据状态",defaultValue = "DISABLED",required = true,summary = "枚举可选项举例")
36      private DataStatusEnum dataStatus;
37
38      @Field(displayName = "店铺标志")
39      private List<PetShopOptionEnum> options;
40
41      @Field(displayName = "一年内新店")
42      private Boolean oneYear;
43
44      @Field.Float
45      @Field(displayName = "店内员工平均年龄")
46      private BigDecimal averageAge;
47
48      @Field.Text
49      @Field(displayName = "描述")
50      private String description;
51
52      @Field.Html
53      @Field(displayName = "html描述")
54      private String descHtml;
55
56      @Field.Date(type = DateTypeEnum.DATE,format = DateFormatEnum.DATE)
57      @Field(displayName = "店庆")
58      private Date anniversary;
59
60  //      前端未提供默认支持，演示不了。这个留在我们自定义前端页面时演示
61  //      @Field.Date(type = DateTypeEnum.YEAR,format = DateFormatEnum.YEAR)
62  //      @Field(displayName = "开店年份")
63  //      private Date shopYear;
64
65      @Field.Money
66      @Field(displayName = "收入",priority = 1)
67      private BigDecimal income;
68
69      @Field.many2many(relationFields = {"petShopId"},referenceFields = {"petTalentId"},throughClass =PetShopRelPetTalent.class)
70      @Field(displayName = "推荐达人",summary = "推荐该商品的达人们")
71      private List<PetTalent> petTalents;
72
73  }
```

Step2. 重启看效果。

（1）进入宠物商品—编辑页面，可以看到多了推荐达人字段，单击"添加"选择"达人记录"。至于达人数据管理这里就不赘述了，只要为达人模型配一个菜单入口进去新增就好了，如图3-121所示。

图 3-121　进入宠物商品—编辑页面

（2）进入宠物商品—列表页面，就可以看到达人字段所选的记录，如图 3-122 所示。

图 3-122　进入宠物商品—列表页面

（3）宠物商店的效果参考宠狗商品的编辑路径。
（4）查看数据库表，对应的表和字段都已经生成了，如图 3-123 所示。

图 3-123　数据库表对应的表和字段都已生成

**4. O2M 关系类型（举例）**

O2M 跟 M2O 和 M2M 不一样，O2M 的交互方式新建关联模型记录一并提交，而 M2O 和 M2M 是通过添加方式选择已有关联模型记录。

Step1. 把 PetShopProxyA 的 items 字段提取到父类 PetShop 中去。

为什么要把 PetShopProxyA 的 items 提取到 PetShop 中去呢？因为代理模型新增的字段都是非存储字段，如果需要保存，需要自己覆盖模型的 create 和 update 方法手工保存关联对象。这个在 3.3.2 节介绍代理模型时介绍过。

（1）去除 PetShopProxyA 的 items 字段，代码如下。

```
1  package pro.shushi.pamirs.demo.api.proxy;
2
3  import pro.shushi.pamirs.demo.api.model.PetShop;
4  import pro.shushi.pamirs.meta.annotation.Model;
5  import pro.shushi.pamirs.meta.enmu.ModelTypeEnum;
6
7  @Model.model(PetShopProxyA.MODEL_MODEL)
8  @Model.Advanced(type = ModelTypeEnum.PROXY)
9  @Model(displayName = "宠物店铺代理模型A",summary="宠物店铺代理模型A")
10 public class PetShopProxyA extends PetShop {
11     public static final String MODEL_MODEL="demo.PetShopProxyA";
12
13 }
```

（2）在 PetShop 中增加 items 字段，代码如下。

```
15 @Model.model(PetShop.MODEL_MODEL)
16     @Model(displayName = "宠物店铺",summary="宠物店铺",labelFields ={"shopName"} )
17     @Model.Code(sequence = "DATE_ORDERLY_SEQ",prefix = "P",size=6,step=1,initial = 10000,format = "yyyyMMdd")
18 public class PetShop extends AbstractDemoIdModel {
19         public static final String MODEL_MODEL="demo.PetShop";
20
21         @Field(displayName = "店铺编码")
22         private String code;
23
24         @Field(displayName = "店铺编码2")
25         @Field.Sequence(sequence = "DATE_ORDERLY_SEQ",prefix = "C",size=6,step=1,initial = 10000,format = "yyyyMMdd")
26         private String codeTwo;
27
28         @Field(displayName = "店铺名称",required = true,immutable=true)
29         private String shopName;
30
31         @Field(displayName = "开店时间",required = true)
32         private Time openTime;
33
34         @Field(displayName = "闭店时间",required = true)
35         private Time closeTime;
36
37         @Field.Enum
38         @Field(displayName = "数据状态",defaultValue = "DRAFT",required = true,summary = "枚举可选项举例")
39         private DataStatusEnum dataStatus;
40
41         @Field(displayName = "店铺标志")
42         private List<PetShopOptionEnum> options;
43
44         @Field(displayName = "一年内新店")
45         private Boolean oneYear;
46
47         @Field.Float
48         @Field(displayName = "店内员工平均年龄")
49         private BigDecimal averageAge;
50
51         @Field.Text
```

```
52          @Field(displayName = "描述")
53          private String description;
54
55          @Field.Html
56          @Field(displayName = "html描述")
57          private String descHtml;
58
59          @Field.Date(type = DateTypeEnum.DATE,format = DateFormatEnum.DATE)
60          @Field(displayName = "店庆")
61          private Date anniversary;
62
63          @Field.Date(type = DateTypeEnum.YEAR,format = DateFormatEnum.YEAR)
64          @Field(displayName = "开店年份")
65          private Date publishYear;
66
67          @Field.Money
68          @Field(displayName = "收入",priority = 1)
69          private BigDecimal income;
70
71
72          @Field.many2many(relationFields = {"petShopId"},referenceFields = {"petTa
    lentId"},throughClass =PetShopRelPetTalent.class)
73          @Field(displayName = "推荐达人",summary = "推荐该商品的达人们")
74          private List<PetTalent> petTalents;
75
76
77          @Field.one2many
78          @Field(displayName = "商品列表")
79          @Field.Relation(relationFields = {"id"},referenceFields = {"shopId"})
80          private List<PetItem> items;
81
82      }
```

Step2. 重启应用查看效果。

前文中 PetShopProxyB 代理模型扩展了 catItems 萌猫商品列表字段，因为该字段是代理模型的非存储字段。但上一步操作中原本 PetShopProxyB 从 PetShopProxyA 继承的非存储字段 items，转而变成了从 PetShop 继承的存储字段。那么我们试下 catItems 和 items 都添加记录的效果吧。

（1）单击"菜单商店管理 B"进入 PetShopProxyB 代理模型的管理页，选择商店记录，单击"编辑"进入商店管理 B—编辑页，商品列表字段处新增两个商品分别是萌猫商品 1005，宠狗商品 1006，如图 3-124 和图 3-125 所示。

图 3-124　单击"菜单商店管理 B"进入 PetShopProxyB 代理模型的管理页

图 3-125 新增商品信息填写

（2）萌猫商品列表处字段新增两个商品分别是萌猫商品 001 和代理模型萌猫商品，如图 3-126 和图 3-127 所示。

图 3-126 萌猫商品列表

图 3-127 新增代理模型萌猫商品

（3）保存查看商品管理页面，发现只是创建了萌猫商品 1005 和宠狗商品 1006，而没有代理模型萌猫商品，如图 3-128 所示。

图 3-128　保存查看商品管理页面

**5. O2O 关系类型**

O2O 在前端体验上，跟 M2O 一样都是下拉单选，但有一个区别是目标对象只能被一个操作对象绑定。

**6. O2M，O2O 字段的提交策略（该版本还未支持）**

注解的 onUpdate 和 onDelete 属性指定在删除模型或更新模型关系字段值时，对关联模型进行的相应操作。操作包括 RESTRICT、NO ACTION、CASCADE 和 SET NULL，默认值为 SET NULL。

● RESTRICT：是指模型与关联模型有关联记录的情况下，引擎会阻止模型关系字段的更新或删除模型记录。

● NO ACTION：是指不做约束（这里与数据库约束的定义不相同）。

● CASCADE：表示在更新模型关系字段或者删除模型时，级联更新关联模型对应记录的关联字段值或者级联删除关联模型对应记录。

● SET NULL：表示在更新模型关系字段或者删除模型的时候，关联模型的对应关联字段将被 SET NULL（该字段值允许为 null 的情况下，若不允许为 null，则引擎阻止对模型的操作）。

## 3.4　Oinone 以函数为内在

函数（Function）是 Oinone 可管理的执行逻辑单元，跟模型绑定则对应模型的方法：

（1）描述满足数学领域函数定义，含有三个要素：定义域 A、值域 C{f(x)，x 属于 A} 和对应法则 f。其中核心是对应法则 f，它是函数关系的本质特征。

（2）满足面向对象原则，可设置不同开放级别，本地与远程智能切换。

### 3.4.1　构建第一个 Function

在前文中的 3.3.3 节、3.3.2 节涉及包括在 Action 中自定义（Action 背后都对应一个 Function）和 queryPage 进行 Function 覆盖，还是独立抽取的公共逻辑，Function 作为 Oinone 的可管理的执行逻辑单元，是无处不在的。这也是为什么说 Oinone 以函数为内在的原因。

**1. 构建第一个 Function**

数据管理器和数据构造器是 Oinone 为模型自动赋予的 Function，是内在数据管理能力。模型其他 Function 都需要用以下四种方式主动定义。

1）伴随模型新增函数（举例）

它跟模型的 Java 类定义在一起，复用模型的命名空间。

Step1. 为 PetShop 增加一个名为 sayHello 的 Function，代码如下。

```java
package pro.shushi.pamirs.demo.api.model;
…… //import

@Model.model(PetShop.MODEL_MODEL)
@Model(displayName = "宠物店铺",summary="宠物店铺",labelFields ={"shopName"} )
@Model.Code(sequence = "DATE_ORDERLY_SEQ",prefix = "P",size=6,step=1,initial = 10
000,format = "yyyyMMdd")
public class PetShop extends AbstractDemoIdModel {
    public static final String MODEL_MODEL="demo.PetShop";
    …… //省略其他代码
    @Function(openLevel = FunctionOpenEnum.API)
    @Function.Advanced(type=FunctionTypeEnum.QUERY)
    public PetShop sayHello(PetShop shop){
        PamirsSession.getMessageHub().info("Hello:"+shop.getShopName());
        return shop;
    }
}
```

Step2. 重启看效果。

用 graphQL 工具 Insomnia 查看效果。

（1）用 Insomnia 模拟登录：

① 创建一个 login 请求，用于保存 login 请求，为后续模拟登录保留快捷方式，如图 3-129 所示。

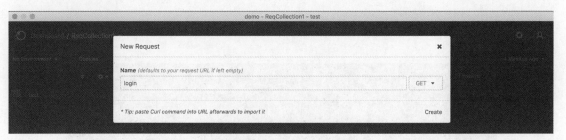

图 3-129　创建一个 login 请求

② 下面为登录请求的 GraphQL，请在 post 输入框中输入。如果请求输入框提示错误，可以单击 schema 的"Refresh Schema"来刷新文档，代码如下。

```
mutation {
  pamirsUserTransientMutation {
    login(user: {login: "admin", password: "admin"}) {
      broken
      errorMsg
      errorCode
      errorField
    }
  }
}
```

③ 单击 Send 按钮，我们可以看到登录成功的反馈信息，如图 3-130 所示。

图 3-130  登录成功的反馈信息

（2）用 Insomnia 模拟访问 PetShop 的 sayHello 方法（代码如下），gql 的返回中，我们可以看到两个核心返回（如图 3-131 所示）：

一是方法正常返回的 shopName；

二是 "PamirsSession.getMessageHub().info ("Hello:"+shop.getShopName())" 代码执行的结果，在 messages 中有一个消息返回，更多消息机制请见"框架之信息传递"。

```
1  query{
2    petShopQuery{
3      sayHello(shop:{shopName:"cpc"}){
4        shopName
5      }
6    }
7  }
```

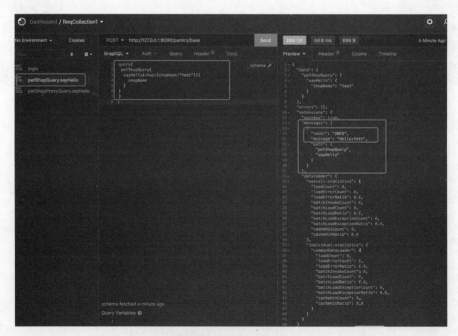

图 3-131  代码执行结果

(3)用 Insomnia 模拟访问 PetShopProxy 的 sayHello 方法。

效果同用 Insomnia 模拟访问 PetShop 的 sayHello 方法,体现 Function 的继承特性。

2)独立新增函数绑定到模型(举例)

用独立方法定义类,并采用 Model.Model 或 Fun 注解,但是 value 都必须是模型的编码,如 @Model.Model (PetShop.MODEL_MODEL) 或 @Fun (PetShop.MODEL_MODEL)。

Step1. 提取 PetShop 的 sayHello 方法独立到 PetShopService 中。

(1)注释掉 PetShop 的 sayHello 方法,代码如下。

```
1  package pro.shushi.pamirs.demo.api.model;
2  …… //import
3
4  @Model.model(PetShop.MODEL_MODEL)
5  @Model(displayName = "宠物店铺",summary="宠物店铺",labelFields ={"shopName"} )
6  @Model.Code(sequence = "DATE_ORDERLY_SEQ",prefix = "P",size=6,step=1,initial = 10
   000,format = "yyyyMMdd")
7  public class PetShop extends AbstractDemoIdModel {
8      public static final String MODEL_MODEL="demo.PetShop";
9      …… //省略其他代码
10 //    @Function(openLevel = FunctionOpenEnum.API)
11 //    @Function.Advanced(type=FunctionTypeEnum.QUERY)
12 //    public PetShop sayHello(PetShop shop){
13 //        PamirsSession.getMessageHub().info("Hello:"+shop.getShopName());
14 //        return shop;
15 //    }
16 }
```

(2)新增 PetShopService 接口类。

接口的方法上要加上 @Function 注解,这样模块依赖 api 包的时候,会自动注册远程服务的消费者,代码如下。

```
7  @Fun(PetShop.MODEL_MODEL)
8  //@Model.model(PetShop.MODEL_MODEL)
9  public interface PetShopHelloService {
10     @Function
11     PetShop sayHello(PetShop shop);
12 }
```

(3)新增 PetShopServiceImpl 实现类,代码如下。

```
12 @Fun(PetShop.MODEL_MODEL)
13 //@Model.model(PetShop.MODEL_MODEL)
14 @Component
15 public class PetShopHelloServiceImpl implements PetShopHelloService {
16
17     @Override
18     @Function(openLevel = FunctionOpenEnum.API)
19     @Function.Advanced(type= FunctionTypeEnum.QUERY)
20     public PetShop sayHello(PetShop shop) {
21         PamirsSession.getMessageHub().info("Hello:"+shop.getShopName());
22         return shop;
23     }
24 }
```

Step2. 重启看效果。

具备继承特性。

3)独立新增函数只做公共逻辑单元(举例)

只能 Java 后端访问,不生成 GraphQL 的 schema,即使配置 @Function (openLevel = FunctionOpenEnum.API),也相当于 FunctionOpenEnum.REMOTE。如同 3.3.4 节中的 PetCatItemQueryService 的作用,提取公共的逻辑,并且可管理。

Step1. 修改 PetShopService 和 PetShopServiceImpl 的命名空间，代码如下。

修改 PetShopService 的命名空间，代码如下。

```
 7    @Fun(PetShopHelloService.FUN_NAMESPACE)
 8    public interface PetShopHelloService {
 9
10        String FUN_NAMESPACE = "demo.PetShopHelloService";
11
12        @Function
13        PetShop sayHello(PetShop shop);
14    }
```

修改 PetShopServiceImpl 的命名空间，代码如下。

```
12    @Fun(PetShopHelloService.FUN_NAMESPACE)
13    @Component
14    public class PetShopHelloServiceImpl implements PetShopHelloService {
15
16        @Override
17        @Function(openLevel = FunctionOpenEnum.API)
18        @Function.Advanced(type= FunctionTypeEnum.QUERY)
19        public PetShop sayHello(PetShop shop) {
20            PamirsSession.getMessageHub().info("Hello:"+shop.getShopName());
21            return shop;
22        }
23    }
```

Step2. 重启看效果。

刷新 GraphQL schema，原先的 post 请求输入框会报错，单击"提交"结果也会报 GraphQL 未定义，如图 3-132 所示。

图 3-132　刷新 GraphQL schema 看效果

4）伴随 ServerAction 新增函数

ServerAction 我们前面也多次提到过，比如在介绍模型类型中的代理模型和传输模型时都定义过 ServerAction，其背后都默认定义了一个 Function。如 PetShopBatchUpdate 模型在 PetShopBatchUpdateAction 类中定义了一个 conform 的 ServerAction，背后就定义了一个 namespace 为 demo.PetShopBatchUpdate，fun 为 conform 的 Function，而且开放级别为 API。

**2. Java 同名不同参数方法（不建议）**

Java 的同名不同参数，在很多远程调用框架如 dubbo 也是不支持的，在 Oinone 中也需要特殊处理，要以不同的 name 和 fun 来区别。

Step1. 为 PetShop 定义两个同名方法，并加上 Function 注解。

我们把 PetShop 模型下的 sayHello 函数恢复一下，并增加一个同名方法 sayHello，但注解上 @Function (name = "sayHello2") 和 @Function.fun ("sayHello2")。修改完以后 sayHello 和 sayHello2 都能在 Insomnia 通过 GQL 来访问，代码如下。

```java
1  package pro.shushi.pamirs.demo.api.model;
2  ...... //import
3
4  @Model.model(PetShop.MODEL_MODEL)
5  @Model(displayName = "宠物店铺",summary="宠物店铺",labelFields ={"shopName"} )
6  @Model.Code(sequence = "DATE_ORDERLY_SEQ",prefix = "P",size=6,step=1,initial = 10
   000,format = "yyyyMMdd")
7  public class PetShop extends AbstractDemoIdModel {
8      public static final String MODEL_MODEL="demo.PetShop";
9      ...... //省略其他代码
10     @Function(openLevel = FunctionOpenEnum.API)
11     @Function.Advanced(type= FunctionTypeEnum.QUERY)
12     public PetShop sayHello(PetShop shop){
13         PamirsSession.getMessageHub().info("Hello:"+shop.getShopName());
14         return shop;
15     }
16     @Function(name = "sayHello2",openLevel = FunctionOpenEnum.API)
17     @Function.Advanced(type= FunctionTypeEnum.QUERY)
18     @Function.fun("sayHello2")
19     public PetShop sayHello(PetShop shop, String s) {
20         PamirsSession.getMessageHub().info("Hello:"+shop.getShopName()+",s:"+s);
21         return shop;
22     }
23  }
```

Step2. 重启看效果，如图3-133所示。

图3-133　重启看效果

**3. 配置**

非模型但带有函数的类必须使用 @Fun 注解来标识当前类为非模型带函数的类。如果需要提供远程服务，需要在 API 包中声明注解了 @Fun 注解的函数接口，并在接口的方法上加上 @Function 注解。

**1）函数配置**

函数定义可以无返回值，也允许定义无参函数。如果入参为原始类型请使用对应封装类型声明。命名空间和函数编码相同的覆盖函数有且仅有一个生效，在模型类中定义优先级高于在模型类外定义，在模型类中越靠后优先级越高。

可以使用 @Model.Model、@Fun 注解函数的命名空间。先取 @Model.Model 注解值，在本类查找注解，如果本类未配置或注解值为空则在父类或接口上查找；若为空则取 @Fun 注解值，先在本类查找注解，如果本类未配置或注解值为空则在父类或接口上查找；若皆为空则取全限定类名。

可以使用 @Function.fun 注解配置函数编码。取函数编码先在本类方法查找注解，如果本类方法未配置或注解值为空则在父类或接口方法上查找，若皆为空则取方法名。

如果接口或者父类配置了命名空间和函数编码并且有多个实现类或继承类，实现方法使用默认的方法名作为函数编码，则会导致多个实现方法函数编码冲突，需要使用 @Function.fun 为每个实现类的对应方法配置唯一的函数编码。但大多数场景一个接口只有一个实现类。

推荐为函数声明接口，并在接口上进行注解（@Fun 或 @Model.Model，@Function），函数实现接口即可；如果需要为函数开启远程服务，必须为函数声明接口并注解，接口须放在 API 工程中。系统会根据函数开放级别是否 REMOTE 来自动注册服务提供者和消费者。

命名空间和函数编码的注解方式适用于所有函数。

2）函数命名规范

函数命名规范见表 3-20。

表 3-20 函数命名规范

| 模型属性 | 默认取值规范 | 命名规则规范 |
|---|---|---|
| namespace | 先取 @Model.Model 注解值，先在本类查找注解，如果本类未配置或注解值为空则在父类或接口上查找；若为空则取 @Fun 注解值，先在本类查找注解，如果本类未配置或注解值为空则在父类或接口上查找；若皆为空则取全限定类名 | 长度必须小于或等于 128 个字符 |
| name | 默认使用 Java 方法名 | 仅支持数字、字母<br>必须以字母开头<br>长度必须小于或等于 128 个字符<br>不能以 get、set 为开头作为函数名称 |
| fun | 默认使用 name 属性 | 长度必须小于或等于 128 个字符 |
| summary | 默认使用 displayName 属性 | 不能使用分号<br>长度必须小于或等于 500 个字符 |
| description | NULL，注解无法定义 | 长度必须小于或等于 65535 个字符 |
| openLevel | 函数的开放等级默认值：<br>{FunctionOpenEnum.LOCAL, FunctionOpenEnum.REMOTE} | FunctionOpenEnum 枚举值<br>LOCAL (2L，"本地调用""本地调用"),<br>REMOTE (4L，"远程调用""远程调用"),<br>API (8L，"开放接口""开放接口") |
| Advanced.displayName | 默认使用 name 属性 | 长度必须小于或等于 128 个字符 |
| Advanced.timeout | 超时时间默认为 5 秒，远程调用时配置生效 | |

### 3.4.2 函数的开放级别与类型

**1. 函数开放级别**

我们在日常开发中通常会因为安全性，为方法定义不同的开放层级，或者通过应用分层把需要对 Web 开放的接口统一定义在一个独立的应用中。Oinone 也提供类似的策略，所有逻辑都通过 Function 来归口统一管理，所以在 Function 时可以定义其开放级别有 API、REMOTE、LOCAL 三种类型，配置可多选。

1）四种自定义新增方式与开放级别的对应关系

四种自定义新增方式与开放级别的对应关系见表 3-21。

表 3-21　四种自定义新增方式与开放级别的对应关系

| 函数 | 本地调用（LOCAL） | 远程调用（REMOTE） | 开放（API） |
| --- | --- | --- | --- |
| 伴随模型新增函数 | 支持 | 支持"默认" | 支持 |
| 独立新增函数绑定到模型 | 支持 | 支持"默认" | 支持 |
| 独立新增函数只做公共逻辑单元 | 支持 | 支持"默认" | — |
| 伴随 ServerAction 新增函数 | — | — | 必选 |

2）远程调用（REMOTE）

如果函数的开放级别为本地调用，则不会发布远程服务和注册远程服务消费者。

（1）非数据管理器函数。

提供者：如果函数定义在当前部署包的启动应用中，则主动发布远程服务提供者。

消费者：如果函数定义在部署依赖包中但未在当前部署包的启动应用中，则系统会默认注册远程消费者。发布注册的远程服务使用命名空间和函数编码进行路由。

所以非数据管理器函数的消费者并不需要感知该服务是本地提供还是远程提供。而服务提供者也不需要手动注册远程服务。

（2）数据管理器类函数。

提供者：如果数据管理器函数所在模型定义在当前部署包的启动应用中，则系统会主动发布数据管理器的远程服务作为数据管理器的远程服务提供者。

消费者：如果模型定义在部署依赖包中但未在当前部署包的启动应用中，则系统会主动注册数据管理器的远程服务消费者。

所以数据管理器类函数的消费者与服务提供者并不需要感知函数的远程调用。

**2. 函数类型**

函数的类型语义分为增、删、改、查，在编程模式下目前用于 Function 为 API 级别，生成 GraphQL 的 Schema 时放在 query 或 mutation。"查"放在 query，其余放在 mutation。在无代码编辑器里还用页面分类管理。

### 3.4.3　函数的相关特性

以下从 Oinone 函数拥有的三方面特性展开介绍：

一是面向对象，继承与多态；

二是面向切面编程，拦截器；

三是 SPI 机制，扩展点。

**1. 面向对象—继承与多态**

1）继承

我们在 3.4.1 节中的伴随模型新增函数和独立类新增函数绑定到模型都是为父模型 PetShop 新增了 sayHello 的 Function。同样其子模型都具备 sayHello 的 Function。所以我们是通过 Function 的 namespace 来做依据的，子模型会继承以母模型的编码为 namespace 的 Function。

2）多态（举例）

Oinone 的多态只提供覆盖功能，不提供重载，因为 Oinone 相同 name 和 fun 的情况下不会去识别参数个数和类型。

Step1. 为 PetShop 新增 hello 函数，代码如下。

```java
package pro.shushi.pamirs.demo.api.model;
…… //import

@Model.model(PetShop.MODEL_MODEL)
@Model(displayName = "宠物店铺",summary="宠物店铺",labelFields ={"shopName"} )
@Model.Code(sequence = "DATE_ORDERLY_SEQ",prefix = "P",size=6,step=1,initial = 10000,format = "yyyyMMdd")
public class PetShop extends AbstractDemoIdModel {
    public static final String MODEL_MODEL="demo.PetShop";
    …… //省略其他代码
    @Function(openLevel = FunctionOpenEnum.API)
    @Function.Advanced(type= FunctionTypeEnum.QUERY)
    public PetShop sayHello(PetShop shop){
        PamirsSession.getMessageHub().info("Hello:"+shop.getShopName());
        return shop;
    }
    @Function(name = "sayHello2",openLevel = FunctionOpenEnum.API)
    @Function.Advanced(type= FunctionTypeEnum.QUERY)
    @Function.fun("sayHello2")
    public PetShop sayHello(PetShop shop, String s) {
        PamirsSession.getMessageHub().info("Hello:"+shop.getShopName()+",s:"+s);
        return shop;
    }

    @Function(openLevel = FunctionOpenEnum.API)
    @Function.Advanced(type= FunctionTypeEnum.QUERY)
    public PetShop hello(PetShop shop){
        PamirsSession.getMessageHub().info("Hello:"+shop.getShopName());
        return shop;
    }
}
```

Step2. 为 PetShopProxyB 新增对应的三个函数。

其中 PetShopProxyB 新增的 hello 函数，在 Java 中重载了 hello，在代码中 new PetShopProxyB() 是可以调用父类的 sayHello 单参方法，也可以调用本类的双参方法。但在 Oinone 的体系中对于 PetShopProxyB 只有一个可识别的 Function，就是双参的 sayHello，代码如下。

```java
@Model.model(PetShopProxyB.MODEL_MODEL)
@Model.Advanced(type = ModelTypeEnum.PROXY,inherited ={PetShopProxy.MODEL_MODEL,PetShopProxyA.MODEL_MODEL} )
@Model(displayName = "宠物店铺代理模型B",summary="宠物店铺代理模型B")
public class PetShopProxyB extends PetShop {

    public static final String MODEL_MODEL="demo.PetShopProxyB";

    @Field.one2many
    @Field(displayName = "萌猫商品列表")
    @Field.Relation(relationFields = {"id"},referenceFields = {"shopId"})
    private List<PetCatItem> catItems;

    @Function(openLevel = FunctionOpenEnum.API)
    @Function.Advanced(type= FunctionTypeEnum.QUERY)
    public PetShop sayHello(PetShop shop){
        PamirsSession.getMessageHub().info("PetShopProxyB Hello:"+shop.getShopName());
        return shop;
    }

    @Function(name = "sayHello2",openLevel = FunctionOpenEnum.API)
    @Function.Advanced(type= FunctionTypeEnum.QUERY)
    @Function.fun("sayHello2")
    public PetShop sayHello(PetShop shop,String hello){
```

```
38              PamirsSession.getMessageHub().info("PetShopProxyB say:"+hello+","+shop.ge
    tShopName());
39              return shop;
40          }
41          @Function(openLevel = FunctionOpenEnum.API)
42          @Function.Advanced(type= FunctionTypeEnum.QUERY)
43          public PetShop hello(PetShop shop,String hello){
44              PamirsSession.getMessageHub().info("PetShopProxyB hello:"+hello+","+shop.
    getShopName());
45              return shop;
46          }
47      }
```

Step3. 重启看效果。

（1）查看 petShopQuery 的 sayHello、sayHello2、hello 三个函数，结果正常，代码如下。

```
 1 ▸ query{
 2 ▸    petShopQuery{
 3 ▸      sayHello(shop:{shopName:"cpc"}){
 4          shopName
 5        }
 6 ▸      sayHello2(shop:{shopName:"cpc"},s:"sss"){
 7          shopName
 8        }
 9 ▸      hello(shop:{shopName:"cpc"}){
10          shopName
11        }
12      }
13 ▸    petShopProxyBQuery{
14 ▸      sayHello(shop:{shopName:"cpc"}){
15          shopName
16        }
17 ▸      sayHello2(shop:{shopName:"cpc"},hello:"sss"){
18          shopName
19        }
20 ▸      hello(shop:{shopName:"cpc"},hello:"hello"){
21          shopName
22        }
23      }
24    }
```

（2）查看 petShopProxyBQuery 的 sayHello、sayHello2、hello 三个函数结果都被重载了，而且 hello 函数传不传参数 hello，都是调用的本类的双参函数，如图 3-134 所示。

图 3-134　查看 petShopProxyBQuery 的三个函数结果都已被重载

3）模型的默认函数重写

对默认函数的重写，本书很多处都有涉及，大家可以自行测试，如图 3-22 所示。

表 3-22　模型的默认函数重写

| 名称 | 重写示例 | 说明 |
| --- | --- | --- |
| 创建 | @Action.Advanced (name = FunctionConstants.create, managed = true)<br>@Action (displayName = " 确定 ", summary = " 创建 ", bindingType = ViewTypeEnum.FORM)<br>@Function (name = FunctionConstants.create)<br>@Function.fun (FunctionConstants.create)<br>public PetType create (PetType data) | 这里以宠物品种为例罗列出来，在方法上以注解的方式增加"重写示例"列的内容 |
| 修改 | @Function.Advanced (type = FunctionTypeEnum.UPDATE)<br>@Action.Advanced (name = FunctionConstants.update, managed = true)<br>@Action (displayName = " 确定 ", summary = " 修改 ", bindingType = ViewTypeEnum.FORM)<br>@Function (name = FunctionConstants.update)<br>@Function.fun (FunctionConstants.update)<br>public PetType update (PetType data) | |
| 查询 | @Function.Advanced (type = FunctionTypeEnum.QUERY)<br>@Function.fun (FunctionConstants.queryPage)<br>@Function (openLevel = {FunctionOpenEnum.API})<br>public Pagination&lt;PetType&gt; queryPage (Pagination&lt;PetType&gt; page, IWrapper&lt;PetType&gt; queryWrapper) | |
| 删除 | @Function.Advanced (type = FunctionTypeEnum.DELETE)<br>@Function.fun (FunctionConstants.deleteWithFieldBatch)<br>@Function (name = FunctionConstants.delete)<br>@Action (displayName = " 删除 ", contextType = ActionContextTypeEnum.SINGLE_AND_BATCH)<br>public PetType delete (PetType data) | |
| 构造 | @Function (openLevel = FunctionOpenEnum.API)<br>@Function.Advanced (type= FunctionTypeEnum.QUERY) | 在 3.3.2 节中"新增 PetShopBatchUpdateAction 类"提到覆盖数据构造器（construct），接收从宠物商店列表多选带过来的数据参数，非 PetShopBatchUpdate 本模型参数不能放第一个，用 List&lt;PetShopProxy&gt; petShopList 来接收，进行数据组装逻辑处理，对应数据也是由 Pet Shop Batch Update 来承载返回给 Pet Shop Batch Update 的 Form 编辑页 |

**2. 面向切面——拦截器**

1）拦截器

拦截器为平台中满足条件的函数以非侵入方式，根据优先级扩展函数执行前和执行后的逻辑。

使用方法上的 @Hook 注解可以标识方法为拦截器。前置扩展点需要实现 HookBefore 接口；后置扩展点需要实现 HookAfter 接口。传入的参数包含当前拦截函数定义与该函数的参数。拦截器可以根据函数定义与传入的参数增加处理逻辑。

拦截器分为前置拦截器和后置拦截器，前者的出入参为所拦截函数的入参，后者的出入参为所拦截函数的出参。可以使用 @Hook 注解或 Hook 模型的非必填字段 module、Model、fun、函数类型、active 筛选出对当前拦截方法有效的拦截器。若未配置任何过滤属性，拦截器将对所有函数生效。

根据拦截器的优先级 priority 属性可以对拦截器的执行顺序进行调整。priority 数字越小，越先执行。

2）前置拦截（举例）

增加一个前置拦截，对 PetShop 的 sayHello 函数进行前置拦截，修改函数的入参的 shopName 属性，在其前面增加 "hookbefore:" 字符串。并查看效果。

Step1. 新增 PetShopSayHelloHookBefore 实现 HookBefore 接口，代码如下。

为 run 方法增加 @Hook 注解。

（1）配置 module={DemoModule.MODULE_MODULE}，这里 module 代表的是执行模块，该 Hook 只匹配由 DemoModule 模块为发起入口的请求。

（2）配置 Model={PetShop.MODEL_MODEL}，该 Hook 只匹配 PetShop 模型。

（3）配置 fun={"sayHello"}，该 Hook 只匹配函数编码为 sayHello 的函数。

```
10     @Component
11   public class PetShopSayHelloHookBefore implements HookBefore {
12         @Override
13         @Hook(module = {DemoModule.MODULE_MODULE},model = {PetShop.MODEL_MODEL},fun =
    {"sayHello"})
14         public Object run(Function function, Object... args) {
15             if(args!=null && args[0]!=null){
16                 PetShop arg = (PetShop)args[0];
17                 arg.setShopName("hookbefore:"+ arg.getShopName());
18             }
19             return args;
20         }
21
22     }
```

Step2. 重启查看效果。

用 graphQL 工具 Insomnia 查看效果，如果访问提示未登录，则请先登录。参考 3.4.1 节。

（1）用 http://127.0.0.1:8090/pamirs/base 访问，结果会发现 PetShopSayHelloHookBefore 不起作用。是因为本次请求是以 base 模块作为发起模块，而我们用 module={DemoModule.MODULE_MODULE} 声明了该 Hook 只匹配由 DemoModule 模块为发起入口的请求，如图 3-135 所示。

图 3-135　用 http://127.0.0.1:8090/pamirs/base 访问

（2）用 http://127.0.0.1:8090/pamirs/demoCore 访问，前端是以模块名作为访问入口，不是模块编码，这里大家要注意，如图 3-136 所示。

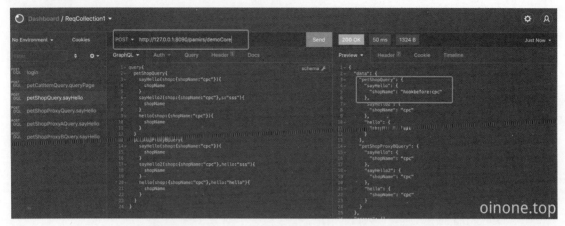

图 3-136　用 http://127.0.0.1:8090/pamirs/demoCore 访问

（3）用 http://127.0.0.1:8090/pamirs/demoCore 访问，更换到 petShop 的子模型 petShopProxy 来访问 sayHello 函数，结果我们发现是没有效果的。因为配置 Model={PetShop.MODEL_MODEL}，该 Hook 只匹配 PetShop 模型。

3）后置拦截（举例）

增加一个后置拦截，对 PetShop 的 sayHello 函数进行后置拦截，修改函数的返回结果的 shopName 属性，在其后面增加 "hookAfter:" 字符串。并查看效果。

Step1. 新增 PetShopSayHelloHookAfter 实现 HookAfter 接口，代码如下。

为 run 方法增加 @Hook 注解。

（1）配置 Model={PetShop.MODEL_MODEL}，该 Hook 只匹配 PetShop 模型。

（2）配置 fun={"sayHello"}，该 Hook 只匹配函数编码为 sayHello 的函数。

```
 9      @Component
10    public class PetShopSayHelloHookAfter implements HookAfter {
11
12        @Override
13        @Hook(model = {PetShop.MODEL_MODEL},fun = {"sayHello"})
14        public Object run(Function function, Object ret) {
15            if (ret == null) {
16                return null;
17            }
18            PetShop result =null;
19            if (ret instanceof Object[]) {
20                Object[] rets = (Object[])((Object[])ret);
21                if (rets.length == 1) {
22                    result = (PetShop)rets[0];
23                }
24            } else {
25                result = (PetShop)ret;
26            }
27            result.setShopName(result.getShopName()+":hookAfter");
28            return result;
29        }
30    }
```

Step2. 重启查看效果。

（1）用 http://127.0.0.1:8090/pamirs/base 访问，结果我们会发现 PetShopSayHelloHookAfter 是起作用的。PetShopSayHelloHookBefore 没有配置模块过滤，如图 3-137 所示。

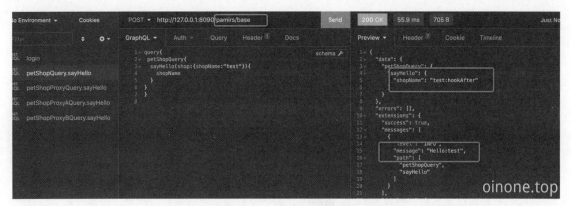

图 3-137　用 http://127.0.0.1:8090/pamirs/base 访问

（2）用 http://127.0.0.1:8090/pamirs/demoCore 访问，结果我们会发现 PetShopSayHelloHookAfte 和 PetShopSayHelloHookBefore 同时起作用，如图 3-138 所示。

图 3-138　http://127.0.0.1:8090/pamirs/demoCore 访问

（3）我们会发现 HookAfter 只对结果做了修改，所以 message 中可以看到 hookbefore，但看不到 hookAfter。

4）注意点

（1）不管前置拦截器，还是后置拦截器都可以配置多个拦截器，根据拦截器的优先级 priority 属性可以对拦截器的执行顺序进行调整。priority 数字越小，越先执行。小伙伴们可以自行尝试。

（2）拦截器必须是 jar 依赖，不然执行会报错。特别是如果配置了一个没有过滤条件的拦截器，就要非常小心。

（3）模块启动 yml 文件可以过滤不需要执行的 hook，具体配置见 4.1.1 节。

（4）调用入口不是由前端发起，而是在后端编程中直接调用，默认不会生效，如果要生效请参考 4.1.9 节。

**3. SPI 机制—扩展点**

扩展点结合拦截器的设计，Oinone 可以点、线、面一体化管理 Function。

扩展点用于扩展函数逻辑。扩展点类似于 SPI（Service Provider Interface）机制，是一种服务发现机制，这一机制为函数逻辑的扩展提供了可能。

1）构建第一个扩展点

（1）自定义扩展点（举例）。

在我们日常开发中，随着对业务理解的深入，往往还会在一些逻辑中预留扩展点，以便日后应对不同需求时可以灵活替换某一小块逻辑。

在 3.3.4 节中的 PetCatItemQueryService，是独立新增函数，只做公共逻辑单元。现在我们给它的实现类增加一个扩展点。在 PetCatItemQueryServiceImpl 的 queryPage 方法中原本会先查询 PetCatType 列表，我们这里假设这个逻辑随着业务发展未来会发生变化，我们可以预先预留"查询萌猫类型扩展点"。

Step1. 新增扩展点定义 PetCatItemQueryCatTypeExtpoint，代码如下。

（1）扩展点命名空间：在接口上用 @Ext 声明扩展点命名空间。会优先在本类查找 @Ext，若为空则往接口向上做遍历查找，返回第一个查找到的 @Ext.value 注解值，使用该值再获取函数的命名空间；如果未找到，则返回扩展点全限定类名。所以我们这里扩展点命名空间为 pro.shushi.pamirs.demo.api.extpoint.PetCatItemQueryCatTypeExtpoint。

（2）扩展点技术名称：先取 @ExtPoint.name，若为空则取扩展点接口方法名。所以我们这里技术名为 queryCatType。

```java
10  @Ext
11  public interface PetCatItemQueryCatTypeExtpoint {
12
13      @ExtPoint(displayName = "查询萌猫类型扩展点")
14      List<PetCatType> queryCatType();
15
16  }
```

Step2. 修改 PetCatItemQueryServiceImpl（用 Ext、run 模式调用），代码如下。

修改 queryPage，增加扩展点的使用代码。扩展点的使用有以下两种方式：

方式一，使用命名空间和扩展点名称调用 Ext.run (namespace, fun，参数 )；

方式二，使用函数式接口调用 Ext.run ( 函数式接口，参数 )。

我们这里用了第二种方式：

（1）用 PetCatItemQueryCatTypeExtpoint 的全限定类名作为扩展点的命名空间（namespace）；

（2）用 queryCatType 的方法名作为扩展点的技术名称（name）；

（3）根据 "namespace+name" 去找到匹配扩展点实现，并根据规则是否匹配，以及优先级唯一确定一个扩展点实现去执行逻辑。

```java
1   package pro.shushi.pamirs.demo.core.service;
2
3   ……省略依赖包
4
5   @Fun(PetCatItemQueryService.FUN_NAMESPACE)
6   @Component
7   public class PetCatItemQueryServiceImpl implements PetCatItemQueryService {
8
9       @Override
10      @Function
11      public Pagination<PetCatItem> queryPage(Pagination<PetCatItem> page, IWrapper
    <PetCatItem> queryWrapper) {
12
13      //   List<PetCatType> typeList = new PetCatType().queryList();
14           List<PetCatType> typeList = Ext.run(PetCatItemQueryCatTypeExtpoint::query
    CatType, new Object[]{});
15           if(!CollectionUtils.isEmpty(typeList)) {
16               List<Long> typeIds = typeList.stream().map(PetCatType::getId).collect
    (Collectors.toList());
17               queryWrapper = Pops.<PetCatItem>f(queryWrapper).from(PetCatItem.MODEL
    _MODEL).in(PetCatItem::getTypeId, typeIds).get();
18           }
```

```
19          return new PetCatItem().queryPage(page,queryWrapper);
20      }
21
22      ……省略其他函数
23  }
24
```

Step3. 新增扩展点实现 PetCatItemQueryCatTypeExtpointOne，代码如下。

（1）扩展点命名空间要与扩展点定义一致，用 @Ext (PetCatItemQueryCatTypeExtpoint.class)；

（2）@ExtPoint.Implement 声明是在 @Ext 声明的命名空间下，且技术名为 queryCatType 的扩展点实现。

```
11  @Ext(PetCatItemQueryCatTypeExtpoint.class)
12  public class PetCatItemQueryCatTypeExtpointOne implements PetCatItemQueryCatTypeExtpoint {
13
14      @Override
15      @ExtPoint.Implement(displayName = "查询萌猫类型扩展点的默认实现")
16      public List<PetCatType> queryCatType() {
17          PamirsSession.getMessageHub().info("走的是第一个扩展点");
18          List<PetCatType> typeList = new PetCatType().queryList();
19          return typeList;
20      }
21  }
```

Step4. 重启看效果。

（1）萌猫商品—列表页面的逻辑没有变化，说明 typeList 从扩展点中取到了，如图 3-139 所示。

图 3-139　萌猫商品—列表页面的逻辑无变化

（2）用 Insomnia 直接发起 GraphQL 请求，返回结果里可以明确知道这是扩展点实现 PetCatItemQueryCatTypeExtpointOne 执行的结果，如图 3-140 所示。

图 3-140　用 Insomnia 直接发起 GraphQL 请求

Step5. 自行测试扩展点的优先级，代码如下。

附上第二个扩展点实现的代码，快去试试吧。

```
11  @Ext(PetCatItemQueryCatTypeExtpoint.class)
12  public class PetCatItemQueryCatTypeExtpointTwo implements PetCatItemQueryCatTypeExtpoint {
13
14      @Override
15      @ExtPoint.Implement(priority = 95,displayName = "查询萌猫类型扩展点的实现,优先级取胜")
16      public List<PetCatType> queryCatType() {
17          PamirsSession.getMessageHub().info("走的是第二个扩展点");
18          List<PetCatType> typeList = new PetCatType().queryList();
19          return typeList;
20      }
21  }
```

（2）默认扩展点（举例）。

由前端直接发起调用 Oinone 后端 Function（能被前端直接发起的 Function 前提是 namespace 挂在模型上），当前端通过 GraphQL 发起对函数的请求时，Oinone 都会默认执行的三个内置扩展点分别是前置扩展点、覆盖扩展点和后置扩展点。

默认扩展点与函数的关联关系。扩展点扩展的函数与扩展点通过扩展点的命名空间和技术名称关联。扩展点与所扩展函数的命名空间一致。前置扩展点、重载扩展点和后置扩展点的技术名称的规则是所扩展函数的函数编码 fun 加上 "Before" "Override" 和 "After" 后缀；方法体内调用扩展点直接使用接口调用，所以技术名称可以任意定义，只需要在同一命名空间下唯一即可。

我们在 3.3.4 节中有提到过通过实现扩展点来保证子模型与父模型数据同步。此例我们替换 PetShop 的 sayHello 函数。

Step1. 新增扩展点定义 PetShopSayhelloOverrideExtpoint，代码如下。

```
7   @Ext(PetShop.class)
8   public interface PetShopSayhelloOverrideExtpoint {
9
10      @ExtPoint(displayName = "覆盖PetShop的sayHello执行逻辑")
11      public PetShop sayHelloOverride(PetShop shop);
12
13  }
```

Step2. 新增扩展点实现 PetShopSayhelloOverrideExtpointImpl，代码如下。

```
9   @Ext(PetShop.class)
10  public class PetShopSayhelloOverrideExtpointImpl implements PetShopSayhelloOverrideExtpoint {
11
12      @ExtPoint.Implement(displayName = "覆盖PetShop的sayHello执行逻辑")
13      public PetShop sayHelloOverride(PetShop shop){
14          PamirsSession.getMessageHub().info("OverrideExtpoint Hello:"+shop.getShopName());
15          return shop;
16      }
17  }
```

Step3. 确保 PetShop 的 sayHello 函数存在。

请查看 3.4.1 节。

Step4. 重启查看效果，如图 3-141 所示。

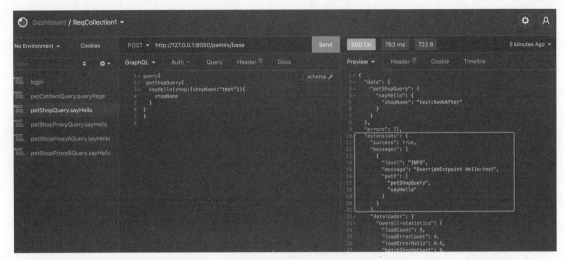

图 3-141　重启查看效果

### 2）小结

Oinone 用默认扩展点为 Function 提供三种默认扩展点，并通过自定义扩展点在 Function 逻辑内部任意插入扩展点，让 Function 作为 Oinone 的逻辑管理单元的可管理性大大提升。同时结合拦截器的设计，Oinone 可以点、线、面一体化管理 Function。

注：默认扩展点，不是由前端发起而是由后端编程调用，默认不会生效，如果要生效请参考 4.1.9 节。

## 3.5　Oinone 以交互为外在

交互组件（UI Component）：

（1）用组件化的方式统一管理菜单、布局和视图。

（2）用 Action 做衔接，来勾绘出模块的前端交互拓扑，描述所有可操作行为。

Oinone 的页面路由串联规则是以菜单为导航，用 Action 做衔接，用 View 做展示。查看模块的前端交互的简易逻辑图，方便大家理解如用菜单、视图、Action 来勾绘出模块的前端交互拓扑。本节重点介绍以下几点。

（1）如何定义菜单、视图、行为以及其初始化。

（2）xml 配置：视图配置（包括字段联动）、字段配置、布局配置、动作配置。

（3）前端组件自定义。

### 3.5.1　构建第一个 Menu

在前面章节中我们已对菜单有所熟悉，菜单就好比地图、导航，没有地图、导航就无法畅游模块并进行相关业务操作。在 3.3.4 节中介绍"多表继承"的部分就有菜单的初始化，这里将展开介绍初始化 Menu 的两种方式：注解式、数据初始化式。

**1. 注解式（举例）**

Step1. 分析现有菜单注解，代码如下。

用 @UxMenus 声明 DemoMenus 为菜单初始化入口，同时该类在 DemoModule 配置扫描路径中，那么通过 DemoMenus 初始化的菜单都挂在 demo_core 这个模块上。

如果采用这种模式，建议同一个模块的菜单都只配置在一处。

```
13    @UxMenus
14  ▾ public class DemoMenus implements ViewActionConstants {
15        @UxMenu("商店管理")@UxRoute(PetShopProxy.MODEL_MODEL) class PetShopProxyMenu{}
16        @UxMenu("商店管理A")@UxRoute(PetShopProxyA.MODEL_MODEL) class PetShopProxyAMenu{}
17        @UxMenu("商店管理B")@UxRoute(PetShopProxyB.MODEL_MODEL) class PetShopProxyBMenu{}
18        @UxMenu("商品管理")@UxRoute(PetItem.MODEL_MODEL) class ItemMenu{}
19        @UxMenu("宠狗商品")@UxRoute(PetDogItem.MODEL_MODEL) class DogItemMenu{}
20        @UxMenu("萌猫商品")@UxRoute(PetCatItem.MODEL_MODEL) class CatItemMenu{}
21        @UxMenu("宠物品种")@UxRoute(PetType.MODEL_MODEL) class PetTypeMenu{}
22        @UxMenu("萌猫品种")@UxRoute(PetCatType.MODEL_MODEL) class CatTypeMenu{}
23        @UxMenu("宠狗品种")@UxRoute(PetDogType.MODEL_MODEL) class DogTypeMenu{}
24        @UxMenu("宠物达人")@UxRoute(PetTalent.MODEL_MODEL) class PetTalentMenu{}
25    }
```

Step2. 改造现有菜单注解。

（1）菜单的层级关系通过 @UxMenu 的嵌套进行描述。

（2）菜单单击效果有三种，分别对应不同的 Action 类型。

① 通过 @UxRoute 定义一个与菜单绑定的 viewAction，@UxMenu ("创建商店")@UxRoute (value = PetShop.MODEL_MODEL,viewName = "redirectCreatePage",viewType = ViewTypeEnum.FORM)，其中 viewName 代表视图的 name（其默认值为 redirectListPage，也就是跳转到列表页），value 代码视图所属模型的编码，viewType 代表 view 类型（其默认值为 ViewTypeEnum.TABLE）。

② @UxServer 定义一个与菜单绑定的 serverAction，@UxMenu ("UxServer")@UxServer (Model = PetCatItem.MODEL_MODEL,name = "uxServer")，其中 name 代表 serverAction 的 name，Model 或 value 代码 serverAction 所属模型的编码。

③ @UxLink 定义一个与菜单绑定的 UrlAction，@UxMenu ("Oinone 官网")@UxLink (value = "http://www.oinone.top",openType= ActionTargetEnum.OPEN_WINDOW)，其中 value 为跳转 url，openType 为打开方式默认为 ActionTargetEnum.ROUTER，打开方式有以下几种：

● ROUTER ("router"，"页面路由"，"页面路由")；
● DIALOG ("dialog"，"页面弹窗"，"页面弹窗")；
● DRAWER ("drawer"，"打开抽屉"，"打开抽屉")；
● OPEN_WINDOW ("openWindow"，"打开新窗口"，"打开新窗口")。

（1）配合菜单演示，PetCatItemAction 增加一个 uxServer 的 ServerAction，代码如下。

```
1   package pro.shushi.pamirs.demo.core.action;
2
3   ……包引用
4
5   @Model.model(PetCatItem.MODEL_MODEL)
6   @Component
7 ▾ public class PetCatItemAction extends DataStatusBehavior<PetCatItem> {
8       ……省略其他代码
9
10      @Action(displayName = "uxServer")
11 ▾    public PetCatItem uxServer(PetCatItem data){
12          PamirsSession.getMessageHub().info("uxServer");
13          return data;
14      }
15
16  }
```

（2）新的菜单初始化代码如下。

```
17  @UxMenus public class DemoMenus implements ViewActionConstants {
18
19      @UxMenu("商店") class ShopMenu{
20          @UxMenu("UxServer")@UxServer(model = PetCatItem.MODEL_MODEL,name = "uxSer
    ver") class ShopSayHelloMenu{
21          }
22          @UxMenu("创建商店")@UxRoute(value = PetShop.MODEL_MODEL,viewName = "redirec
    tCreatePage",viewType = ViewTypeEnum.FORM) class ShopCreateMenu{
23          }
24          @UxMenu("商店管理")@UxRoute(PetShopProxy.MODEL_MODEL) class ShopProxyMenu{
25          }
26          @UxMenu("商店管理A")@UxRoute(PetShopProxyA.MODEL_MODEL) class ShopProxyAMen
    u{
27          }
28          @UxMenu("商店管理B")@UxRoute(PetShopProxyB.MODEL_MODEL) class ShopProxyBMen
    u{
29          }
30
31      }
32      @UxMenu("商品") class ItemPMenu{
33          @UxMenu("商品管理")@UxRoute(PetItem.MODEL_MODEL) class ItemMenu{
34          }
35          @UxMenu("宠狗商品")@UxRoute(PetDogItem.MODEL_MODEL) class DogItemMenu{}
36          @UxMenu("萌猫商品")@UxRoute(PetCatItem.MODEL_MODEL) class CatItemMenu{}
37      }
38      @UxMenu("品种")@UxRoute(PetType.MODEL_MODEL) class PetTypePMenu{
39          @UxMenu("宠物品种")@UxRoute(PetType.MODEL_MODEL) class PetTypeMenu{}
40          @UxMenu("萌猫品种")@UxRoute(PetCatType.MODEL_MODEL) class CatTypeMenu{}
41          @UxMenu("宠狗品种")@UxRoute(PetDogType.MODEL_MODEL) class DogTypeMenu{}
42      }
43      @UxMenu("宠狗达人")@UxRoute(PetTalent.MODEL_MODEL) class PetTalentMenu{}
44      @UxMenu("友情链接")class FlinkManagement{
45          @UxMenu("Oinone官网")@UxLink(value = "http://www.oinone.top",openType= Act
    ionTargetEnum.OPEN_WINDOW) class SsLink{}
46          @UxMenu("百度") @UxLink(value = "http://www.baidu.com",openType= ActionTar
    getEnum.OPEN_WINDOW) class BaiduLink{}
47      }
48  }
```

Step3. 重启看效果。

我们会发现菜单按照我们预先设定的效果进行组织和展示,如图3-142所示。

图3-142　菜单已按照我们预先设定的效果进行组织和展示

### 2. 数据初始化式（不推荐）

在模块启动生命周期中,调用InitializationUtil工具中的createViewActionMenu。在前面的学习中我们在DemoModuleMetaDataEditor这个类中用InitializationUtil工具初始化过viewAction和View。大家可以自行回忆,温故知新。

### 3.5.2 构建第一个 View

虽然我们有小眼睛（表格视图中的眼睛符号）可以让用户自定义展示字段和排序喜好，以及通过权限控制行、列展示，但在日常业务开发中还是会对页面进行调整，以满足业务方对交互友好和便捷性的要求。本小节先介绍页面结构与逻辑，再完成自定义 view 的 Template 和 Layout，以及整个母版的 Template 和 Layout。

**1. 整体介绍**

1）页面的构成讲解

页面交互拓扑图如图 3-143 所示。

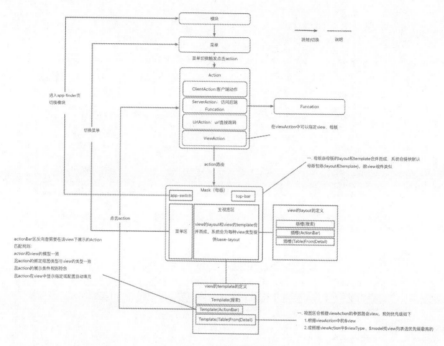

图 3-143　页面交互拓扑图

注：页面逻辑交互拓扑图说明。

（1）模块作为主切换入口。

（2）模块决定菜单列表。

（3）菜单切换触发单击 action。

（4）前端根据 Mask、View 进行渲染。

①Mask 是母版确定了主题、非主内容分发区域所使用组件和主内容分发区域联动方式的页面模板。全局、应用、视图动作、视图都可以通过 Mask 属性指定母版。

②Mask 和 View 都是由 layout 定义和 template 定义合并而成，系统会提供默认母版，以及为每种视图提供默认 layout。

③layout 与 template 通过插槽进行匹配。

（5）Action 根据不同类型分别访问后端服务、url 跳转、页面路由、发起客户端动作等。

（6）Action 路由可以指定 Mask，视图组件的 layout、template。

①当 layout 没有指定的时候则用系统默认的。

②当 template 没有指定的时候，且视图组件相同类型有多条记录时，根据优先级选取。

(7) Mask 和视图组件的 layout 优先级：视图组件 > 视图动作 > 应用 > 全局。

2）默认母版以及各类视图组件

（1）母版布局。

① 默认母版基础布局 base-layout，代码如下。

```
1  <mask layout="default">
2      <header slot="header"/>
3      <container slot="main" name="main">
4          <sidebar slot="sidebar"/>
5          <container slot="content"/>
6      </container>
7      <footer slot="footer"/>
8  </mask>
```

② 母版 template，代码如下。

```
1  <mask name="defaultMask">
2      <template slot="header">
3          <container name="appBar">
4              <element widget="logo"/>
5              <element widget="appFinder"/>
6          </container>
7          <container name="operationBar">
8              <element widget="notification"/>
9              <element widget="dividerVertical"/>
10             <element widget="languages"/>
11         </container>
12         <element widget="userProfile"/>
13     </template>
14     <template slot="sidebar">
15         <element widget="navMenu"/>
16     </template>
17     <template slot="content">
18         <element widget="breadcrumb"/>
19         <element widget="mainView"/>
20     </template>
21 </mask>
```

注：上例中因为名称为 main 的插槽不需要设置更多的属性，所以在 template 中缺省了 main 插槽的 template 标签。

③ 最终可执行视图，代码如下。

```
1  <mask name="defaultMask">
2      <header>
3          <container name="appBar">
4              <element widget="logo"/>
5              <element widget="appFinder"/>
6          </container>
7          <container name="operationBar">
8              <element widget="notification"/>
9              <element widget="dividerVertical"/>
10             <element widget="languages"/>
11         </container>
12         <element widget="userProfile"/>
13     </header>
14     <container name="main">
15         <sidebar name="sidebar">
16             <element widget="navMenu"/>
17         </sidebar>
18         <container name="content">
19             <element widget="breadcrumb"/>
20             <element widget="mainView"/>
21         </container>
22     </container>
23     <footer/>
24 </mask>
```

（2）表格视图布局。

① 默认表格视图基础布局 base-layout，代码如下。

```
<view type="table">
    <view type="search">
        <element widget="search" slot="search">
            <xslot name="fields" slotSupport="field" />
        </element>
    </view>
    <pack widget="fieldset">
        <element widget="actionBar" slot="actions" slotSupport="action" />
        <element widget="table" slot="table">
            <xslot name="fields" slotSupport="field" />
            <element widget="actionsColumn" slot="actionsColumn">
                <xslot name="rowActions" slotSupport="action" />
            </element>
        </element>
    </pack>
</view>
```

注：table 标签的子标签为 column 组件，如果 field 填充到元数据插槽，fields 没有 column 组件将自动包裹 column 组件。

② 表格视图 template，代码如下。

```
<view type="table" model="xxx" name="tableViewExample">
    <template slot="search">
        <field data="name"/>
    </template>
    <template slot="actions">
        <action name="create"/>
    </template>
    <template slot="fields">
        <field data="id"/>
        <field data="name"/>
        <field data="code"/>
    </template>
    <template slot="rowActions">
        <action name="delete"/>
        <action name="update"/>
    </template>
</view>
```

③ 最终可执行视图，代码如下。

```
<view type="table" model="xxx" name="tableViewExample">
    <view type="search">
        <element widget="search">
            <field data="name"/>
        </element>
    </view>
    <action-bar>
        <action name="create"/>
    </action-bar>
    <table>
        <column>
            <field data="id"/>
        </column>
        <column>
            <field data="name"/>
        </column>
        <column>
            <field data="code"/>
        </column>
        <column>
            <action name="delete"/>
            <action name="update"/>
        </column>
    </table>
</view>
```

(3)表单视图布局。

① 默认表单视图基础布局 base-layout,代码如下。

```
1  <view type="form">
2      <element widget="actionBar" slot="actions" slotSupport="action"/>
3      <element widget="form" slot="form">
4          <xslot name="fields" slotSupport="pack,field"/>
5      </element>
6  </view>
```

② 表单视图 template,代码如下。

```
1  <view type="form" model="xxx" name="viewExample">
2      <template slot="actions">
3          <action name="submit"/>
4      </template>
5      <template slot="fields">
6          <pack widget="group">
7              <field data="id"/>
8              <field data="name" widget="string"/>
9              <field data="code"/>
10         </pack>
11         <pack widget="tabs">
12             <pack widget="tab" title="商品列表">
13                 <field data="items" />
14             </pack>
15             <pack widget="tab" title="子订单列表">
16                 <field data="orders" />
17             </pack>
18         </pack>
19     </template>
20 </view>
```

注:tabs 标签的子标签为 tab,如果 dsl 填充到 layout,没有 tab 标签将自动包裹 tab 标签。

③ 最终可执行视图,代码如下。

```
1  <view type="form" model="xxx" name="viewExample">
2      <action-bar>
3          <action name="submit"/>
4      </action-bar>
5      <form>
6          <group>
7              <field data="id"/>
8              <field data="name" widget="string"/>
9              <field data="code"/>
10         </group>
11         <tabs>
12             <tab title="商品列表">
13                 <field data="items" />
14             </tab>
15             <tab title="子订单列表">
16                 <field data="orders" />
17             </tab>
18         </tabs>
19     </form>
20 </view>
```

(4)详情视图布局。

① 默认详情视图基础布局 base-layout,代码如下。

```
1  <view type="detail">
2      <element widget="actionBar" slot="actions" slotSupport="action"/>
3      <element widget="detail" slot="detail">
4          <xslot name="fields" slotSupport="pack,field"/>
5      </element>
6  </view>
```

② 详情视图 template,代码如下。

```
1  <view name="viewExample">
2      <template slot="actions">
3          <action name="back"/>
4      </template>
5      <template slot="fields">
6          <pack widget="group">
7              <field data="id"/>
8              <field data="name" widget="string"/>
9              <field data="code"/>
10         </pack>
11         <pack widget="tabs">
12             <pack widget="tab" title="商品列表">
13                 <field data="items" />
14             </pack>
15             <pack widget="tab" title="子订单列表">
16                 <field data="orders" />
17             </pack>
18         </pack>
19     </template>
20 </view>
```

③ 最终可执行视图,代码如下。

```
1  <view type="detail" model="xxx" name="viewExample">
2      <action-bar>
3          <action name="back"/>
4      </action-bar>
5      <detail>
6          <group>
7              <field data="id"/>
8              <field data="name" widget="string"/>
9              <field data="code"/>
10         </group>
11         <tabs>
12             <tab title="商品列表">
13                 <field data="items" />
14             </tab>
15             <tab title="子订单列表">
16                 <field data="orders" />
17             </tab>
18         </tabs>
19     </detail>
20 </view>
```

**2. 构建 View 的 Template**

我们在很多时候需要自定义模型的管理页面,而不是直接使用默认页面,比如字段的展示与隐藏,Action 是否在这个页面上出现,搜索条件自定义,等等,本小节带您一起学习如何自定义 View 的 Template。

1)自定义 View 的 Template

在使用默认 layout 的情况下,我们来做几个自定义视图 Template,并把文件放到指定目录下,如图 3-144 所示。

图 3-144 自定义 View 的 Template

2）第一个 Table

Step1. 自定义 PetTalent 的列表。

（1）我们先通过数据库查看默认页面定义，找到 base_view 表，过滤条件设置为 Model ='demo.PetTalent'，可以看到该模型下对应的所有 view，这些是系统根据该模型的 ViewAction 对应生成的默认视图，找到类型为"表格（type = TABLE）"的记录，查看 template 字段（如图 3-145 所示），代码如下。

图 3-145 找到 base_view 表

```
1  <view name="tableView" cols="1" type="TABLE" enableSequence="false">
2    <template slot="actions" autoFill="true"/>
3    <template slot="rowActions" autoFill="true"/>
4    <template slot="fields">
5      <field invisible="true" data="id" label="ID" readonly="true"/>
6      <field data="name" label="达人"/>
7      <field data="dataStatus" label="数据状态">
8        <options>
9          <option name="DRAFT" displayName="草稿" value="DRAFT" state="ACTIVE"/>
10         <option name="NOT_ENABLED" displayName="未启用" value="NOT_ENABLED" state="ACTIVE"/>
11         <option name="ENABLED" displayName="已启用" value="ENABLED" state="ACTIVE"/>
12         <option name="DISABLED" displayName="已禁用" value="DISABLED" state="ACTIVE"/>
13       </options>
14     </field>
15     <field data="createDate" label="创建时间" readonly="true"/>
16     <field data="writeDate" label="更新时间" readonly="true"/>
17     <field data="createUid" label="创建人id"/>
18     <field data="writeUid" label="更新人id"/>
19   </template>
20   <template slot="search" autoFill="true" cols="4"/>
21 </view>
```

（2）对比 view 的 template 定义与页面差异，从页面上看给 view 的定义少了创建人 id 和更新人

id，因为这两个字段元数据定义 invisible 属性（如图 3-146、图 3-147 所示）。

① 当 XML 里面没有配置，则用元数据覆盖。

② 当 XML 里面配置了，则不会用元数据覆盖。

在下一步中我们只要 view 的 DSL 中给这两个字段加上 invisible="false" 就可以展示出来了。

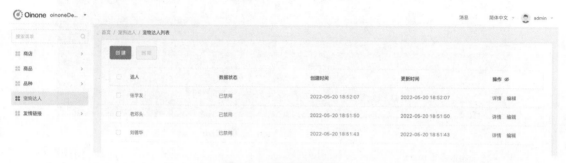

图 3-146　查看列表展示

```
AbstractModel.class
Decompiled .class file, bytecode version: 52.0 (Java 8)
    private Date createDate;
    @Base
    @pro.shushi.pamirs.meta.annotation.Field.Advanced(
        columnDefinition = "DATETIME NOT NULL DEFAULT CURRENT_TIMESTAMP ON UPDATE CURRENT_TIMESTAM
        batchStrategy = FieldStrategyEnum.NEVER
    )
    @Field(
        displayName = "更新时间",
        priority = 210L
    )
    private Date writeDate;
    @Base
    @Field(
        displayName = "创建人id",
        priority = 220L,
        invisible = true
    )
    private Long createUid;
    @Base
    @Field(
        displayName = "更新人id",
        priority = 230L,
        invisible = true
    )
    private Long writeUid;

    public AbstractModel() {
    }
```

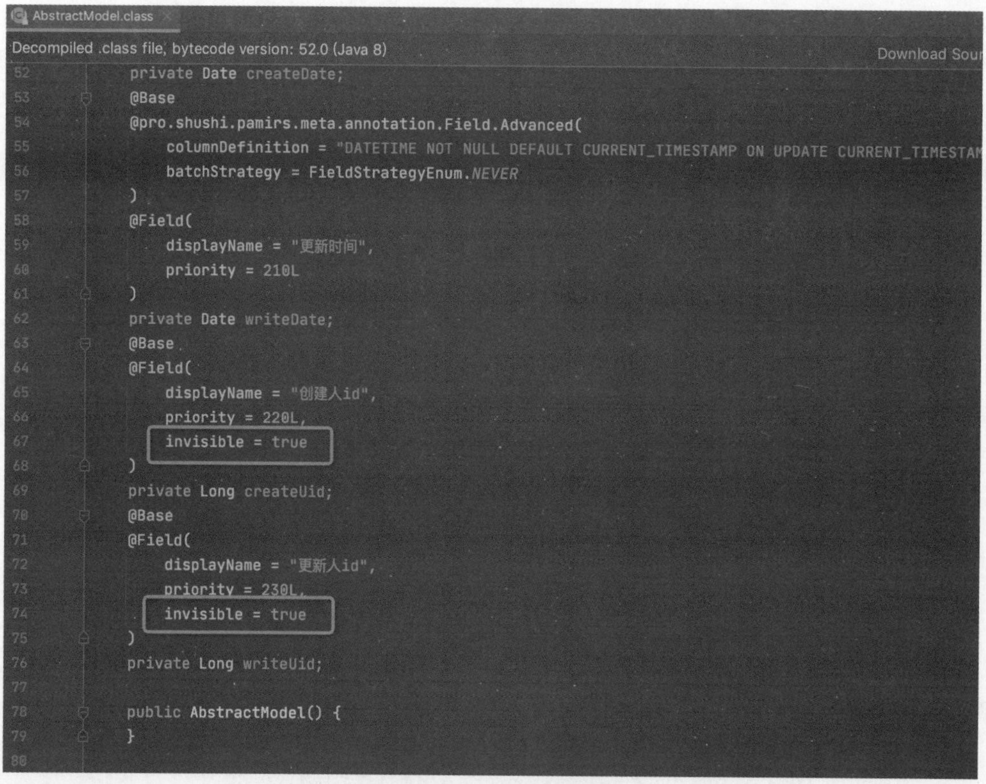

图 3-147　invisible 属性

（3）新建 pet_talent_table.xml 文件放到对应的 pamirs/views/demo_core/template 目录下，内容如下：

① 对比默认视图，在自定义视图时需要额外增加属性 Model="demo.PetTalent"。

② name 设置为“tableView”，系统重启后会替换掉 base_view 表中 Model 为“demo.PetTalent”，name 为“tableView”，type 为“TABLE”的数据记录。

name 不同但 type 相同，且 viewAction 没有指定时，根据优先级 priority 进行选择。小伙伴们可以尝试修改 name="tableView1"，并设置 priority 为 1，默认生成的优先级为 10，越小越优先。

③ createUid 和 writeUid 字段，增加 invisible=“false”的属性定义，代码如下。

```xml
1  <view name="tableView" model="demo.PetTalent" cols="1" type="TABLE" enableSequenc
   e="false">
2    <template slot="actions" autoFill="true"/>
3    <template slot="rowActions" autoFill="true"/>
4    <template slot="fields">
5      <field invisible="true" data="id" label="ID" readonly="true"/>
6      <field data="name" label="达人"/>
7      <field data="dataStatus" label="数据状态">
8        <options>
9          <option name="DRAFT" displayName="草稿" value="DRAFT" state="ACTIVE"/>
10         <option name="NOT_ENABLED" displayName="未启用" value="NOT_ENABLED" state
   ="ACTIVE"/>
11         <option name="ENABLED" displayName="已启用" value="ENABLED" state="ACTIV
   E"/>
12         <option name="DISABLED" displayName="已禁用" value="DISABLED" state="ACTIV
   E"/>
13       </options>
14     </field>
15     <field data="createDate" label="创建时间" readonly="true"/>
16     <field data="writeDate" label="更新时间" readonly="true"/>
17     <field data="createUid" label="创建人id" invisible="false"/>
18     <field data="writeUid" label="更新人id" invisible="false"/>
19   </template>
20   <template slot="search" autoFill="true" cols="4"/>
21 </view>
```

Step2. 重启应用看效果。

重启应用看效果，如图3-148所示。

图3-148　重启应用看效果

3）第一个Form

Step1. 自定义PetTalent的编辑页，代码如下。

（1）我们先通过数据库查看默认页面定义，找到base_view表，过滤条件设置为Model ='demo. PetTalent'，可以看到该模型下对应的所有view，这些是系统根据该模型的ViewAction对应生成的默认视图，找到类型为"表单（type = FORM）"的记录，查看template字段。

（2）新建一个pet_talent_form.xml文件放在对应的pamirs/views/demo_core/template目录下，把数据状态下拉选项去除一个"草稿"选项：

① 对比默认视图，在自定义视图时需要额外增加属性Model="demo.PetTalent"；

② name设置为"formView1"，系统重启后会在base_view表中新增一个Model为"demo. PetTalent"，name为"formView1"，type为"FORM"，并设置priority为1的数据记录：

name不同但type相同，且viewAction没有指定时，根据优先级priority进行选择，默认生成的优先级为10，越小越优先。所以在此打开新增或编辑页面默认会路由到我们新配置的view上。

```xml
<view name="formView1" model="demo.PetTalent" cols="2" type="FORM" priority="1">
    <template slot="actions" autoFill="true"/>
    <template slot="fields">
        <pack widget="group" title="基础信息">
            <field invisible="true" data="id" label="ID" readonly="true"/>
            <field data="name" label="达人"/>
            <field data="dataStatus" label="数据状态">
                <options>
                    <!-- option name="DRAFT" displayName="草稿" value="DRAFT" state="ACTIVE"/ -->
                    <option name="NOT_ENABLED" displayName="未启用" value="NOT_ENABLED" state="ACTIVE"/>
                    <option name="ENABLED" displayName="已启用" value="ENABLED" state="ACTIVE"/>
                    <option name="DISABLED" displayName="已禁用" value="DISABLED" state="ACTIVE"/>
                </options>
            </field>
            <field invisible="true" data="createDate" label="创建时间" readonly="true"/>
            <field invisible="true" data="writeDate" label="更新时间" readonly="true"/>
            <field data="createUid" label="创建人id"/>
            <field data="writeUid" label="更新人id"/>
        </pack>
    </template>
</view>
```

Step2. 重启看效果。

重启看效果，如图 3-149 所示。

图 3-149　重启看效果

4）第一个 Detail

Step1. 自定义 PetTalent 的详情页，代码如下。

（1）我们先通过数据库查看默认页面定义，找到 base_view 表，过滤条件设置为 Model ='demo. PetTalent'，可以看到该模型下对应的所有 view，这些是系统根据该模型的 ViewAction 对应生成的默认视图，找到类型为表单（type = DETAIL）的记录，查看 template 字段。

（2）新建一个 pet_talent_detail.xml 文件放在 pamirs-demo-core 的 pamirs/views/demo_core/template 目录下：

① 对比默认视图，在自定义视图时需要额外增加属性 Model="demo.PetTalent"；

② 把分组的 title 从基础信息改成基础信息 1。

```xml
<view name="detailView" cols="2" type="DETAIL" model="demo.PetTalent">
    <template slot="actions" autoFill="true"/>
    <template slot="fields">
        <pack widget="group" title="基础信息1">
            <field invisible="true" data="id" label="ID" readonly="true"/>
            <field data="name" label="达人"/>
            <field data="dataStatus" label="数据状态">
                <options>
                    <option name="DRAFT" displayName="草稿" value="DRAFT" state="ACTIVE"/>
```

```
10          <option name="NOT_ENABLED" displayName="未启用" value="NOT_ENABLED" stat
   e="ACTIVE"/>
11          <option name="ENABLED" displayName="已启用" value="ENABLED" state="ACTIV
   E"/>
12          <option name="DISABLED" displayName="已禁用" value="DISABLED" state="ACT
   IVE"/>
13        </options>
14      </field>
15      <field data="createDate" label="创建时间" readonly="true"/>
16      <field data="writeDate" label="更新时间" readonly="true"/>
17      <field data="createUid" label="创建人id"/>
18      <field data="writeUid" label="更新人id"/>
19    </pack>
20  </template>
21  </view>
```

Step2. 重启看效果。

重启看效果，如图 3-150 所示。

图 3-150　重启看效果

5）第一个 Search

Step1. 自定义 PetTalent 的列表页的搜索项，代码如下。

修改 pet_talent_table.xml 文件，默认情况下 `<template slot="search" autoFill="true" cols="4"/>` 代表着跟表格的字段自动填充搜索字段，我们搜索条件只保留数据状态和创建时间。

```
1 ▾ <view name="tableView" model="demo.PetTalent" cols="1" type="TABLE" enableSequenc
     e="true" baseLayoutName="">
2        <template slot="actions" autoFill="true"/>
3        <template slot="rowActions" autoFill="true"/>
4 ▾      <template slot="fields">
5          <field invisible="true" data="id" label="ID" readonly="true"/>
6          <field data="name" label="达人"/>
7 ▾        <field data="dataStatus" label="数据状态">
8 ▾          <options>
9              <option name="DRAFT" displayName="草稿" value="DRAFT" state="ACTIV
   E"/>
10             <option name="NOT_ENABLED" displayName="未启用" value="NOT_ENABLE
   D" state="ACTIVE"/>
11             <option name="ENABLED" displayName="已启用" value="ENABLED" state
   ="ACTIVE"/>
12             <option name="DISABLED" displayName="已禁用" value="DISABLED" stat
   e="ACTIVE"/>
13           </options>
14         </field>
15         <field data="createDate" label="创建时间" readonly="true"/>
16         <field data="writeDate" label="更新时间" readonly="true"/>
17         <field data="createUid" label="创建人id" invisible="false"/>
18         <field data="writeUid" label="更新人id" invisible="false"/>
19       </template>
20 ▾     <template slot="search" cols="4">
21         <field data="dataStatus" label="数据状态">
22 ▾         <options>
```

```
23                <option name="DRAFT" displayName="草稿" value="DRAFT" state="ACTIV
    E"/>
24                <option name="NOT_ENABLED" displayName="未启用" value="NOT_ENABLE
    D" state="ACTIVE"/>
25                <option name="ENABLED" displayName="已启用" value="ENABLED" state
    ="ACTIVE"/>
26                <option name="DISABLED" displayName="已禁用" value="DISABLED" stat
    e="ACTIVE"/>
27            </options>
28        </field>
29        <field data="createDate" label="创建时间"/>
30    </template>
31 </view>
```

Step2. 重启看效果。

重启看效果，如图 3-151 所示。

图 3-151　重启看效果

6）其他

search 默认查询的是模型的 queryPage 函数，但我们有时候需要替换调用的函数，下一个版本支持。其核心场景为当搜索条件中有非存储字段，如果直接用 queryPage 函数的 rsql 拼接就会报错。本版本临时替代方案见 4.1.4 一节。

**3. 构建 View 的 Layout**

在日常需求中也经常需要调整 Layout 的情况，如出现树表结构（左树右表、级联），则需要通过修改 View 的 Layout 来完成。现在就带您学习一下 Layout 的自定义。

1）第一个表格 Layout

如果我们想去除表格视图区域的搜索区、ActionBar（操作区），为视图自定义一个简单的 Layout 就可以啦。

Step1. 新建一个表格的 Layout，代码如下。

在 views/demo_core/layout 路径下增加一个名为 sample_table_layout.xml 的文件，name 设置为 sampleTableLayout。

```
1  <view type="TABLE" name="sampleTableLayout">
2  <!--    <view type="SEARCH">-->
3  <!--        <pack widget="fieldset">-->
4  <!--            <element widget="search" slot="search" slotSupport="field"/>-->
5  <!--        </pack>-->
6  <!--    </view>-->
7      <pack widget="fieldset" style="height: 100%" wrapperStyle="height: 100%">
8          <pack widget="row" style="height: 100%; flex-direction: column">
9  <!--            <pack widget="col" mode="full" style="flex: 0 0 auto">-->
10 <!--                <element widget="actionBar" slot="actionBar" slotSupport="act
```

```
11   <!--                    <xslot name="actions" slotSupport="action" />-->
12   <!--                </element>-->
13   <!--            </pack>-->
14       <pack widget="col" mode="full" style="min-height: 234px">
15           <element widget="table" slot="table" slotSupport="field">
16               <xslot name="fields" slotSupport="field" />
17               <element widget="rowAction" slot="rowActions" slotSupport="action"/>
18           </element>
19       </pack>
20   </pack>
21  </view>
```

Step2. 修改宠物达人自定义表格 Template，代码如下。

在 view 标签上增加 layout 属性值为 "sampleTableLayout"。

```
1  <view name="tableView" model="demo.PetTalent" cols="1" type="TABLE" enableSequence
      ="true" layout="sampleTableLayout">
2      ……省略其他
3  </view>
```

Step3. 重启看效果。

重启看效果，如图 3-152 所示。

图 3-152　重启看效果

Step4. 修改宠物达人自定义表格 Template。

去除在 view 标签上的 layout 属性配置，让其恢复正常。

2）第一个树表 Layout

这里以"给商品管理页面以树表的方式增加商品类目过滤"为例。

Step1. 增加商品类目模型，代码如下。

增加 PetItemCategory 模型继承 CodeModel，新增两个字段定义 name 和 parent，其中 parent 字段 M2O 关联自身模型，非必填字段（如字段值为空即为一级类目）。

```
7   @Model.model(PetItemCategory.MODEL_MODEL)
8   @Model(displayName = "宠物商品类目",summary="宠物商品类目",labelFields={"name"})
9   public class PetItemCategory extends CodeModel {
10
11      public static final String MODEL_MODEL="demo.PetItemCategory";
12      @Field(displayName = "类目名称",required = true)
13      private String name;
14
15      @Field(displayName = "父类目")
16      @Field.many2one
17      private PetItemCategory parent;
18  }
```

**Step2.** 修改自定义商品模型，代码如下。

为商品模型 PetItem 增加一个 category 字段关联 PetItemCategory。

```
1  @Field(displayName = "类目")
2  @Field.many2one
3  private PetItemCategory category;
```

**Step3.** 新增名为 treeTableLayout 的 Layout。

在 views/demo_core/layout 路径下增加一个名为 tree_table_layout.xml 文件，name 设置为 treeTableLayout，如图 3-153 所示。

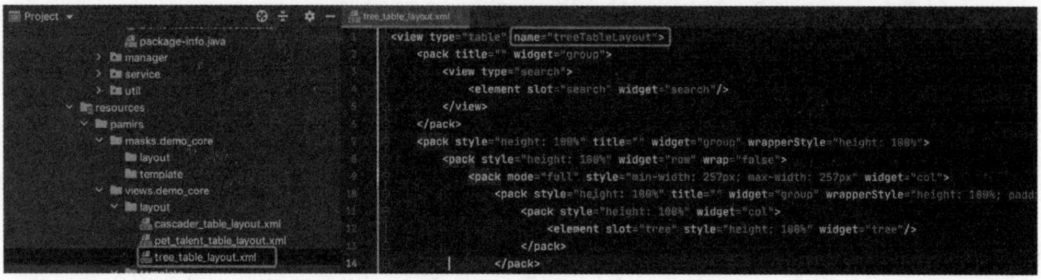

图 3-153　新增名为 treeTableLayout 的 Layout

代码说明如下。

```xml
 1  <view type="TABLE" name="treeTableLayout">
 2    <view type="SEARCH">
 3      <pack widget="fieldset">
 4        <element widget="search" slot="search" slotSupport="field"/>
 5      </pack>
 6    </view>
 7    <pack widget="fieldset" style="height: 100%" wrapperStyle="height: 100%">
 8      <pack widget="row" wrap="false" style="height: 100%">
 9        <pack widget="col" mode="full" style="min-width: 257px; max-width: 257px">
10          <pack widget="fieldset" style="height: 100%" wrapperStyle="height: 100%; padding: 24px 16px">
11            <pack widget="col" style="height: 100%">
12              <element widget="tree" slot="tree" style="height: 100%" />
13            </pack>
14          </pack>
15        </pack>
16        <pack widget="col" mode="full" style="min-width: 400px">
17          <pack widget="row" style="height: 100%; flex-direction: column">
18            <pack widget="col" mode="full" style="width: 100%; flex: 0 0 auto">
19              <element widget="actionBar" slot="actionBar" slotSupport="action">
20                <xslot name="actions" slotSupport="action" />
21              </element>
22            </pack>
23            <pack widget="col" mode="full" style="width: 100%; min-height: 345px">
24              <element widget="table" slot="table" slotSupport="field">
25                <xslot name="fields" slotSupport="field" />
26                <element widget="rowAction" slot="rowActions" slotSupport="action" />
27              </element>
28            </pack>
29          </pack>
30        </pack>
31      </pack>
32    </pack>
33  </view>
```

**Step4.** 自定义商品管理的 Template。

（1）在 views/demo_core/template 路径下增加一个名为 pet_item_table.xml 的文件，如图 3-154 所示。

图 3-154　在 views/demo_core/template 路径下增加一个名为 pet_item_table.xml 的文件

（2）跟普通自定义 template 的区别在于：

① 配置 layout 属性为"treeTableLayout"跟前面 layout 定义一致；

② 配置了 widget 为 tree 的 template 节点，node 可以配置多个，node 配置说明如下：

● Model：模型编码，必填。

● label：数据标题，支持表达式，必填。

● labelFields：数据标题中使用的字段列表，必填。

● references：层级关联字段，第一层无效，其他层必填。模型编码 # 字段。

● selfReferences：自关联字段，模型编码 # 字段。

● search：单击搜索字段，必须使用主表格字段。模型编码 # 字段。

● g.filter，层级过滤条件。模型编码 # 字段。

以上所有使用 # 拼接的属性配置，与 Model 一致的情况下，均可以省略模型编码。

自定义商品管理的 Template 代码如下。

```xml
<view name="tableView" model="demo.PetItem" type="TABLE" cols="1" enableSequence="false" layout="treeTableLayout">
    <template slot="actions" autoFill="true"/>
    <template slot="rowActions" autoFill="true"/>
    <template slot="fields">
      <field invisible="true" priority="5" data="id" label="ID" readonly="true"/>
      <field priority="90" data="code" label="编码"/>
      <field priority="101" data="itemName" label="商品名称"/>
      <field priority="101" data="dataStatus" label="数据状态">
        <options>
          <option name="DRAFT" displayName="草稿" value="DRAFT" state="ACTIVE"/>
          <option name="NOT_ENABLED" displayName="未启用" value="NOT_ENABLED" state="ACTIVE"/>
          <option name="ENABLED" displayName="已启用" value="ENABLED" state="ACTIVE"/>
          <option name="DISABLED" displayName="已禁用" value="DISABLED" state="ACTIVE"/>
        </options>
      </field>
      <field priority="102" data="price" label="商品价格"/>
      <field priority="103" data="shop" label="店铺">
        <options>
          <option references="demo.PetShop" referencesType="STORE" referencesLabelFields="shopName">
            <field name="shopName" data="shopName" ttype="STRING"/>
          </option>
        </options>
      </field>
      <field priority="104" data="shopId" label="店铺id"/>
```

```xml
25      <field priority="105" data="type" label="品种">
26        <options>
27          <option references="demo.PetType" referencesType="STORE" referencesLabelFields="name">
28            <field name="name" data="name" ttype="STRING"/>
29          </option>
30        </options>
31      </field>
32      <field priority="106" data="typeId" label="品种类型"/>
33      <field priority="107" data="petItemDetails" label="详情">
34        <options>
35          <option references="demo.PetItemDetail" referencesType="TRANSIENT" referencesLabelFields="remark">
36            <field name="remark" data="remark" ttype="STRING"/>
37          </option>
38        </options>
39      </field>
40      <field priority="108" data="tags" label="商品标签"/>
41      <field priority="109" data="petTalents" label="推荐达人">
42        <options>
43          <option references="demo.PetTalent" referencesType="STORE" referencesLabelFields="name">
44            <field name="name" data="name" ttype="STRING"/>
45          </option>
46        </options>
47      </field>
48      <field priority="200" data="createDate" label="创建时间" readonly="true"/>
49      <field priority="210" data="writeDate" label="更新时间" readonly="true"/>
50      <field priority="220" data="createUid" label="创建人id"/>
51      <field priority="230" data="writeUid" label="更新人id"/>
52    </template>
53    <template slot="search" autoFill="true" cols="4"/>
54    <template enableSearch="true" slot="tree" style="height: 100%" widget="tree">
55      <nodes>
56  <!--        <node label="activeRecord.name" labelFields="name" model="demo.PetItemCategory" search="demo.PetItem#category" selfReferences="parent"/> -->
57          <node label="activeRecord.name" labelFields="name" model="demo.PetItemCategory" search="demo.PetItem#category" selfReferences="demo.PetItemCategory#parent"/>
58      </nodes>
59    </template>
60  </view>
```

Step5. 为商品类目增加管理入口。

修改 DemoMenus 类，增加类目管理菜单（如图 3-155 所示），代码如下。

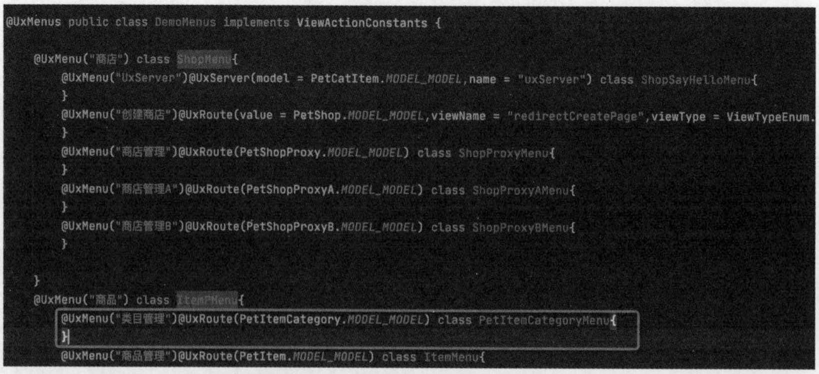

```
@UxMenus public class DemoMenus implements ViewActionConstants {
    @UxMenu("商店") class ShopMenu{
        @UxMenu("UxServer")@UxServer(model = PetCatItem.MODEL_MODEL,name = "uxServer") class ShopSayHelloMenu{
        }
        @UxMenu("创建商店")@UxRoute(value = PetShop.MODEL_MODEL, viewName = "redirectCreatePage", viewType = ViewTypeEnum.
        @UxMenu("商店管理")@UxRoute(PetShopProxy.MODEL_MODEL) class ShopProxyMenu{}
        @UxMenu("商店管理A")@UxRoute(PetShopProxyA.MODEL_MODEL) class ShopProxyAMenu{
        }
        @UxMenu("商店管理B")@UxRoute(PetShopProxyB.MODEL_MODEL) class ShopProxyBMenu{}
    }
    @UxMenu("商品") class ItemPMenu{
        @UxMenu("类目管理")@UxRoute(PetItemCategory.MODEL_MODEL) class PetItemCategoryMenu{
        }
        @UxMenu("商品管理")@UxRoute(PetItem.MODEL_MODEL) class ItemMenu{
```

图 3-155　增加类目管理菜单

```
1  @UxMenu("类目管理")@UxRoute(PetItemCategory.MODEL_MODEL) class PetItemCategoryMenu{
2  }
```

Step6. 重启看效果。

（1）进入类目管理页面，新增商品类目数据，如图3-156所示。

图3-156　进入类目管理页面

（2）进入商品管理页面，找一行数据修改其类目字段，然后再单击左边树看过滤效果，如图3-157所示。

图3-157　进入商品管理页面

3）第一个级联Layout

这里以"给商品管理页面以级联的方式增加商品类目过滤"为例，该例子中左边级联项由多个模型组成。

Step1. 增加商品类目类型模型。

增加PetItemCategoryType模型集成CodeModel，新增两个字段定义name，代码如下。

```
 7  @Model.model(PetItemCategoryType.MODEL_MODEL)
 8  @Model(displayName = "宠物商品类目类型",summary="宠物商品类目类型",labelFields={"name"})
 9  public class PetItemCategoryType extends CodeModel {
10
11      public static final String MODEL_MODEL="demo.PetItemCategoryType";
12      @Field(displayName = "类目类型名称",required = true)
13      private String name;
14
15  }
16
```

Step2. 修改商品类目。

为商品类目模型 PetItemCategory 增加一个 type 字段 m2o 关联 PetItemCategoryType，代码如下。

```
1  @Field(displayName = "类目类型")
2  @Field.many2one
3  private PetItemCategoryType type;
```

Step3. 新增名为 cascaderTableLayout 的 Layout。

在 views/demo_core/layout 路径下增加一个名为 cascader_table_layout.xml 的文件，如图 3-158 所示。

图 3-158　新增名为 cascader_table_layout.xml 的文件

代码如下。

```xml
<view type="TABLE" name="cascaderTableLayout">
    <view type="SEARCH">
        <pack widget="fieldset">
            <element widget="search" slot="search" slotSupport="field"/>
        </pack>
    </view>
    <pack widget="fieldset" style="height: 100%" wrapperStyle="height: 100%; overflow: auto">
        <pack widget="row" wrap="false" style="height: 100%">
            <pack widget="col" mode="full" style="flex: unset">
                <element widget="card-cascader" slot="cardCascader" style="height: 100%" />
            </pack>
            <pack widget="col" mode="full" style="min-width: 564px">
                <pack widget="row" style="height: 100%; flex-direction: column">
                    <pack widget="col" mode="full" style="width: 100%; flex: 0 0 auto">
                        <element widget="actionBar" slot="actionBar" slotSupport="action">
                            <xslot name="actions" slotSupport="action" />
                        </element>
                    </pack>
                    <pack widget="col" mode="full" style="width: 100%; min-height: 345px">
                        <element widget="table" slot="table" slotSupport="field">
                            <xslot name="fields" slotSupport="field" />
                            <element widget="rowAction" slot="rowActions" slotSupport="action"/>
                        </element>
                    </pack>
                </pack>
            </pack>
        </pack>
    </pack>
</view>
```

Step4. 修改商品管理的 Template。

修改在 views/demo_core/template 路径下名为 pet_item_table.xml 的文件：

① 配置 layout 属性为 "cascaderTableLayout" 跟前面 layout 定义一致；

② 配置了 widget 为 card-cascader 的 template 节点，node 可以配置多个，node 配置跟树表的配置一致，代码如下：

```xml
<view name="tableView" model="demo.PetItem"  type="TABLE" cols="1" enableSequence=
    <template slot="actions" autoFill="true"/>
    <template slot="rowActions" autoFill="true"/>
    <template slot="fields">
        <field invisible="true" priority="5" data="id" label="ID" readonly="true"/>
        <field priority="90" data="code" label="编码"/>
        <field priority="101" data="itemName" label="商品名称"/>
        <field priority="101" data="dataStatus" label="数据状态">
            <options>
                <option name="DRAFT" displayName="草稿" value="DRAFT" state="ACTIVE"/>
                <option name="NOT_ENABLED" displayName="未启用" value="NOT_ENABLED" state="
                <option name="ENABLED" displayName="已启用" value="ENABLED" state="ACTIVE"/
                <option name="DISABLED" displayName="已禁用" value="DISABLED" state="ACTIVE
            </options>
        </field>
        <field priority="102" data="price" label="商品价格"/>
        <field priority="103" data="shop" label="店铺">
            <options>
                <option references="demo.PetShop" referencesType="STORE" referencesLabelFi
                    <field name="shopName" data="shopName" ttype="STRING"/>
                </option>
            </options>
        </field>
        <field priority="104" data="shopId" label="店铺id"/>
        <field priority="105" data="type" label="品种">
            <options>
                <option references="demo.PetType" referencesType="STORE" referencesLabelFi
                    <field name="name" data="name" ttype="STRING"/>
                </option>
            </options>
        </field>
        <field priority="106" data="typeId" label="品种类型"/>
        <field priority="107" data="petItemDetails" label="详情">
            <options>
                <option references="demo.PetItemDetail" referencesType="TRANSIENT" referen
                    <field name="remark" data="remark" ttype="STRING"/>
                </option>
            </options>
        </field>
        <field priority="108" data="tags" label="商品标签"/>
        <field priority="109" data="petTalents" label="推荐达人">
            <options>
                <option references="demo.PetTalent" referencesType="STORE" referencesLabel
                    <field name="name" data="name" ttype="STRING"/>
                </option>
            </options>
        </field>
        <field priority="200" data="createDate" label="创建时间" readonly="true"/>
        <field priority="210" data="writeDate" label="更新时间" readonly="true"/>
        <field priority="220" data="createUid" label="创建人id"/>
        <field priority="230" data="writeUid" label="更新人id"/>
    </template>
    <template slot="search" autoFill="true" cols="4"/>
    <template enableSearch="true" slot="cardCascader" style="height: 100%" widget="c
        <nodes>
            <node label="activeRecord.name" title="类目类型" labelFields="name" model
            <node label="activeRecord.name" title="类目" labelFields="name" model="de
        </nodes>
    </template>
</view>
```

Step5. 为商品类目类型增加管理入口，如图3-159所示。

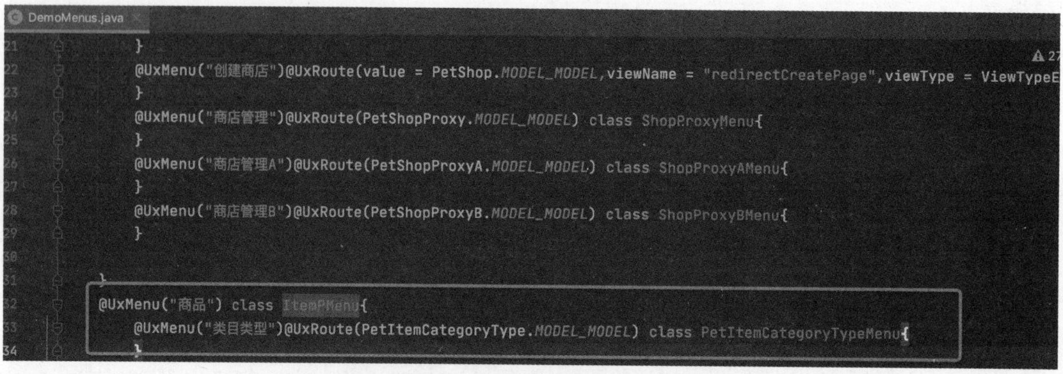

图3-159　为商品类目类型增加管理入口

代码如下。

```
1  @UxMenu("类目类型")@UxRoute(PetItemCategoryType.MODEL_MODEL) class PetItemCategory
   TypeMenu{
2  }
```

Step6. 重启看效果。

（1）进入类目类型管理页面，新增商品类目类型数据，如图3-160所示。

图3-160　进入类目类型管理页面

（2）进入类目管理页面，修改一级类目的类型，如图3-161所示。

图3-161　修改一级类目的类型

（3）进入商品管理页面，找一行数据修改其类目字段，然后再单击左边级联项看过滤效果，如图 3-162 所示。

图 3-162　进入商品管理页面

### 3.5.3　Action 的类型

各类动作我们都碰到过，但都没有展开讲过。这里我们来系统介绍 Oinone 涉及的所有 Action 类型。

**1. 动作类型**

（1）服务器动作 ServerAction：类似于 Spring MVC 的控制器 Controller，通过模型编码和动作名称路由，定义存储模型或代理模型将为该模型自动生成动作名称为 consturct、queryOne、queryPage、create、update、delete、deleteWithFieldBatch 的服务器动作。定义传输模型将为该模型自动生成动作名称为 consturct 的服务器动作。

（2）窗口动作 ViewAction：站内跳转，通过模型编码和动作名称路由，系统将为存储模型和代理模型自动生成动作名称为 redirectDetailPage 的跳转详情页窗口动作，动作名称为 redirectListPage 的跳转列表页窗口动作，动作名称为 redirectCreatePage 的跳转新增页窗口动作，以及动作名称为 redirectUpdatePage 的跳转更新页窗口动作。

（3）跳转动作 UrlAction：外链跳转。

（4）客户端动作 ClientAction：调用客户端函数。

**2. 默认动作**

（1）如果在 UI 层级，有开放新增语义函数，则会默认生成新增的窗口动作 ViewAction，跳转到新增页面；

（2）如果在 UI 层级，有开放更新语义函数，则会默认生成修改的窗口动作 ViewAction，跳转到更新页面；

（3）如果在 UI 层级，有开放删除语义函数，则会默认生成删除的客户端动作 ClientAction，弹出删除确认对话框。

**3. 第一个服务器动作 ServerAction**

1）回顾第一个 ServerAction

第一个 ServerAction 是在 3.3.2 节出现的，再来看一下当时的定义代码。

```java
 1  package pro.shushi.pamirs.demo.core.action;
 2
 3  ……引用类
 4
 5  @Model.model(PetShopProxy.MODEL_MODEL)
 6  @Component
 7  public class PetShopProxyAction extends DataStatusBehavior<PetShopProxy> {
 8
 9      @Override
10      protected PetShopProxy fetchData(PetShopProxy data) {
11          return data.queryById();
12      }
13      @Action(displayName = "启用")
14      public PetShopProxy dataStatusEnable(PetShopProxy data){
15          data = super.dataStatusEnable(data);
16          data.updateById();
17          return data;
18      }
19
20  ……其他代码
21
22  }
```

（1）@Action 注解将创建服务器动作，并 @Model.Model 绑定；

（2）自定义 ServerAction 请勿使用 get、set、unset 开头命名方法或 toString 命名方法。

2）ServerAction 之校验（举例）

Step1. 为动作配置校验表达式，代码如下。

使用 @Validation 注解为 PetShopProxyAction 的 dataStatusEnable 服务端动作进行校验表达式配置。

```java
 1  package pro.shushi.pamirs.demo.core.action;
 2
 3  ……引用类
 4
 5  @Model.model(PetShopProxy.MODEL_MODEL)
 6  @Component
 7  public class PetShopProxyAction extends DataStatusBehavior<PetShopProxy> {
 8
 9      @Override
10      protected PetShopProxy fetchData(PetShopProxy data) {
11          return data.queryById();
12      }
13      @Validation(ruleWithTips = {
14              @Validation.Rule(value = "!IS_BLANK(data.code)", error = "编码为必填项"),
15              @Validation.Rule(value = "LEN(data.shopName) < 128", error = "名称过长，不能超过128位"),
16      })
17      @Action(displayName = "启用")
18      public PetShopProxy dataStatusEnable(PetShopProxy data){
19          data = super.dataStatusEnable(data);
20          data.updateById();
21          return data;
22      }
23
24  ……其他代码
25
26  }
```

注：

（1）ruleWithTips 可以声明多个校验规则及错误提示；

（2）IS_BLANK 和 LEN 为内置文本函数，更多内置函数见"函数之内置函数与表达式"；

（3）当内置函数不满足时参考"Action 之校验"。

Step2. 重启看效果。

在"商店管理"页面单击"启用"得到了预期返回错误信息，显示"错误 编码为必填项"，如图 3-163 所示。

图 3-163　在"商店管理"页面单击"启用"得到了预期返回错误信息

3）ServerAction 之前端展示规则（举例）

既然后端对 ServerAction 发起提交做了校验，那能不能在前端就不展示了呢？当然可以，我们现在就来试一下。

Step1. 配置 PetShopProxyAction 的 dataStatusEnable 的前端出现规则，代码如下。

用注解 @Action.Advanced(invisible="!(activeRecord.code !== undefined && !IS_BLANK(activeRecord.code))") 来表示，注意这里配对 invisible 是给前端识别的，所以写法上跟后端的校验有些不一样，但一些函数是前后端一致实现的，如内置函数 IS_BLANK，activeRecord 在前端用于表示当前记录。

```
package pro.shushi.pamirs.demo.core.action;

……引用类

@Model.model(PetShopProxy.MODEL_MODEL)
@Component
public class PetShopProxyAction extends DataStatusBehavior<PetShopProxy> {

    @Override
    protected PetShopProxy fetchData(PetShopProxy data) {
        return data.queryById();
    }
    @Validation(ruleWithTips = {
            @Validation.Rule(value = "!IS_BLANK(data.code)", error = "编码为必填项"),
            @Validation.Rule(value = "LEN(data.name) < 128", error = "名称过长，不能超过128位"),
    })
    @Action(displayName = "启用")
    @Action.Advanced(invisible="!(activeRecord.code !== undefined && !IS_BLANK(activeRecord.code))")
    public PetShopProxy dataStatusEnable(PetShopProxy data){
        data = super.dataStatusEnable(data);
        data.updateById();
        return data;
    }
    ……其他代码
}
```

Step2. 重启看效果。

我们发现店铺编码为空的记录，没有了启用的操作按钮，如图 3-164 所示。

图 3-164　店铺编码为空的记录，没有了启用的操作按钮

4）ServerAction 配置说明

（1）常用配合。

① contextType 设置动作上下文类型。

● SINGLE（默认）——单行，常用于列表页（展示在每行末尾的操作栏中）和表单页（展示在页面上方）。

● BATCH——多行，常用于列表页（展示在表格上方按钮区）。

● SINGLE_AND_BATCH——单行或多行，常用于列表页（展示在表格上方按钮区）。

● CONTEXT_FREE——上下文无关，常用于列表页（展示在表格上方按钮区）。

② bindingType 设置按钮所在页面类型（以下仅说明常用类型，详见 ViewTypeEnum）。

● TABLE——列表页。

● GALLERY——画廊。

● FORM——表单页。

● DETAIL——详情页。

● CUSTOM——自定义页。

（2）注解大全。

@Action
├── displayName 显示名称
├── summary 摘要摘要
├── contextType 动作上下文，可选项详见 ActionContextTypeEnum
├── bindingType 所在页面类型，可选项详见 ViewTypeEnum
├── Advanced 更多配置
│   ├── name 技术名称，默认 Java 方法名
│   ├── args 参数，默认 Java 参数
│   ├── type 方法类型，默认 UPDATE，可选项详见 FunctionTypeEnum
│   ├── language 方法实现语言，默认 Java，可选项详见 FunctionLanguageEnum
│   ├── invisible 隐藏规则
│   ├── bindingView 绑定特定视图
│   └── priority 展示顺序

## 4. 第一个窗口动作 ViewAction

1）回顾第一个 ViewAction

第一个 ViewAction 是在 3.3.2 节出现的，再来看一下当时的定义代码。

```java
1   package pro.shushi.pamirs.demo.core.init;
2
3   ……引用类
4
5   @Component
6   public class DemoModuleMetaDataEditor implements MetaDataEditor {
7       @Override
8       public void edit(AppLifecycleCommand command, Map<String, Meta> metaMap) {
9           InitializationUtil util = InitializationUtil.get(metaMap, DemoModule.MODULE_MODULE,DemoModule.MODULE_NAME);
10          if(util==null){
11              return;
12          }
13          ……其他代码
14          //初始化自定义前端行为
15          viewActionInit(util);
16          ……其他代码
17      }
18      private void viewActionInit(InitializationUtil util){
19          util.createViewAction("demo_petShop_batch_update","批量更新数据状态", PetShopProxy.MODEL_MODEL,
20                  InitializationUtil.getOptions(ViewTypeEnum.TABLE), PetShopBatchUpdate.MODEL_MODEL,ViewTypeEnum.FORM, ActionContextTypeEnum.SINGLE_AND_BATCH
21                  , ActionTargetEnum.DIALOG,null,null);
22      }
23      ……其他代码
24  }
```

2)createViewAction 参数详解(建议更换为注解方式)

createViewAction 参数详解见表 3-23。

表 3-23 createViewAction 参数详解

| 参数名 | 类型 | 说明 |
| --- | --- | --- |
| name | String | 技术名称,唯一要求 |
| displayName | String | 展示名称 |
| originModel | String | ViewAction 的绑定模型 |
| originViewTypes | List<ViewTypeEnum> | ViewAction 在绑定模型的哪些视图类型上展示 |
| targetModel | String | ViewAction 跳转到的目标模型 |
| targetViewType | ViewTypeEnum | ViewAction 跳转到的目标模型的什么类型视图 |
| contextType | ActionContextTypeEnum | ViewAction 在绑定模型视图上设置展示的动作上下文类型<br>SINGLE(默认)——单行,常用于列表页(展示在每行末尾的操作栏中)和表单页(展示在页面上方)<br>BATCH——多行,常用于列表页(展示在表格上方按钮区)<br>SINGLE_AND_BATCH——单行或多行,常用于列表页(展示在表格上方按钮区)<br>CONTEXT_FREE——上下文无关,常用于列表页(展示在表格上方按钮区) |
| pageTarget | ActionTargetEnum | 页面打开方式<br>DIALOG:页面弹窗<br>DRAWER:打开抽屉 |
| resViewName | String | ViewAction 跳转到的目标模型的指定视图的名称,该视图的类型需要跟 targetViewType 一致 |
| title | String | 页面标题 |
| 非 createViewAction 方法的参数,但是 ViewAction 模型有的属性,可在调用 createViewAction 以后,通过 setXX 属性方法来设置对应属性字段值 |||

续表

| 参数名 | 类型 | 说明 |
|---|---|---|
| priority | Intget | 展示顺序 |
| bindingView | String | 绑定特定视图 |
| invisible | String | 隐藏规则 |
| laod | String | 对于特殊模型的窗口动作进行定制化的加载方式 |
| filter | String | 代表后端过滤，是一定会加上的过滤条件，用户无感知 |
| domain | String | 代表前端过滤，是默认会加上的过滤条件，用户可以去除该搜索条件 |

注解见：@UxRouteButton (action = @UxAction (),value = @UxRoute ())

3）ViewAction 高级参数——Load（下个版本支持）

一般用于以下场景：

（1）对于特殊模型的窗口动作进行定制化的加载方式。

（2）不同模型间跳转时，可根据上一个模型的数据内容加载另一个模型的数据内容。

使用初始化工具类设置 Load 函数，代码如下。

```
1  util.modifyViewAction(TestModele.MODEL_MODEL, InitializationUtil.DEFAULT_CREATE,
2          viewAction -> viewAction.setLoad("createPageLoad"));
```

注：所示修改窗口动作方法将 TestModel 模型的默认创建页的加载函数从 construct 函数改为了 createPageLoad 函数。

替代方案：构建模型子类，通过子类来重写 construct 方法。

4）ViewAction 高级参数 filter 和 domain（举例）。

filter 当前版本支持，domain 下一个版本支持，之所以放一起讲是因为这是过滤的两种形态。

（1）filter 代表后端过滤，是一定会加上的过滤条件，用户无感知。

（2）domain 代表前端过滤，是默认会加上的过滤条件，用户可以去除该搜索条件。

Step1. 修改自定义 pet_talent_table.xml 查询条件增加 name 字段，代码如下。

```
1   <view name="tableView" model="demo.PetTalent" cols="1" type="TABLE" enableSequenc
e="true" layout="petTalentTableLayout">
2   ……其他代码
3   <template slot="search"  cols="4">
4       <field data="name" label="达人"/>
5       <field data="dataStatus" label="数据状态">
6           <options>
7               <option name="DRAFT" displayName="草稿" value="DRAFT" state="ACTIVE"/>
8               <option name="NOT_ENABLED" displayName="未启用" value="NOT_ENABLED" state="ACTIVE"/>
9               <option name="ENABLED" displayName="已启用" value="ENABLED" state="ACTIVE"/>
10              <option name="DISABLED" displayName="已禁用" value="DISABLED" state="ACTIVE"/>
11          </options>
12      </field>
13      <field data="createDate" label="创建时间"/>
14  </template>
15  </view>
```

Step2. 为宠物达人模型的两个菜单入口分别配置 filter 和 domain，代码如下。

把"宠物达人"菜单调整为三个菜单"宠物达人1""宠物达人2""宠物达人3"，分别设置 filter 和 domain。

(1) 修改以菜单 "宠物达人 1" 为入口的 ViewAction 的 filter：

@UxRoute (filter = "name =like=' 老 '") 字符串要符合 RSQL。

(2) 修改以菜单 "宠狗达人 2" 为入口的 ViewAction 的 domain：

@UxRoute (domain = "name =like=' 老 '") 字符串要符合 RSQL。

(3) 修改以菜单 "宠狗达人 3" 跟函数结合，设置时间默认过滤条件：

① @UxRoute (domain = "createDate =ge= '${ADD_DAY (NOW_STR(),-7)}' and createDate =lt= '${NOW_STR ()}'") 字符串要符合 RSQL。

② createDate =ge= '${ADD_DAY (NOW_STR (),-7)}' and createDate =lt= '${NOW_STR ()}'：

● 用到函数需要用 ${} 装饰；

● 更多函数参见 4.1.12 节。

(4) domain 的操作符需要跟页面搜索字段定义的操作符一致，比如 name 字符串字段搜索默认操作符是 =like=，如果配置成其他则无效。

```
1  @UxMenu("宠物达人1")@UxRoute(value = PetTalent.MODEL_MODEL,filter = "name =like=
   '老'") class PetTalentMenu{}
2  @UxMenu("宠物达人2")@UxRoute(value = PetTalent.MODEL_MODEL,domain = "name =like=
   '老'") class PetTalent2Menu{}
3  @UxMenu("宠物达人3")@UxRoute(value = PetTalent.MODEL_MODEL,domain = "createDate =ge
   = '${ADD_DAY(NOW_STR(),-7)}' and createDate =lt= '${NOW_STR()}'") class PetTalent3
   Menu{}
```

Step3. 重启看效果。

因为 "宠物达人 1" 对应 ViewAction 加的是 Filter，所以只能看到达人名称带 "老" 字的记录，如图 3-165 所示。

图 3-165　"宠物达人 1" 搜索记录

因为 "宠物达人 2" 对应 ViewAction 加的是 Domain，前端搜索栏里 "达人" 字段搜索条件为 "老" 字，达人名称带 "老" 字的记录，可以手工删除再次搜索全部数据，如图 3-166 所示。

图 3-166　"宠物达人 2" 搜索记录

因为"宠物达人3"对应ViewAction加的是Domain，前端搜索栏里"创建时间"字段带近7天过滤条件，可以手工删除再次搜索全部数据，如图3-167所示。

图3-167　"宠物达人3"搜索记录

**5. 第一个跳转动作 UrlAction**

（1）回顾第一个UrlAction，代码如下。

在"构建第一个Menu"一节中用 @UxMenu.url 定义了一个百度的菜单，该菜单背后就是一个普通的UrlAction。

```
@UxMenu("Oinone官网")@UxLink(value = "http://www.oinone.top",openType= ActionTarget
Enum.OPEN_WINDOW) class SsLink{}
```

（2）URL 计算表达式（暂不支持），代码如下。

```
@UxMenu("百度") @UxLink(value = "http://www.baidu.com?wd=${activeRecord.technicalNa
me}",openType= ActionTargetEnum.OPEN_WINDOW) class BaiduLink{}
```

（3）Compute 函数（暂不支持），代码如下。

```
@UxMenu("百度") @UxLink(value = "http://www.baidu.com",openType= ActionTargetEnum.
OPEN_WINDOW,context = {@Prop(name="wd",value= "activeRecord.technicalName"),compu
te="computeSearchUrl"}) class BaiduLink{}

@Model.model(PetTalent.MODEL_MODEL)
public class PetTalentAction {

    @Function(openLevel = FunctionOpenEnum.API, summary = "计算搜索Url")
    @Function.Advanced(type = FunctionTypeEnum.QUERY)
    public String computeSearchUrl(TestModel data) {
        return "https://www.baidu.com/s?wd=" + data.getName();
    }
}
```

注：代码所示创建链接动作含义为以新标签页方式跳转至指定URL，该URL值来自调用后端computeSearchUrl方法的返回值。

**6. 第一个客户端动作 ClientAction**

1）基础客户端动作（举例）

给批量修改店铺状态弹出页面增加"自定义返回"和"自定义关闭"按钮，这里注意contextType只能配置为"ActionContextTypeEnum.SINGLE"，因为Form默认只展示contextType为SINGLE的Action。

Step1. 在 PetShopBatchUpdateAction 增加 @UxClientButton 注解，代码如下。

```
1  @Model.model(PetShopBatchUpdate.MODEL_MODEL)
2  @UxClientButton(action = @UxAction(name = "demo_back_test", label = "自定义返回",con
   textType = ActionContextTypeEnum.SINGLE,bindingType = ViewTypeEnum.FORM),value = @
   UxClient(value = "$$internal_GotoListTableRouter"))
3  @UxClientButton(action = @UxAction(name = "demo_close_test", label = "自定义关闭",co
   ntextType = ActionContextTypeEnum.SINGLE,bindingType = ViewTypeEnum.FORM),value =
   @UxClient(value = "$$internal_DialogCancel"))
4  @Component
5  public class PetShopBatchUpdateAction {
6  }
```

Step2. 重启看效果。

重启看效果，如图 3-168 所示。

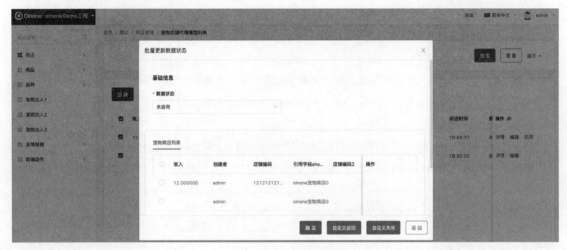

图 3-168　重启看效果

2）前端动作之组合动作（举例）

通过自定义 View 的 Template 来设置组合动作，同时学习自定义视图时如何设置需要展示 Action，还是以批量修改店铺状态弹出页面为例子，我们只展示一个组合动作按钮"组合动作"，它包含表单校验、提交、关闭并刷新主视图等动作。

Step1. 自定义弹出框 View。

（1）在 views/demo_core/template 路径下增加一个名为 pet_shop_batch_update_form.xml 的文件。

（2）再通过数据库查看默认页面定义，找到 base_view 表，过滤条件设置为 Model ='demo.PetShopBatchUpdate'，可以看到该模型下对应的所有 view，这些是系统根据该模型的 ViewAction 对应生成的默认视图，找到类型为"表单（type = FORM）"的记录，查看 template 字段，复制给 pet_shop_batch_update_form.xml 文件。

（3）把"<template slot="actions" autoFill="true"/>"替换成以下内容：

```
1  <template slot="actions" >
2      <action actionType="composition" label="组合动作">
3          <action actionType="client" name="$$internal_ValidateForm" />
4          <action name="conform" />
5          <action actionType="client" name="$$internal_GotoListTableRouter" />
6      </action>
7  </template>
```

Step2. 重启看效果。

这个页面只保留了"组合动作"一个按钮,其他没有配置的按钮就会隐藏掉,如图 3-169 所示。

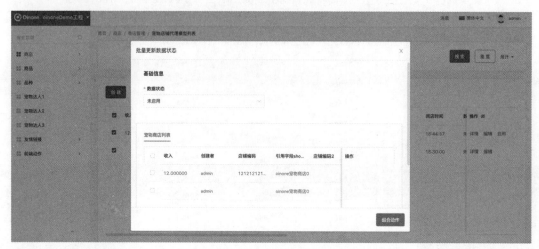

图 3-169　未配置的按钮均会被隐藏

3）平台默认前端动作

有元数据定义名的建议最好用元数据名,比如平台默认前端导入动作"\$\$internal_GotoListImportDialog"可以用"internalGotoListImportDialog"来替代,这样的好处是管理一致,比如权限等功能设置,代码如下。

```
1  <!-- <action name="$$internal_GotoListImportDialog" label="导入" /> 可以用下面方式替
       代,可以用于权限 -->
2  <action name="internalGotoListImportDialog" label="导入" />
```

平台默认前端动作见表 3-24。

表 3-24　平台默认前端动作

| name | 适用场景 | 功能描述 | 元数据定义名 |
| --- | --- | --- | --- |
| \$\$internal_GotoListTableRouter | 通用 | 返回上一个页面 | internalGotoListTableRouter |
| \$\$internal_DeleteOne | 关系字段表格 | 删除绑定的表格的选中数据 | internalDeleteOne |
| \$\$internal_DialogCancel | 弹窗 | 关闭当前弹窗 | |
| \$\$internal_ReloadData | 弹窗 | 刷新当前主视图组件 | |
| \$\$internal_ListInsertOneAndCloseDialog | 通用 | 打开一个创建弹窗 | |
| \$\$internal_GotoM2MListDialog | 多对多关系表格 | 打开 m2m 表格的创建弹窗 | |
| \$\$internal_GotoO2MCreateDialog | 一对多关系表格 | 打开 o2m 表格的创建弹窗 | |
| \$\$internal_GotoO2MEditDialog | 一对多关系表格 | 打开 o2m 表格的编辑弹窗 | |
| \$\$internal_ListInsertOneAndBackToList | 表单 | 校验表单数据,提交到后端,返回表格(新建) | |
| \$\$internal_ListUpdateOneAndBackToList | 表单 | 校验表单数据,提交到后端,返回表格(更新) | |
| \$\$internal_ValidateForm | 表单 | 校验当前表单 | |
| \$\$internal_GotoListExportDialog | 一般用于表格 | 数据导出的 action | internal_GotoListImportDialog |
| \$\$internal_GotoListImportDialog | 一般用于表格 | 数据导入的 action | internalGotoListImportDialog |

## 3.5.4 Ux 注解详解

我们默认视图已经基本可以用了，但实际业务中还是会有一些不大不小的自定义需求，写自定义视图又太麻烦，现在学习一种更加轻量的模式：后端研发可以通过注解来配置视觉交互。该系列注解以 Ux 开头，例如 @UxHomepage、@UxMenu、@UxAction、@UxView、@UxWidget 等。

视图 XML 的配置优先级大于代码上的注解，也就是代码上的注解影响的是默认展示逻辑。

### 1. Ux 家族图谱

我们先通过家族图谱做个简单了解（如图 3-170 所示），脑海里有一个印象当有需要的时候知道能不能做，深入了解还需要大家多多动手去尝试。

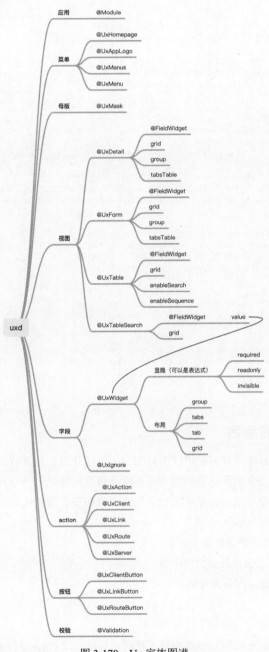

图 3-170　Ux 家族图谱

**2. 默认视图后端配置举例**

在下面的代码片段中 UxTable、UxForm、UxDetail、UxTableSearch 都有涉及，几个特殊点做些解释，其他的留大家自行测试，代码如下。

● Group 分组的配置逻辑：为了不让一个分组内的字段不断地写 Group，采取了第一个字段写了 Group，到下一个出现的 Group 之间的字段都自动归为一个 Group。

● 搜索整体不展示可以用 "@UxTable (enableSearch = false)" 配置在模型的类上。

● 字段搜索用 "UxTableSearch" 配置在模型的字段上，其特殊逻辑是只要你配了一个字段，系统就不自动补充了，例子中表格页的搜索栏只会留下店铺名称和店铺编码。

```
1    ……其他代码
2    //@UxTable(enableSearch = false),整体不支持搜索
3    public class PetShop extends AbstractDemoIdModel {
4        public static final String MODEL_MODEL="demo.PetShop";
5
6        @Field(displayName = "店铺编码")
7        @UxForm.FieldWidget(@UxWidget(group = "Form基础数据"))//Form分组
8        @UxTableSearch.FieldWidget(@UxWidget())//支持搜索
9        private String code;
10
11       @Field(displayName = "店铺编码2")
12       @Field.Sequence(sequence = "DATE_ORDERLY_SEQ",prefix = "C",size=6,step=1,initial = 10000,format = "yyyyMMdd")
13       private String codeTwo;
14
15       @UxTableSearch.FieldWidget(@UxWidget())//支持搜索
16       @UxTable.FieldWidget(@UxWidget(invisible = "true"))//表格中不展示支持搜索
17       @Field(displayName = "店铺名称",required = true,immutable=true)
18       private String shopName;
19
20       @Field(displayName = "一年内新店")
21       @UxForm.FieldWidget(@UxWidget(widget = "Switch",group = "Form基础数据"))//Switch,Checkbox可以切换着看,字段可选widget参考【字段的配置】一文
22       private Boolean oneYear;
23
24
25       @Field(displayName = "开店时间",required = true)
26       @UxDetail.FieldWidget(@UxWidget(invisible = "true"))//详情不展示
27       private Time openTime;
28
29       @Field(displayName = "闭店时间",required = true)
30       @UxDetail.FieldWidget(@UxWidget(invisible = "true"))//详情不展示
31       private Time closeTime;
32       …… 其他代码
33   }
```

### 3.5.5 设计器的结合

在页面开发的时候，直接通过前端组件和视图 xml 进行开发虽然开放性很大，但我们经常会忘记视图的配置属性，同时用 xml 配置的页面因为缺少设计数据，导致无法直接在设计器中复制，自定义页面得从头设计。今天就带大家一起来学习如何结合无代码设计器来完成页面开发，并把设计后的页面元数据装载为标准产品的一部分。

Step1. 安装 Docker（详情参见附录）。

如果没有 Docker 的话，请自行到官网下载。

Step2. 下载 Docker 镜像，并导入镜像。

（1）镜像下载。

① 镜像下载用户与密码，代码如下。

如需商业版镜像，需要加入 Oinone 商业版本伙伴专属群，向 Oinone 技术支持获取临时用户名与密

码,镜像会定时更新并通知大家。

```
1  docker login --username=cr_temp_user registry.cn-zhangjiakou.aliyuncs.com
```

② 前端镜像,代码如下。

```
1  docker pull registry.cn-zhangjiakou.aliyuncs.com/oinone/designer-frontend:3.0.2
2
```

③ 后端镜像,代码如下。

```
1  docker pull registry.cn-zhangjiakou.aliyuncs.com/oinone/designer-backend:3.0.2
```

(2) 本地结构说明。

下载结构包并解压(下载详情可见附录),如图 3-171 所示。

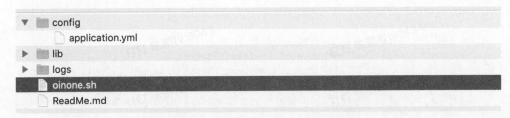

图 3-171　下载结构包 Oinone-op-ds 并解压

① config 是放 application.yml 的目录,可以在 application.yml 配置需要启动的自有模块同时修改对应其他中间件配置项。

② lib 是放自有模块的 jar 包以及其对应的依赖包,比如 pamirs-demo-api-1.0.0-SNAPSHOT.jar 和 pamirs-demo-core-1.0.0-SNAPSHOT.jar。

③ logs 是运行时系统日志目录。

Step3. application.yml 配置示例,代码如下。

application.yml 文件用我们 demo 工程的 application.yml 来替换同时增加对应配置(主要是合并 bootstrap.yml 和增加启动模块),注意点有:

(1) 要把 192.168.1.173 换成非 127.0.0.1 的机器分配 IP,通过 ifconfig(mac)或 ipconfig(windows)命令查询;

(2) 数据库换成自己 demo 工程的数据库,demo6_v3 --> demo;

(3) 增加 demo_core 模块。

```
1   dubbo:
2     #dubbo的配置
3   server:
4     address: 0.0.0.0
5     port: 8091
6     sessionTimeout: 3600
7   spring:
8     redis:
9     #redis的配置
10  logging:
11      #日志的配置
12  pamirs:
13    framework:
```

```yaml
14      system:
15        system-ds-key: base
16        system-models: base.WorkerNode
17      data:
18        default-ds-key: pamirs
19        ds-map:
20          base: base
21      gateway:
22        statistics: true
23        show-doc: true
24      meta:
25        dynamic: false
26    persistence:
27      global:
28        auto-create-database: true
29        auto-create-table: true
30    datasource:
31      pamirs:
32        driverClassName: com.mysql.cj.jdbc.Driver
33        type: com.alibaba.druid.pool.DruidDataSource
34        url: jdbc:mysql://127.0.0.1:3306/demo?useSSL=false&allowPublicKeyRetrieval=true&useServerPrepStmts=true&cachePrepStmts=true&useUnicode=true&characterEncoding=utf8&serverTimezone=Asia/Shanghai&autoReconnect=true&allowMultiQueries=true
35        username: root
36        password: oinone
37        #...........
38        asyncInit: true
39      #...........
40    boot:
41      init: true
42      sync: true
43      modules:
44        - base
45        #-其他module
46        - demo_core
47      tenants:
48        - pamirs
49    auth:
50        #权限过滤配置
51    eip:
52        #eip配置
```

**Step4. 检查中间件。**

我们前面中间件所绑定的 IP 都是 127.0.0.1，因为我们这里使用了 docker 来访问，需要让中间件支持真正机器分配 IP 访问。

**Step4.1. 检查 MySQL。**

确保 yml 文件中配置的用户可以通过机器 IP 来访问，例子中我们使用的是 root 用户，按以下步骤检查。

第一步：

```
1  mysql -u root -p
2  use mysql
```

第二步：

```
oinonedeMacBook-Pro:归档 oinone$ mysql -u root -p
Enter password:
Welcome to the MySQL monitor.  Commands end with ; or \g.
Your MySQL connection id is 12
Server version: 8.0.26 MySQL Community Server - GPL

Copyright (c) 2000, 2021, Oracle and/or its affiliates.

Oracle is a registered trademark of Oracle Corporation and/or its
affiliates. Other names may be trademarks of their respective
owners.

Type 'help;' or '\h' for help. Type '\c' to clear the current input statement.

mysql> use mysql
Reading table information for completion of table and column names
You can turn off this feature to get a quicker startup with -A

Database changed
```

第三步：

```
1  select User,authentication_string,Host from user;
```

第四步：

```
mysql> select User,authentication_string,Host from user
    -> ;
+------------------+------------------------------------------------------------------------+-----------+
| User             | authentication_string                                                  | Host      |
+------------------+------------------------------------------------------------------------+-----------+
| mysql.infoschema | $A$005$THISISACOMBINATIONOFINVALIDSALTANDPASSWORDTHATMUSTNEVERBRBEUSED  | localhost |
| mysql.session    | $A$005$THISISACOMBINATIONOFINVALIDSALTANDPASSWORDTHATMUSTNEVERBRBEUSED  | localhost |
| mysql.sys        | $A$005$THISISACOMBINATIONOFINVALIDSALTANDPASSWORDTHATMUSTNEVERBRBEUSED  | localhost |
| root             | *AEE671527C21CDDCAA7467CECCDBEF016A0D2007                              | localhost |
```

第五步：

```
1  update user set Host='%' where user='root';
2  flush privileges;
```

Step4.2. 检查 redis。

找到 redis 安装目录，编辑 redis.conf，bind 从 127.0.0.1 改成 0.0.0.0 或对应本机 IP，重启 redis，代码如下。

```
# IF YOU ARE SURE YOU WANT YOUR INSTANCE TO LISTEN TO ALL THE INTERFACES
# JUST COMMENT THE FOLLOWING LINE.
# ~~~~~~~~~~~~~~~~~~~~~~~~~~~~~~~~~~~~~~~~~~~~~~~~~~~~~~~~~~~~~~~~~~~~~~~~
bind 0.0.0.0

# Protected mode is a layer of security protection, in order to avoid that
```

如果 redis 访问有问题，可以尝试在启动命令中增加 "--protected-mode no" 参数：

```
1  nohup redis-server --protected-mode no &
```

Step4.3. 检查 RocketMq。

检查 broker 节点配置 IP，如果有配置不能用 127.0.0.1。

Step4.4. 检查 ZooKeeper。

如果本地搭集群方式需要检查 IP，如果有配置不能用 127.0.0.1。

Step5. 启动 docker。

启动 docker：

（1）打开 Oinone.sh 文件，修改脚本变量 configDir 和 Ip，代码如下。

```bash
#!/bin/bash
configDir=yml文件所在位置
Ip=本地ip非127.0.0.1

docker run -d --name designer-backend -p 8091:8091  -v $configDir/config/:/opt/pamirs/ext  -v $configDir/logs:/opt/pamirs/logs -v $configDir/lib:/opt/pamirs/outlib registry.cn-zhangjiakou.aliyuncs.com/oinone/designer-backend:3.0.2

docker run -d --name designer-frontend -p 80:80 --add-host localhost.oinone.top:$Ip registry.cn-zhangjiakou.aliyuncs.com/oinone/designer-frontend:3.0.2
```

（2）执行脚本，代码如下。

```
sh oinone.sh
```

Step6. 体验并设计页面。

输入 http://localhost/ 访问，通过 App Finder 切换时多了设计器一项，如图 3-172 所示。

图 3-172　通过 App Finder 切换时多了设计器一项

Step7. 导出元数据，并固化元数据到产品中。

Step7.1. 通过接口获取界面设计器设计页面的元数据。

（1）调用接口定义，代码如下。

```
mutation {
    uiDesignerExportReqMutation {
        export(data: { module: "demo_core" }) {
            jsonUrl
        }
    }
}
```

（2）返回结果，代码如下。

```
{
    "data": {
        "uiDesignerExportReqMutation": {
            "export": {
                "jsonUrl": "https://pamirs.oss-cn-hangzhou.aliyuncs.com/kubernetes/upload/kailas/lowcode/designer-pre/demo_corexxxx.json"
            }
        }
    }
}
```

jsonUrl 为导出结果，单击 ULR 下载导出文件。

Step7.2. 第三步元数据导入到项目中。

（1）pom 依赖，代码如下。

```
1    <dependency>
2        <groupId>pro.shushi.pamirs.designer</groupId>
3        <artifactId>pamirs-designer-install-common</artifactId>
4        <version>3.0.1-SNAPSHOT</version>
5    </dependency>
```

（2）将第二步下载后的文件放入项目中（注意文件放置的位置）。放置在工程的 resources 下面，如图 3-173 所示。

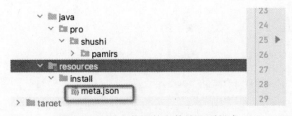

图 3-173　将下载后的文件放入项目中

（3）项目启动过程中，将文件中的数据导入（通常放在 core 模型的 init 包下面），代码如下。

```
12  @Slf4j
13  @Component
14  public class DemoAppInstall implements MetaDataEditor, SystemBootAfterInit {
15
16      //private static final String INSTALL_DATA_PATH = "install/meta.json";
17      private static final String INSTALL_DATA_PATH = "导出文件所在路径";
18
19      @Override
20      public void edit(AppLifecycleCommand command, Map<String, Meta> metaMap) {
21          log.info("开始安装demo应用-元数据");
22          InitializationUtil util = InitializationUtil.get(metaMap, DemoModule.MODU
    LE_MODULE/***改成自己的Module*/, DemoModule.MODULE_NAME/***改成自己的Module*/);
23          DesignerInstallHelper.mateInitialization(util, INSTALL_DATA_PATH);
24      }
25
26      @Override
27      public boolean init(AppLifecycleCommand command) {
28          //do nothing
29          return true;
30      }
31
32      @Override
33      public int priority() {
34          return 1;
35      }
36  }
```

### 3.5.6　DSL 配置

因为默认视图很难满足客户的个性化需求，所以日常开发中 view 的配置是避免不了的。本系列比较全面地介绍 View 配置的各个方面，涉及视图、字段、动作、布局等。

**1. 字段的配置**

1) 字段组件类型

字段类型可以配置哪些组件名？本节把 Oinone 平台默认支持的所有 widget 都进行了罗列，方便大家查阅。

2）字段组件匹配规则

字段组件没有严格地按组件名（widget）、字段类型（ttype）、视图组件类型（viewType）限定，而是按以下匹配规则：

① 按组件名最优先匹配。

② 按最大匹配原则。

字段类型、视图组件类型。每个属性权重一分。

③ 按后注册优先原则。

3）通用属性

字段通用属性如图 3-174 所示。

| 属性 | 属性描述 | 属性名称 | 默认值 | 类型 |
| --- | --- | --- | --- | --- |
| 标题 | 字段的标题名称 | label | — | string |
| 占位提示 | 一个字段的描述信息，常用于说明当前字典的范围、注意事项等 | placeholder | — | string |
| 描述说明 | 组件描述信息 | hint | — | string |
| 数据校验 | 与表单其他数据联动校验 | validator | — | 表达式 |
| 数据校验不通过提示 | 数据校验不通过提示 | validatorMessage | 校验失败 | string |
| 只读 | 字段的状态，可见，不可编辑 | readonly | false | boolean或者表达式 |
| 隐藏 | 字段的状态，不可见，不可编辑 | invisible | false | boolean或者表达式 |
| 必填 | 字段是否必填 | required | false | boolean或者表达式 |
| 禁用 | 字段是否禁用 | disabled | false | boolean或者表达式 |
| 宽度 | 属性在页面中的宽度 | colSpan | 01/02 | |
| 标题排列方式 | 标题排列方式：水平，横向 | layout | vertical | vertical \| horizontal |
| 默认值 | 默认值，多值逗号分隔 | defaultValue | - | 根据不同ttype有不同的默认值类型 |

图 3-174 字段通用属性

4）字段组件大全

组件名为"—"表明不需要指定，是该字段类型默认的组件名，见表 3-25。

表 3-25　字段组件大全

| 组件名称 | 组件名 | 对应字段类型 | 属性 | 属性描述 | 属性名称 | 默认值 | 类型 |
|---|---|---|---|---|---|---|---|
| 单行文本 | — | STRING | 通用属性 | | | | |
| | | | 文本类型 | 密码：password；文本：text | type | text | string |
| | | | 最小长度 | 输入框填写数据时最少输入的长度值 | minLength | — | number |
| | | | 最大长度 | 输入框填写数据时最多输入的长度值 | maxLength | — | number |
| | | | 输入格式 | 单行文本组件特有的属性，通过规则校验内容，提供一些常用的，也支持自定义校验正则 | pattern | — | 正则表达式 |
| | | | 输入格式不通过 | 输入格式不通过提示语 | tips | 校验失败 | string |
| | | | 显示计数器 | 设置输入框是否显示字数计数器 | showCount | false | boolean |
| | | | 显示清除按钮 | 设置输入框是否有一键清除的按钮及功能 | allowClear | true | true |
| | | | 支持前缀 | 开启前缀 | showPrefix | false | boolean |
| | | | 支持后缀 | 开启后缀 | showSuffix | false | boolean |
| | | | 前缀类型 | ICON：图标；TEXT：文本 | prefixType | — | string |
| | | | 后缀类型 | ICON：图标；TEXT：文本 | suffixType | — | string |
| | | | 前缀内容 | 文本内容或者图标引用名 | prefix | — | string |
| | | | 后缀内容 | 文本内容或者图标引用名 | suffix | — | string |
| | | | 前缀存储 | 前缀存储，仅前缀类型为文本可用 | prefixStore | — | boolean |
| | | | 后缀存储 | 后缀存储，仅后缀类型为文本可用 | suffixStore | — | boolean |
| 多行文本 | — | TEXT | 通用属性 | | | | |
| | | | 最小长度 | 输入框填写数据时最少输入的长度值 | minLength | — | number |
| | | | 最大长度 | 输入框填写数据时最多输入的长度值 | maxLength | — | number |
| | | | 显示计数器 | 设置输入框是否显示字数计数器 | showCount | false | boolean |
| | | | 显示清除按钮 | 设置输入框是否有一键清除的按钮及功能 | allowClear | true | true |

续表

| 组件名称 | 组件名 | 对应字段类型 | 属性 | 属性描述 | 属性名称 | 默认值 | 类型 |
|---|---|---|---|---|---|---|---|
| 整数 | — | INTEGER | 通用属性 | | | | |
| | | | 最大值 | 最大值 | max | — | number |
| | | | 最小值 | 最小值 | min | — | number |
| | | | 支持前缀 | 开启前缀 | showPrefix | false | boolean |
| | | | 支持后缀 | 开启后缀 | showSuffix | false | boolean |
| | | | 前缀类型 | ICON：图标；TEXT：文本 | prefixType | — | string |
| | | | 后缀类型 | ICON：图标；TEXT：文本 | suffixType | — | string |
| | | | 前缀内容 | 文本内容或者图标引用名 | prefix | — | string |
| | | | 后缀内容 | 文本内容或者图标引用名 | suffix | — | string |
| | | | 显示千分位 | 显示千分位 | showThousandth | false | boolean |
| | | | 显示清除按钮 | 设置输入框是否有一键清除的按钮及功能 | allowClear | true | true |
| 小数 | — | FLOAT | 通用属性 | | | | |
| | | | 最大值 | 最大值 | max | — | number |
| | | | 最小值 | 最小值 | min | — | number |
| | | | 支持前缀 | 开启前缀 | showPrefix | false | boolean |
| | | | 支持后缀 | 开启后缀 | showSuffix | false | boolean |
| | | | 前缀类型 | ICON：图标；TEXT：文本 | prefixType | — | string |
| | | | 后缀类型 | ICON：图标；TEXT：文本 | suffixType | — | string |
| | | | 前缀内容 | 文本内容或者图标引用名 | prefix | — | string |
| | | | 后缀内容 | 文本内容或者图标引用名 | suffix | — | string |
| | | | 显示千分位 | 显示千分位 | showThousandth | false | boolean |
| | | | 保留小数位数 | 保留小数位数 | precision | — | number |
| | | | 显示清除按钮 | 设置输入框是否有一键清除的按钮及功能 | allowClear | true | true |
| 下拉单选 | M2M、O2O、ENUM不需要配置BOOLEAN为Select | M2M O2O BOOLEAN ENUM | 通用属性 | | | | |
| | | | 选项配置 | 选项配置使用 &lt;option name="" displayName="" invisible=""&gt; 节点可配置多个选项，关系型下拉不支持 | options | — | |
| | | | 选项字段 | 关系型下拉选项展示标题，例如"activeRecord.name" | optionLabel | 模型数据标题 | 表达式 |

续表

| 组件名称 | 组件名 | 对应字段类型 | 属性 | 属性描述 | 属性名称 | 默认值 | 类型 |
|---|---|---|---|---|---|---|---|
| | | | 搜索字段 | 关系下拉搜索使用字段，例如"name, code"，字段名用逗号分隔 | searchFields | name | string |
| | | | 过滤条件 | 关系型下拉数据过滤，例如"activeRecord.name == '小明'" | domain | — | 表达式 |
| 下拉多选 | — | O2M、M2M、ENUM | 通用属性 | | | | |
| | | | 选项配置 | 选项配置使用<option name="" displayName="" invisible=""> 节点可配置多个选项，关系型下拉不支持 | options | — | |
| | | | 选项字段 | 关系型下拉选项展示标题，例如"activeRecord.name" | optionLabel | 模型数据标题 | 表达式 |
| | | | 搜索字段 | 关系下拉搜索使用字段，例如"name, code"，字段名用逗号分隔 | searchFields | name | string |
| | | | 过滤条件 | 关系型下拉数据过滤，例如"activeRecord.name == '小明'" | domain | — | 表达式 |
| | | | 最多选择个数 | 多选最多选择的个数 | maxNumber | — | number |
| | | | 最少选择个数 | 多选最少选择的个数 | minNumber | — | number |
| 开关 | — | BOOLEAN | 通用属性 | | | | |
| | | | 默认值 | 默认值 | defaultValue | false | boolean |
| 单选框 | Radio | BOOLEAN、ENUM | 通用属性 | | | | |
| | | | 选项配置 | 选项配置使用<option name="" displayName="" invisible=""> 节点可配置多个选项，关系型下拉不支持 | options | — | |
| | | | 排列方式 | 单选框的排列方式 | orientation | horizontal | vertical \| horizontal |

续表

| 组件名称 | 组件名 | 对应字段类型 | 属性 | 属性描述 | 属性名称 | 默认值 | 类型 |
|---|---|---|---|---|---|---|---|
| 复选框 | Checkbox | ENUM, multi=true | 通用属性 | | | | |
| | | | 选项配置 | 选项配置使用 <option name="" displayName="" invisible=""> 节点可配置多个选项，关系型下拉不支持 | options | — | — |
| | | | 排列方式 | 单选框的排列方式 | orientation | horizontal | vertical \| horizontal |
| 富文本 | — | HTML | 通用属性 | | | | |
| 年份 | — | Year | 通用属性 | | | | |
| 日期 | — | Date | 通用属性 | | | | |
| | | | 日期格式 | CHINESE: 2019 年 04 月 06 日<br>CHINESE_YEAR_MONTH: 2019 年 04 月<br>HYPHEN_YEAR_MONTH: 2019—04<br>SLASH_YEAR_MONTH: 2019/04<br>HYPHEN: 2019—04—06<br>SLASH: 2019/04/06 | dateFormat | CHINESE | CHINESE CHINESE_YEAR_MONTH HYPHEN_YEAR_MONTH SLASH_YEAR_MONTH HYPHEN SLASH |
| 日期时间 | — | DateTime | 通用属性 | | | | |
| | | | 时间格式 | COLON_NORMAL: HH:mm:ss<br>COLON_SHORT: HH:mm<br>AP_COLON_NORMAL: A hh:mm:ss<br>AP_COLON_SHORT: A hh:mm | timeFormat | COLON_NORMAL | COLON_NORMAL COLON_SHORT AP_COLON_NORMAL AP_COLON_SHORT |
| | | | 日期格式 | CHINESE: 2019 年 04 月 06 日<br>CHINESE_YEAR_MONTH: 2019 年 04 月<br>HYPHEN_YEAR_MONTH: 2019—04<br>SLASH_YEAR_MONTH: 2019/04<br>HYPHEN: 2019—04—06<br>SLASH: 2019/04/06 | dateFormat | CHINESE | CHINESE CHINESE_YEAR_MONTH HYPHEN_YEAR_MONTH SLASH_YEAR_MONTH HYPHEN SLASH |

续表

| 组件名称 | 组件名 | 对应字段类型 | 属性 | 属性描述 | 属性名称 | 默认值 | 类型 |
|---|---|---|---|---|---|---|---|
| 时间 | — | Time | 通用属性 | | | | |
| | | | 时间格式 | COLON_NORMAL: HH:mm:ss COLON_SHORT: HH:mm AP_COLON_NORMAL: A hh:mm:ss AP_COLON_SHORT: A hh:mm | timeFormat | COLON_NORMAL | COLON_NORMAL COLON_SHORT AP_COLON_NORMAL AP_COLON_SHORT |
| 颜色选择器 | ColorPicker | STRING | 通用属性 | | | | |
| 文件上传 | Upload | STRING multi=true | 通用属性 | | | | |
| | | | 最大上传文件个数 | 最大上传文件个数 | limit | — | number |
| | | | 最大上传文件体积 | 最大上传文件体积，单位为 MB | limitSize | — | number |
| | | | 限制上传文件类型 | 限制上传文件类型，例如"jpg,mp4"，用逗号分隔 | limitFileType | — | string |
| 图片上传 | UploadImg | STRING multi=true | 通用属性 | | | | |
| | | | 最大上传文件个数 | 最大上传文件个数 | limit | — | number |
| | | | 最大上传文件体积 | 最大上传文件体积，单位为 MB | limitSize | — | number |
| | | | 限制上传文件类型 | 限制上传文件类型，例如"jpg,mp4"，逗号分隔 | limitFileType | — | string |
| 标签 | — | STRING multi=true | 数量限制 | 限制输入标签的个数 | limit | — | number |
| 表单 | Form | M2O、O2O | 通用属性 | | | | |
| | | | 标题排列方式 | 标题排列方式：水平，横向，控制整个表单的排列方式 | layout | vertical | vertical \| horizontal |
| 表格 | Table | O2M、M2M | 通用属性 | | | | |
| | | | 操作列显示数量 | 表格操作列显示数量，1~5 可选 | inlineActiveCount | 3 | 1, 2, 3, 4, 5 |
| | | | 默认分页条数 | 表格分页默认分页条数 | defaultPageSize | 30 | 10, 15, 30, 50, 100, 200 |
| Iframe 网页 | Iframe | STRING | 通用属性 | | | | |
| 超链接 | Hyperlinks | STRING | 通用属性 | | | | |
| | | | 链接文字 | 超链接的展示标题 | | | |

续表

| 组件名称 | 组件名 | 对应字段类型 | 属性 | 属性描述 | 属性名称 | 默认值 | 类型 |
|---|---|---|---|---|---|---|---|
| 时间范围 | — | Year、Date、DateTime、Time | 通用属性 | | | | |
| 货币 | — | CURRENCY | 最大值 | 最大值 | max | — | number |
| | | | 最小值 | 最小值 | min | — | number |
| | | | 支持前缀 | 开启前缀 | showPrefix | false | boolean |
| | | | 支持后缀 | 开启后缀 | showSuffix | false | boolean |
| | | | 前缀类型 | ICON：图标；TEXT：文本 | prefixType | — | string |
| | | | 后缀类型 | ICON：图标；TEXT：文本 | suffixType | — | string |
| | | | 前缀内容 | 文本内容或者图标引用名 | prefix | — | string |
| | | | 后缀内容 | 文本内容或者图标引用名 | suffix | — | string |
| | | | 显示千分位 | 显示千分位 | showThousandth | false | boolean |
| | | | 保留小数位数 | 保留小数位数 | precision | — | number |
| | | | 显示清除按钮 | 设置输入框是否有一键清除的按钮及功能 | allowClear | true | true |
| 手机号 | — | PHONE | 最小长度 | 输入框填写数据时最少输入的长度值 | minLength | — | number |
| | | | 最大长度 | 输入框填写数据时最多输入的长度值 | maxLength | — | number |
| | | | 输入格式 | 单行文本组件特有的属性，通过规则校验内容，提供一些常用的，也支持自定义校验正则 | pattern | — | 正则表达式 |
| | | | 输入格式不通过 | 输入格式不通过提示语 | tips | 校验失败 | string |
| | | | 显示计数器 | 设置输入框是否显示字数计数器 | showCount | false | boolean |
| | | | 显示清除按钮 | 设置输入框是否有一键清除的按钮及功能 | allowClear | true | true |
| | | | 支持前缀 | 开启前缀 | showPrefix | false | boolean |
| | | | 支持后缀 | 开启后缀 | showSuffix | false | boolean |
| | | | 前缀类型 | ICON：图标；TEXT：文本 | prefixType | — | string |
| | | | 后缀类型 | ICON：图标；TEXT：文本 | suffixType | — | string |
| | | | 前缀内容 | 文本内容或者图标引用名 | prefix | — | string |

续表

| 组件名称 | 组件名 | 对应字段类型 | 属性 | 属性描述 | 属性名称 | 默认值 | 类型 |
|---|---|---|---|---|---|---|---|
| 手机号 | — | PHONE | 后缀内容 | 文本内容或者图标引用名 | suffix | — | string |
| | | | 前缀存储 | 前缀存储，仅前缀类型为文本可用 | prefixStore | — | boolean |
| | | | 后缀存储 | 后缀存储，仅后缀类型为文本可用 | suffixStore | — | boolean |
| 邮箱 | — | EMAIL | 最小长度 | 输入框填写数据时最少输入的长度值 | minLength | — | number |
| | | | 最大长度 | 输入框填写数据时最多输入的长度值 | maxLength | — | number |
| | | | 输入格式 | 单行文本组件特有的属性，通过规则校验内容，提供一些常用的，也支持自定义校验正则 | pattern | — | 正则表达式 |
| | | | 输入格式不通过 | 输入格式不通过提示语 | tips | 校验失败 | string |
| | | | 显示计数器 | 设置输入框是否显示字数计数器 | showCount | false | boolean |
| | | | 显示清除按钮 | 设置输入框是否有一键清除的按钮及功能 | allowClear | true | true |
| | | | 支持前缀 | 开启前缀 | showPrefix | false | boolean |
| | | | 支持后缀 | 开启后缀 | showSuffix | false | boolean |
| | | | 前缀类型 | ICON：图标；TEXT：文本 | prefixType | — | string |
| | | | 后缀类型 | ICON：图标；TEXT：文本 | suffixType | — | string |
| | | | 前缀内容 | 文本内容或者图标引用名 | prefix | — | string |
| | | | 后缀内容 | 文本内容或者图标引用名 | suffix | — | string |
| | | | 前缀存储 | 前缀存储，仅前缀类型为文本可用 | prefixStore | — | boolean |
| | | | 后缀存储 | 后缀存储，仅后缀类型为文本可用 | suffixStore | — | boolean |

5）字段可选的系统组件名

字段可选系统组件名见表 3-26。

表 3-26　字段可选系统组件名

| 组件名称<br>* 代表统配组件名 | 支持字段类型 | 支持视图类型<br>* 代表统配视图组件类型 | 说明 |
|---|---|---|---|
| Table | | | |
| *<br>[TableO2MFieldWidget] | OneToMany | ViewType.Table | 表格默认 o2m 组件 |

续表

| 组件名称<br>* 代表统配组件名 | 支持字段类型 | 支持视图类型<br>* 代表统配视图组件类型 | 说明 |
|---|---|---|---|
| * | ManyToMany | ViewType.Table | 表格默认 m2m 组件 |
| UploadImg<br>[TableM2MUploadImgFieldWidget] | ManyToMany | ViewType.Table | 图片上传<br>往往用于跟 PamirsFile 对象关联的时候 |
| UploadImg<br>TableM2OUploadImgFieldWidget | ManyToOne | ViewType.Table | 图片上传<br>往往用于跟 PamirsFile 对象关联的时候 |
| Image<br>[TableStringImageFieldWidget] | String | ViewType.Table | 把字段值当 url 展示图片 |
| *<br>[TableBooleanFieldWidget] | Boolean | ViewType.Table | 表格默认的 Boolean 组件 |
| *<br>[TableEnumFieldWidget] | Enum | ViewType.Table | 表格默认的 Enum 组件 |
| *<br>[TableFloatFieldWidget] | Float | ViewType.Table | 表格默认的 Float 组件 |
| *<br>[TableIntegerFieldWidget] | Integer | ViewType.Table | 表格默认的 Integer 组件 |
| *<br>[TableLongFieldWidget] | Long | ViewType.Table | 表格默认的 Long 组件 |
| *<br>[TableStringFieldWidget] | String | ViewType.Table | 表格默认的 String 组件 |
| *<br>[TableYearFieldWidget] | Year | ViewType.Table | 表格默认的 Year 组件 |
| Search | | | |
| *<br>[SearchBooleanSelectFieldWidget] | Boolean | ViewType.Search | 搜索默认的 Boolean 组件 |
| *<br>[SearchRangeDateTimeFieldWidget] | DateTime | ViewType.Search | 搜索默认的 DateTime 组件 |
| Form | | | |
| Checkbox<br>[FormBooleanCheckboxFieldWidget] | Boolean | ViewType.Form,<br>ViewType.Search,<br>ViewType.Detail | |
| *<br>[FormBooleanFieldWidget] | Boolean | ViewType.Form, ViewType.Search, ViewType.Detail | Form\Search\Detail 默认的 Boolean 组件，Radio 形式 |
| Switch | Boolean | ViewType.Form, ViewType.Search, ViewType.Detail | |
| *<br>[FormMoneyFieldWidget] | Currency | ViewType.Form, ViewType.Search, ViewType.Detail | Form\Search\Detail 默认的 Currency 组件 |
| *<br>[FormDateFieldWidget] | Date | ViewType.Form, ViewType.Detail | Form\Detail 默认的 Date 组件 |
| *<br>[FormDateTimeFieldWidget] | DateTime | ViewType.Form, ViewType.Detail | Form\Detail 默认的 Date 组件 |

续表

| 组件名称<br>* 代表统配组件名 | 支持字段类型 | 支持视图类型<br>* 代表统配视图组件类型 | 说明 |
|---|---|---|---|
| RangeDatePickerTime | DateTime | ViewType.Form, ViewType.Detail | |
| *<br>[FormEnumFieldWidget] | Enum | ViewType.Form, ViewType.Search, ViewType.Detail | Form\Search\Detail 默认的 Enum 组件 |
| *<br>[FormFloatFieldWidget] | Float | ViewType.Form, ViewType.Search, ViewType.Detail | Form\Search\Detail 默认的 Float 组件 |
| RichText | HTML | * | 所有 viewType 的 HTML 字段都可以配 RichText |
| *<br>[FormIntegerFieldWidget] | Integer | ViewType.Form, ViewType.Search, ViewType.Detail | Form\Search\Detail 默认的 Integer\Currency 组件 |
| Select | ManyToMany | ViewType.Form, ViewType.Search | |
| *<br>[FormM2MTableFieldWidget] | ManyToMany | ViewType.Form, ViewType.Detail | Form\Detail 默认的 ManyToMany 组件 |
| Upload | ManyToMany | ViewType.Form, ViewType.Detail | |
| UploadImg | ManyToMany | ViewType.Form, ViewType.Detail | |
| Form<br>[FormM2OFormFieldWidget] | ManyToOne | ViewType.Form, ViewType.Detail | |
| Select、* | ManyToOne | ViewType.Form, ViewType.Search, ViewType.Detail, ViewType.Table | Form\Search\Detail\Table 默认的 ManyToOne 组件 |
| SSConstructSelect<br>[FormM2OConstructSelectFieldWidget | ManyToOne | ViewType.Form, ViewType.Search, ViewType.Detail | |
| Upload<br>[FormM2OUploadFieldWidget] | ManyToOne | ViewType.Form, ViewType.Detail | |
| UploadImg | ManyToOne | ViewType.Form, ViewType.Detail | |
| *<br>[TableM2OFieldWidget] | ManyToOne | ViewType.Table | Table 默认的 ManyToOne 组件 |
| *<br>FormMapFieldFormFieldWidget | Map | ViewType.Form, ViewType.Detail | Form\Detail 默认的 Map 组件 |
| Select<br>[FormO2MFieldSelectWidget] | OneToMany | ViewType.Form, ViewType.Search | |
| Table<br>[FormO2MFieldWidget] | OneToMany | ViewType.Form, ViewType.Detail | |
| *<br>[FormRelatedFieldWidget] | Related | ViewType.Form, ViewType.Search, ViewType.Detail | Form\Search\Detail 默认的 Related 组件 |
| Email | String | ViewType.Form, ViewType.Search, ViewType.Detail | |
| *<br>[FormStringFieldWidget] | String | ViewType.Form, ViewType.Search, ViewType.Detail | |

| 组件名称<br>* 代表统配组件名 | 支持字段类型 | 支持视图类型<br>* 代表统配视图组件类型 | 说明 |
| --- | --- | --- | --- |
| Password | String | ViewType.Form, ViewType.Search, ViewType.Detail | |
| Phone | String | ViewType.Form, ViewType.Search, ViewType.Detail | |
| VerificationCode | String | ViewType.Form | |
| DomainExpGenerator | Text | ViewType.Form | |
| *<br>[FormTextFieldWidget] | Text | ViewType.Form, ViewType.Search, ViewType.Detail | Form\Search\Detail 默认的 Text 组件 |
| *<br>[FormTimeFieldWidget] | Time | ViewType.Form, ViewType.Detail | |
| *<br>[FormYearPickerControlWidget] | Year | ViewType.Form, ViewType.Search | |
| Detail | | | |
| *<br>[DetailBooleanFieldWidget] | Boolean | ViewType.Detail | Detail 默认的 Boolean 组件 |
| *<br>[DetailCurrencyFieldWidget] | Currency | ViewType.Detail | Detail 默认的 Currency 组件 |
| *<br>[DetailDateTimeFieldWidget] | DateTime | ViewType.Detail | Detail 默认的 DateTime 组件 |
| *<br>[DetailEnumFieldWidget] | Enum | ViewType.Detail | Detail 默认的 Enum 组件 |
| *<br>[DetailFloatFieldWidget] | Float | ViewType.Detail | Detail 默认的 Float 组件 |
| *<br>[DetailHtmlFieldWidget] | HTML | ViewType.Detail | Detail 默认的 HTML 组件 |
| *<br>[DetailIdFieldWidget] | ID | ViewType.Detail | Detail 默认的 ID 组件 |
| *<br>[DetailIntegerFieldWidget] | Integer | ViewType.Detail | Detail 默认的 Integer 组件 |
| Select<br>[DetailM2MSelectFieldWidget] | ManyToMany | ViewType.Detail | |
| UploadImg | ManyToMany | ViewType.Detail | |
| Select | OneToMany | ViewType.Detail | |
| *<br>[DetailStringFieldWidget] | String | ViewType.Detail | Detail 默认的 String 组件 |
| *<br>[DetailTextFieldWidget] | Text | ViewType.Detail | Detail 默认的 Text 组件 |
| *<br>[DetailTimeFieldWidget] | Time | ViewType.Detail | Detail 默认的 Time 组件 |
| *<br>[DetailYearFieldWidget] | Year | ViewType.Detail | Detail 默认的 Year 组件 |

6）字段属性配置

（1）常规字段 field 通用属性。

Table 字段通用属性见表 3-27。

表 3-27　Table 字段通用属性

| 配置项 | 可选值 | 默认值 | 作用 |
| --- | --- | --- | --- |
| label | string | displayName | 展示中文名称 |
| widget | string | | 见上方的 widget，字段组件类型 |
| independentlyEditable | true，false | false | 是否启用单字段编辑 |

Form 字段通用属性见表 3-28。

表 3-28　Form 字段通用属性

| 配置项 | 可选值 | 默认值 | 作用 |
| --- | --- | --- | --- |
| 基础校验 | | | |
| pattern | | | 前端正则校验 |
| maxValue | | | 最大值 |
| minValue | | | 最小值 |
| maxLength | | | 最大长度 |
| minLength | | | 最小长度 |
| required | true，false，string | false | 是否必填 |
| 字段状态：自读、是否可见 | | | |
| invisible | true，false，string | false | 是否隐藏 |
| readonly | true，false，string | false | 是否只读 |
| 字段数据处理：数据源过滤、计算、默认值 | | | |
| defaultValue | * | | 前端默认值 |
| domain | string | | rsql 表达式 |
| compute | string | | 计算值 |
| 其他配置 | | | |
| label | string | displayName | 展示中文名称<br>false 为不展示 label |
| hint | string | | 一些字段的说明性文字 |
| widget | string | | 见上方的 widget，字段组件类型 |
| colSpan | number | 1 | 字段在表单一行中所占比例<br>具体看布局的配置 |
| mutil | true，false | list 的默认 true<br>其他为 false | 是否支持多值<br>针对枚举、o2m、m2m、m2o |

关系字段扩展配置：关系字段展示的形态更多，在通用和 form 的配置基础上，我们也增加了一些场景的配置形态，满足更多的业务场景，见表 3-29。

表 3-29 关系字段扩展配置

| 配置项 | 可选值 | 默认值 | 作用 |
|---|---|---|---|
| Select 类组件配置 | | | |
| labelField | string | | 展示字段以 separator 连接 |
| separator | string | | |
| searchField | string | | 搜索所用的字段 |

搜索字段通用属性见表 3-30。

表 3-30 搜索字段通用属性

| 配置项 | 可选值 | 默认值 | 作用 |
|---|---|---|---|
| mutil | true, false | list 默认为 true；其他为 false | 是否支持多值 针对枚举、o2m、m2m、m2o |
| label | string | displayName | 展示中文名称 false 为不展示 label |
| domain | string | | rsql 表达式 |
| blank | true, false | false | 是否加入自选搜索字段 |
| operator | 见 "rsql 操作符" | 根据 ttype 默认推断 | 搜索 filter 特有 |

**举例 mutil 和 operator**

我们经常会碰到 mutil 和 operator 两种场景，它们非常有用：

① mutil：单选的枚举值要在搜索框中选择多个值进行过滤，如订单状态枚举，我们在搜索已下单未支付和已支付未发货的两种状态时就非常有必要；

② operator：默认 ttype 为 String，搜索的时候都是以 like 为操作符，但对于一些数据量比较大的表，我们希望使用 = 或者单边 like（注：默认形式为 like="%××%"，单边 like="××%" 或者 like="%××"。）就可以派上用场了。

Step1. 新增 PetTalentSexEnum 枚举类，代码如下。

```
 6    @Dict(dictionary = PetTalentSexEnum.DICTIONARY,displayName = "萌猫体型")
 7    public class PetTalentSexEnum extends BaseEnum<PetTalentSexEnum,Integer> {
 8
 9        public static final String DICTIONARY ="demo.PetTalentSexEnum";
10
11        public final static PetTalentSexEnum MAN =create("MAN",1,"男","男");
12        public final static PetTalentSexEnum FEMAL =create("FEMAL",2,"女","女");
13    }
14
```

Step2. 新增 PetFile 模型，代码如下。

```
 6    @Model.model(PetFile.MODEL_MODEL)
 7    @Model(displayName = "文件",summary="文件",labelFields = {"url"})
 8    public class PetFile extends AbstractDemoIdModel{
 9        public static final String MODEL_MODEL="demo.PetFile";
10
11        @Field(displayName = "图片路径")
12        private String url;
13
14    }
15
```

Step3. 修改 PetTalent 模型，代码如下。

```java
10    @Model.model(PetTalent.MODEL_MODEL)
11    @Model(displayName = "宠物达人",summary="宠物达人")
12 ▸  public class PetTalent extends AbstractDemoIdModel{
13        public static final String MODEL_MODEL="demo.PetTalent";
14
15        @Field(displayName = "达人")
16        private String name;
17
18        @Field.one2many
19        @Field(displayName = "达人图片")
20        private List<PetFile> picList;
21
22        @Field.many2many(relationFields = {"petTalentId"},referenceFields = {"petShopId"},throughClass =PetShopRelPetTalent.class)
23        @Field(displayName = "推荐宠物商店")
24        private List<PetShop> petShops;
25
26        @Field.Enum
27        @Field(displayName = "性别")
28        private PetTalentSexEnum petTalentSex;
29
30        @Field.many2one
31        @Field(displayName = "创建者",required = true)
32        @Field.Relation(relationFields = {"createUid"},referenceFields = {"id"})
33        private PamirsUser creater;
34
35    }
36
```

Step4. 修改"宠物达人"表格视图 Template 中的 search 部分，代码如下。

```xml
1
2 ▸  <template slot="search"  cols="4">
3        <field data="name" label="达人" operator="==" />
4        <field data="petTalentSex" multi="true"/>
5        <field data="petShops" />
6 ▸      <field data="dataStatus" label="数据状态" multi="true">
7 ▸          <options>
8                <option name="DRAFT" displayName="草稿" value="DRAFT" state="ACTIVE"/>
9                <option name="NOT_ENABLED" displayName="未启用" value="NOT_ENABLED" state="ACTIVE"/>
10               <option name="ENABLED" displayName="已启用" value="ENABLED" state="ACTIVE"/>
11               <option name="DISABLED" displayName="已禁用" value="DISABLED" state="ACTIVE"/>
12           </options>
13       </field>
14       <field data="createDate" label="创建时间"/>
15   </template>
16
```

Step5. 重启看效果，如图 3-175 所示。

我们看到发起的请求中 name 变成了 ==，而性别可以下拉多选并以 or 拼接。在 3.5.3 节中"ViewAction 高级参数 filter 和 domain（举例）"的宠物达人 2 菜单针对 name 配置 domain = "name =like=' 老 '" 将无效。

图 3-175　查看宠物达人 2 菜单

（2）视图中字段间的联动。

字段联动需求主要在于：数据处理、展示处理和数据校验等字段间相互动态作用，其他复杂的联动需要自定义字段的 widget，如平台提供的 SSConstructSelect。

数据处理：通过字段的数据处理相关属性配置来完成如 domain、compute。

展示处理：通过字段的数据展示相关属性配置来完成如 invisible、readonly。

数据校验：通过字段的基础校验相关属性配置来完成如 required。

domain 的举例：我们经常会在 o2m 和 m2m 的 filter 或 field 中设置 domain 来过滤数据，但是如何在设置过滤条件时用上其他字段的值呢？在 4.1.19 节中我们在过滤条件中用上了后端占位符，这个在日常开发中也很有用。下面给出"过滤条件时用上其他字段的值"的例子。

Step1. 修改"宠物达人"表格视图 Template 中的 search 部分。

将 petShops 字段的 domain 设置成 createUid == ${activeRecord.creater.id}，表示该字段的可选范围取决于 creater 字段，activeRecord 为前端内置关键字，获取当前操作记录，代码如下。

```
1  <template slot="search" cols="4">
2    <field data="name" label="达人" operator="==" />
3    <field data="petTalentSex" multi="true" label="达人性别"/>
4    <field data="creater" />
5    <field data="petShops" label="宠物商店" domain="createUid == ${activeRecord.creater.id}"/>
6    <field data="dataStatus" label="数据状态" multi="true">
7      <options>
8        <option name="DRAFT" displayName="草稿" value="DRAFT" state="ACTIVE"/>
9        <option name="NOT_ENABLED" displayName="未启用" value="NOT_ENABLED" state="ACTIVE"/>
10       <option name="ENABLED" displayName="已启用" value="ENABLED" state="ACTIVE"/>
11       <option name="DISABLED" displayName="已禁用" value="DISABLED" state="ACTIVE"/>
12     </options>
13   </field>
14   <field data="createDate" label="创建时间"/>
15 </template>
```

Step2. 重启看效果，如图 3-176、图 3-177 所示。

图 3-176　查看"宠物达人"列表页（1）

图 3-177　查看"宠物达人"列表页（2）

compute 的举例：字段的值通过 compute 计算而来。

Step1. 为 PetTalent 增加一个 nick 字段，代码如下。

```
1  @Field(displayName = "昵称")
2  private String nick;
```

Step2. 修改 PetTalent 的 Form 视图增加下面代码，代码如下。

```
1  <field data="nick" compute="activeRecord.name"/>
```

Step3. 重启看效果。

我们发现昵称的值会跟着达人字段变化而变化，如图 3-178 所示。

图 3-178　昵称的值会跟着达人字段变化而变化

invisible、readonly、required 的配置方式均与 compute 类似，感兴趣的读者可自行体验。

（3）通过自定义组件进行联动（SSConstructSelect）。

目前对于复杂的联动需要自定义 widget，在值变化时提交数据到后端，然后将后端返回值对其他字段的赋值操作。

这里举例 SSConstructSelect 组件，它在值变化的时候会调用后端 constructMirror 的函数，并把对象返回的值进行赋值。

Step1. 新增 PetTalentAction 类，并增加一个 constructMirror 函数，代码如下。

```
10     @Model.model(PetTalent.MODEL_MODEL)
11     @Component
12     public class PetTalentAction {
13         @Function(openLevel = FunctionOpenEnum.API)
14         @Function.Advanced(type= FunctionTypeEnum.QUERY)
15         public PetTalent constructMirror(PetTalent data){
16             return data.setName("oinone");
17         }
18     }
19
```

Step2. 在 PetTalent 的 form 视图指定 widget，代码如下。

widget "SSConstructSelect" 有两个属性配置：

① submitFields 提交后端请求时会带上字段列表，默认为：当前配置字段。

② responseFields 请求返回值后会影响的字段列表，默认为：影响所有字段。

```
1  <field data="creater" widget="SSConstructSelect" submitFields="creater,name" responseFields="name"/>
```

Step3. 重启看效果。

选择创建者，后端会把达人改成 Oinone，而昵称又通过前端 compute 计算为 Oinone，如图 3-179 所示。

图 3-179　查看展示效果

（4）前端上下文关键字。

前端上下文关键字见表 3-31。

表 3-31　前端上下文关键字

| 关键字 | 说明 | 举例说明 |
| --- | --- | --- |
| activeRecord | 当前对象 | 字段间联动之 domain 的举例 |
| rootRecord | 根对象 | 在视图嵌套的情况下，子视图需要用到父视图数据，则可以通过 rootRecord 来获取字段数据 |
| openerRecord | 打开者对象 | 在弹出框时需要用到打开者对象数据时，则可以用 openerRecord 来获取父窗口对象数据 |
| scene | 当前页面的 viewAction.name | |

**2. 视图的配置**

视图的大致配置我们在 3.5.2 节已经介绍过，这里主要介绍视图层的基本属性配置，这些配置会透传给视图内的组件 Widget，组件会根据配置内容做出不同的呈现样式。

1）视图的配置

（1）Table 的配置见表 3-32。

表 3-32　Table 的配置

| 配置项 | 可选值 | 默认值 | 作用 |
| --- | --- | --- | --- |
| activeCount | number | 5 | 表格上方动作区默认展示操作的数量，超过个数的操作将被折叠收起 |
| inlineActiveCount | number | 3 | 表格最右侧操作列默认展示操作的数量，超过个数的操作将被折叠收起 |
| defaultPageSize | number | 30 | 表格默认分页条数 |

（2）Form/Detail 的配置见表 3-33。

表 3-33　Form/Detail 的配置

| 配置项 | 可选值 | 默认值 | 作用 |
|---|---|---|---|
| direction | horizontal/vertical（大小写不明感） | vertical | 表单标题排列方式 |

2）Table 的配置项举例

Step1. 修改宠物达人的表格视图。

在宠物达人的自定义表格视图的 Template 文件中增加三个属性配置 activeCount="1"、inlineActiveCount="1"、defaultPageSize="1"，代码如下。

```
1  <view name="tableView" model="demo.PetTalent" cols="1" activeCount="1" inlineActiv
    eCount="1" defaultPageSize="1"  type="TABLE"  enableSequence="true"  >
2  </view>
```

Step2. 重启看效果，如图 3-180 所示。

图 3-180　重启看效果

3）Form 的配置举例

Step1. 修改宠物达人的表单视图。

在宠物达人的自定义表格视图的 Template 文件中增加一个属性配置 direction = "horizontal"，代码如下。

宠物达人此前增加了一些字段，利用默认视图把新增字段也展示出来。还是通过数据库查看默认页面定义，找到 base_view 表，过滤条件设置为 Model ='demo.PetTalent' and name ='formView'，查看 template 字段，把里面涉及新增字段复制到 pet_talent_form.xml 文件中。

```
1  <view name="formView1" model="demo.PetTalent" cols="2" type="FORM" priority="1"
     direction = "horizontal">
2  </view>
```

Step2. 重启看效果，如图 3-181 所示。

图 3-181　重启看效果

**3. 布局的配置**

布局是将页面拆分成一个一个的小单元，按照上下中左右进行排列。

1）前端布局

在前端领域中，布局可以分为 Float、Flex、Grid 三大块，Float 可以说得上是上古时期的布局了，如今市面已经很少见了，除了一些古老的网站。

目前，平台主要支持通过配置 XML 的 cols 和 span 进行布局。平台也同样支持自由布局，合理地使用 row、col、containers 和 container 四个布局容器相关组件，可以实现各种类型的布局样式，换句话说，平台实现的自由布局功能是 Flex 和 Grid 的结合体。

这里主要讲解 Flex 和 Grid 布局，以及目前新的模板布局实现的思路。

2）Flex 布局

Flex 布局采用的是一维布局，所谓的一维布局就是只有一个方向，没有体积、面积，比如一条直线，它适合做局部布局，就像我们原来的顶部菜单、面包屑导航，以及现在的主视图字段配置，如图 3-182、图 3-183 和图 3-184 所示。

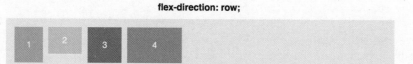

图 3-182　Flex 布局（1）

图 3-183　Flex 布局（2）

图 3-184　Flex 布局（3）

从上图可以看看出，Flex 布局只能在 $X$ 轴和 $Y$ 轴进行转换，它无法对上下左右四个方向同时处理，因为它没"面积"的概念。所以它最适合做局部布局。

（1）优点。

可以看出，目前主流的浏览器都支持 Flex 布局，所以 Flex 兼容性强，如果你想对局部做布局处理，Flex 是最好选择，如图 3-185 所示。

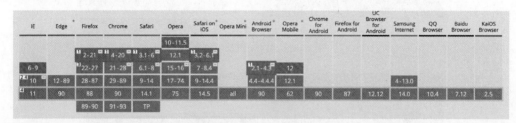

图 3-185　Flex 兼容性

（2）缺陷。

刚刚也提到了，用户想要的布局是千奇百怪的，如果他想要的布局在现有的功能中无法实现怎么办？让用户放弃？还是说服他使用现在的布局？

3）Grid 布局

Grid 布局系统采用的是二维布局，二维布局有四个方向：上、下、左、右。它只有面积没有体积。

从图 3-186、图 3-187 及图 3-188 中的代码可以看出：Grid 天然地支持整体布局，它甚至可以移行换位，它的强大之处可以颠覆你对布局的认知。

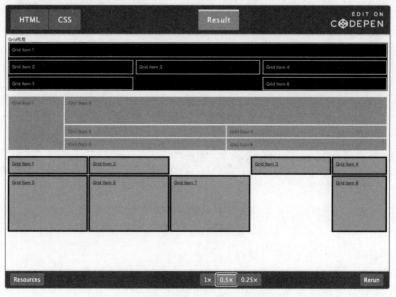

图 3-186　Grid 布局概览

图 3-187　HTML 模式布局

图 3-188　CSS 模式布局

虽然 Grid 很强大，但是它有一个致命的缺陷，那就是兼容性。

从图 3-189 可以看出，IE10 以下、Chrome56 以下、Firefox51 以下等等都不支持该属性。

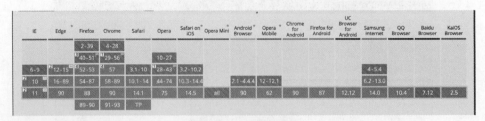

图 3-189　Grid 的兼容性

4）平台布局

（1）名词解释。

① 布局容器相关组件：row、col、containers 和 container 这四个组件的统称。

② 基础布局属性：包括 cols、colSpan、span 和 offset。

（2）概述。

一般地，大多数组件配置了基本的布局属性，包括 cols 和 colSpan。

① cols：将内容区中的每行拆分的列数；当前元素未配置该属性时，将取离当前元素最近父元素属性。默认为 1。

② colSpan：当前元素相对于父元素在每行中所占列比例；优先于 span。默认为 FULL。
- FULL：1；
- HALF：1/2；
- THIRD：1/3；
- TWO_THIRDS：2/3；
- QUARTER：1/4；
- THREE_QUARTERS：3/4。

③ span：当前元素相对于父元素在每行中所占列数。默认为 cols。

④ offset：当前元素相对于原所在列的偏移列数。

规定默认栅格数为 24，基础布局属性均是相对于默认栅格数而言的。

在使用基础布局属性时，应尽可能保证公式（24 / cols）* span 以及以下列举的公式，其计算结果为整数，否则可能出现不符合预期的结果。

在使用 colSpan 时，span 的计算公式为 span = cols * colSpan。

在使用 offset 时，offset 的计算公式为 offset =（24 / cols）* offset。

（3）平台内置组件。

下面列举了平台内置组件在 Form 和 Detail 视图中使用时所支持的布局相关属性，包括可能影响部分区域显隐的相关属性。

表头说明：

① 标签：xml 配置标签。

② 组件名称：xml 属性；多个值属于别称。如 widget="form"。

③ 配置项：xml 属性名称。

④ 可选值：声明支持的配置项属性类型，不可识别或不可解析时将采用默认值。

⑤ 作用：对配置项的简单描述，部分布局属性在上方进行了描述，不再赘述。

平台内置组件见表 3-34。

表 3-34 平台内置组件

| 标签 | 组件名称 | 配置项 | 可选值 | 默认值 | 作用 |
| --- | --- | --- | --- | --- | --- |
| view | | cols | number | 1 | |
| element | form | cols | number | 1 | |
| | | colSpan | enum | FULL | |
| | | span | number | cols | |
| | | offset | number | | |
| | detail | cols | number | 1 | |
| | | colSpan | enum | FULL | |
| | | span | number | cols | |
| | | offset | number | | |
| | DateTimeRangePicker | colSpan | enum | FULL | |
| | | span | number | cols | |
| | | offset | number | | |

续表

| 标签 | 组件名称 | 配置项 | 可选值 | 默认值 | 作用 |
|---|---|---|---|---|---|
| element | DateRangePicker | colSpan | enum | FULL | |
| | | span | number | cols | |
| | | offset | number | | |
| | TimeRangePicker | colSpan | enum | FULL | |
| | | span | number | cols | |
| | | offset | number | | |
| | YearRangePicker | colSpan | enum | FULL | |
| | | span | number | cols | |
| | | offset | number | | |
| field | 任意 | colSpan | enum | FULL | |
| | | span | number | cols | |
| | | offset | number | | |
| pack | fieldset/group | cols | number | 1 | |
| | | colSpan | enum | FULL | |
| | | span | number | cols | |
| | | offset | number | | |
| | | title | string | 分组 | 标题；空字符串或不填，则隐藏标题区 |
| | tabs | cols | number | 1 | |
| | | colSpan | enum | FULL | |
| | | span | number | cols | |
| | | offset | number | | |
| | tab | cols | number | 1 | |
| | | title | string | 选项页 | 标题；必填；空字符串或不填则显示默认值 |
| | row | cols | number | 1 | |
| | | align | top/middle/bottom | | 垂直对齐方式 |
| | | justify | start/end/center/space-around/space-between | | 水平对齐方式 |
| | | gutter | number,number?<br>示例：<br>24 = 24,24<br>12,12 | 24,24 | 水平/垂直间距 |
| | | wrap | boolean | true | 是否允许换行 |
| | col | cols | number | 1 | |
| | | colSpan | enum | FULL | |
| | | span | number | cols | |
| | | offset | number | | 偏移单元格数 |
| | | mode/widthType | manual/full | manual | 列模式；手动/自动填充；使用自动填充时，将忽略 span、offset 属性 |

续表

| 标签 | 组件名称 | 配置项 | 可选值 | 默认值 | 作用 |
|---|---|---|---|---|---|
| pack | containers | cols | number | 1 | |
| | | align | top/middle/bottom | | 垂直对齐方式 |
| | | justify | start/end/center/space-around/space-between | | 水平对齐方式 |
| | | gutter | number,number?<br>示例：24 = 24,24<br>12,12 | 0,24 | 水平/垂直间距 |
| | | wrap | boolean | true | 是否允许换行 |
| | container | cols | number | 1 | |
| | | colSpan | enum | FULL | |
| | | span | number | cols | |
| | | offset | number | | 偏移单元格数 |
| | | align | top/middle/bottom | | 垂直对齐方式 |
| | | justify | start/end/center/space-around/space-between | | 水平对齐方式 |
| | | gutter | number,number?<br>示例：<br>24 = 24,24<br>12,12 | 0,24 | 水平/垂直间距 |
| | | wrap | boolean | true | 是否允许换行 |
| | | mode/widthType | manual/full | full | 列模式：手动/自动填充；使用自动填充时，将忽略 span、offset 属性 |

（4）基础布局

基础布局提供了在不使用任何布局容器相关组件的情况下，仅使用 cols、span、offset 这三个属性控制行列的布局能力。

其本质上是 flex 布局的扩展，但依旧无法脱离 flex 布局本身的限制，即元素始终是自上而下、自左向右紧凑的。

下面将使用 fieldset 和 tabs/tab 组件来介绍各个属性在实际场景中的使用。

示例 1：默认撑满一行，代码如下：

```
1  <pack widget="fieldset" title="示例1">
2    <field data="code" widget="Input" label="编码" placeholder="请输入" required="true" />
3    <field data="name" widget="Input" label="名称" placeholder="请输入" required="true" />
4  </pack>
```

结果如图 3-190 所示。

图 3-190　示例 1 结果展示

示例2：一行两列，cols=2； colSpan=HALF/span=1。代码如下：

```
1  <pack widget="fieldset" title="示例2" cols="2">
2      <field data="code" widget="Input" colSpan="HALF" label="编码" placeholder="请输入" required="true" />
3      <field data="name" widget="Input" colSpan="HALF" label="名称" placeholder="请输入" required="true" />
4  </pack>
5  -------------------------------- 或 --------------------------------
6  <pack widget="fieldset" title="示例2" cols="2">
7      <field data="code" widget="Input" span="1" label="编码" placeholder="请输入" required="true" />
8      <field data="name" widget="Input" span="1" label="名称" placeholder="请输入" required="true" />
9  </pack>
```

结果如图 3-191 所示。

图 3-191　示例 2 结果展示

示例 3：使用 offset 实现中间空一个字段空间的布局。

注：offset 的作用有限，offset 最优实践的前提是在同一行中进行偏移，要实现特殊布局功能，请使用自由布局相关布局能力。

```
1  <pack widget="fieldset" title="示例3" cols="3">
2      <field data="code" widget="Input" span="1" label="编码" placeholder="请输入" required="true" />
3      <field data="name" widget="Input" span="1" offset="1" label="名称" placeholder="请输入" required="true" />
4  </pack>
```

结果如图 3-192 所示。

图 3-192　示例 3 结果展示

示例 4：属性 cols 就近取值，代码如下：

```
1  <!-- 所有tab将使用cols="2"属性 -->
2  <pack widget="tabs" cols="2">
3      <pack widget="tab" title="示例4-1">
4          <field data="code" widget="Input" span="1" label="编码" placeholder="请输入" required="true" />
5      </pack>
6      <pack widget="tab" title="示例4-2">
7          <field data="name" widget="Input" span="1" label="名称" placeholder="请输入" required="true" />
8      </pack>
9  </pack>
```

结果如图 3-193、图 3-194 所示。

图 3-193　示例 4 结果展示（1）

图 3-194 示例 4 结果展示（2）

tab 属性代码如下。

```
1   <!-- 所有tab将使用cols="2"属性 -->
2   <pack widget="tabs" cols="2">
3       <!-- 特指该tab使用cols="1"属性，其他tab继续使用tabs配置的cols="2"属性 -->
4       <pack widget="tab" title="示例4-1" cols="1">
5           <field data="code" widget="Input" span="1" label="编码" placeholder="请输入" required="true" />
6       </pack>
7       <pack widget="tab" title="示例4-2">
8           <field data="name" widget="Input" span="1" label="名称" placeholder="请输入" required="true" />
9       </pack>
10  </pack>
```

（5）自由布局。

自由布局提供了无法通过基础布局能力实现的其他布局能力，总的来说，自由布局是 grid 布局和 flex 布局的结合，它既拥有 grid 布局对页面进行单元格拆/合的能力，在每个单元格中，又能使用 flex 布局进行紧凑排列。

下面将使用 fieldset 组件介绍各个属性在实际场景中的使用。

row/col：两个组件共同形成行和列，在一行中拆分成 24 个栅格，每个列的跨度不超过 24。当一行中，所有列的跨度和超过 24 时，将会自动换行。

containers/row/container：三个组件共同形成一个二维网格，以此实现 grid 布局的基本能力。每个单元格（container）中使用 flex 布局。

以下示例为了体现布局效果，可能会出现重复字段定义，业务上在使用时需要避免这种定义。

示例 1：仅使用 1/2 左侧空间，代码如下。

小贴士：在使用 row 和 col 组合时，如果在一个 col 中有且仅有一个子元素，则 col 可以缺省。col 相关属性可以配置在该子元素上。

```
1   <pack widget="fieldset" title="示例1">
2       <pack widget="row" cols="2">
3           <!-- 此处显式定义col组件，field标签上的基础布局属性将失效 -->
4           <pack widget="col" span="1">
5               <field data="code" widget="Input" label="编码" placeholder="请输入" required="true" />
6           </pack>
7       </pack>
8       <pack widget="row" cols="2">
9           <pack widget="col" span="1">
10              <field data="name" widget="Input" label="名称" placeholder="请输入" required="true" />
11          </pack>
12      </pack>
13  </pack>
14  -------------------------------- 或 --------------------------------
15  <pack widget="fieldset" title="示例1">
16      <pack widget="row" cols="2">
17          <!-- 此处缺省col组件，field标签上的基础布局属性可生效 -->
18          <field data="code" widget="Input" span="1" label="编码" placeholder="请输入" required="true" />
19      </pack>
20      <pack widget="row" cols="2">
21          <field data="name" widget="Input" span="1" label="名称" placeholder="请输入" required="true" />
22      </pack>
23  </pack>
```

结果如图 3-195 所示。

图 3-195　示例 1 结果展示

示例 2：使用布局容器实现中间空一个字段空间的布局，代码如下。

```
<pack widget="fieldset" title="示例2">
    <pack widget="containers">
        <pack widget="row">
            <pack widget="container">
                <field data="code" widget="Input" label="编码" placeholder="请输入" required="true" />
            </pack>
            <pack widget="container"></pack>
            <pack widget="container">
                <field data="name" widget="Input" label="名称" placeholder="请输入" required="true" />
            </pack>
        </pack>
    </pack>
</pack>
```

结果如图 3-196 所示。

图 3-196　示例 2 结果展示

示例 3：一行 5 列（基础布局属性公式无法计算出整数的情况），代码如下。

```
<pack widget="fieldset" title="示例3">
    <pack widget="containers">
        <pack widget="row">
            <pack widget="container">
                <field data="code" widget="Input" label="编码" placeholder="请输入" required="true" />
            </pack>
            <pack widget="container">
                <field data="name" widget="Input" label="名称" placeholder="请输入" required="true" />
            </pack>
            <pack widget="container">
                <field data="code" widget="Input" label="编码" placeholder="请输入" required="true" />
            </pack>
            <pack widget="container">
                <field data="code" widget="Input" label="编码" placeholder="请输入" required="true" />
            </pack>
            <pack widget="container">
                <field data="code" widget="Input" label="编码" placeholder="请输入" required="true" />
            </pack>
        </pack>
    </pack>
</pack>
```

结果如图 3-197 所示。

图 3-197　示例 3 结果展示

示例 4：共 2 行，其中 1 行为 3 列，另 1 行为 2 列。

该示例使用任何一种组件组合都可以实现，结果一致。

```
1  <!-- 使用containers/row/container -->
2  <pack widget="fieldset" title="示例4">
3      <pack widget="containers">
4          <pack widget="row">
5              <pack widget="container">
6                  <field data="code" widget="Input" label="编码" placeholder="请输入" required="true" />
7              </pack>
8              <pack widget="container">
9                  <field data="name" widget="Input" label="名称" placeholder="请输入" required="true" />
10             </pack>
11             <pack widget="container">
12                 <field data="code" widget="Input" label="编码" placeholder="请输入" required="true" />
13             </pack>
14         </pack>
15         <pack widget="row">
16             <pack widget="container">
17                 <field data="code" widget="Input" label="编码" placeholder="请输入" required="true" />
18             </pack>
19             <pack widget="container">
20                 <field data="code" widget="Input" label="编码" placeholder="请输入" required="true" />
21             </pack>
22         </pack>
23     </pack>
24 </pack>
25 ------------------------------ 或 ------------------------------
26 <!-- 使用row/col，其中col缺省 -->
27 <pack widget="fieldset" title="示例4">
28     <pack widget="row" cols="3">
29         <field data="code" widget="Input" span="1" label="编码" placeholder="请输入" required="true" />
30         <field data="name" widget="Input" span="1" label="名称" placeholder="请输入" required="true" />
31         <field data="code" widget="Input" span="1" label="编码" placeholder="请输入" required="true" />
32     </pack>
33     <pack widget="row" cols="2">
34         <field data="code" widget="Input" span="1" label="编码" placeholder="请输入" required="true" />
35         <field data="code" widget="Input" span="1" label="编码" placeholder="请输入" required="true" />
36     </pack>
37 </pack>
```

结果如图 3-198 所示。

示例4

* 编码
请输入

* 名称
请输入

* 编码
请输入

* 编码
请输入

* 编码
请输入

图 3-198　示例 4 结果展示

示例 5：布局容器的垂直居中。左侧容器高度被子元素撑开，右侧容器在垂直方向居中。代码如下。

```
1  <pack widget="fieldset" title="示例5">
2      <pack widget="containers">
3          <pack widget="row">
4              <pack widget="container">
5                  <field data="code" widget="Input" label="编码" placeholder="请输入"
6                  <field data="name" widget="Input" label="名称" placeholder="请输入"
7              </pack>
8              <pack widget="container" align="middle">
9                  <field data="code" widget="Input" label="编码" placeholder="请输入"
10             </pack>
11         </pack>
12     </pack>
13 </pack>
```

结果如图 3-199 所示。

示例5

* 编码
请输入

* 编码
请输入

* 名称
请输入

图 3-199　示例 5 结果展示

（6）举例。

这里拿 PetTalent 举例，仿造教程上面效果，除了例子中的效果，自己可以做更多的尝试。

Step1. 修改 PetTalent 的 form 视图为给 creater 字段增加一个属性配置 offset="1"，代码如下。

```
1  <field data="creater" widget="SSConstructSelect" offset="1" submitFields="creater,
   name" responseFields="name"/>
2
```

Step2. 重启看效果，如图 3-200 所示。

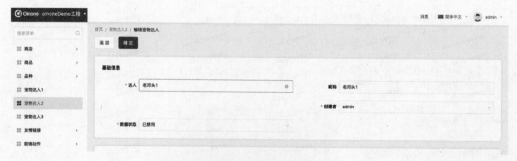

图 3-200　实际展示效果

Step3. 修改 PetTalent 的 form 视图为给基础信息增加一个属性 cols="4"，给 name 字段增加一个属性 span="2"，给 creater 和 nick 字段增加一个属性 span="1"，代码如下。

```
1  <pack widget="group" title="基础信息" cols="4">
2      <field invisible="true" data="id" label="ID" readonly="true"/>
3      <field data="name" label="达人" required ="true" span="2"/>
4      <field data="nick" compute="activeRecord.name" readonly = "true" span="1"/>
5      <field data="creater" widget="SSConstructSelect" submitFields="creater,name" responseFields="name" span="1"/>
6      <field data="dataStatus" label="数据状态" >
7          <options>
8              <!-- option name="DRAFT" displayName="草稿" value="DRAFT" state="ACTIVE"/ -->
9              <option name="NOT_ENABLED" displayName="未启用" value="NOT_ENABLED" state="ACTIVE"/>
10             <option name="ENABLED" displayName="已启用" value="ENABLED" state="ACTIVE"/>
11             <option name="DISABLED" displayName="已禁用" value="DISABLED" state="ACTIVE"/>
12         </options>
13     </field>
14     <field invisible="true" data="createDate" label="创建时间" readonly="true"/>
15     <field invisible="true" data="writeDate" label="更新时间" readonly="true"/>
16     <field data="createUid" label="创建人id"/>
17     <field data="writeUid" label="更新人id"/>
18 </pack>
```

Step4. 重启看效果，如图 3-201 所示。

图 3-201　查看实际效果

**4. 动作的配置**

在 3.5.3 节中已介绍了 Action 的几种类型和组合动作。

1）通用配置

动作的通用配置见表 3-35。

表 3-35　动作的通用配置

| 配置项 | 可选值 | 默认值 | 作用 |
|---|---|---|---|
| name | | | 动作名称 |
| label | | | 显示名称 |
| icon | | | 图标 |
| type | primary<br>default<br>link | primary | 按钮类型样式，支持主要样式、次要样式以及链接样式 |
| bizStyle | default<br>success<br>warning<br>danger<br>info | default | 按钮业务样式，支持成功（green）、警告（yellow）、危险（red）、信息（grey）四种样式 |

续表

| 配置项 | 可选值 | 默认值 | 作用 |
|---|---|---|---|
| invisible | true<br>false<br>condition | false | 展示规则，有简单的 true/false 显隐，也支持复杂的表达式 |
| disabled | true<br>false<br>condition | 根据动作上下文类型进行自动推断 | 是否禁用<br>自动推断规则：<br>当上下文类型为"单行"时，相当于使用表达式 LIST_COUNT (context.activeRecords) != 1<br>当上下文类型为"多行"时，相当于使用表达式 LIST_COUNT (context.activeRecords) <= 1<br>当上下文类型为"单行"或"多行"时，相当于使用表达式 LIST_COUNT (context.activeRecords) == 0 |
| disabledTitle | string | 根据动作上下文类型进行自动推断 | 禁用悬浮提示 |

2）二次确认配置

二次确认框默认支持两种模式，对话框和气泡框。

（1）对话框如图 3-202 所示。

图 3-202　对话框提示

（2）气泡框如图 3-203 所示。

图 3-203　气泡框警告

（3）配置项见表 3-36。

表 3-36　配置项

| 配置项 | 可选值 | 默认值 | 作用 | 备注 |
|---|---|---|---|---|
| confirm | string |  | 二次确认提示文字 | 配置后开启二次确认 |
| confirmType | POPPER（气泡提示框）<br>MODAL（对话框） | POPPER | 确认框类型 |  |
| confirmPosition | TM（按钮上方）<br>BM（按钮下方）<br>LM（按钮左侧）<br>RM（按钮右侧） | BM | 确认框位置 | 气泡框该配置生效 |

续表

| 配置项 | 可选值 | 默认值 | 作用 | 备注 |
|---|---|---|---|---|
| enterText | | 确定 | 确定按钮文字 | |
| cancelText | | 取消 | 取消按钮文字 | |

3）弹出层动作配置（窗口动作 ViewAction）

目前平台对于弹出层支持了两种展示形式：弹窗（modal/dialog）和抽屉（drawer）。

支持两种配置方式：内嵌视图配置和引用已有页面。内嵌视图配置优先于引用已有页面。

（1）内嵌视图配置。

该配置对于弹窗和抽屉均适用。代码如下。

```
1  <action name="窗口动作名称" label="创建">
2      <view model="模型编码" type="form">
3          <template slot="form" widget="form">
4              <field data="id" invisible="true" />
5              <field data="code" label="编码" widget="Input" />
6              <field data="name" label="名称" widget="Input" />
7          </template>
8          <template slot="footer">
9              <action name="$$internal_DialogCancel" label="关闭" type="default" />
10             <action name="create" label="确定" />
11         </template>
12     </view>
13 </action>
```

（2）引用已有页面配置。

该配置对于弹窗和抽屉均适用。代码如下。

```
1  <view model="模型编码" type="form">
2      <template slot="form" widget="form">
3          <field data="id" invisible="true" />
4          <field data="code" label="编码" widget="Input" />
5          <field data="name" label="名称" widget="Input" />
6      </template>
7      <template slot="footer">
8          <action name="$$internal_DialogCancel" label="关闭" type="default" />
9          <action name="create" label="确定" />
10     </template>
11 </view>
```

```
1  <action name="窗口动作名称" label="创建" resViewName="$viewName$" />
```

（3）弹窗。

当窗口动作的路由方式（target）为 dialog 时，内嵌视图 / 引用页面将以弹窗形式展示在页面上。

弹窗配置项见表 3-37。

表 3-37　弹窗配置项

| 配置项 | 可选值 | 默认值 | 作用 |
|---|---|---|---|
| title | string | 动作名称 | 标题名称 |
| width | string/number/enum<br>small（560px）<br>medium（890px）<br>large（1200px） | medium | 宽度 |

示例：配置标题名称为自定义创建弹窗，宽度为 70% 的弹窗，代码如下。

```
1  <!-- 内嵌视图配置 -->
2  <action name="窗口动作名称" label="创建">
3    <template slot="default" title="自定义创建弹窗" width="70%">
4      <view model="模型编码" type="form">
5        <template slot="form" widget="form">
6          <field data="id" invisible="true" />
7          <field data="code" label="编码" widget="Input" />
8          <field data="name" label="名称" widget="Input" />
9        </template>
10       <template slot="footer">
11         <action actionType="client" name="$$internal_DialogCancel" label="关闭" type="default" />
12         <action name="create" label="确定" />
13       </template>
14     </view>
15   </template>
16 </action>
17
18 <!-- 引用已有页面配置 -->
19 <action name="窗口动作名称" label="创建" resViewName="引用页面名称">
20   <template slot="default" title="自定义创建弹窗" width="70%" />
21 </action>
```

（4）抽屉。

当窗口动作的路由方式（target）为 drawer 时，内嵌视图/引用页面将以弹窗形式展示在页面上。抽屉配置项见表 3-38。

表 3-38　抽屉配置项

| 配置项 | 可选值 | 默认值 | 作用 | 备注 |
| --- | --- | --- | --- | --- |
| title | string | 动作名称 | 标题名称 | |
| placement | top（上）<br>right（右）<br>bottom（下）<br>left（左） | right | 抽屉打开位置 | |
| width | string/number/enum<br>small（20%）<br>medium（40%）<br>large（80%） | small | 宽度 | 打开位置为 left 和 right 时生效 |
| height | string/number/enum<br>small（20%）<br>medium（40%）<br>large（80%） | small | 宽度 | 打开位置为 top 和 bottom 时生效 |

示例：配置从下方打开，高度为 large 的抽屉，代码如下。

```
1  <!-- 内嵌视图配置 -->
2  <action name="窗口动作名称" label="创建">
3    <template slot="default" placement="bottom" height="large">
4      <view model="模型编码" type="form">
5        <template slot="form" widget="form">
6          <field data="id" invisible="true" />
7          <field data="code" label="编码" widget="Input" />
8          <field data="name" label="名称" widget="Input" />
9        </template>
10       <template slot="footer">
11         <action actionType="client" name="$$internal_DialogCancel" label="关闭" type="default" />
12         <action name="create" label="确定" />
13       </template>
14     </view>
15   </template>
16 </action>
17
18 <!-- 引用已有页面配置 -->
19 <action name="窗口动作名称" label="创建" resViewName="引用页面名称">
20   <template slot="default" placement="bottom" height="large" />
21 </action>
```

（5）组合动作配置。

具体例子参见 3.5.3 节中介绍的"前端动作之组合动作"。

① 服务器动作串行，代码如下。

```xml
<action actionType="composition" label="组合动作">
    <action name="服务器动作1" />
    <action name="服务器动作2" />
</action>
```

② 服务器动作与跳转动作组合执行，代码如下。

```xml
<action actionType="composition" label="组合动作">
    <action name="服务器动作1" />
    <action name="跳转动作1" />
</action>
```

③ 后端动作与前端动作组合，代码如下。

```xml
<action actionType="composition" label="组合动作">
    // 校验表单
    <action actionType="client" name="$$internal_ValidateForm" />
    <action name="服务器动作" />
    // 返回上一级页面
    <action actionType="client" name="$$internal_GotoListTableRouter" />
</action>
```

（6）展示规则。

在 3.5.3 节中介绍 "ServerAction 之前端展示规则（举例）" 时用到 invisible 这个属性定义。这个在 xml 中也可以配置，而且前端的优先级高于后端。

可以通过 invisible 属性配置一个表达式来使动作根据数据记录条件显示、隐藏。

如果 invisible 所引用的动作的 bindingType 是列表型视图，要用 context.activeRecords 获取当前选中记录，如果是对象型视图或列表型视图的行内显示，则需用 context.activeRecord，代码如下。

```xml
<!-- 列表型视图 -->
<action name="a" invisible="context.activeRecords && context.activeRecords[0].a === true" />
<!-- 对象型视图或列表型视图的行内显示 -->
<action name="b" invisible="context.activeRecord && context.activeRecord.b === true" />
```

4）举例

（1）Table 视图动作配置。

示例：不允许删除编码为 "5" 的数据。

当配置 disabled 时，自动推断规则将会失效，需要按需配置。下方的配置保留了单行或多行的自动推断规则。

Table 视图动作配置代码如下。

```xml
<view model="模型编码" type="table">
    <template slot="actionBar" widget="actionBar">
        <action name="delete" label="删除" disabled="LIST_COUNT(context.activeRecords) == 0 || LIST_CONTAINS(LIST_FIELD_VALUES(context.activeRecords, '', 'code'), '5')" refreshData="true" />
    </template>
    <template slot="fields">
        <field data="id" invisible="true" />
        <field data="code" label="编码" widget="Input" />
        <field data="name" label="名称" widget="Input" />
    </template>
</view>
```

结果展示如下。

① 未选中时的结果如图 3-204 所示。

图 3-204　未选中时

② 选中编码不是"5"的数据的结果如图 3-205 所示。

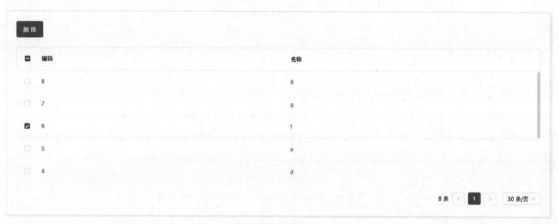

图 3-205　选中编码不是"5"的数据

在 3.5.3 节中，我们介绍了 Action 的几种类型，以及组合动作。

③ 选中编码包含"5"的数据时的结果如图 3-206 所示。

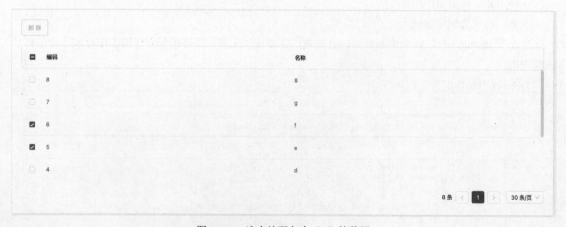

图 3-206　选中编码包含"5"的数据

（2）form 视图动作配置。

示例：创建和编辑使用同一个视图配置，代码如下。

```xml
<view model="模型编码" type="form">
    <template slot="actionBar" widget="actionBar">
        <action actionType="client" name="$$internal_GotoListTableRouter" label="返回" type="default" />
        <action name="create" label="创建" invisible="!IS_NULL(activeRecord.id)" validateForm="true" goBack="true" />
        <action name="update" label="更新" invisible="IS_NULL(activeRecord.id)" validateForm="true" goBack="true" />
    </template>
    <template slot="fields">
        <pack widget="fieldset" title="基础信息">
            <field data="id" invisible="true" />
            <field data="code" label="编码" widget="Input" />
            <field data="name" label="名称" widget="Input" />
        </pack>
    </template>
</view>

<!-- 使用组合动作实现validateForm和goBack属性 -->
<view model="模型编码" type="form">
    <template slot="actionBar" widget="actionBar">
        <action actionType="client" name="$$internal_GotoListTableRouter" label="返回" type="default" />
        <action actionType="composition" label="创建" invisible="!IS_NULL(activeRecord.id)">
            <action actionType="client" name="$$internal_ValidateForm" />
            <action name="create" />
            <action actionType="client" name="$$internal_GotoListTableRouter" />
        </action>
        <action actionType="composition" label="更新" invisible="IS_NULL(activeRecord.id)">
            <action actionType="client" name="$$internal_ValidateForm" />
            <action name="update" />
            <action actionType="client" name="$$internal_GotoListTableRouter" />
        </action>
    </template>
    <template slot="fields">
        <pack widget="fieldset" title="基础信息">
            <field data="id" invisible="true" />
            <field data="code" label="编码" widget="Input" />
            <field data="name" label="名称" widget="Input" />
        </pack>
    </template>
</view>
```

结果展示如下。

① 使用上下文无关的跳转动作进入该视图时的结果如图 3-207 所示。

图 3-207　使用上下文无关的跳转动作进入该视图时

② 使用"单行"的跳转动作进入该视图时的结果如图 3-208 所示。

图 3-208　使用"单行"的跳转动作进入该视图时

5）实战

Step1. 修改宠物商品代理表格视图的 Template。

启用和禁用服务器动作根据状态分别显示其中一个，代码如下。

```
<view name="tableView1" type="TABLE" cols="2" enableSequence="false" model='demo.PetShopProxy' priority="1" >
    <template slot="actions" autoFill="true"/>
    <template slot="rowActions">
        <action name="dataStatusEnable" label="启用" invisible="activeRecord.dataStatus == 'ENABLED'" />
        <action name="dataStatusDisable" label="禁用" invisible="activeRecord.dataStatus == 'DISABLED'" />
    </template>
    <template slot="fields">
        <field priority="1" data="income" label="收入"/>
        <field priority="101" data="code" label="店铺编码"/>
        <field priority="102" data="relatedShopName" label="引用字段shopName"/>
        <field priority="102" data="codeTwo" label="店铺编码2"/>
        <field priority="103" data="createrId" label="引用创建者Id"/>
        <field priority="110" data="description" label="描述"/>
        <field priority="111" data="description1" label="描述"/>
        <field priority="112" data="descHtml" label="html描述"/>
        <field priority="113" data="anniversary" label="店庆"/>
        <field priority="114" data="publishYear" label="开店年份"/>
        <!-- 表格其他字段-->
    </template>
</view>
```

Step2. 重启看效果。

"已启用"状态时只显示"禁用"按钮，如图 3-209 所示。

图 3-209　"已启用"状态时只显示"禁用"按钮

"已禁用"状态时只显示"启用"按钮,如图3-210所示。

图 3-210　"已禁用"状态时只显示"启用"按钮

## 3.5.7　前端组件自定义(初级篇)

在日常开发过程中因为个性化的业务与交互诉求,会出现原有组件无法满足的情况,在这种情况下一般需要前端研发介入进行新组件开发,然后交由后端研发进行对应的视图配置。在此我们主要介绍如何开发组件以及组件匹配规则。

在3.5.6节中简单提到过"字段的匹配规则",里面介绍了平台默认提供的组件在匹配规则中用到了viewType、ttype、widget,实际上平台匹配规则会更丰富,还是拿字段组件来说,平台支持字段id(id)、字段名(name)、视图类型(viewType)、视图Id(viewId)、视图名称(viewName)、字段类型(ttype)、组件名(widget)、模型名(Model,前端用模型名,而非编码)等。

组件名按最优权重匹配,其他按最大匹配原则、后注册优先原则,我们就可以将它们灵活应用于不同场景。表3-39为平台提供的组件匹配机制。

表 3-39　组件匹配机制

| 名称 | Token | 描述 | 筛选(匹配)条件 |
| --- | --- | --- | --- |
| Application | ROOT_TOKEN | 静态 Token 由平台提供 Application 为应用入口 | 无 |
| View | ViewWidget.Token 构造 | 动态 Token,为当前视图组件 | id,name,type,Model,widget |
| Field | FieldWidget.Token 构造 | 动态 Token,当前字段组件 | id,name,viewType,viewId,viewName,ttype,widget,Model |
| Action | ActionWidget.Token 构造 | 动态 Token,当前的动作组件 | id,name,viewType,actionType,Model,target |
| Group | GroupWidget.Token 构造 | 动态 Token,xml 节点中的 Group 节点对应的组件 | widget |
| Adapter | BaseAdapter.Token 构造 | 动态 Token,用于转换业务数据类型 | from,to |

注:在本节例子中如果标志(后端),是需要后端配合。Oinone采用特色业务前后端分离的开发模式,只有在组件不满足需求或者特色业务组件开发的时候进行前端专业开发工作。这样既保留前后端分离架构带来的好处,同时减少了业务开发过程中前后端不必要的沟通工作,极大地提升了效率。

**1. 自定义主题**

在页面交互中,样式的变化也是前端的核心工作之一。接下来介绍自定义主题相关的功能。

1）自定义主题

（1）在项目 src 目录下，新建 theme.ts 文件。如图 3-211 所示。

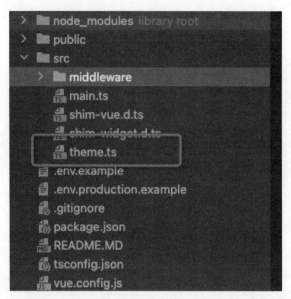

图 3-211　在项目 src 目录下，新建 theme.ts 文件

（2）在 theme.ts 内部定义主题名称和 css 变量名（替换主色为例，将主色系替换成黑色），代码如下。

```
1  export const themeName = 'OinoneTheme';
2  export const themeCssVars = {
3    'primary-color': 'black',
4    'primary-color-hover': 'black',
5    'primary-color-rgb': '0, 0, 0',
6    'primary-color-focus': 'black',
7    'primary-color-active': 'black',
8    'primary-color-outline': 'black',
9  };
```

（3）在 main.ts 注册，代码如下。

```
1  import { registerTheme, VueOioProvider } from '@kunlun/dependencies'; // 引入注册主
   题组件
2  import { themeName, themeCssVars } from './theme'; // 引入theme.ts
3  registerTheme(themeName, themeCssVars);// 注册
4  VueOioProvider(
5  {
6    http: {
7      url: location.origin,
8      callback: interceptor
9    },
10   browser: {
11     title: 'Oinone - 构你想象!',
12     favicon: 'https://pamirs.oss-cn-hangzhou.aliyuncs.com/pamirs/image/default_
   favicon.ico'
13   },
14   theme: [themeName] // 定义的themeName传入provider中
15  },
16  []
17 );
```

（4）刷新页面看效果，如图 3-212 所示。

图 3-212　刷新页面看效果

（5）完整代码如下。

```
import { VueOioProvider, registerTheme } from '@kunlun/dependencies';
import { themeName, themeCssVars } from './theme';
registerTheme(themeName, themeCssVars);

VueOioProvider(
  {
    http: {
      url: location.origin,
      callback: interceptor
    },
    browser: {
      title: 'Oinone - 构你想象!',
      favicon: 'https://pamirs.oss-cn-hangzhou.aliyuncs.com/pamirs/image/default_favicon.ico'
    },
    theme: [themeName]
  },
  []
);
```

2）自定义组件如何使用主题变量

以自定义页面为例自定义表单。

（1）theme.ts 添加变量，代码如下。

```
export const themeName = 'OinoneTheme';
export const themeCssVars = {
  'primary-color': 'black',
  'primary-color-hover': 'black',
  'primary-color-rgb': '0, 0, 0',
  'primary-color-focus': 'black',
  'primary-color-active': 'black',
  'primary-color-outline': 'black',
  'custom-theme-var': 'blue' // 自定义变量
};
```

（2）修改 petForm.vue，代码如下。

```vue
<template>
  <div class="petFormWrapper">
    <form :model="formState" @finish="onFinish">
      <a-form-item label="宠物名称" id="name" name="name" :rules="[{ required: true, message: '请输入宠物名称！' }]">
        <a-input v-model:value="formState.name" @input="onNameChange" />
      </a-form-item>

      <a-form-item label="宠物年龄" id="age" name="age" :rules="[{ required: true, message: '请输入宠物年龄！' }]">
        <a-input-number v-model:value="formState.age" />
        <a-button @click="reloadData" class="custom-theme-button">重新渲染数据</a-button>
      </a-form-item>

      <a-form-item
        label="宠物年龄"
        id="birthday"
        name="birthday"
        :rules="[{ required: true, message: '请输入宠物生日！' }]"
      >
        <a-date-picker v-model:value="formState['birthday']" value-format="YYYY-MM-DD" />
      </a-form-item>
    </form>
  </div>
</template>

<script lang="ts">
import { defineComponent, reactive } from 'vue';
import { Form } from 'ant-design-vue';

export default defineComponent({
  props: ['onFieldChange', 'reloadData'],
  components: { Form },
  setup(props) {
    const formState = reactive({
      name: '',
      age: '',
      birthday: ''
    });

    const onFinish = () => {
      console.log(formState);
    };

    const onNameChange = (event) => {
      props.onFieldChange('name', event.target.value);
    };

    const reloadData = async () => {
      await props.reloadData();
    };

    return {
      formState,
      reloadData,
      onNameChange,
      onFinish
    };
  }
});
</script>
<style lang="scss">
.petFormWrapper {
  .custom-theme-button {
    background-color: var(--oio-custom-theme-var); // 使用主题变量
    color: white;
  }
}
</style>
```

(3) 页面效果如图 3-213 所示。

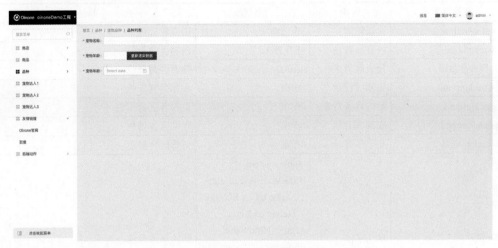

图 3-213 页面效果

3）平台内置变量

（1）平台内置变量基础见表 3-40。

表 3-40 平台内置变量基础

| 变量名 | 变量 | 默认值 | 描述 |
| --- | --- | --- | --- |
| primary-color | 主色 | #035DFF | 主要色系，直接使用或者用它的透明度色系 |
| primary-color-rgb | 主色 RGB | 3, 93, 255 | 主要色系，改变主色时需要同步更改。与 primary-color 的区别是，作用场景是主色的透明度百分比，比如 background: rgba (var (--oio-primary-color-rgb), 0.1) |
| primary-color-hover | 主色悬停 | #3F84FF | 主色悬停色 |
| primary-color-focus | 主焦点 | #3F84FF | 主焦点色 |
| primary-color-active | 主激活 | #024CDE | 主激活色 |
| primary-color-outline | 主轮廓 | #035DFF | 主轮廓色 |
| success-color | 成功 | #6DD400 | 成功色，用于表述正向反馈 |
| success-color-hover | 成功悬停 | #90DE3D | 成功悬停色 |
| success-color-active | 成功激活 | #6BBB00 | 成功激活色 |
| success-color-outline | 成功轮廓色 | #6DD400 | 成功轮廓色 |
| waring-color | 警告 | #F7B500 | 警告色 |
| waring-color-hover | 警告悬停 | #F9C73D | 警告悬停色 |
| waring-color-active | 警告激活 | #D99200 | 警告激活色 |
| waring-color-outline | 警告轮廓 | #F7B500 | 警告轮廓色 |
| info-color | 通知 | #035DFF | 通知色 |
| info-color-hover | 通知悬停 | #3F84FF | 通知悬停色 |
| info-color-active | 通知激活 | #024CDE | 通知激活色 |
| info-color-outline | 通知轮廓 | #035DFF | 通知轮廓色 |
| error-color | 错误 | #E02020 | 错误色 |
| error-color-hover | 错误悬停 | #E75555 | 错误悬停色 |
| error-color-active | 错误激活 | #C51C26 | 错误激活色 |

续表

| 变量名 | 变量 | 默认值 | 描述 |
|---|---|---|---|
| error-color-outline | 错误轮廓 | #E02020 | 错误轮廓色 |
| body-background | 内容区背景 | #F3F7FA | 内容区背景色 |
| search-background | 搜索区背景 | #ffffff | 搜索区背景色 |
| header-background | 头部区背景 | #ffffff | 头部区背景色 |
| main-background | 主内容区背景 | #ffffff | 主内容区背景色 |
| footer-background | 底部区背景 | #ffffff | 底部区背景色 |
| menu-background | 菜单区背景 | #ffffff | 菜单区背景色 |
| font-family | 默认字体 | -apple-system, BlinkMacSystemFont,'Segoe UI', Roboto,'Helvetica Neue', Arial,'Noto Sans', sans-serif,'Apple Color Emoji','Segoe UI Emoji','Segoe UI Symbol','Noto Color Emoji'; | 默认字体设置 |
| text-color | 主文字 | rgba（0,0,0,0.85） | 主文字色 |
| text-color-rgb | 主文字RGB | 38, 38, 38 | 主文字RGB |
| text-color-secondary | 次要文字 | rgba（0,0,0,0.65） | 次要文字色 |
| font-size | 基础文字 | 14px | 基础文字 |
| font-size-lg | 大号文字 | 16px | 大号文字 |
| font-size-sm | 小号文字 | 12px | 小号文字 |
| font-weight | 默认加粗 | 400 | 默认加粗 |
| font-weight-thick | 加粗 | bold | 加粗 |
| font-weight-bold | 加粗系数 | 700 | 加粗系数 |
| line-height | 基础行高 | 18px | 基础行高 |
| line-height-lg | 大号行高 | 20px | 大号行高 |
| line-height-sm | 中号行高 | 16px | 中号行高 |
| line-height-xs | 小号行高 | 14px | 小号行高 |
| line-height-xxs | 最小号行高 | 12px | 最小号行高 |
| border-radius | 默认圆角 | 4px | 默认圆角 |
| border-radius-lg | 大号圆角 | 8px | 大号圆角 |
| border-radius-sm | 小号圆角 | 2px | 小号圆角 |
| border-color | 边框颜色 | #e3e7ee | 边框颜色 |
| border-width | 边框宽度 | 1px | 边框宽度 |
| border-style | 边框风格 | solid | 边框风格 |
| padding-lg | 大号内间距 | 24px | 大号内间距 |
| padding-md | 中号内间距 | 16px | 中号内间距 |
| padding-sm | 小号内间距 | 12px | 小号内间距 |
| padding-xs | 迷你内间距 | 8px | 迷你内间距 |
| padding-xxs | 最小号内间距 | 4px | 最小号内间距 |
| margin-lg | 大号外间距 | 24px | 大号外间距 |

续表

| 变量名 | 变量 | 默认值 | 描述 |
| --- | --- | --- | --- |
| margin-md | 中号外间距 | 16px | 中号外间距 |
| margin-sm | 小号外间距 | 12px | 小号外间距 |
| margin-xs | 迷你外间距 | 8px | 迷你外间距 |
| margin-xxs | 最小号外间距 | 4px | 最小号外间距 |
| height | 基础高度 | 40px | 基础高度 |
| height-lg | 大号高度 | 54px | 大号高度 |
| height-sm | 小号高度 | 32px | 小号高度 |
| readonly-color | 只读色 | #262626 | 只读色 |
| readonly-bg | 只读背景色 | #fcfcfc | 只读背景色 |
| readonly-active-bg | 只读激活色 | #fcfcfc | 只读激活色 |
| readonly-border-color | 只读边框色 | #e3e7ee | 只读边框色 |
| disabled-color | 禁用色 | rgba（0, 0, 0, 0.25） | 禁用色 |
| disabled-bg | 禁用背景色 | #fcfcfc | 禁用背景色 |
| disabled-active-bg | 禁用激活色 | #fcfcfc | 禁用激活色 |
| disabled-border-color | 禁用边框色 | #d9d9d9 | 禁用边框色 |

（2）平台内置变量按钮见表 3-41。

表 3-41　平台内置变量按钮

| 变量名 | 变量 | 默认值 | 描述 |
| --- | --- | --- | --- |
| background | 按钮背景色 | #ffffff | 按钮背景色 |
| border-width | 边框宽度 | 1px | 边框宽度 |
| border-style | 边框风格 | solid | 边框风格 |
| border-color | 边框颜色 | #d9d9d9 | 边框颜色 |
| border-radius | 圆角 | 4px | 圆角 |
| outline | 轮廓 | none | 轮廓 |
| shadow | 阴影 | none | 阴影 |
| text-color | 文字颜色 | rgba（0,0,0,0.85） | 文字颜色 |
| background-active | 背景激活 | #ffffff | 背景激活 |
| border-width-active | 激活边框宽度 | 1px | 激活边框宽度 |
| border-style-active | 边框激活风格 | solid | 激活边框风格 |
| border-color-active | 激活边框颜色 | #d9d9d9 | 激活边框颜色 |
| border-radius-active | 激活圆角 | 4px | 激活圆角 |
| outline-active | 激活轮廓 | none | 激活轮廓 |
| shadow-active | 激活阴影 | none | 激活阴影 |
| text-color-active | 文字激活色 | rgba（0,0,0,0.85） | 文字激活色 |
| background-focus | 聚焦背景色 | #ffffff | 聚焦背景色 |
| border-width-focus | 聚焦边框宽度 | 1px | 聚焦边框宽度 |
| border-style-focus | 聚焦边框风格 | solid | 聚焦边框风格 |
| border-color-focus | 聚焦边框颜色 | #d9d9d9 | 聚焦边框颜色 |
| border-radius-focus | 聚焦圆角 | 4px | 聚焦圆角 |
| outline-focus | 聚焦轮廓 | none | 聚焦轮廓 |
| shadow-focus | 聚焦阴影 | none | 聚焦阴影 |

续表

| 变量名 | 变量 | 默认值 | 描述 |
| --- | --- | --- | --- |
| text-color-focus | 聚焦文字颜色 | rgba（0,0,0,0.85） | 聚焦文字颜色 |
| background-hover | 悬停背景色 | #ffffff | 悬停背景色 |
| border-width-hover | 悬停边框宽度 | 1px | 悬停边框宽度 |
| border-style-hover | 悬停边框风格 | solid | 悬停边框风格 |
| border-color-hover | 悬停边框颜色 | #d9d9d9 | 悬停边框颜色 |
| border-radius-hover | 悬停圆角 | 4px | 悬停圆角 |
| outline-hover | 悬停轮廓色 | none | 悬停轮廓色 |
| shadow-hover | 悬停阴影色 | none | 悬停阴影色 |
| text-color-hover | 悬停文字颜色 | rgba（0,0,0,0.85） | 悬停文字颜色 |
| border-color-danger | 悬停危险边框色 | #E02020 | 悬停危险边框色 |
| background-visited | 访问后背景色 | #ffffff | 访问后背景色 |
| border-width-visited | 访问后边框宽度 | 1px | 访问后边框宽度 |
| border-style-visited | 访问后边框风格 | solid | 访问后边框风格 |
| border-color-visited | 访问后边框颜色 | #d9d9d9 | 访问后边框颜色 |
| border-radius-visited | 访问后圆角 | 4px | 访问后圆角 |
| outline-visited | 访问后轮廓 | none | 访问后轮廓 |
| shadow-visited | 访问后阴影 | none | 访问后阴影 |
| text-color-visit | 访问后文字颜色 | rgba（0,0,0,0.85） | 访问后文字颜色 |
| primary-background | 主背景色 | #035DFF | 主背景色 |
| primary-border-width | 主边框宽度 | 1px | 主边框宽度 |
| primary-border-style | 主边框风格 | solid | 主边框风格 |
| primary-border-color | 主边框颜色 | #035DFF | 主边框颜色 |
| primary-border-radius | 主圆角 | 4px | 主圆角 |
| primary-outline | 主轮廓 | none | 主轮廓 |
| primary-shadow | 主阴影 | none | 主阴影 |
| primary-text-color | 主文字颜色 | #ffffff | 主文字颜色 |
| default-text-color | 次要按钮文字颜色 | #035DFF | 次要按钮文字颜色 |
| default-border-width | 次要按钮边框宽度 | 1px | 次要按钮边框宽度 |
| primary-background-danger | 危险背景色 | #E02020 | 危险背景色 |
| primary-border-color-danger | 危险按钮边框色 | #E02020 | 危险边框色 |
| primary-text-color-danger | 危险按钮文字色 | #ffffff | 危险按钮文字色 |
| default-text-color-danger | 次要危险按钮文字颜色 | #E02020 | 危险按钮次要色 |
| default-background-danger | 次要危险按钮背景颜色 | #ffffff | 次要危险按钮背景颜色 |
| default-border-width-danger | 次要危险按钮边框宽度 | 1px | 次要危险按钮边框宽度 |
| default-border-color-danger | 次要危险按钮边框颜色 | #E02020 | 次要危险按钮边框颜色 |
| link-background-danger | 文字危险按钮背景色 | none | 文字危险按钮背景色 |
| link-border-color-danger | 文字危险按钮边框颜色 | none | 文字危险按钮边框颜色 |
| primary-background-info | 通知按钮背景颜色 | #8c8c8c | 通知按钮背景颜色 |
| primary-border-color-info | 通知按钮边框颜色 | #8c8c8c | 通知按钮边框颜色 |
| primary-text-color-info | 通知按钮文字颜色 | #ffffff | 通知按钮文字颜色 |
| default-text-color-info | 次要通知按钮文字颜色 | #8c8c8c | 次要通知按钮文字颜色 |
| default-background-info | 次要通知按钮背景颜色 | #ffffff | 次要通知按钮背景颜色 |

续表

| 变量名 | 变量 | 默认值 | 描述 |
|---|---|---|---|
| default-border-width-info | 次要通知按钮边框宽度 | 1px | 次要通知按钮边框宽度 |
| default-border-color-info | 次要通知按钮边框颜色 | #8c8c8c | 次要通知按钮边框颜色 |
| link-background-info | 通知文字按钮背景颜色 | none | 通知文字按钮背景色 |
| link-border-color-info | 通知文字按钮边框颜色 | none | 通知文字按钮边框颜色 |
| link-text-color-info | 通知文字按钮文字颜色 | #8c8c8c | 通知文字按钮文字颜色 |
| primary-background-warning | 警告按钮背景颜色 | #F7B500 | 警告按钮背景颜色 |
| primary-border-color-warning | 警告按钮边框颜色 | #F7B500 | 警告按钮边框颜色 |
| primary-text-color-warning | 警告按钮文字颜色 | #ffffff | 警告按钮文字颜色 |
| default-text-color-warning | 次要警告按钮文字颜色 | #F7B500 | 次要警告按钮文字颜色 |
| default-background-warning | 次要警告按钮背景颜色 | #F7B500 | 次要警告按钮背景颜色 |
| default-border-width-warning | 次要警告按钮边框宽度 | 1px | 次要警告按钮边框宽度 |
| default-border-color-warning | 次要警告按钮边框颜色 | #F7B500 | 次要警告按钮边框颜色 |
| link-background-warning | 文字警告按钮背景色 | none | 文字警告按钮背景色 |
| link-border-color-warning | 文字警告按钮边框颜色 | none | 文字警告按钮边框颜色 |
| link-text-color-warning | 文字警告按钮文字颜色 | #F7B500 | 文字警告按钮文字颜色 |
| primary-background-success | 成功按钮背景色 | #6DD400 | 成功按钮背景色 |
| primary-border-color-success | 成功按钮边框颜色 | #6DD400 | 成功按钮边框颜色 |
| primary-text-color-success | 成功按钮文字颜色 | #ffffff | 成功按钮文字颜色 |
| default-text-color-success | 次要成功按钮文字颜色 | #6DD400 | 次要成功按钮文字颜色 |
| default-background-success | 次要成功按钮背景颜色 | #ffffff | 次要成功按钮背景颜色 |
| default-border-width-success | 次要成功按钮边框宽度 | 1px | 次要成功按钮边框宽度 |
| default-border-color-success | 次要成功按钮边框颜色 | #6DD400 | 次要成功按钮边框颜色 |
| link-background-success | 文字成功按钮背景颜色 | none | 文字成功按钮背景颜色 |
| link-border-color-success | 文字成功按钮边框颜色 | none | 文字成功按钮边框颜色 |
| link-text-color-success | 文字成功按钮文字颜色 | #6DD400 | 文字成功按钮文字颜色 |
| primary-background-active | 主激活色 | #E02020 | 主激活色 |
| primary-border-width-active | 主激活边框色 | #024CDE | 主激活边框色 |
| primary-border-style-active | 主激活边框风格 | solid | 主激活边框风格 |
| primary-border-color-active | 主激活边框颜色 | 1px | 主激活边框颜色 |
| primary-border-radius-active | 主激活圆角 | 4px | 主激活圆角 |
| primary-outline-active | 主激活轮廓 | none | 主激活轮廓 |
| primary-shadow-active | 主激活阴影 | none | 主激活阴影 |
| primary-text-color-active | 主激活文字颜色 | #ffffff | 主激活文字颜色 |
| primary-background-focus | 主焦点背景色 | #E02020 | 主焦点背景色 |
| primary-border-width-focus | 主焦点边框宽度 | 1px | 主焦点边框宽度 |
| primary-border-style-focus | 主焦点边框风格 | solid | 主焦点边框风格 |
| primary-border-color-focus | 主焦点边框颜色 | #024CDE | 主焦点边框颜色 |
| primary-border-radius-focus | 主焦点圆角 | 4px | 主焦点圆角 |
| primary-outline-focus | 主焦点轮廓 | none | 主焦点轮廓 |
| primary-shadow-focus | 主焦点阴影 | none | 主焦点阴影 |
| primary-text-color-focus | 主悬停文字色 | #E02020 | 主悬停文字色 |
| primary-background-hover | 主悬停背景 | #024CDE | 主悬停背景 |

续表

| 变量名 | 变量 | 默认值 | 描述 |
|---|---|---|---|
| primary-border-width-hover | 主悬停边框宽度 | solid | 主悬停边框宽度 |
| primary-border-style-hover | 主悬停边框风格 | 1px | 主悬停边框风格 |
| primary-border-color-hover | 主悬停边框颜色 | 4px | 主悬停边框颜色 |
| primary-border-radius-hover | 主悬停边框圆角 | none | 主悬停边框圆角 |
| primary-outline-hover | 主悬停轮廓 | none | 主悬停轮廓 |
| primary-shadow-hover | 主悬停阴影 | #E02020 | 主悬停阴影 |
| primary-text-color-hover | 主悬停文字色 | #024CDE | 主悬停文字色 |
| primary-background-visited | 主访问后背景色 | #035DFF | 主访问后背景色 |
| primary-border-width-visited | 主访问后边框宽度 | #024CDE | 主访问后边框宽度 |
| primary-border-style-visited | 主访问后边框风格 | solid | 主访问后边框风格 |
| primary-border-color-visited | 主访问后边框颜色 | 1px | 主访问后边框颜色 |
| primary-border-radius-visited | 主访问后边框圆角 | 4px | 主访问后边框圆角 |
| primary-outline-visited | 主访问后轮廓 | none | 主访问后轮廓 |
| primary-shadow-visited | 主访问后阴影 | none | 主访问后阴影 |
| primary-text-color-visit | 主访问后文字颜色 | #E02020 | 主访问后文字颜色 |
| link-background | 链接按钮背景色 | transparent | 链接按钮背景色 |
| link-border-width | 链接按钮边框宽度 | none | 链接按钮边框宽度 |
| link-border-style | 链接按钮边框风格 | none | 链接按钮边框风格 |
| link-border-color | 链接按钮文字颜色 | none | 链接按钮文字颜色 |
| link-border-radius | 链接按钮圆角 | none | 链接按钮圆角 |
| link-outline | 链接轮廓 | none | 链接轮廓 |
| link-shadow | 链接按钮阴影 | none | 链接按钮阴影 |
| link-text-color | 链接按钮文字颜色 | #E02020 | 链接按钮文字颜色 |
| link-background-active | 链接按钮激活背景色 | transparent | 链接按钮激活背景色 |
| link-border-width-active | 链接按钮激活边框宽度 | none | 链接按钮激活边框宽度 |
| link-border-style-active | 链接按钮激活边框风格 | none | 链接按钮激活边框风格 |
| link-border-color-active | 链接按钮激活边框颜色 | none | 链接按钮激活边框颜色 |
| link-border-radius-active | 链接按钮激活圆角 | none | 链接按钮激活圆角 |
| link-outline-active | 链接按钮激活轮廓 | none | 链接按钮激活轮廓 |
| link-shadow-active | 链接按钮激活阴影 | none | 链接按钮激活阴影 |
| link-text-color-active | 链接按钮激活文字颜色 | #E02020 | 链接按钮激活文字颜色 |
| link-background-focus | 链接按钮聚焦背景色 | transparent | 链接按钮聚焦背景色 |
| link-border-width-focus | 链接按钮聚焦边框宽度 | none | 链接按钮聚焦边框宽度 |
| link-border-style-focus | 链接按钮聚焦边框风格 | none | 链接按钮聚焦边框风格 |
| link-border-color-focus | 链接按钮聚焦边框颜色 | none | 链接按钮聚焦边框颜色 |
| link-border-radius-focus | 链接按钮聚焦圆角 | none | 链接按钮聚焦圆角 |
| link-outline-focus | 链接按钮聚焦轮廓 | none | 链接按钮聚焦轮廓 |
| link-shadow-focus | 链接按钮聚焦阴影 | none | 链接按钮聚焦阴影 |
| link-text-color-focus | 链接按钮聚焦文字颜色 | #E02020 | 链接按钮聚焦文字颜色 |
| link-background-hover | 链接按钮悬停背景色 | transparent | 链接按钮悬停背景色 |
| link-border-width-hover | 链接按钮悬停边框宽度 | none | 链接按钮悬停边框宽度 |
| link-border-style-hover | 链接按钮悬停边框风格 | none | 链接按钮悬停边框风格 |

续表

| 变量名 | 变量 | 默认值 | 描述 |
| --- | --- | --- | --- |
| link-border-color-hover | 链接按钮悬停边框颜色 | none | 链接按钮悬停边框颜色 |
| link-border-radius-hover | 链接按钮悬停圆角 | none | 链接按钮悬停圆角 |
| link-outline-hover | 链接按钮悬停轮廓 | none | 链接按钮悬停轮廓 |
| link-shadow-hover | 链接按钮悬停阴影 | none | 链接按钮悬停阴影 |
| link-text-color-hover | 链接按钮悬停文字颜色 | #E02020 | 链接按钮悬停文字颜色 |
| link-background-visited | 链接按钮选中后背景 | transparent | 链接按钮选中后背景 |
| link-border-width-visited | 链接按钮选中后边框宽度 | none | 链接按钮选中后边框宽度 |
| link-border-style-visited | 链接按钮选中后边框风格 | none | 链接按钮选中后边框风格 |
| link-border-color-visited | 链接按钮选中后边框颜色 | none | 链接按钮选中后边框颜色 |
| link-border-radius-visited | 链接按钮选中后圆角 | none | 链接按钮选中后圆角 |
| link-outline-visited | 链接按钮选中后轮廓 | none | 链接按钮选中后轮廓 |
| link-shadow-visited | 链接按钮选中后阴影 | none | 链接按钮选中后阴影 |
| link-text-color-visit | 链接按钮选中后文字颜色 | #E02020 | 链接按钮选中后文字颜色 |

（3）平台内置变量输入框见表3-42。

表3-42 平台内置变量输入框

| 变量名 | 变量 | 默认值 | 描述 |
| --- | --- | --- | --- |
| background | 背景色 | #ffffff | 背景色 |
| border-width | 边框宽度 | 1px | 边框宽度 |
| border-color | 边框颜色 | #e3e7ee | 边框颜色 |
| border-radius | 圆角 | 4px | 圆角 |
| outline | 轮廓 | none | 轮廓 |
| shadow | 阴影 | none | 阴影 |
| text-color | 文字颜色 | rgba（0,0,0,0.85） | 文字颜色 |
| line-height | 行高 | 22px | 行高 |
| counter-background | 计数区域背景色 | #ffffff | 计数区块背景色 |
| counter-color | 计数器文字颜色 | rgba（0, 0, 0, 0.25） | 计数器文字颜色 |
| counter-font-size | 计数器文字大小 | 12px | 计数器文字大小 |
| background-hover | 悬停背景颜色 | #ffffff | 悬停背景颜色 |
| border-width-hover | 悬停边框宽度 | 1px | 悬停边框宽度 |
| border-style-hover | 悬停边框风格 | solid | 悬停边框风格 |
| border-color-hover | 悬停边框颜色 | #3F84FF | 悬停边框颜色 |
| border-radius-hover | 悬停边框圆角 | 4px | 悬停边框圆角 |
| outline-hover | 悬停轮廓 | none | 悬停轮廓 |
| shadow-hover | 悬停阴影 | none | 悬停阴影 |
| text-color-hover | 悬停文字颜色 | rgba（0,0,0,0.85） | 悬停文字颜色 |
| background-focus | 激活背景色 | #ffffff | 激活背景色 |
| border-width-focus | 激活边框宽度 | 1px | 激活边框宽度 |
| border-style-focus | 激活边框风格 | solid | 激活边框风格 |
| border-color-focus | 激活边框颜色 | #3F84FF | 激活边框颜色 |
| border-radius-focus | 激活圆角 | 4px | 激活圆角 |
| outline-focus | 激活轮廓 | none | 激活轮廓 |

| 变量名 | 变量 | 默认值 | 描述 |
|---|---|---|---|
| shadow-focus | 激活阴影 | 0px 0px 0px 2px rgba（3,93,255,0.1） | 激活阴影 |
| text-color-focus | 激活文字颜色 | rgba（0,0,0,0.85） | 激活文字颜色 |
| readonly-border-color | 只读边框颜色 | #e3e7ee | 只读边框颜色 |
| error-border-color | 异常边框颜色 | #ff4d4f | 异常时边框颜色 |
| disabled-border-color | 只读边框颜色 | #d9d9d9 | 只读边框颜色 |

（4）平台内置变量分页器见表 3-43。

表 3-43 平台内置变量分页器

| 变量名 | 变量 | 默认值 | 描述 |
|---|---|---|---|
| height | 高度 | 32px | 高度 |
| item-width | 分页器宽度 | 32px | 分页器宽度 |
| item-height | 分页器高度 | 30px | 分页器高度 |

（5）平台内置变量下拉框见表 3-44。

表 3-44 平台内置变量下拉框

| 变量名 | 变量 | 默认值 | 描述 |
|---|---|---|---|
| background | 背景色 | #ffffff | 背景色 |
| border-width | 边框宽度 | 1px | 边框宽度 |
| border-color | 边框颜色 | #e3e7ee | 边框颜色 |
| border-radius | 圆角 | 4px | 圆角 |
| outline | 轮廓 | none | 轮廓 |
| shadow | 阴影 | none | 阴影 |
| text-color | 文字颜色 | rgba（0,0,0,0.85） | 文字颜色 |
| background-hover | 悬停背景颜色 | #ffffff | 悬停背景颜色 |
| border-width-hover | 悬停边框宽度 | 1px | 悬停边框宽度 |
| border-style-hover | 悬停边框风格 | solid | 悬停边框风格 |
| border-color-hover | 悬停边框颜色 | #3F84FF | 悬停边框颜色 |
| border-radius-hover | 悬停边框圆角 | 4px | 悬停边框圆角 |
| outline-hover | 悬停轮廓 | none | 悬停轮廓 |
| shadow-hover | 悬停阴影 | none | 悬停阴影 |
| text-color-hover | 悬停文字颜色 | rgba（0,0,0,0.85） | 悬停文字颜色 |
| background-focus | 激活背景色 | #ffffff | 激活背景色 |
| border-width-focus | 激活边框宽度 | 1px | 激活边框宽度 |
| border-style-focus | 激活边框风格 | solid | 激活边框风格 |
| border-color-focus | 激活边框颜色 | #3F84FF | 激活边框颜色 |
| border-radius-focus | 激活圆角 | 4px | 激活圆角 |
| outline-focus | 激活轮廓 | none | 激活轮廓 |
| shadow-focus | 激活阴影 | 0px 0px 0px 2px rgba（3,93,255,0.1） | 激活阴影 |

续表

| 变量名 | 变量 | 默认值 | 描述 |
|---|---|---|---|
| text-color-focus | 激活文字颜色 | rgba（0,0,0,0.85） | 激活文字颜色 |
| readonly-border-color | 只读边框颜色 | #e3e7ee | 只读边框颜色 |
| error-border-color | 异常边框颜色 | #ff4d4f | 异常时边框颜色 |
| disabled-border-color | 只读边框颜色 | #d9d9d9 | 只读边框颜色 |
| item-readonly-radius | 选项只读圆角 | 4px | 选项只读圆角 |

（6）平台内置变量多行文本见表 3-45。

表 3-45 平台内置变量多行文本

| 变量名 | 变量 | 默认值 | 描述 |
|---|---|---|---|
| background | 背景色 | #ffffff | 背景色 |
| border-width | 边框宽度 | 1px | 边框宽度 |
| border-color | 边框颜色 | #e3e7ee | 边框颜色 |
| border-radius | 圆角 | 4px | 圆角 |
| outline | 轮廓 | none | 轮廓 |
| shadow | 阴影 | none | 阴影 |
| text-color | 文字颜色 | rgba（0,0,0,0.85） | 文字颜色 |
| background-hover | 悬停背景颜色 | #ffffff | 悬停背景颜色 |
| border-width-hover | 悬停边框宽度 | 1px | 悬停边框宽度 |
| border-style-hover | 悬停边框风格 | solid | 悬停边框风格 |
| border-color-hover | 悬停边框颜色 | #3F84FF | 悬停边框颜色 |
| border-radius-hover | 悬停边框圆角 | 4px | 悬停边框圆角 |
| outline-hover | 悬停轮廓 | none | 悬停轮廓 |
| shadow-hover | 悬停阴影 | none | 悬停阴影 |
| text-color-hover | 悬停文字颜色 | rgba（0,0,0,0.85） | 悬停文字颜色 |
| background-focus | 激活背景色 | #ffffff | 激活背景色 |
| border-width-focus | 激活边框宽度 | 1px | 激活边框宽度 |
| border-style-focus | 激活边框风格 | solid | 激活边框风格 |
| border-color-focus | 激活边框颜色 | #3F84FF | 激活边框颜色 |
| border-radius-focus | 激活圆角 | 4px | 激活圆角 |
| outline-focus | 激活轮廓 | none | 激活轮廓 |
| shadow-focus | 激活阴影 | 0px 0px 0px 2px rgba（3, 93, 255,0.1） | 激活阴影 |
| text-color-focus | 激活文字颜色 | rgba（0,0,0,0.85） | 激活文字颜色 |

（7）平台内置变量文件上传见表 3-46。

表 3-46　平台内置变量文件上传

| 变量名 | 变量 | 默认值 | 描述 |
| --- | --- | --- | --- |
| background | 背景色 | #ffffff | 背景色 |
| border-width | 边框宽度 | 1px | 边框宽度 |
| border-color | 边框颜色 | #e3e7ee | 边框颜色 |
| border-radius | 圆角 | 4px | 圆角 |
| outline | 轮廓 | none | 轮廓 |
| shadow | 阴影 | none | 阴影 |
| text-color | 文字颜色 | rgba（0,0,0,0.85） | 文字颜色 |
| background-hover | 悬停背景颜色 | #ffffff | 悬停背景颜色 |
| border-width-hover | 悬停边框宽度 | 1px | 悬停边框宽度 |
| border-style-hover | 悬停边框风格 | solid | 悬停边框风格 |
| border-color-hover | 悬停边框颜色 | #3F84FF | 悬停边框颜色 |
| border-radius-hover | 悬停边框圆角 | 4px | 悬停边框圆角 |
| outline-hover | 悬停轮廓 | none | 悬停轮廓 |
| shadow-hover | 悬停阴影 | none | 悬停阴影 |
| text-color-hover | 悬停文字颜色 | rgba（0,0,0,0.85） | 悬停文字颜色 |
| background-focus | 激活背景色 | #ffffff | 激活背景色 |
| border-width-focus | 激活边框宽度 | 1px | 激活边框宽度 |
| border-style-focus | 激活边框风格 | solid | 激活边框风格 |
| border-color-focus | 激活边框颜色 | #3F84FF | 激活边框颜色 |
| border-radius-focus | 激活圆角 | 4px | 激活圆角 |
| outline-focus | 激活轮廓 | none | 激活轮廓 |
| shadow-focus | 激活阴影 | 0px 0px 0px 2px rgba（3, 93, 255,0.1） | 激活阴影 |
| text-color-focus | 激活文字颜色 | rgba（0,0,0,0.85） | 激活文字颜色 |

**2. 自定义组件——字段**

字段的自定义是前端最常见的开发需求，Oinone 平台目前管理页面组件库用的是 antd，组件开发大部分就是基于 antd 组件，用 "vue+ts" 的语法进行封装，并注册为系统组件。

1）自定义字段组件的三部曲

（1）新建 XField.vue 来定义一个 vue 原生组件，代码如下。

```
1  <template>
2      <div>{{ value }}</div>
3  </template>
4  <script lang="ts">
5  import { defineComponent } from 'vue';
6  export default defineComponent({
7      props: ['value'],
8  });
9  </script>
10 <style lang="scss">
11     //样式写在这里
12 </style>
```

（2）新建 XFieldWidget.ts 封装成 Oinone 的组件，依赖 @kunlun/dependencies 包提供的 SPI 机制实现，匹配规则见 3.5.7 节的开篇介绍。代码如下。

```
1  import { FormFieldWidget, ModelFieldType, ViewType, SPI } from '@kunlun/dependenc
   ies';
2  import XField from './XField.vue';
3
4  @SPI.ClassFactory(
5    FormFieldWidget.Token({
6      viewType: [ViewType.Detail],
7      ttype: ModelFieldType.String,
8      name: 'XField'
9    })
10 )
11 export class XFieldWidget extends FormFieldWidget {
12   public initialize(props) {
13     super.initialize(props);
14     this.setComponent(XField);
15     return this;
16   }
17 }
18
```

（3）注册代码如下。

```
1  export * from './XFieldWidget';
```

2）第一个自定义字段组件

接下来一步一步地自定义一个千分位展示的金额字段组件，感受一下字段组件开发过程。

Step1.（前端）新建 currency 目录。

在开始前回顾一下我们的前端工程结构，我们在前端工程的 src/field 建一个 currency 目录，把自定义组件放到这个目录下，如图 3-214 所示。

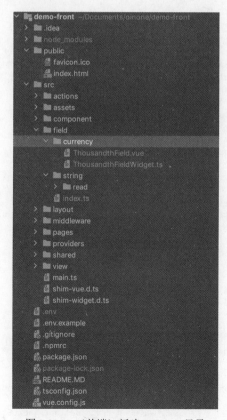

图 3-214 （前端）新建 currency 目录

Step2.（前端）新建一个 vue 原生组件 ThousandthField.vue。

（1）使用 antd 的 a-input 组件。

通过 value="valueStr"，让组件 value 跟 valueStr 进行绑定。

通过 onThousandthFieldChange、onThousandthFieldFocus、onThousandthFieldBlur 分别监听组件的值变化、聚焦、失焦。

（2）定义 vue 组件。

① props 定义接收从父组件传递过来的数据。

② setup (props)：

● transformThousandthStr 定义千分位转换逻辑；

● watch 对 valueStr 第一次赋值；

● onThousandthFieldChange，值变化时修改真正的 value；

● onThousandthFieldFocus，获取焦点的时候把 valueStr 转换为 value；

● onThousandthFieldBlur，失焦时把 valueStr 转化成千分位形式。

定义 vue 组件代码如下。

```ts
<template>
  <div class="form-input">
    <a-input
      class="oio-input"
      :value="valueStr"
      :placeholder="placeholder"
      :disabled="disabled"
      @change="onThousandthFieldChange"
      @focus="onThousandthFieldFocus"
      @blur="onThousandthFieldBlur"
    />
  </div>
</template>

<script lang="ts">
import { defineComponent, ref, watch,onRenderTriggered } from 'vue';
export default defineComponent({
  props: ['disabled', 'placeholder', 'value', 'change', 'blur', 'focus'],
  setup(props) {
    const transformThousandthStr = (num) => {
      if (num) {
        const reg = /\d{1,3}(?=(\d{3})+$)/g;
        if (num && num.toString().indexOf(".") == -1) {
          return (num + "").replace(reg, "$&,");
        } else {
          return num
            .toString()
            .replace(/(\d)(?=(\d{3})+\.)/g, function($0, $1) {
              return $1 + ",";
            });
        }
      }
    };
    var reg=/\d{1,3}(?=(\d{3})+$)/g
    const valueStr = ref<string>();
    const flag = ref<boolean>(true);
    //直接获取时，可能首次无法获取到值，需要使用watch的方法
    watch(
      () => props.value,
      (newVal) => {
        if(flag.value){
          if(props.value!= null && props.value != undefined && props.value != ''){
            valueStr.value = transformThousandthStr(props.value);
            flag.value =false;
```

```
45                  }
46              }
47          },
48          { immediate: true, deep: true },
49      );
50
51      const onThousandthFieldChange = (event) => {
52          valueStr.value=event.target.value;
53          props.change(event.target.value);
54      };
55      const onThousandthFieldFocus = (event) => {
56          //获取焦点时,恢复数字展示
57          props.focus && props.focus();
58          if(props.value){
59              valueStr.value = props.value;
60          }
61      };
62
63      const onThousandthFieldBlur = (event) => {
64          //失去焦点时赋值
65          props.blur && props.blur();
66          if(event.target.value == null || event.target.value == undefined ||event.target.value == ''){
67              props.change(0);
68              valueStr.value = '0';
69          }else{
70              props.change(event.target.value );
71              //失去焦点时用千分位展示
72              valueStr.value = transformThousandthStr(props.value);
73          }
74      };
75
76      return { valueStr, onThousandthFieldChange,onThousandthFieldBlur,onThousandthFieldFocus };
77      },
78  });
79  </script>
```

Step3.(前端)新建 ThousandthFieldWidget.ts 封装成 Oinone 组件。

(1)继承 FormFieldWidget。

(2)initialize 为组件初始化方法。

(3)通过 @SPI.ClassFactory (FieldWidget.Token ({ viewType:[ViewType.Form,ViewType.Detail,ViewType.Search],ttype: [ModelFieldType.Currency],widget: 'ThousandthFieldWidget' })) 来定义组件匹配规则:

① viewType:ViewType.Form,ViewType.Detail,ViewType.Search。

② ttype:ModelFieldType.Currency(对应后端 @Field.Money)。

③ widget:ThousandthFieldWidget。

代码如下。

```
1   import { FieldWidget, FormFieldWidget, ModelFieldType, SPI, ViewType, Widget } from '@kunlun/dependencies';
2   import ThousandthField from './ThousandthField.vue';
3
4   @SPI.ClassFactory(
5       FieldWidget.Token({
6           viewType: [ViewType.Form, ViewType.Detail, ViewType.Search],
7           ttype: [ModelFieldType.Currency],
8           widget: 'ThousandthFieldWidget'
9       })
10  )
11  export class ThousandthFieldWidget extends FormFieldWidget {
```

```
12    public initialize(props) {
13        super.initialize(props);
14        this.setComponent(ThousandthField);
15        return this;
16    }
17
18    @Widget.Method()
19    public change(value) {
20        super.change(value);
21    }
22
23    @Widget.Method()
24    public get disabled() {
25        return this.isReadonly();
26    }
27  }
28
```

Step4.（前端）注册 ThousandthFieldWidget。

修改 field 目录下 index.ts，代码如下。

```
1  export * from './string/read/DetailStringUpperCaseFieldWidget';
2  export * from './currency/ThousandthFieldWidget';
```

Step5.（后端）修改 PetTalent 模型。增加一个年收入字段，ttype 定义为 Money，后端的 Money 对应前端 Currency。代码如下。

```
1  @Field.Money
2  @Field(displayName = "年收入")
3  private BigDecimal annualIncome;
```

Step6.（后端）修改 PetTalent 的 Form 视图的 Template 文件，增加 annualIncome 字段，把 widget 指定为 ThousandthFieldWidget。代码如下。

```
1  <field data="annualIncome" widget="ThousandthFieldWidget"/>
2
```

Step7. 重启系统看效果，如图 3-215 所示。

图 3-215　重启系统看效果

3）替换原有默认组件

上面的例子是新增一个名叫 ThousandthFieldWidget 的匹配 ttype 为 Money 的字段组件，如果我们希望把系统中所有的 Money 都替换成这个组件，而不是每个模型每个页面逐一配置，那么根据匹配规则来修改组件代码就可以了。

Step1.（前端）修改 ThousandthFieldWidget.ts 的匹配规则，把原来的 widget: 'ThousandthFieldWidget' 改成 widget: ['ThousandthFieldWidget', '*']。

（1）* 代表通配。

（2）也可以删除 widget: 'ThousandthFieldWidget'，不定义也是通配。

（3）widget: ['ThousandthFieldWidget', '*'] 即代表通配，同时也可以配置 ThousandthFieldWidget。

代码如下。

```typescript
import { FieldWidget, FormFieldWidget, ModelFieldType, SPI, ViewType, Widget } from '@kunlun/dependencies';
import ThousandthField from './ThousandthField.vue';

@SPI.ClassFactory(
  FieldWidget.Token({
    viewType: [ViewType.Form, ViewType.Detail, ViewType.Search],
    ttype: [ModelFieldType.Currency],
    widget: ['ThousandthFieldWidget', '*']
  })
)
export class ThousandthFieldWidget extends FormFieldWidget {
  public initialize(props) {
    super.initialize(props);
    this.setComponent(ThousandthField);
    return this;
  }

  @Widget.Method()
  public change(value) {
    super.change(value);
  }

  @Widget.Method()
  public get disabled() {
    return this.isReadonly();
  }
}
```

Step2.（后端）修改 PetTalent 的 Form 视图，把 annualIncome 字段的 widget 配置去掉（widget="ThousandthFieldWidget"），代码如下。

```xml
<field data="annualIncome" />
```

Step3. 重启看效果，结果同上面 case。

4）获取 xml 属性

在自定义组件过程中，可以针对不同的业务场景进行封装，一些通用的个性化需求可以沉淀在 xml 中。通过不同视图的 xml 配置，满足前端组件的个性化。

xml 中定义 key 和对应的 value 代码如下。

```xml
//xml 中定义key和对应的value
<field name="字段名称" data="字段名称" testDslConfigKey="test" />
```

（1）在组件的 ts 内获取，代码如下。

```typescript
const { testDslConfigKey = '' } = this.getDsl();
```

（2）当需要响应式时，在 ts 内部定义响应式变量赋值，代码如下。

```
1  // 定义响应式变量
2  @Widget.Reactive()
3  private testDslConfigKey: string = '';
4  // 从dsl内获取
5  const { testDslConfigKey = '' } = this.getDsl();
6  // 赋值给响应式变量
7  this.testDslConfigKey = testDslConfigKey;
```

（3）在对应的 vue 文件中获取，代码如下。

```
1  <script lang="ts">
2    import { defineComponent } from 'vue';
3
4    export default defineComponent({
5      props: {
6        testDslConfigKey: {
7          tyep: String,
8          default: '',
9        },
10     }
11   });
12 </script>
```

5）值的提交和获取

在组件定义中，需要继承对应 ttype 的基类，基类中内置了值的提交、获取对应函数和变量。值的提交和获取需要在 ts 文件中操作，字段对应的值默认是响应式变量。

（1）值提交。

在 ts 文件中提交，代码如下。

```
1  // 第一种写法
2  super.change(value);
3  // 第二种写法
4  this.change(value)
```

（2）值获取。

注意，this.value 是异步赋值，如果业务中需要对值进行操作，需要判断空值和处理异常，代码如下。

```
1  // ts文件内this.value即是当前字段的值，是响应式变量
2  this.value
```

**3. 自定义组件——动作**

在日常开发中经常也会碰到自定义一些前端动作的需求。自定义动作有两个核心概念：一是定义前端 Action；二是定义前端 Action 的组件。这个跟其他自定义组件不大一样，比如自定义字段组件，字段本身已经是存在的，但自定义前端 Action 的组件，首先得确保前端 Action 存在，不存在得先定义前端 Action。

1）第一个动作组件

自定义一个全新前端动作组件只要简单的几步，下面例子构建了一个 DoNothingActionWidget 动作组件。

Step1.（前端）Action 目录规范。

在开始前回顾一下前端工程结构，自定义组件放在前端工程的 src/actions 目录下，如图 3-216 所示。

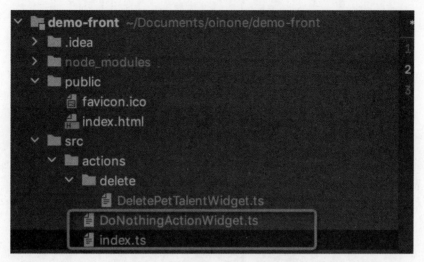

图 3-216 （前端）Action 目录规范

Step2.（前端）新建 DoNothingActionWidget 组件。

（1）定义前端 Action 的组件 DoNothingActionWidget 继承 ActionWidget。

（2）用 @SPI.ClassFactory (ActionWidget.Token ({ name: 'demo.doNothing' })) 来定义组件匹配规则：name：demo.doNothing

（3）用 Action.registerAction 定义前端 Action：

① * //* 为模型通配符，可指定模型。

② displayName: ' 啥也没干 ', // 动作展示名。

③ name: 'demo.doNothing', // 动作技术名称。

④ id: 'demo.doNothing', // 动作组件 id。

⑤ contextType: ActionContextType.ContextFree, // 设置动作上下文类型。

⑥ bindingType: [ViewType.Table] // 设置按钮所在页面类型。

代码如下。

```
1   import {Action,ActionContextType,ActionWidget,executeConfirm,IClientAction,SPI,ViewType,Widge} from '@kunlun/dependencies';
2
3   @SPI.ClassFactory(ActionWidget.Token({ name: 'demo.doNothing' }))
4   export class DoNothingActionWidget extends ActionWidget {
5     @Widget.Method()
6     public async clickAction() {
7       const confirmRs = executeConfirm('oinone第一个自定义Action，啥也没干');
8     }
9   }
10
11    //定义动作元数据
12    Action.registerAction('*', {
13      displayName: '啥也没干',
14      name: 'demo.doNothing',
15      id: 'demo.doNothing',
16      contextType: ActionContextType.ContextFree,
17      bindingType: [ViewType.Table]
18    } as IClientAction);
```

Step3.（前端）注册 DoNothingActionWidget 组件。

修改 action 目录下的 index.ts，代码如下。

```
1   export * from './DoNothingActionWidget';
```

Step4.（后端）修改 PetTalent 表格视图的 Template，代码如下。

```
1  <!-- <template slot="actions" autoFill="true"/> slot里面有aciton定义就不再自动填充其他Action-->
2  <template slot="actions" autoFill="true">
3    <action actionType="client" name="demo.doNothing" label="第一个自定义Action" />
4    <action name="delete" label="删除" />
5    <action name="redirectCreatePage" label="创建" />
6  </template>
```

Step5. 重启看效果，如图 3-217 所示。

图 3-217　重启看效果

2）覆盖平台动作（举例一）

字段的 Widget 可以替换，Action 的 Widget 同样可以替换，现在就来写一个自定义动作 Widget 来覆盖特定动作的前端逻辑。

Step1.（前端）新建 DeletePetTalentWidget.ts，代码如下。

在 src/action/delete 目录下新建 DeletePetTalentWidget.ts 文件，用于给 PetTalent 模型的列表的删除加上二次确认。

```
1  import {Action,ActionWidget,callFunction,DeleteOneActionWidget,executeConfirm,getModel,IServerAction,ModelDefaultActionName,RELOAD_VIEW,SPI,ViewType,Widget,WidgetSubjection} from '@kunlun/dependencies';
2
3  @SPI.ClassFactory(ActionWidget.Token({ model: 'demo.PetTalent', name: ModelDefaultActionName.delete }))
4  export class DeletePetTalentWidget extends DeleteOneActionWidget {
5    @Widget.SubContext(RELOAD_VIEW)
6    protected reload$!: WidgetSubjection<boolean | string>;
7
8    @Widget.Method()
9    public executeAction(action: IServerAction) {
10     const res = callFunction(
11       this.action.model!,
12       action as IServerAction,
13       this.activeRecords as Record<string, unknown>[]
14     );
15     res.then(() => {
16       this.action$.next({ action });
17     });
18     return res;
19   }
20
21   @Widget.Method()
22   public _clickAction(action: IServerAction): unknown | void | Promise<void> | Promise<unknown> {
23     return this.validateConfirm().then((res) => {
24       if (res) {
25         return this.executeAction(action);
26       }
```

```
27            });
28        }
29
30        @Widget.Method()
31        public async clickAction() {
32            const confirmRs = await executeConfirm('是否确定要删除');
33            if (confirmRs) {
34                const model = await getModel(this.action.model as string);
35                const deleteAction = (model.serverActionList || []).find((a) => a.name === 'delete')! as IServerAction;
36                try {
37                    await this._clickAction(deleteAction);
38                    this.reload$.subject.next(true);
39                } catch (error) { }
40            }
41        }
42    }
```

Step2.（前端）注册 DeletePetTalentWidget 组件，修改 action 目录下的 index.ts。代码如下。

```
1    export * from './delete/DeletePetTalentWidget';
2    export * from './DoNothingActionWidget';
```

Step3. 重启看效果，如图 3-218 所示。

图 3-218　重启看效果

3）覆盖平台动作（举例二）

再举一个例子，在创建页面中，"确认"按钮的自定义动作 Widget 来覆盖原有动作的前端逻辑。

Step1. 新建 CreatePetServer.ts，代码如下。

给 PetTalent 模型的表单添加一个自定义提交的动作。

```
1    import {ActionWidget,callFunction,IServerAction,getModel,ModelDefaultActionName,RELOAD_VIEW,SPI,Widget,WidgetSubjection} from '@kunlun/dependencies';
2
3    @SPI.ClassFactory(ActionWidget.Token({ model: 'demo.PetTalent', name: ModelDefaultActionName.create }))
4    export class CreatePetServer extends ActionWidget {
5        @Widget.SubContext(RELOAD_VIEW)
6        protected reload$!: WidgetSubjection<boolean | string>;
7
8        @Widget.Method()
9        public async executeAction(action: IServerAction) {
10           const data = this.activeRecords[0];
11           const res = await callFunction(this.action.model!, action as IServerAction, d
```

```
ata, undefined, undefined, {});
12       return res;
13     }
14
15     @Widget.Method()
16     public _clickAction(action: IServerAction): unknown | void | Promise<void> | Promise<unknown> {
17       return this.executeAction(action);
18     }
19
20     @Widget.Method()
21     public async clickAction() {
22       const model = await getModel(this.action.model as string);
23       const action = (model.serverActionList || []).find((a) => a.name === 'create')! as IServerAction;
24       try {
25         await this._clickAction(action);
26         history.back();
27         this.reload$.subject.next(true);
28       } catch (error) {}
29     }
30   }
31
```

Step2.（前端）注册 CreatePetServer，修改 action 目录下的 index.ts。代码如下。

```
1  export * from './delete/DeletePetTalentWidget';
2  export * from './DoNothingActionWidget';
3  export * from './CreatePetServer';
```

**4. 自定义视图——表格**

了解了字段、动作组件的自定义，再学习核心内容区的自定义也就比较轻松了，基本思路跟字段组件自定义是一样的。

1）自定义字段组件的三部曲

（1）新建 XView.vue 定义一个 vue 原生组件，代码如下。

```
1  <template>
2    <div class="container">
3        这是你的第一个viewWidget，这个widget教会你如何自定义一个viewWidget<br/>
4        内容区就是你发挥想象的地方
5    </div>
6  </template>
7
8  <script lang="ts">
9  import { defineComponent } from 'vue';
10
11 export default defineComponent({
12   props: [],
13 });
14 </script>
```

（2）新建 XViewWidget.ts 封装成 Oinone 的组件，代码如下。依赖 @kunlun/core 包提供的 SPI 机制实现，匹配规则见本节内容的开篇介绍。

```typescript
1  import { Entity, getModel, getModelByUrl, IModel, SPI, ViewWidget, Widget } from
   '@kunlun/dependencies';
2  import XView from './XView.vue';
3
4  @SPI.ClassFactory(ViewWidget.Token({ widget: 'demo.X' }))
5  //视图的自定义组件必须继承自ViewWidget基类
6  export class XViewWidget extends ViewWidget {
7    @Widget.Reactive()
8    private loading = false;
9
10   private modelInstance!: IModel;
11
12   public setBusy(busy: boolean) {
13     this.loading = busy;
14   }
15
16   //重写了父类的initialize方法控制页面渲染
17   public initialize(props) {
18     super.initialize(props);
19     this.setComponent(XView);
20     return this;
21   }
22
23   //页面加载时获取元数据
24   public async $$mounted() {
25     super.mounted();
26     const modelModel = getModelByUrl();
27     this.modelInstance = await getModel(modelModel);
28   }
29
30   //重写数据提交方法,一般针对form
31   public async submit() {
32     //数据提交
33     return {};
34   }
35   //重写数据获取方法
36   public async fetchData(data?: Entity[], options?, variables?: Record<string, un
   known>) {
37     //获取数据
38     return {};
39   }
40  }
41
```

(3) 注册,代码如下。

```typescript
1  export * from './XViewWidget';
```

2) 第一个自定义 View 组件

接下来为 PetTalent 模型一步一步地自定义一个 Card 模式展示的表格 view 组件,感受一下视图组件的开发过程。

Step1. (前端) 新建 pet 目录。

在开始前回顾一下前端工程结构,找到前端工程的 src/view 目录,新建 pet 目录把自定义组件放到这个目录下,如图 3-219 所示。

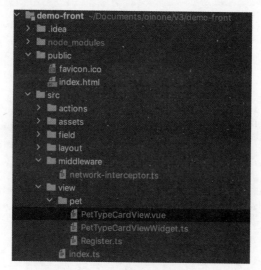

图 3-219　（前端）新建 pet 目录

Step2.（前端）新建一个 vue 原生组件 PetTypeCardView.vue，代码如下。

```vue
<template>
  <div class="card" style="display: flex;flex-wrap: wrap;">
    <div v-for="(card, index) in dataSource">
      <a-card :title="card.name" style="width: 300px">
        <p>品种: {{ card.name }}</p>
      </a-card>
    </div>
  </div>
  <div class="table-container-page">
    <a-pagination
      :show-total="(total) => `${total} 条`"
      show-size-changer
      :current="pagination.current"
      :pageSize="pagination.pageSize"
      :total="pagination.total"
      @change="onChange"
      @showSizeChange="onChange"
      :page-size-options="['15', '30', '50', '100']"
    >
      <template #buildOptionText="props">
        <span>{{ props.value }}条/页</span>
      </template>
    </a-pagination>
  </div>
</template>

<script lang="ts">
import { defineComponent } from 'vue';

export default defineComponent({
  props: {
    dataSource: {
      tyep: Array,
      default: [],
    },
    pagination: {
      type: Object,
      default: {
        current: 1,
        pageSize: 10,
        total: 0
      }
    }
  }
});
</script>
```

Step3.（前端）新建 PetTypeCardViewWidget.ts 封装成 Oinone 组件。

(1) 继承 TableWidget，复用表格能力。

(2) initialize 为组件初始化方法。

(3) 通过 @SPI.ClassFactory (ViewWidget.Token ({ type: [ViewType.Custom,ViewType.Table], widget: 'PetTypeCard'})) 来定义组件匹配规则：

① viewType：ViewType.Custom,ViewType.Table。

② widget：PetTypeCard。

代码如下：

```ts
import { Condition, CustomWidget, getModel, getModelByUrl, IModel,
IQueryPageOption, queryPage, SPI, TableWidgetV3 } from '@kunlun/dependencies';
import PetTalentCardView from './PetTypeCardView.vue';

@SPI.ClassFactory(CustomWidget.Token({ widget: 'PetTypeCard' }))
export class PetTypeCardViewWidget extends TableWidgetV3 {
    public initialize(props) {
        super.initialize({
            ...props,
            tableFieldNames: []
        });
        this.setComponent(PetTalentCardView);
        return this;
    }

    private modelInstance!: IModel;

    public async fetchData(content: Record<string, unknown>[] = [],
options: IQueryPageOption = {}, variables: Record<string, unknown> = {}) {
        const reqData = content[0] || {};
        const condition = new Condition('1==1');
        condition.and(this.buildSearchConditions(reqData));
        const fields = this.modelInstance.modelFields;
        const res = await queryPage(this.modelInstance!.model, { condition, ...option
s }, fields, variables, {
            maxDepth: 1
        });
        this.setPagination({ total: res.totalElements, pageSize: res.size, current: o
ptions.currentPage || 1 });
        const data = res.content || [];

        this.loadData(data);
        this.setContent(content);
        this.setOptions(options);

        return this.dataSource;
    }

    public async $$mounted() {
        super.mounted();
        const modelModel = getModelByUrl();
        this.modelInstance = await getModel(modelModel);
        this.fetchData();
    }
}
```

Step4.（前端）自定义一个 Layout 并注册为只影响宠物品种模型的 Table 视图。

新建 Register.ts 文件，通过 registerLayout 函数注册为只影响宠物品种模型的 Table 视图，如果

moduleName 和 Model 不配置则会影响所有模型 Table 视图的默认 Layout。代码如下。

```
import {ViewType, registerLayout} from '@kunlun/dependencies';

export const install = () => {
  registerLayout(`<view type="TABLE">
    <pack widget="fieldset">
      <view type="SEARCH">
        <element widget="search" slot="search" slotSupport="field" />
      </view>
    </pack>
    <pack widget="fieldset" style="height: 100%" wrapperStyle="height: 100%">
      <pack widget="row" style="height: 100%; flex-direction: column">
        <pack widget="col" mode="full" style="flex: 0 0 auto">
          <element widget="actionBar" slot="actionBar" slotSupport="action">
            <xslot name="actions" slotSupport="action" />
          </element>
        </pack>
        <pack widget="col" mode="full">
          <custom widget="PetTypeCard" slotSupport="field">
          </custom>
        </pack>
      </pack>
    </pack>
  </view>`, {
    viewType: ViewType.Table,
    moduleName: 'DemoCore',
    model: 'demo.PetType'
  });
}
install();
```

Step5.（前端）注册 PetTalentCardViewWidget，修改 view 目录下 index.ts。代码如下。

```
export * from './pet/PetTypeCardViewWidget';
export * from './pet/Register';
```

Step6.（前端）重启系统看效果，如图 3-220 所示。

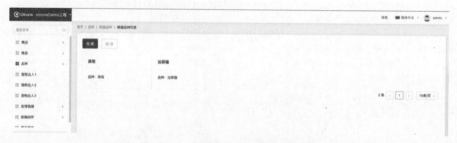

图 3-220　（前端）重启系统看效果

我们可以看到品种列表的表格已经发生变化。

3）第一个自定义的大数据表格

Setp1.（前端）新建 virtual-table 目录，如图 3-221 所示。

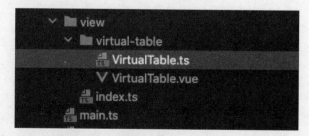

图 3-221　（前端）新建 virtual-table 目录

Step2.（前端）新建一个 vue 原生组件 VirtualTable.vue。新增打印、导出、搜索、聚合等功能，若需要其他复杂功能，请查阅第三方文档自行定制。代码如下。

```ts
<template>
  <div class="virtual-table">
    <div class="search">模糊搜索：<vxe-input type="text" @blur="(e) => onSearchChange(e, 'shopName')"></vxe-input></div>
    <div>
      <vxe-button @click="printEvent2">打印</vxe-button>
      <vxe-button @click="exportDataEvent">导出</vxe-button>
    </div>
    <vxe-table border show-overflow show-header-overflow ref="xTable" height="600" width="100%" :print-config="demo1.tablePrint" :row-config="{ isCurrent: true, isHover: true, useKey: true }" :column-config="{ resizable: true }" :export-config="{}" :loading="demo1.loading" :edit-config="{ trigger: 'click', mode: 'cell' }" :sort-config="{ trigger: 'cell' }">
      <vxe-column type="seq" width="100" fixed="left"></vxe-column>
      <vxe-column field="code" title="店铺编码" width="120" sortable fixed="left"></vxe-column>
      ...
      <vxe-column field="petShopGm" title="店长" width="120" fixed="right"></vxe-column>
    </vxe-table>
  </div>
</template>

<script lang="ts">
import { defineComponent, reactive, ref, watch } from 'vue';
import { Input as VXEInput, VXETable, VxeTableInstance, VxeButtonEvents } from 'vxe-table';
export default defineComponent({
  props: ['dataSource', 'loadData'],
  components: { VXETable, VXEInput },
  setup(props) {
    const demo1 = reactive({
      loading: false,
      tableData: [],
      // 打印配置
      tablePrint: {
        // 自定义打印的样式示例
        style: `
          .vxe-table {
            color: #000000; // 修改表格默认颜色
            font-size: 12px; // 修改表格默认字体大小
            font-family: "Microsoft YaHei",微软雅黑,"MicrosoftJhengHei",华文细黑,STHeiti,MingLiu; // 修改表格默认字体
          }
        `
      }
    });

    const xTable = ref<VxeTableInstance>();
    const printEvent1: VxeButtonEvents.Click = () => {
      const $table = xTable.value;
      $table!.print();
    };
    const printSelectEvent: VxeButtonEvents.Click = () => {
      const $table = xTable.value;
      $table!.print({
        data: $table!.getCheckboxRecords()
      });
    };
    const printEvent2: VxeButtonEvents.Click = () => {
      const $table = xTable.value;
      $table!.openPrint();
    };
```

```
55      const exportDataEvent: VxeButtonEvents.Click = () => {
56        const $table = xTable.value;
57        $table!.exportData({ type: 'csv' });
58      };
59      const loadList = () => {
60        demo1.loading = true;
61        const $table = xTable.value;
62        // 使用函数式加载
63        if ($table) {
64          $table.reloadData(demo1.tableData).then(() => {
65            demo1.loading = false;
66          });
67        } else {
68          demo1.loading = false;
69        }
70      };
71      watch(
72        () => props.dataSource,
73        () => {
74          if (props.dataSource.length) {
75            demo1.tableData = props.dataSource;
76            loadList();
77          }
78        },
79        { deep: true }
80      );
81      const onValueChange = (e, index, name) => {
82        (demo1.tableData[index] as any)[name] = e.value;
83        loadList();
84      };
85      const onSearchChange = (e, name) => {
86        let condition = '';
87        if (name === 'shopName') {
88          condition = condition + `shopName=like='%${e.value}%'`;
89        }
90        props.loadData(condition);
91      };
92      return {
93        xTable, demo1, loadList, printEvent1, printEvent2, printSelectEvent,exportDataEvent, onValueChange, onSearchChange
94      };
95    }
96  });
97  </script>
98
```

Step3.（前端）新建 VirtualTable.ts 封装成 Oinone 组件。

（1）继承 ViewWidget。

（2）initialize 为组件初始化方法。

（3）通过 @SPI.ClassFactory (CustomWidget.Token ({ widget:'VirtualTable'})) 定义组件匹配规则：widget：VirtualTable。

代码如下。

```
1   import {Condition, getModel,getModelByUrl,IModel,IQueryPageOption,queryPage,SPI,
2     Widget,ViewWidget, Entity, CustomWidget, http,} from '@kunlun/dependencies';
3   import VirtualTableVue from './VirtualTable.vue';
4
5   @SPI.ClassFactory(CustomWidget.Token({ widget: 'VirtualTable' }))
6   export class VirtualTableWidget extends ViewWidget {
7     public initialize(props) {
8       super.initialize(props);
9       this.setComponent(VirtualTableVue);
10      return this;
```

```
11      }
12      private modelInstance!: IModel;
13      @Widget.Reactive()
14      private dataSource: Entity[] = []
15      @Widget.Method()
16      public async loadData(
17        condition?: string
18      ) {
19        const res = await this.customQueryPage(condition);
20        this.dataSource = res.content as Entity[];
21      }
22      // pageSize: -1 请求全部数据
23      private async customQueryPage(condition: string = '') {
24        const query = `{
25          petShopProxyAQuery {
26            queryPage(page: {currentPage: 1, size: -1, groupBy: "code"}, queryWrapper: {rsql: "1==1 ${condition && `and ${condition}`}"}) {
27              content {
28                income
29                id
30                petShopGmId
31                code
32                codeTwo
33                createUid
34                writeUid
35              }
36              size
37              totalPages
38              totalElements
39            }
40          }
41        }`
42
43        const result = await http.query('DemoCore', query);
44        return result['data']['petShopProxyAQuery']['queryPage']
45      }
46
47      public async $$mounted() {
48        super.mounted();
49        const modelModel = getModelByUrl();
50        this.modelInstance = await getModel(modelModel);
51        this.loadData();
52      }
53    }
54
```

Step4.（前端）自定义一个 Layout 并注册为只影响宠物店铺代理模型 A 模型的 Table 视图。

新建 Register.ts 文件，通过 registerLayout 函数注册为只影响宠物店铺代理模型 A 的 Table 视图，如果 moduleName 和 Model 不配置则会影响所有的模型 Table 视图的默认 Layout。代码如下。

```
1   import { registerLayout, ViewType } from '@kunlun/dependencies'
2
3   export const install = () => {
4     registerLayout(`<view type="TABLE">
5   <pack widget="fieldset" style="height: 100%" wrapperStyle="height: 100%">
6     <pack widget="row" style="height: 100%; flex-direction: column">
7       <pack widget="col" mode="full">
8         <custom widget="VirtualTable" slotSupport="field">
9         </custom>
10      </pack>
11    </pack>
12  </pack>
13  </view>`, {
14    viewType: ViewType.Table,
15    moduleName: 'DemoCore',
16    model: 'demo.PetShopProxyA'
17  })
18  }
```

```
19
20    install();
```

Step5.（前端）注册 VirtualTable，修改 view 目录下 index.ts。代码如下。

```
1  export * from './virtual-table/VirtualTable'
2  export * from './virtual-table/Register'
```

Step6.（前端）重启系统看效果，如图 3-222 所示。

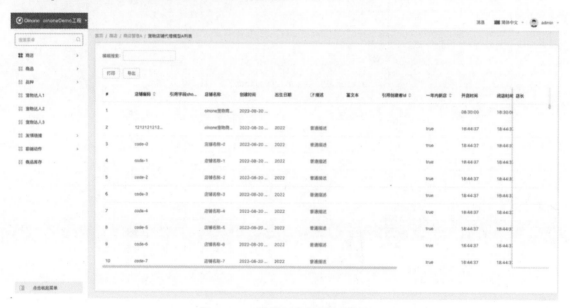

图 3-222　（前端）重启系统看效果

**5. 自定义视图——表单**

前面讲到如何自定义宠物达人表格，在中后台管理系统中，表单也是重要的组成部分。

Step1.（前端）新建 petForm 目录。

新建 petForm 目录，把自定义组件放到 src/view 目录下，如图 3-223 所示。

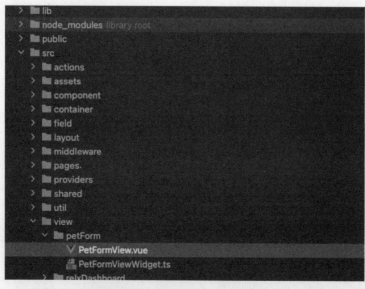

图 3-223　新建 petForm 目录，把自定义组件放到 src/view 目录下

Step2.（前端）新建一个 vue 原生组件 PetForm.vue，代码如下。

```ts
<template>
  <div class="petFormWrapper">
    <form :model="formState" @finish="onFinish">
      <a-form-item label="品种种类" id="name" name="kind" :rules="[{ required: true, message: '请输入品种种类!', trigger: 'focus' }]">
        <a-input v-model:value="formState.kind" @input="(e) => onNameChange(e, 'kind')" />
      </a-form-item>

      <a-form-item label="品种名" id="name" name="name" :rules="[{ required: true, message: '请输入品种名!', trigger: 'focus' }]">
        <a-input v-model:value="formState.name" @input="(e) => onNameChange(e, 'name')" />
      </a-form-item>

      <a-button @click="reloadData">重新渲染数据</a-button>
    </form>
  </div>
</template>

<script lang="ts">
import { defineComponent, reactive } from 'vue';
import { Form } from 'ant-design-vue';

export default defineComponent({
  props: ['onChange', 'reloadData'],
  components: { Form },
  setup(props) {
    const formState = reactive({
      kind: '',
      name: '',
    });

    const onFinish = () => {
      console.log(formState);
    };

    const onNameChange = (event, name) => {
      props.onChange(name, event.target.value);
    };

    const reloadData = async () => {
      await props.reloadData();
    };

    return {
      formState,
      reloadData,
      onNameChange,
      onFinish
    };
  }
});
</script>
```

Step3.（前端）新建 PetFormViewWidget.ts 封装成 Oinone 组件。

（1）继承 FormWidget，复用表单能力。

（2）initialize 为组件初始化方法。

（3）通过 @SPI.ClassFactory (CustomWidget.Token ({ widget:'PetForm'})) 定义组件匹配规则：

widget：PetForm。

代码如下。

```typescript
import { constructOne, FormWidget, queryOne, SPI, ViewWidget, Widget, IModel, get
ModelByUrl, getModel, getIdByUrl, FormWidgetV3, CustomWidget, CallChaining } from
 '@kunlun/dependencies';
import PetFormView from './PetForm.vue';

@SPI.ClassFactory(CustomWidget.Token({ widget: 'PetForm' }))
export class PetFormViewWidget extends FormWidgetV3 {
  public initialize(props) {
    super.initialize(props);
    this.setComponent(PetFormView);
    return this;
  }

  /**
   * 数据提交
   * @protected
   */
  @Widget.Reactive()
  @Widget.Inject()
  protected callChaining: CallChaining | undefined;

  private modelInstance!: IModel;

  /**
   * 重要！！！！
   * 当字段改变时修改formData
   * */
  @Widget.Method()
  public onFieldChange(fieldName: string, value) {
    this.setDataByKey(fieldName, value);
  }

  /**
   * 表单编辑时查询数据
   * */
  public async fetchData(content: Record<string, unknown>[] = [], options: Record
<string, unknown> = {}, variables: Record<string, unknown> = {}) {
    this.setBusy(true);
    const context: typeof options = { sourceModel: this.modelInstance.model, ...o
ptions };
    const fields = this.modelInstance?.modelFields;
    try {
      const id = getIdByUrl();
      const data = (await queryOne(this.modelInstance.model, (content[0] || { id
 }) as Record<string, string>, fields, variables, context)) as Record<string, unkn
own>;

      this.loadData(data);
      this.setBusy(false);
      return data;
    } catch (e) {
      console.error(e);
    } finally {
      this.setBusy(false);
    }
  }

  /**
   * 新增数据时获取表单默认值
   * */
  @Widget.Method()
  public async constructData(content: Record<string, unknown>[] = [], options: Re
cord<string, unknown> = {}, variables: Record<string, unknown> = {}) {
```

```
57        this.setBusy(true);
58        const context: typeof options = { sourceModel: this.modelInstance.model, ...o
    ptions };
59        const fields = this.modelInstance.modelFields;
60        const reqData = content[0] || {};
61        const data = await constructOne(this.modelInstance!.model, reqData, fields, v
    ariables, context);
62        return data as Record<string, unknown>;
63      }
64
65      @Widget.Method()
66      private async reloadData() {
67        const data = await this.constructData();
68        // 覆盖formData
69        this.setData(data);
70      }
71
72      @Widget.Method()
73      public onChange(name, value) {
74        this.formData[name] = value;
75      }
76
77      protected async mounted() {
78        super.mounted();
79        const modelModel = getModelByUrl();
80        this.modelInstance = await getModel(modelModel);
81        this.fetchData();
82        // 数据提交钩子函数！！！
83        this.callChaining?.callBefore(() => {
84          return this.formData;
85        });
86      }
87    }
88
```

**Step4.**（前端）自定义一个 Layout 并注册为只影响宠物品种模型的 Form 视图。

新建 Register.ts 文件，通过 registerLayout 函数注册为只影响宠物品种模型的 Form 视图，如果 moduleName 和 Model 不配置则会影响所有的模型 Form 视图的默认 Layout。代码如下。

```
1   import { registerLayout, ViewType } from '@kunlun/dependencies'
2
3   export const install = () => {
4     registerLayout(`<view type="FORM">
5       <element widget="actionBar" slot="actionBar" slotSupport="action">
6         <xslot name="actions" slotSupport="action" />
7       </element>
8       <custom widget="PetForm">
9       </custom>
10    </view>`, {
11      viewType: ViewType.Form,
12      moduleName:'DemoCore',
13      model: 'demo.PetType'
14    });
15  }
16
17  install();
```

**Step5.**（前端）注册 PetFormViewWidget，修改 view 目录下的 index.ts。代码如下。

```
1   export * from './petForm/PetFormViewWidegt';
2   export * from './petForm/Register';
```

**Step6.**（前端）重启系统。

前端刷新页面，单击"宠物品种"的"创建"按钮进入表单，如图 3-224 所示。

图 3-224　前端刷新页面，单击"宠物品种"的"创建"按钮进入表单

**6. 自定义 mask**

这里讲述如何替换已有的 mask 和添加自己的 mask 组件。以菜单放到 header 区域为例。

1）自定义 mask 组件

Step1.（前端）新建 mask 目录。

新建 mask 目录，把自定义组件放到 src/mask 目录下，如图 3-225 所示。

图 3-225　新建 mask 目录，把自定义组件放到 src/mask 目录下

Step2.（前端）新建一个 vue 原生组件 CustomMaskComponent.vue，代码如下。

```
1  <template>
2    <div class="custom-mask-component" :key="menus.length">
3      <div class="menu-area oio-scrollbar">
4        <div class="menu-content">
5          <a-menu class="oinone-menu" mode="horizontal" :selectedKeys="selectKeys"
    v-model:openKeys="innerOpenKeys" @click="onMenuSelected" @openChange="openChang
    e">
6          <template v-for="item in menus" :key="item.key">
7            <template v-if="!item.children || !item.children.length">
8              <a-menu-item :key="item.key" :title="item.title">
9                <template #icon>
10                 <oio-icon :icon="item.menu.icon || DEFAULT_MENU_ICON" />
11                </template>
12                {{ item.title }}
13              </a-menu-item>
14            </template>
15            <template v-else>
```

```
16              <menu-item :menu-info="item" :key="item.key" @handleSubMenuSelected
   ="handleSubMenuSelected" />
17            </template>
18          </template>
19        </a-menu>
20      </div>
21    </div>
22  </div>
23 </template>
24 <script lang="ts">
25 import { defineComponent, nextTick, onMounted, ref } from 'vue';
26 import { OioIcon } from '@kunlun/vue-ui-common';
27 import MenuItem from './MenuItem.vue';
28
29 export default defineComponent({
30   props: ['mode', 'module', 'onSelect', 'onClick', 'openChange', 'selectKeys', 'o
   penKeys', 'menus', 'translate'],
31   components: {
32     MenuItem,
33     OioIcon
34   },
35   setup(props) {
36     const DEFAULT_MENU_ICON = 'oinone-menu-caidanmoren';
37     const onMenuSelected = ({ key }) => {
38       if (props.onClick) {
39         props.onClick(key);
40       }
41     };
42
43     onMounted(async () => {
44       await nextTick();
45     });
46
47     const innerOpenKeys = ref([] as string[]);
48     const tempInnerOpenKeys = ref([] as string[]);
49
50     const handleSubMenuSelected = (key) => {
51       if (props.onClick) {
52         props.onClick(key);
53       }
54     };
55
56     return {
57       onMenuSelected,
58       handleSubMenuSelected,
59       DEFAULT_MENU_ICON,
60       innerOpenKeys,
61       tempInnerOpenKeys
62     };
63   }
64 });
65 </script>
66
67 <style lang="scss">
68 .custom-mask-component {
69   display: flex;
70   align-items: center;
71   flex: 1;
72   justify-content: start;
73   .ant-menu-horizontal {
```

MenuItem 组件的代码如下。

```
 1  <template>
 2    <a-sub-menu popupClassName="inline-menu-popup" class="oio-scrollbar" :key="menuInfo.key" @click="handleSubMenuClick(menuInfo)">
 3      <template #icon v-if="!menuInfo.menu.parent">
 4        <oio-icon :icon="menuInfo.menu.icon || DEFAULT_MENU_ICON" />
 5      </template>
 6      <template #title>
 7        <span :title="menuInfo.title">{{ menuInfo.title }}</span>
 8      </template>
 9      <template v-for="item in menuInfo.children" :key="item.key">
10        <template v-if="!item.children || !item.children.length">
11          <a-menu-item :key="item.key" :title="item.title">
12            {{ item.title }}
13          </a-menu-item>
14        </template>
15        <template v-else>
16          <MenuItem :menu-info="item" :key="item.key" />
17        </template>
18      </template>
19
20      <template #expandIcon="context">
21        <oio-icon
22          v-if="menuInfo.children"
23          icon="oinone-menu-caidanxiala"
24          color="rgba(var(--oio-text-color-rgb), 0.75)"
25          :style="{
26            transform: context.isOpen ? 'rotate(360deg)' : 'rotate(270deg)',
27            'margin-right': '-20px',
28            transition: 'all 0.3s ease'
29          }"
30        />
31      </template>
32    </a-sub-menu>
33  </template>
34  <script lang="ts">
35  import { computed, defineComponent } from 'vue';
36  import { OioIcon } from '@kunlun/vue-ui-common';
37
38  export default defineComponent({
39    name: 'MenuItem',
40    props: {
41      menuInfo: {
42        type: Object,
43        default: () => ({})
44      },
45      selectKeys: {
46        type: Array,
47        default: () => []
48      },
49      collapsed: {
50        type: Boolean
51      }
52    },
```

```
53      emits: ['handleSubMenuSelected'],
54      components: {
55        OioIcon
56      },
57      setup(props, { emit }) {
58        const DEFAULT_MENU_ICON = 'oinone-menu-caidanmoren';
59        const menuId = computed(() => {
60          return props.menuInfo?.menu?.id;
61        });
62
63        const isSelected = computed(() => {
64          const { selectKeys = [] } = props;
65          return menuId.value === selectKeys[0];
66        });
67
68        const handleSubMenuClick = (e) => {
69          if (!e.children || !e.children.length) {
70            emit('handleSubMenuSelected', e.key);
71          }
72        };
73
74        return {
75          DEFAULT_MENU_ICON,
76          menuId,
77          isSelected,
78          handleSubMenuClick
79        };
80      }
81    });
82  </script>
```

Step3.（前端）新建 CustomMaskComponent.ts 封装成 Oinone 组件。

通过 @SPI.ClassFactory (ViewWidget.Token ({ widget:'CustomMaskComponent'})) 来定义组件匹配规则：

widget：CustomMaskComponent。

代码如下。

```
1  import { ActionType, DslNodeWidget, Entity, executeUrlAction, IMenu, IModule, resolveMenu, Router, SPI, Subscription, useCurrentContextService, useMatched, useRouter, ViewWidget, Widget } from '@kunlun/dependencies';
2  import { OioNotification } from '@kunlun/vue-ui-antd';
3
4  import CustomMaskComponent from './CustomMaskComponent.vue';
5
6  @SPI.ClassFactory(
7    ViewWidget.Token({
8      widget: 'CustomMaskComponent'
9    })
10 )
11 export default class CustomMaskComponentWidget extends DslNodeWidget {
12   public initialize(props) {
13     super.initialize(props);
14     this.setComponent(CustomMaskComponent);
15     return this;
16   }
17   @Widget.Reactive()
18   private menus: any[] = [];
19
20   @Widget.Reactive()
21   private collapsed = false;
22
```

```ts
23      @Widget.Inject('mode')
24      @Widget.Reactive()
25      private mode: 'horizontal' | 'inline' = 'inline';
26
27      private router!: Router;
28
29      private contextSub?: Subscription;
30
31      private onModuleChange(module) {
32        if (!module) {
33          return;
34        }
35
36        this.module = module;
37        this.menus.splice(0);
38        this.menus = resolveMenu(module.allMenus);
39        this.sortMenu(this.menus);
40      }
41
42      public mounted() {
43        const { getCurrentContext, moduleMap } = useCurrentContextService();
44
45        this.router = useRouter().router;
46        // 监听路由切换
47        this.contextSub = getCurrentContext()
48          .pluck('module')
49          .subscribe(async (moduleName) => {
50            const module = moduleMap.get(moduleName);
51            if (module) {
52              this.onModuleChange(JSON.parse(JSON.stringify(module)));
53            }
54          });
55      }
56
57      @Widget.Reactive()
58      private module: IModule | null = null;
59
60      // 菜单点击
61      @Widget.Method()
62      private onClick(menuId: string) {
63        const menu = this.module!.allMenus.find((m) => m.id === menuId);
64        if (menu) {
65          const { viewAction, serverAction, urlAction, actionType } = menu;
66          if (!viewAction && !serverAction && !urlAction) {
67            OioNotification.warning('请注意', '该菜单未配置页面或动作');
68            return;
69          }
70          if (actionType === ActionType.Server && serverAction) {
71            const requestMapping = menu.mapping || {};
72            const menuContext = menu.context || {};
73          } else if (actionType === ActionType.URL && urlAction) {
74            executeUrlAction(urlAction);
75          } else if (actionType === ActionType.View && viewAction) {
76            const { model, domain } = viewAction;
77            const extraParam = {} as Entity;
78            const viewType = viewAction.resView?.type || viewAction.viewType;
79            this.selectKeys = [menuId];
80            const {
81              page: { language = '' }
82            } = useMatched().matched.segmentParams;
83
84            const params = {
```

```
 85              path: 'page',
 86              extra: { preserveParameter: false },
 87              parameters: {
 88                module: viewAction.resModuleName || this.module!.name,
 89                model,
 90                viewType,
 91                action: viewAction.name,
 92                menu: JSON.stringify({ selectedKeys: this.selectKeys, openKeys: this.openKeys }),
 93                scene: viewAction.name,
 94                menuRandomKey: Date.now(),
 95                ...extraParam
 96              }
 97            } as any;
 98
 99            if (language) {
100              (params.parameters as Record<string, string>).language = language;
101            }
102
103            this.router.push({
104              segments: [params]
105            });
106          }
107        }
108      }
109
110      // 菜单排序
111      private sortMenu(menus: any[]) {
112        menus.sort((a, b) => {
113          if (a.menu.priority === b.menu.priority) {
114            const indexA = this.module!.allMenus.findIndex((m) => m.id === a.menu.id);
115            const indexB = this.module!.allMenus.findIndex((m) => m.id === b.menu.id);
116            return indexA - indexB;
117          }
118          return Number(a.menu.priority) - Number(b.menu.priority);
119        });
120        menus.forEach((m) => {
121          if (m.children) {
122            this.sortMenu(m.children);
123          }
124        });
125      }
126
127      @Widget.Reactive()
128      private selectKeys: string[] = [];
129
130      @Widget.Reactive()
131      private openKeys: string[] = [];
132
133      @Widget.Method()
134      private openChange(openKeys: string[]) {
135        const latestOpenKey = openKeys.find((key) => this.openKeys.indexOf(key) === -1);
136        if (this.rootSubmenuKeys.indexOf(latestOpenKey!) === -1) {
137          this.openKeys = openKeys;
138        } else {
139          this.openKeys = latestOpenKey ? [latestOpenKey] : [];
140        }
141      }
142
143      private get rootSubmenuKeys() {
144        return this.menus.map((m) => m.menu.id);
145      }
146
147      // 菜单选中
148      @Widget.Reactive()
149      private onSelect(menuId: string) {
150        const menu = this.module!.allMenus.find((m) => m.id === menuId)!;
151        const openKeys: string[] = [];
152        let tmp = menu;
153        while (tmp.parent) {
```

```
155          // eslint-disable-next-line @typescript-eslint/no-loop-func
156          tmp = this.module!.allMenus.find((m) => m.id === tmp.parent!.id)!;
157        }
158        openKeys.push(tmp.id);
159        this.openChange(openKeys.reverse());
160        this.onClick(menuId as any);
161      }
162    }
```

Step4.（前端）注册 CustomMaskComponent。

（1）修改 mask 目录下的 index.ts，代码如下。

```
1    export * from './custom-mask-component/CustomMaskComponent';
```

（2）main.ts 引入，代码如下。

```
1    import './mask';
```

Step5. 修改 mask，代码如下。

```
1    import { registerMask } from '@kunlun/dependencies';
2    /**
3     * @param maskTpl mask模板
4     * @param model mask对应模型 可缺省
5     * @param actionName mask对应的action 可缺省
6     * @description 当model和actionName都缺省时，mask即为全局的模板
7     */
8    registerMask(`<mask>
9      <header>
10       <widget widget="app-switcher" />
11       <block flex="1">
12         <widget widget="CustomMaskComponent" />
13       </block>
14       <block>
15         <widget widget="notification" />
16         <widget widget="divider" />
17         <widget widget="language" />
18         <widget widget="divider" />
19         <widget widget="user" />
20       </block>
21     </header>
22     <container>
23       <content>
24         <widget widget="breadcrumb" />
25         <block height="100%" width="100%">
26           <widget width="100%" widget="main-view" />
27         </block>
28       </content>
29     </container>
30    </mask>`, '', '')
```

Step6. 前端刷新页面，如图 3-226 所示。

图 3-226　前端刷新页面

2）默认 Mask

（1）系统内置 Mask。

系统内置的 Mask xml，多标签栏默认放在顶部，基于以下几点考虑。

① Oinone Mask 替换范围是页面、应用、整体；

② 页面、应用定制灵活度高。

系统内置 Mask，代码如下。

```
<mask>
    <multi-tabs />
    <header>
        <widget widget="app-switcher" />
        <block>
            <widget widget="notification" />
            <widget widget="divider" />
            <widget widget="language" />
            <widget widget="divider" />
            <widget widget="user" />
        </block>
    </header>
    <container>
        <sidebar>
            <widget widget="nav-menu" height="100%" />
        </sidebar>
        <content>
            <breadcrumb />
            <block height="100%" width="100%" >
                <widget width="100%" widget="main-view" />
            </block>
        </content>
    </container>
```

页面效果如图 3-227 所示。

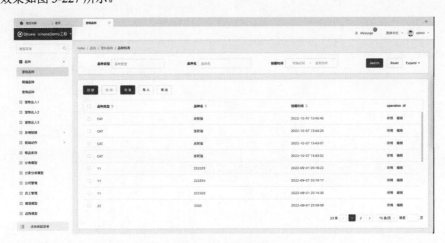

图 3-227　页面效果

（2）替换多标签栏位置。

当只有单应用时，大部分场景多标签栏放在顶部栏下方，本示例给出了多标签栏的另一种展示形式的页面配置，有更多展示方式，可自由调整多 tab 栏的位置，配合 block 和 content 任意组合。

代码如下。

```
<mask>
  <header>
    <widget widget="app-switcher" />
    <block>
      <widget widget="notification" />
      <widget widget="divider" />
      <widget widget="language" />
      <widget widget="divider" />
      <widget widget="user" />
    </block>
  </header>
  <container>
    <sidebar>
      <widget widget="nav-menu" height="100%" />
    </sidebar>
    <block height="100%" width="100%" flex-direction="column" alignContent="flex-start" flex-wrap="nowrap">
      <multi-tabs />
      <content>
        <block height="100%" width="100%">
          <widget width="100%" height="100%" widget="main-view" />
        </block>
      </content>
    </block>
  </container>
</mask>
```

页面效果如图 3-228 所示。

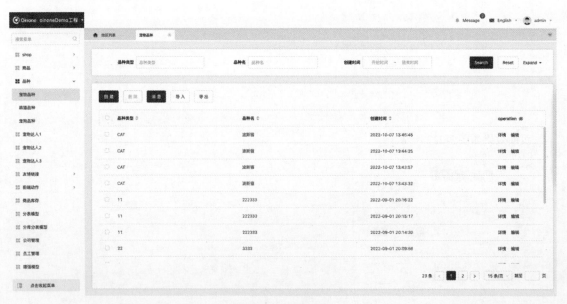

图 3-228　页面效果

# 第 4 章  Oinone 的高级特性

本章中主要介绍元数据的详细构成以及相关高级特性，比如事务、异步、网络协议、分布式等。
1. 后端高级特性
2. 前端高级特性
3. Oinone 分布式体验及进阶体验
4. 研发辅助：结构性代码及 SQL 优化

## 4.1 后端高级特性

了解 Oinone 的基础知识后基本上可以胜任业务代码的开发，但对于构建一个完整的应用，作为技术专家或者架构师需要考虑的方面还有很多，本章将详细讲解构建应用所需知识，让您能成为那个可以带领小伙伴飞的人。

### 4.1.1 模块之 yml 文件结构详解

本节是对 demo 的 boot 工程的 application-*.yml 文件关于 Oinone 相关配置的扩充讲解，大家可以先通读有个印象，以备不时之需。

在 3.2 节中，构建了第一个 Module，并通过启动前后应用，直观地感受到我们自己建的 demo 模块。在上述过程中想必大家都了解到 Oinone 的 boot 工程是专门用来做应用启动管理的，它完全没有任何业务逻辑，只决定启动哪些模块、启动方式以及相关配置。它跟 Spring Boot 的一个普通工程没有什么差异。所有我们只要看 application-*.yml 文件，Oinone 提供了哪些特殊配置就能窥探一二。

这里主要介绍 pamirs 路径下的核心以及常用的配置项。

**1. pamirs.boot**

1）pamirs.boot.init

（1）描述：启动加载程序时，判断是否启动元数据、业务数据和基础设施的加载与更新程序，在应用启动时同时对模块进行生命周期管理。

（2）true ## 标准版，只支持 true。

2）pamirs.boot.sync

（1）描述：同步执行加载程序，启动时对模块进行生命周期管理采用同步方式。

（2）true ## 标准版，只支持 true。

3）pamirs.boot.modules

（1）描述：启动模块列表。这里只有 base 模块是必需的。为了匹配前端模板，在 demo 的例子中加入了其他几个通用业务模块。当然这些通用业务模块也可以大大降低开发难度以及提升业务系统的设计质量。

（2）- base：Oinone 的基础模块。

（3）- common：Oinone 的一些基础辅助功能。

（4）- sequence：序列的能力。

（5）- resource：基础资源。

（6）- user：基础用户。

（7）- auth：权限。

（8）- message：消息。

（9）- international：国际化。

（10）- business：商业关系。

（11）- file：文件，demo 里没有默认加入，如果要开发导入 / 导出相关功能，可以对应引入该模块。

（12）-……还有很多通用业务模块以及这些模块的详细介绍，将在第 6 章展开介绍。

**4）pamirs.boot.mode**

dev：不走缓存，可以直接修改元数据。特别是在页面设计的时候，可以修改 base_view 表直接生效，不需要重启系统。

pamirs.boot.mode 配置举例如图 4-1 所示。

```
1  pamirs:
2    boot:
3      init: true
4      sync: true
5      modules:
6        - base
7        - common
8        - sequence
9        - resource
10       - user
11       - auth
12       - message
13       - international
14       - business
15       - demo_core
```

图 4-1　pamirs.boot.mode 配置举例

**2. pamirs.boot.profile 与 pamirs.boot.options**

pamirs.boot.option 在 pamirs.boot.options 中可以自定义可选项，也可以根据 pamirs.boot.profile 属性来指定这些可选项，pamirs.boot.profile 属性的默认值为 CUSTOMIZE。只有 pamirs.boot.profile=CUSTOMIZE 时，才能在 pamirs.boot.options 中自定义可选项，见表 4-1。

表 4-1　pamirs.boot.options 列表

| 可选项 | 说明 | 默认值 | AUTO | READONLY | PACKAGE | DDL |
| --- | --- | --- | --- | --- | --- | --- |
| reloadModule | 是否加载存储在数据库中的模块信息 | false | true | true | true | true |
| checkModule | 校验依赖模块是否安装 | false | true | true | true | true |
| loadMeta | 是否扫描包读取模块元数据 | true | true | false | true | true |
| reloadMeta | 是否加载存储在数据库中元数据 | false | true | true | true | true |
| computeMeta | 是否重算元数据 | true | true | false | true | true |
| editMeta | 编辑元数据，是否支持编程式编辑元数据 | true | true | false | true | true |
| diffMeta | 差量减计算元数据 | false | false | false | false | false |
| refresh SessionMeta | 刷新元数据缓存 | true | true | true | true | true |
| rebuildHttpApi | 刷新重建前后端协议 | true | true | true | false | false |
| diffTable | 差量追踪表结构变更 | false | false | false | true | false |
| rebuildTable | 更新重建表结构 | true | true | false | true | false |
| printDDL | 打印重建表结构 DDL | false | false | false | false | true |
| publishService | 发布服务，是否发布远程服务 | true | true | true | false | false |
| updateModule | 分布式模块管理 | false | true | false | true | false |

续表

| 可选项 | 说明 | 默认值 | AUTO | READONLY | PACKAGE | DDL |
|---|---|---|---|---|---|---|
| updateMeta | 初始化与更新元数据，是否将元数据的变更写入数据库 | false | true | false | true | false |
| updateData | 初始化与更新内置业务数据，是否将内置业务数据的变更写入数据库 | true | true | false | true | false |
| params | 扩展参数 | 可自定义 | 可自定义 | 可自定义 | 可自定义 | 可自定义 |

可以在启动日志中查看当前服务启动可选项，如图 4-2 所示。

```
Boot Options: {
    "installEnum":"READONLY",
    "upgradeEnum":"AUTO",
    "profile":"CUSTOMIZE",
    "options":{
        "reloadModule":false,
        "checkModule":false,
        "loadMeta":true,
        "reloadMeta":true,
        "computeMeta":true,
        "editMeta":true,
        "diffMeta":true,
        "refreshSessionMeta":true,
        "rebuildHttpApi":true,
        "diffTable":true,
        "rebuildTable":true,
        "printDDL":false,
        "publishService":true,
        "updateModule":false,
        "updateMeta":false,
        "updateData":true
    }
}
```

图 4-2　在启动日志中查看当前服务启动可选项

### 3. pamirs.meta

1）pamirs.meta.metaPackages

（1）描述：自定义元模型的所在包路径。标准版只能是 pro.shushi.pamirs.trigger.Model。

（2）pamirs.meta 配置举例，代码如下。

```
1    pamirs:
2      meta:
3        metaPackages:
4          - pro.shushi.pamirs.trigger.model
```

### 4. pamirs.framework

1）pamirs.framework.gateway：graphql 的可选项配置

（1）pamirs.framework.gateway.show-doc：是否对外提供 gql 的 scheme 文档查询能力，在 3.2.1 节中之所以可以用 Insomnia 开查看后端的文档，是因为这里要配置为 true。

（2）pamirs.framework.gateway.statistics：用于收集 DataLoader 执行过程中的状态，比如缓存命中多少次，已经加载了多少个对象，有多少次错误等。

2）pamirs.framework.hook

（1）pamirs.framework.hook.ignoreAll：默认为 false，为 true 忽略掉所有 hook 函数。

（2）pamirs.framework.hook.excludes：排除掉部分 hook 函数。

（3）pamirs.framework.data：这个经常会使用到，大家一定要了解。

（4）pamirs.framework.data.default-ds-key：模块的默认数据库 key，对应 pamirs.datasource 配置。

（5）pamirs.framework.data.ds-map：为模块指定数据库 key，对应 pamirs.datasource 配置。在 demo 中如果要为 demo_core 这个模块配置独立数据库，可以在这里配置，如 demo_core: demo，并在 pamirs.datasource 配置 key 为 demo 的数据源。

3）pamirs.framework.system

（1）pamirs.framework.system.system-ds-key：元数据系统对应的数据源，对应 pamirs.datasource 配置。

（2）pamirs.framework.system.system-Models：视为元数据模型，一起放到 system-ds-key 库。

（3）pamirs.framework.system 配置举例，代码如下。

```
pamirs:
  framework:
    system:
      system-ds-key: base
      system-models:
        - base.WorkerNode
    data:
      default-ds-key: pamirs
      ds-map:
        base: base
    gateway:
      statistics: true
      show-doc: true
  #hook 如下配置
  #hook:
    #excludes:
      #- pro.shushi.pamirs.core.common.hook.QueryPageHook4TreeAfter
      #- pro.shushi.pamirs.user.api.hook.UserQueryPageHookAfter
      #- pro.shushi.pamirs.user.api.hook.UserQueryOneHookAfter
```

### 5. pamirs.dialect.ds

（1）描述：pamirs.datasource 中数据源的方言信息，以 key 为对应。

（2）子参数有：type（默认：MySQL）；version（默认：8.0）；majorVersion（默认：8）。

（3）pamirs.dialect.ds 配置举例，代码如下。

```
pamirs:
  dialect: #MySQL8.0可不配置
    ds:
      base: # pamirs.datasource中数据源的方言信息，以key为对应
        type: MySQL
        version: 8.0
        majorVersion: 8
      pamirs: # pamirs.datasource中数据源的方言信息，以key为对应
        type: MySQL
        version: 8.0
        majorVersion: 8
```

### 6. pamirs.datasource

（1）描述：安装模块所需要的数据源配置。

（2）pamirs.datasource 配置举例，代码如下。

```yaml
pamirs:
  datasource:
    pamirs:
      driverClassName: com.mysql.cj.jdbc.Driver
      type: com.alibaba.druid.pool.DruidDataSource
      url: jdbc:mysql://127.0.0.1:3306/demo?useSSL=false&allowPublicKeyRetrieval=true&useServerPrepStmts=true&cachePrepStmts=true&useUnicode=true&characterEncoding=utf8&serverTimezone=Asia/Shanghai&autoReconnect=true&allowMultiQueries=true
      username: root
      password: oinone
      initialSize: 5
      maxActive: 200
      minIdle: 5
      maxWait: 60000
      timeBetweenEvictionRunsMillis: 60000
      testWhileIdle: true
      testOnBorrow: false
      testOnReturn: false
      poolPreparedStatements: true
      asyncInit: true
    base:
      driverClassName: com.mysql.cj.jdbc.Driver
      type: com.alibaba.druid.pool.DruidDataSource
      url: jdbc:mysql://127.0.0.1:3306/demo_base?useSSL=false&allowPublicKeyRetrieval=true&useServerPrepStmts=true&cachePrepStmts=true&useUnicode=true&characterEncoding=utf8&serverTimezone=Asia/Shanghai&autoReconnect=true&allowMultiQueries=true
      username: root
      password: oinone
      initialSize: 5
      maxActive: 200
      minIdle: 5
      maxWait: 60000
      timeBetweenEvictionRunsMillis: 60000
      testWhileIdle: true
      testOnBorrow: false
      testOnReturn: false
      poolPreparedStatements: true
      asyncInit: true
```

**7. pamirs.sharding**

（1）描述：Oinone 的分库分表配置，当使用 pamirs-trigger-bridge-tbschedule 工程开启内置 schedule 功能时必须配置。

（2）pamirs.sharding 配置举例，demo 参考 4.1.24 节，代码如下。

```yaml
pamirs:
  sharding:
    define:
      data-sources:
        ds: pamirs
        pamirsSharding: pamirs          #申明pamirsSharding库对应的pamirs数据源
        testShardingDs:                 #申明testShardingDs库对应的testShardingDs_0\1数据源
          - testShardingDs_0
          - testShardingDs_1
      models:
        "[trigger.PamirsSchedule]":
          tables: 0..13
        "[demo.ShardingModel]":
```

```yaml
14               tables: 0..7
15               table-separator: _
16           "[demo.ShardingModel2]":
17             ds-nodes: 0..1        #申明testShardingDs库对应的建库规则
18             ds-separator: _
19             tables: 0..7
20             table-separator: _
21     rule:
22       pamirsSharding: #配置pamirsSharding库的分库分表规则
23         actual-ds:
24           - pamirs #申明pamirsSharding库对应的pamirs数据源
25         sharding-rules:
26           # Configure sharding rule, 以下配置跟sharding-jdbc配置一致
27           - tables:
28               demo_core_sharding_model:
29                 actualDataNodes: pamirs.demo_core_sharding_model_${0..7}
30                 tableStrategy:
31                   standard:
32                     shardingColumn: user_id
33                     shardingAlgorithmName: table_inline
34             shardingAlgorithms:
35               table_inline:
36                 type: INLINE
37                 props:
38                   algorithm-expression: demo_core_sharding_model_${(Long.valueOf(user_id) % 8)}
39         props:
40           sql.show: true
41       testShardingDs: #配置testShardingDs库的分库分表规则
42         actual-ds: #申明testShardingDs库对应的pamirs数据源
43           - testShardingDs_0
44           - testShardingDs_1
45         sharding-rules:
46           # Configure sharding rule, 以下配置跟sharding-jdbc配置一致
47           - tables:
48               demo_core_sharding_model2:
49                 actualDataNodes: testShardingDs_${0..1}.demo_core_sharding_model2_${0..7}
50                 databaseStrategy:
51                   standard:
52                     shardingColumn: user_id
53                     shardingAlgorithmName: ds_inline
54                 tableStrategy:
55                   standard:
56                     shardingColumn: user_id
57                     shardingAlgorithmName: table_inline
58             shardingAlgorithms:
59               table_inline:
60                 type: INLINE
61                 props:
62                   algorithm-expression: demo_core_sharding_model2_${(Long.valueOf(user_id) % 8)}
63               ds_inline:
64                 type: INLINE
65                 props:
66                   algorithm-expression: testShardingDs_${(Long.valueOf(user_id) % 2)}
67         props:
68           sql.show: true
```

**8. pamirs.mapper**

（1）库配置：可以通过 YAML 的 "pamirs.mapper.<global 或者 ds>" 配置项进行库配置。如果未配

置，系统会采用默认值，见表 4-2。

表 4-2 库配置

| 配置项 | 默认值 | 描述 |
| --- | --- | --- |
| databaseFormat | %s | 库名格式化 |
| tableFormat | %s | 表名格式化 |
| tablePattern | %s | 动态表名表达式 |

（2）表配置：可以通过 YAML 的 "pamirs.mapper.<global" 或者 "ds>.table-info" 配置项或者 @Model.Persistence 注解进行表配置。注解优先级大于 YAML 配置文件配置。如果未配置，系统会采用默认值，见表 4-3。

表 4-3 表配置

| 配置项 | 默认值 | 描述 |
| --- | --- | --- |
| logicDelete | true | 是否逻辑删除 |
| logicDeleteColumn | is_delete | 逻辑删除字段名 |
| logicDeleteValue | REPLACE (unix_timestamp (NOW (6)),'.','') | 逻辑删除值 |
| logicNotDeleteValue | 0 | 非逻辑删除值 |
| optimisticLocker | false | 是否开启乐观锁 |
| optimisticLockerColumn | opt_version | 乐观锁字段名 |
| keyGenerator | AUTO_INCREMENT | 主键自增规则 |
| underCamel | true | 驼峰下画线转换 |
| capitalMode | false | 大小写转换 |
| columnFormat | `%s` | 列名格式化，%s 将替换为列名 |
| aliasFormat | `%s` | 字段别名格式化，%s 将替换为字段别名 |
| charset | utf8mb4 | 字符集 |
| collate | bin | 排序字符集 |

（3）pamirs.mapper 配置举例，代码如下。

```
1  pamirs:
2    mapper:
3      static-model-config-locations:
4        - pro.shushi.pamirs
5      batch: collectionCommit #batch方法的批量提交模式
6      batch-config:
7        "[base.Field]":
8          write: 2000
9        "[base.Function]":
10         read: 500
11         write: 2000
12     global:
13       table-info:
14         logic-delete: true
15         logic-delete-column: is_deleted
16         logic-delete-value: REPLACE(unix_timestamp(NOW(6)),'.','')
17         logic-not-delete-value: 0
18         optimistic-locker: false
19         optimistic-locker-column: opt_version
20         key-generator: DISTRIBUTION
```

```
21       table-pattern: '${module}_%s'
22    #可以为指定数据配置
23    #ds:
24    # pamirs:
25    #   table-info:
26    #     logic-delete: true
27    #     logic-delete-column: is_deleted
28    #     logic-delete-value: REPLACE(unix_timestamp(NOW(6)),'.','')
29    #     logic-not-delete-value: 0
30    #     optimistic-locker: false
31    #     optimistic-locker-column: opt_version
32    #     key-generator: DISTRIBUTION
33    #     table-pattern: '${module}_%s'
```

### 9. pamirs.persistence

（1）描述：自动建库、建表。

（2）pamirs.persistence 配置举例，代码如下。

```
1  pamirs:
2    persistence:
3      global:
4        auto-create-database: true
5        auto-create-table: true
```

### 10. pamirs.plus

（1）描述：mybatisplus 的代理配置。

（2）pamirs.plus 配置举例，代码如下。

```
1  pamirs:
2    plus:
3      configuration:
4        map-underscore-to-camel-case: false
5        cache-enabled: false
```

### 11. pamirs.event

（1）pamirs.event.enabled：启用 RocketMQ 功能，不启用的情况下无法使用任何功能，使用详见 4.1.21 节。

（2）pamirs.event.rocket-mq.namesrv-addr：RocketMQ 链接地址字符串。

（3）pamirs.event.rocket-mq.aliyun：阿里云版本适配。

（4）pamirs.event.schedule.enabled：启动异步任务功能，不启用无法使用任何功能。

（5）pamirs.event 配置举例，代码如下。

```
1   pamirs:
2     event:
3       enabled: false
4       schedule:
5         enabled: false
6         ownSign: base
7         auto-init: true
8         auto-create-config-file: true
9       rocket-mq:
10        namesrv-addr: 127.0.0.1:9876
11        # 标识发送消息和消费消息的机器的IP地址，默认为RemotingUtil.getLocalAddress()
```

```
12        client-ip:
13        # 标识发送消息和消费消息的机器的实例名称地址，默认为DEFAULT
14        instance-name:
15        # 标识发送消息和消费消息的机器的实例ID，无默认值
16        namespace:
17        # 为每一个topic添加一个固定前缀
18        topic-prefix:
19        # 是否启用vip netty通道以发送消息
20        vip-channel-enabled: false
21        # 是否启用消息轨迹追踪（该属性对于阿里云MQ无效）
22        enable-trace: false
23        # 消息轨迹追踪的topic名称
24        trace-topic-name:
25        # Pamirs Event工厂
26        event-factory: pro.shushi.pamirs.framework.connectors.event.rocketmq.DefaultRocketMQEventFactory
27        # 阿里云MQ相关配置
28        aliyun:
29          # 使用阿里云MQ
30          enabled: false
31          # 阿里云MQ AccessKey
32          access-key:
33          # 阿里云MQ SecretKey
34          secret-key:
35          # 阿里云MQ InstanceId
36          instance-id:
37          # 阿里云MQ GroupId
38          group-id:
```

### 12. pamirs.auth

（1）pamirs.auth.ModelFilter：通过该配置项，来配置模型的所有对应方法，都不需要经过权限系统的控制。这种配置方式其实有安全漏洞，下个版本中会删去这种配置方法，不建议使用。

（2）pamirs.auth.funFilter：配置模型的特定query方法不需要过权限控制。

（3）pamirs.auth 配置举例，代码如下。

```
1   pamirs:
2     auth:
3       model-filter:
4         - user.PamirsUserTransient
5         - auth.ResourcePermission
6         - auth.AuthGroup
7         - auth.AuthRole
8         - base.View
9         - resource.ResourceCountry
10        - pamirs.web.WebMenu
11        - pamirs.web.WebRenderTransient
12        - pamirs.message.MessageCenter
13        - resource.major.ResourceMajorConfig
14      fun-filter:
15        - namespace: user.PamirsUserTransient
16          fun: login #登录
17        - namespace: user.PamirsUserTransient
18          fun: loginByVerificationCode #手机号验证码登录
19        - namespace: user.PamirsUserTransient
20          fun: loginVerificationCode #手机号登录验证码
21        - namespace: user.PamirsUserTransient
22          fun: signUpVerificationCode #手机号注册验证码
```

### 13. pamirs.file

（1）描述：导入/导出相关配置，使用方法详见6.1节。

（2）pamirs.file 配置举例，代码如下。

```yaml
pamirs:
  file:
    auto-upload-logo: false
    import-property:
      default-each-import: false # 默认逐行导入
      max-error-length: 100 # 默认最大收集错误行数
    export-property:
      default-clear-export-style: false # 默认使用csv导出
      csv-max-support-length: 1000000 # csv导出最大支持1000000行

#文件导入/导出还依赖CDN相关配置
cdn:
  oss:
    name: 阿里云
    type: OSS
    bucket: pamirs
    uploadUrl: oss-cn-hangzhou.aliyuncs.com
    downloadUrl: oss-cn-hangzhou.aliyuncs.com
    accessKeyId: #自行修改值
    accessKeySecret: #自行修改值
    mainDir: upload/demo/test/
    validTime: 3600000
    timeout: 600000
    active: true
    referer:
    localFolderUrl:
```

### 14. pamirs.channel

（1）描述：消息渠道模块的相关配置。

（2）pamirs.channel 配置举例，代码如下。

```yaml
pamris:
  channel:
    zkServers: 127.0.0.1:2181
```

### 15. pamirs.zookeeper

（1）描述：zk 配置的代理。

（2）pamirs.zookeeper 配置举例，代码如下。

```yaml
pamris:
  zookeeper:
    zkConnectString: 127.0.0.1:2181
    zkSessionTimeout: 60000
    rootPath: /demo
```

### 16. pamirs.eip

（1）描述：集成相关配置，使用方法详见 6.2 节。

（2）pamirs.eip 配置举例，代码如下。

```yaml
pamirs:
  eip:
    open-api:
      enabled: false
      standalone:
        host: 127.0.0.1
        port: 8081
        aes-key:
      routes:
        pamirs:
          host: 127.0.0.1
          port: 8081
          aes-key:
```

高德地图接口的 key 配置，代码如下。

```
1  pamirs:
2    eip:
3      map:
4        gd:
5          key: xxxxxx
```

### 17. pamirs.elastic

（1）描述：es 搜索引擎地址，使用方法详见 4.1.25 节。

（2）pamirs.elastic 配置举例，代码如下。

```
1  pamirs:
2    elastic:
3      url: 127.0.0.1:9200
```

### 18. pamirs.load

Oinone 默认模式下元数据都是以 DB 为准，当 A 模块依赖 B 模块，但 A 模块与 B 模块没有部署在一起，同时元数据的 Base 库也不是一个时，A 模块的元数据库没有 B 模型，会导致系统出现模型定义找不到的错误。此时需要 A 模块启动内存元数据模式，因为 A 模块依赖 B 模块，在 A 模块的元数据计算时，同时有 B 模块的接口包，在内存中会扫描 B 模块的模型，所以我们要启动内存模式。应对分布式开发场景设计，更多内容请参见 4.4 节。

pamirs.load 配置举例，代码如下。

```
1  pamirs:
2    load:
3      sessionMode: true
```

## 4.1.2 模块之启动指令

不同启动指令的组合可以满足不同场景需求，下面列举了几个常规组合方式，务必把这几种模式都尝试一遍，会更有体验感。

本节讲解 Oinone 模块的几种启动方式，它们能灵活地应对企业市场的不同场景需求，为 OP（本地化部署）、SaaS 和研发提供个性化支撑，体现了 Oinone 单体与分布式灵活切换的特点。

### 1. 部署参数

部署参数见表 4-4。

表 4-4 部署参数

| 参数 | 名称 | 默认值 | 说明 |
| --- | --- | --- | --- |
| -Plifecycle | 生命周期部署指令 | RELOAD | 可选项：无 /INSTALL/PACKAGE/RELOAD/DDL<br>安装 -install 为 AUTO；upgrade 为 FORCE<br>打包 -install 为 AUTO；upgrade 为 FORCE；profile 为 PACKAGE<br>重启 -install、upgrade、profile 为 READONLY<br>打印变更 DDL-install 为 AUTO；upgrade 为 FORCE；profile 为 DDL |

如果在启动命令中配置了部署参数，可不再设置服务参数和可选项参数。如图 4-3 所示为在启动命令中添加部署参数的示例。

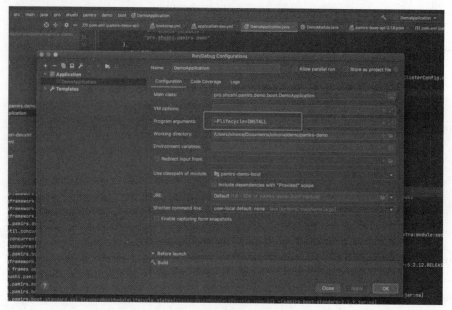

图 4-3 在启动命令中添加部署参数的示例

**2. 使用场景**

**场景一**："DDL（1）+RELOAD（N）"应对专有 DBA。

很多公司数据库是由专门的 DBA 来管理的，不允许应用直接变更数据库相关配置、表结构、初始化数据。而 Oinone 是基于元数据驱动的，任何模型、行为的变化都会自动转化成对物理存储的改变与元数据变化。

Oinone 为了适用企业 OP 场景，特别增加了 DDL 模式。把发布上线分为两个步骤。

（1）用 DDL 模式输出涉及数据库的变更与元数据初始化的脚本，交由客户公司 DBA 审批，并执行。

（2）用 RELOAD 模式，进行正常的应用重启工作，不进行安装、升级以及数据库物理变革等操作。

应用启动/关闭自动 DDL 配置，代码如下。

```
1  #应用启动/关闭自动DDL配置
2  pamirs.boot.profile: CUSTOMIZE
3  pamirs.boot.options.rebuildTable: false
4  pamirs.persistence.global.auto-create-database: false
5  pamirs.persistence.global.auto-create-table: false
```

**场景二**："PACKAGE（1）+RELOAD（N）"提升多机器实例效率。在机器规模相对大的场景中我们会碰到以下问题。

（1）元数据差量计算、数据库变更、元数据变化保存都非常费时，如果每台机器都来一遍是非常费时费力的。

（2）分布式下多机器如果并发进行 INSTALL，会导致数据库修改表结构、元数据变化保存锁死。

所以，我们可以选择一台机器用 PACKAGE，其他机器采用 RELOAD 模式，做到合理规避问题，提升应用发布效率。

**场景三**：INSTALL 应对开发模式。

研发在本地开发模式下 INSTALL 是最有效率的，把所需依赖模块一并启动和调试。

上线如果要用 INSTALL 需要注意，要逐台进行。当然也可以改进为"INSTALL（1）+RELOAD（N）"模式。

**3. 启动命令解读**

（1）查看启动命令。

可以在启动日志中查看当前所用启动命令，如图 4-4 所示。

```
2022-08-17 20:34:22.213  INFO 88676 --- [           main] s.p.f.c.d.d.DataSourceAutoRefreshManager : sharding dataSource trigger with application started.1668739662199
2022-08-17 20:34:22.271  INFO 88676 --- [onPool-worker-3] p.s.p.b.c.initial.PamirsBootMainInitial  : 
Boot Options: {
  "installEnum":"AUTO",
  "upgradeEnum":"FORCE",
  "profile":"AUTO",
  "options":{
    "reloadModule":true,
    "checkModule":true,
    "loadMeta":true,
    "reloadMeta":true,
    "computeMeta":true,
    "editMeta":true,
    "diffMeta":true,
    "refreshSessionMeta":true,
    "rebuildHttpApi":true,
    "diffTable":true,
    "rebuildTable":true,
    "printDDL":false,
    "publishService":true,
    "updateModule":true,
    "updateMeta":true,
    "updateData":true
  }
}
2022-08-17 20:34:23.572 DEBUG 88676 --- [onPool-worker-3] p.s.p.f.c.d.m.G.selectListByPage         : ==>  Preparing: SELECT COUNT(1) FROM `base_module` WHERE `is_deleted` = 0
```

图 4-4  在启动日志中查看当前所用启动命令

（2）生命周期管理：-Plifecycle。

Java -jar <your jar name>.jar -Plifecycle=RELOAD

除了通过启动 YAML 中 pamirs.boot 属性来设置集成方式，还可以在应用启动命令中使用 Plifecycle 参数来快捷控制集成方式和模块生命周期管理。-Plifecycle 参数的可选项有 RELOAD、INSTALL、CUSTOM_INSTALL、PACKAGE、DDL。

启动命令优先级高于 YAML 中 pamirs.boot 的 install、upgrade 和 profile 属性。如果不使用 -Plifecycle 参数，则使用 YAML 中 pamirs.boot 的 install、upgrade 和 profile 属性配置。若 YAML 中未配置，则采用默认值。

Plifecycle 可选项与启动项见表 4-5。

表 4-5  Plifecycle 可选项与启动项对应表

| 启动配置项 | 默认值 | RELOAD | INSTALL | CUSTOM_INSTALL | PACKAGE | DDL |
|---|---|---|---|---|---|---|
| install | AUTO | READONLY | AUTO | AUTO | AUTO | AUTO |
| upgrade | AUTO | READONLY | FORCE | FORCE | FORCE | FORCE |
| profile | CUSTOMIZE | READONLY | AUTO | CUSTOMIZE | PACKAGE | DDL |

profile 属性请参考 4.1.1 节中的"服务启动可选项"内容。只有 pamirs.boot.profile=CUSTOMIZE 时，在 pamirs.boot.options 中自定义的可选项才生效。

（3）自动建表：-PbuildTable。

Java -jar <your jar name>.jar -PbuildTable=NEVER

-PbuildTable 参数用于设置自动构建表结构的方式。如果不使用该参数，则 Options 属性的默认值请参考 4.1.1 节中的"服务启动可选项"内容。-PbuildTable 参数可选项有：

NEVER：不自动构建表结构，会将 pamirs.boot.options 中的 diffTable 和 rebuildTable 属性设置为 false。

EXTEND：增量构建表结构，会将 pamirs.boot.options 中的 diffTable 属性设置为 false，rebuildTable 属性设置为 true。

DIFF：差量构建表结构，会将 pamirs.boot.options 中的 diffTable 和 rebuildTable 属性设置为 true。

（4）模块在线：-PmoduleOnline。

Java -jar <your jar name>.jar -PmoduleOnline=CHECK

-PmoduleOnline 参数用于设置模块在线的方式。如果不使用该参数，则 profile 属性的默认值请参考"服务启动可选项"。-PmoduleOnline 参数可选项有：

NEVER：不读取存储在数据库中的模块信息，会将 pamirs.boot.options 中的 reloadModule 和 checkModule 属性设置为 false。

READ：读取存储在数据库中的模块信息，会将 pamirs.boot.options 中的 checkModule 属性设置为 false，reloadModule 属性设置为 true。

CHECK：读取存储在数据库中的模块信息并校验依赖模块是否已安装，会将 pamirs.boot.options 中的 reloadModule 和 checkModule 属性设置为 true。

（5）元数据在线：-PmetaOnline。

Java -jar <your jar name>.jar -PmetaOnline=MODULE

-PmetaOnline 参数用于设置元数据在线的方式，如果不使用该参数，则 profile 属性的默认值请参考 4.1.1 节中的 "服务启动可选项" 内容。-PmetaOnline 参数可选项有：

NEVER：不持久化元数据，会将 pamirs.boot.options 中的 updateModule、reloadMeta 和 updateMeta 属性设置为 false。（不持久化：只在内存中存储，不存储到数据库）

MODULE：只注册模块信息，会将 pamirs.boot.options 中的 updateModule 属性设置为 true，reloadMeta 和 updateMeta 属性设置为 false。

ALL：注册持久化所有元数据，会将 pamirs.boot.options 中的 updateModule、reloadMeta 和 updateMeta 属性设置为 true。（持久化：存储到数据库中）

（6）开放远程服务：-PenableRpc。

-PenableRpc 参数用于设置是否开启远程服务。如果不使用该参数，则 profile 属性的默认值请参考 4.1.1 节中的 "服务启动可选项" 内容。-PenableRpc 参数可选项为 true 和 false。该参数会将参数值设置到 pamirs.boot.options 中的 publishService 属性。

（7）开启 API 服务：-PopenApi。

-PopenApi 参数用于设置是否开启 HTTP API 服务。如果不使用该参数，则 profile 属性的默认值请参考 4.1.1 节中的 "服务启动可选项" 内容。-PopenApi 参数可选项为 true 和 false，该参数会将参数值设置到 pamirs.boot.options 中的 rebuildHttpApi 属性。

（8）开启字段校验：-PcheckField。

-PcheckField 参数用于设置是否开启字段校验。-PcheckField 参数可选项为 true 和 false。由于通常应用的字段数量非常多，会延长系统启动时长，所以默认不会开启字段校验。

（9）启用数据初始化服务：-PinitData。

-PinitData 参数用于设置是否开启数据初始化服务。如果不使用该参数，则 profile 属性的默认值请参考 4.1.1 节中的 "服务启动可选项" 内容。-PinitData 参数可选项为 true 和 false。该参数会将参数值设置到 pamirs.boot.options 中的 updateData 属性。

**4. 不使用自动构建数据库表功能**

Oinone LCDP 默认提供框架的所有服务，所以会自动构建数据库表。如果不需要使用 Oinone 的存储构建服务，可以设置 YAML 文件中关于自动建表的配置，这样就不会动态构建数据库表，可以手动搭建数据库表。

不使用自动构建数据库表功能，代码如下。

```
1  pamirs:
2    boot:
3      options:
4        rebuildTable: false
```

通过配置启动 YAML 中 pamirs.boot.options.rebuildTable 为 false 彻底关闭自动建表功能。

也可以按需启动 YAML 中 pamirs.persistence 配置来关闭部分数据源的自动建表功能，代码如下。

```
1  pamirs:
2    persistence:
3      global:
4        # 是否自动创建数据库的全局配置，默认为true
5        autoCreateDatabase: true
6        # 是否自动创建数据表的全局配置，默认为true
7        autoCreateTable: true
8      <your ds key>:
9        # 是否自动创建数据库的数据源配置，默认为true
10       autoCreateDatabase: true
11       # 是否自动创建数据表的数据源配置，默认为true
12       autoCreateTable: true
```

persistence 配置既可以针对全局也可以分数据源进行配置。

### 4.1.3 模块之生命周期

了解 Oinone 的生命周期过程，对于理解 Oinone 或者开发高级功能都有非常大的帮助。

**1. 生命周期大图**

生命周期如图 4-5 所示。

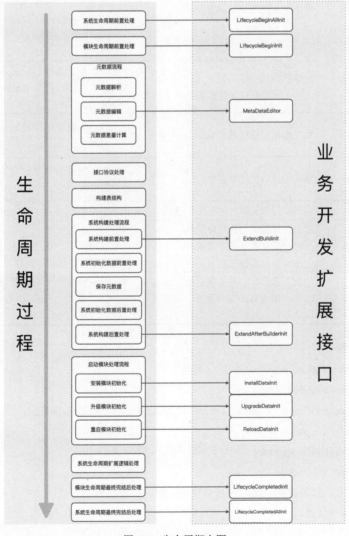

图 4-5 生命周期大图

**2. 平台扩展说明**

平台节点通过 SPI 机制进行扩展，这里不展开，更多详情请见 Oinone 内核揭秘系列文章（可在数式 Oinone 的微信公众号获取）。

**3. 业务扩展说明**

业务扩展说明见表 4-6。

表 4-6  业务扩展说明

| 接口 | 说明 | 使用场景 |
| --- | --- | --- |
| LifecycleBeginAllInit | 系统进入生命周期前置逻辑（注：不能有任何数据库操作） | 系统级别的信息收集上报 |
| LifecycleCompletedAllInit | 系统生命周期完结后置逻辑 | 系统级别的信息收集上报，生命周期过程中的数据或上下文清理 |
| LifecycleBeginInit | 模块进入生命周期前置逻辑（注：不能有任何数据库操作） | 预留，能做的事情比较少 |
| LifecycleCompletedInit | 模块生命周期完结后置逻辑 | 本模块需等待其他模块初始化完毕以后进行初始化。比如：<br>1. 集成模块的初始化<br>2. 权限缓存的初始化<br>…… |
| MetaDataEditor | 元数据编辑（注：不能有任何数据库操作） | 这个在第 3 章中已经多次提及，核心场景是向系统主动注册如 Action、Menu、View 等元数据 |
| ExtendBuildInit | 系统构建前置处理逻辑 | 预留，能做的事情比较少，做一些跟模块无关的事情 |
| ExtendAfterBuilderInit | 系统构建后置处理逻辑 | 预留，能做的事情比较少，做一些跟模块无关的事情 |
| InstallDataInit | 模块在初次安装时的初始化逻辑 | 根据模块启动指令来进行选择执行逻辑，一般用于初始化业务数据。<br>应用启动参数与指令转化逻辑请见 4.1.2 节 |
| UpgradeDataInit | 模块在升级时的初始化逻辑（注：根据启动指令来执行，是否执行一次业务自己控制） | |
| ReloadDataInit | 模块在重启时的初始化逻辑（注：根据启动指令执行，是否执行一次业务自己控制） | |

**4. 常用生命周期举例**

1）Install\Upgrade\Reload 的业务初始化（举例）

Step1. 新建 DemoModuleBizInit。

（1）DemoModuleBizInit 实现 InstallDataInit, UpgradeDataInit, ReloadDataInit。

① InstallDataInit 对应 init；

② UpgradeDataInit 对应 upgrade；

③ ReloadDataInit 对应 reload。

（2）modules 方法代表该初始化类与哪些模块匹配，以模块编码为准。

（3）priority 执行优先级。

代码如下。

```java
package pro.shushi.pamirs.demo.core.init;

import org.springframework.stereotype.Component;
import pro.shushi.pamirs.boot.common.api.command.AppLifecycleCommand;
import pro.shushi.pamirs.boot.common.api.init.InstallDataInit;
import pro.shushi.pamirs.boot.common.api.init.ReloadDataInit;
import pro.shushi.pamirs.boot.common.api.init.UpgradeDataInit;
import pro.shushi.pamirs.demo.api.DemoModule;
import pro.shushi.pamirs.demo.api.enumeration.DemoExpEnumerate;
import pro.shushi.pamirs.meta.common.exception.PamirsException;

import java.util.Collections;
import java.util.List;

@Component
public class DemoModuleBizInit implements InstallDataInit, UpgradeDataInit, ReloadDataInit {
    @Override
    public boolean init(AppLifecycleCommand command, String version) {
        throw PamirsException.construct(DemoExpEnumerate.SYSTEM_ERROR).appendMsg("DemoModuleBizInit: install").errThrow();
        //安装指令执行逻辑
//        return Boolean.TRUE;
    }

    @Override
    public boolean reload(AppLifecycleCommand command, String version) {
        throw PamirsException.construct(DemoExpEnumerate.SYSTEM_ERROR).appendMsg("DemoModuleBizInit: reload").errThrow();
        //重启指令执行逻辑
//        return Boolean.TRUE;
    }

    @Override
    public boolean upgrade(AppLifecycleCommand command, String version, String existVersion) {
        throw PamirsException.construct(DemoExpEnumerate.SYSTEM_ERROR).appendMsg("DemoModuleBizInit: upgrade").errThrow();
        //升级指令执行逻辑
//        return Boolean.TRUE;
    }

    @Override
    public List<String> modules() {
        return Collections.singletonList(DemoModule.MODULE_MODULE);
    }

    @Override
    public int priority() {
        return 0;
    }
}
```

Step2. 重启看效果。

启动指令为 -Plifecycle=INSTALL，转化指令 install 为 AUTO；upgrade 为 FORCE。

因为 DemoModule 已经执行过好多次了，所以会进入 upgrade 逻辑。系统重启的效果跟我们预期的结果一致，确实执行了 DemoModuleBizInit 的 upgrade 方法，如图 4-6 所示。

图 4-6　系统重启执行 DemoModuleBizInit 的 upgrade 方法

2）MetaDataEditor

3.3.2 节介绍传输模型初始化 ViewAction 窗口动作时已有涉及，这里不过多介绍。注意，模块上报元数据只能通过注解或者实现 MetaDataEditor 接口并使用 InitializationUtil 工具来进行，建议用注解方式。

### 4.1.4 模块之元数据详解

本节介绍 Module 相关元数据以及对应代码注解方式。

如您还不了解 Module 的定义，可以先看一下 2.3 节中对 Module 的描述，本节主要带大家了解 Module 元数据构成，能让同学们非常清楚 Oinone 从哪些维度来描述 Module。

**1. 元数据说明**

（1）ModuleDefinition。

ModuleDefinition 的详细信息见表 4-7。

表 4-7　ModuleDefinition 的详细信息

| 元素数据构成 | 含义 | 对应注解 | 备注 |
|---|---|---|---|
| displayName | 显示名称 | @Module (<br>displayName=" ",<br>name=" ",<br>version=" ",<br>category=" ",<br>summary=" ",<br>dependencies={" "," "},<br>exclusions={" "," "},<br>priority=1L) | |
| name | 技术名称 | | |
| latestVersion | 安装版本 | | |
| category | 分类编码 | | |
| summary | 描述摘要 | | |
| moduleDependencies | 依赖模块编码列表 | | |
| moduleExclusions | 互斥模块编码列表 | | |
| priority | 排序 | | |
| module | 模块编码 | @Module.module (" ") | |
| dsKey | 逻辑数据源名 | @Module.Ds (" ") | |
| excludeHooks | 排除拦截器列表 | @Module.Hook (excludes={" "," "}) | |
| website | 站点 | @Module.Advanced (<br>website="http://www.oinone.top",<br>author="Oinone",<br>description="Oinone",<br>application=false,<br>demo=false,<br>web=false,<br>toBuy=false,<br>selfBuilt=true,<br>license=SoftwareLicenseEnum.PEEL1,<br>maintainer="Oinone",<br>contributors="Oinone",<br>url="http://git.com") | |
| author | module 的作者 | | |
| description | 描述 | | |
| application | 是否应用 | | |
| demo | 是否演示应用 | | |
| web | 是否 Web 应用 | | |
| toBuy | 是否需要跳转到 website 去购买 | | |
| selfBuilt | 自建应用 | | |

续表

| 元素数据构成 | 含义 | 对应注解 | 备注 |
| --- | --- | --- | --- |
| license | 许可证 | | 默认 PEEL1<br>可选范围：<br>GPL2<br>GPL2ORLATER<br>GPL3<br>GPL3ORLATER<br>AGPL3<br>LGPL3<br>ORTHEROSI<br>PEEL1<br>PPL1<br>ORTHERPROPRIETARY |
| maintainer | 维护者 | | |
| contributors | 贡献者列表 | | |
| url | 代码库的地址 | | |
| boot | 是否自动安装的引导启动项 | @Boot | 加上该注解表示启动时会自动安装，不管 yml 文件的 modules 是否配置 |
| moduleClazz | 模块定义所在类 | | |
| packagePrefix | 包路径，用于扫描该模块下的其他元数据 | 只有用代码编写的模块才有 | |
| dependentPackagePrefix | 依赖模块列对应的扫描路径 | | |
| state | 状态 | 系统自动计算，无须配置 | |
| metaSource | 元数据来源 | | |
| publishCount | 发布总次数 | | |
| platformVersion | 最新平台版本 | | 本地与中心平台的版本对应。做远程更新时会用到 |
| publishedVersion | 最新发布版本 | | |

（2）UeModule。

是对 ModuleDefinition 的继承，并扩展了与前端交互相关的元数据。详细信息见表 4-8。

表 4-8  UeModule

| 元素数据构成 | 含义 | 对应注解 | 备注 |
| --- | --- | --- | --- |
| homePageModel | 跳转模型编码 | @UxHomepage (UxRoute) | 对应一个 ViewAction，如果 UxRoute 只配置了模型，则默认到该模型的列表页 |
| homePageName | 视图动作或者链接动作名称 | | |
| logo | 图标 | @UxAppLogo (logo=" ") | |

**2．元数据，代码注解方式**

（1）Module。

Module
├── displayName 显示名称
├── name 技术名称

```
├── version 安装版本
├── category 分类编码
├── summary 描述摘要
├── dependencies 依赖模块编码列表
├── exclusions 互斥模块编码列表
├── priority 排序
├── module 模块编码
│   └── value
├── Ds 逻辑数据源名          是 Java 注解中的一种表达方式，自身没有其他属性。
│   └── value
├── Hook 排除拦截器列表
│   └── excludes
├── Advanced 更多配置
│   ├── website 站点
│   ├── author 作者
│   ├── description 描述
│   ├── application 是否为应用
│   ├── demo 是否演示应用
│   ├── web 是否 Web 应用
│   ├── toBuy 是否需要跳转到 website 去购买
│   ├── selfBuilt 是否自建应用
│   ├── license 许可证，枚举默认：PEEL1
│   ├── maintainer 维护者
│   ├── contributors 贡献者
│   └── url 代码库地址
```

（2）相关 Ux 注解。

与模块相关的交互类注解

（3）UxHomepage。

UxHomepage 模块主页
```
└── UxRoute：主页的路由配置
```

（4）UxAppLogo。

UxAppLogo
```
└── logo 图标：应用的图标地址
```

## 4.1.5 模型之持久层配置

**1. 批量操作**

批量操作包括批量创建与批量更新。批量操作的提交类型系统默认值为 batchCommit。

批量提交类型：

● useAffectRows，循环单次单条脚本提交，返回实际影响行数；

● useAndJudgeAffectRows，循环单次单条脚本提交，返回实际影响行数，若实际影响行数与输入不一致，抛出异常；

● collectionCommit，将多个单条更新脚本拼接成一个脚本提交，不能返回实际影响行数；

● batchCommit，使用单条更新脚本批量提交，不能返回实际影响行数。

1）全局配置

全局配置代码如下。

```
1  pamirs:
2    mapper:
3      batch: batchCommit
```

2）运行时配置

非乐观锁模型系统默认采用 batchCommit 提交更新操作；乐观锁模型默认采用 useAndJudgeAffectRows 提交更新操作。也可以使用如下方式在运行时改变批量提交方式。

```
1  Spider.getDefaultExtension(BatchApi.class).run(() -> {
2      更新逻辑
3  }, 批量提交类型枚举);
```

3）运行时校正

如果模型配置了数据库自增主键，而批量新增的提交类型为 batchCommit，则系统将批量提交类型变更为 collectionCommit（如果使用 batchCommit，则需要单条提交以获得正确的主键返回值，性能有所损失）。

如果模型配置了乐观锁，而批量更新的批量提交类型为 collectionCommit 或者 batchCommit，则系统将批量提交类型变更为 useAndJudgeAffectRows。也可以失效乐观锁，让系统不做批量提交类型变更处理。

**2. 乐观锁（举例）**

针对并发修改的数据，往往需要进行并发控制，一般数据库层面有两种并发控制：一种是悲观锁；另一种是乐观锁。Oinone 对乐观锁进行了良好支持。

1）定义方式

乐观锁有如下两种定义方式：

（1）通过快捷继承 VersionModel，构建带有乐观锁、唯一编码且主键为 id 的模型。

（2）可以在字段上使用 @Field.Version 注解来标识该模型更新数据时使用乐观锁。

如果更新的实际影响行数与入参数量不一致，则会抛出异常，错误码为 10150024。如果是批量更新数据，为了返回准确的实际影响行数，批量更新由批量提交改为循环单条数据提交更新，性能有所损失。

2）失效乐观锁

一个模型在某些场景下需要使用乐观锁来更新数据，而另一些场景不需要使用乐观锁来更新数据，则可以使用以下方式在一些场景下失效乐观锁。更多元位指令用法见 4.1.9 节。

失效乐观锁代码如下。

```
1  PamirsSession.directive().disableOptimisticLocker();
2  try{
3      更新逻辑
4  } finally {
5      PamirsSession.directive().enableOptimisticLocker();
6  }
```

3）不抛乐观锁异常

将批量提交类型设置为 useAffectRows 即可，这样可改由外层逻辑对返回的实际影响行数进行自主判断，代码如下。

```
1  Spider.getDefaultExtension(BatchApi.class).run(() -> {
2      更新逻辑，返回实际影响行数
3  }, BatchCommitTypeEnum.useAffectRows);
```

4）构建第一个 VersionModel

Step1. 新建 PetItemInventroy 模型，继承快捷模型 VersionModel，代码如下。

```
1   package pro.shushi.pamirs.demo.api.model;
2
3   import pro.shushi.pamirs.meta.annotation.Field;
4   import pro.shushi.pamirs.meta.annotation.Model;
5   import pro.shushi.pamirs.meta.base.common.VersionModel;
6
7   import java.math.BigDecimal;
8
9   @Model.model(PetItemInventroy.MODEL_MODEL)
10  @Model(displayName = "宠物商品库存",summary="宠物商品库存",labelFields = {"item
    Name"})
11  public class PetItemInventroy extends VersionModel {
12      public static final String MODEL_MODEL="demo.PetItemInventroy";
13
14      @Field(displayName = "商品名称")
15      private String itemName;
16
17      @Field(displayName = "库存数量")
18      private BigDecimal quantity;
19
20  }
```

Step2. 修改 DemoMenu，增加访问入口，代码如下。

```
1   @UxMenu("商品库存")@UxRoute(PetItemInventroy.MODEL_MODEL) class PetItemInvent
    royMenu{}
```

Step3. 重启看效果。

体验一：页面上新增、修改数据库中的数据时字段中的 opt_version 值会自动加 1，如图 4-7～图 4-10 所示。

图 4-7　效果（1）

图 4-8　效果（2）

图 4-9　效果（3）

图 4-10　效果（4）

体验二：同时打开两个页面，依次单击查看库存数量，会发现一个更改成功，一个没有更改成功。但页面都没有报错，只是 update 返回影响行数一个为 1，另一个为 0 而已，如图 4-11、图 4-12 所示。

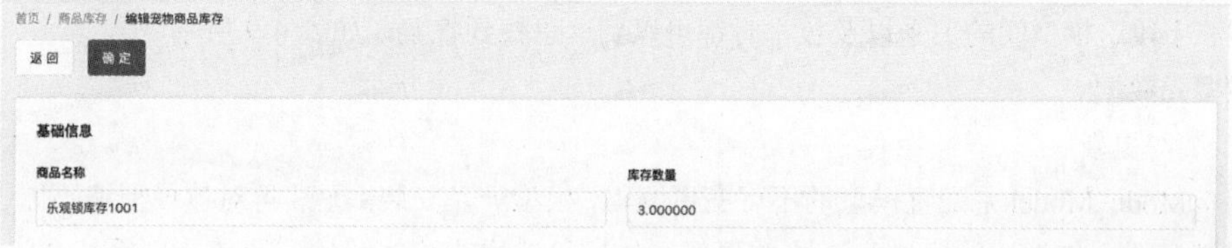

图 4-11 编辑宠物商品库存

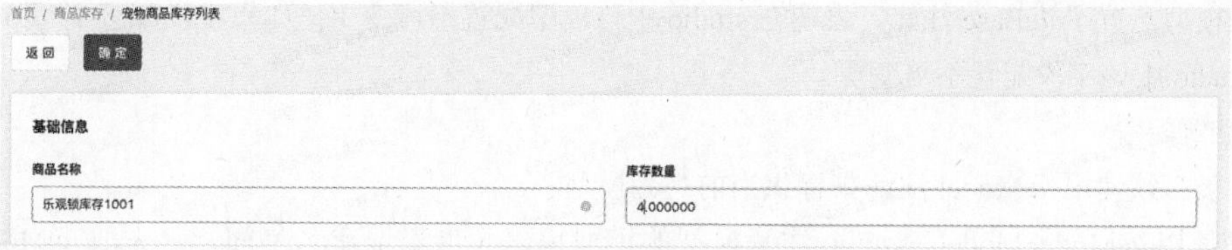

图 4-12 宠物商品库存列表

注：增加了乐观锁，我们在写代码的时候一定要注意，单记录更新操作的时候要去判断返回结果（影响行数），不然没更改成功，程序是不会抛错的，不像 batch 接口默认会报错。

**Step4. 预留任务**：重写 PetItemInventroy 的 update 函数。

留个任务，请各位小伙伴自行测试玩玩，这样会更有体感。

```java
package pro.shushi.pamirs.demo.core.action;

import org.springframework.stereotype.Component;
import pro.shushi.pamirs.demo.api.enumeration.DemoExpEnumerate;
import pro.shushi.pamirs.demo.api.model.PetItemInventroy;
import pro.shushi.pamirs.demo.api.model.PetTalent;
import pro.shushi.pamirs.meta.annotation.Function;
import pro.shushi.pamirs.meta.annotation.Model;
import pro.shushi.pamirs.meta.common.exception.PamirsException;
import pro.shushi.pamirs.meta.constant.FunctionConstants;
import pro.shushi.pamirs.meta.enmu.FunctionOpenEnum;
import pro.shushi.pamirs.meta.enmu.FunctionTypeEnum;

import java.util.ArrayList;
import java.util.List;

@Model.model(PetItemInventroy.MODEL_MODEL)
@Component
public class PetItemInventroyAction {

    @Function.Advanced(type= FunctionTypeEnum.UPDATE)
    @Function.fun(FunctionConstants.update)
    @Function(openLevel = {FunctionOpenEnum.API})
    public PetItemInventroy update(PetItemInventroy data){
        List<PetItemInventroy> inventroys = new ArrayList<>();
        inventroys.add(data);
        //批量更新会自动抛错
        int i = data.updateBatch(inventroys);
        //单记录更新，不自动抛错需要自行判断
//        int i = data.updateById();
//        if(i!=1){
//            throw PamirsException.construct(DemoExpEnumerate.PET_ITEM_INVENTROY_UPDATE_VERSION_ERROR).errThrow();
//        }
        return data;
    }
}
```

### 4.1.6 模型之元数据详解

本节介绍 Model 相关元数据，以及对应代码注解方式。这是 Oinone 的设计精华所在。不知道如何配置模型、字段、模型间的关系以及枚举时都可以到这里找到答案，如表 4-9 所示。

**1. 模型元数据**

1）安装与更新

使用 @Model.Model 来配置模型的不可变更编码。模型一旦安装，无法再对该模型编码值进行修改，之后的模型配置会依据该编码进行查找并更新；如果仍然修改该注解的配置值，则系统会将该模型识别为新模型，存储模型会创建新的数据库表，而原表将会 rename 为废弃表。

如果模型配置了 @Base 注解，表明在 studio 中该模型配置不可变更；如果字段配置了 @Base 注解，表明在 studio 中该字段配置不可变更。

2）注解配置

模型类必须使用 @Model 注解来标识当前类为模型类。

可以使用 @Model.Model、@Fun 注解模型的模型编码（也表示命名空间），先取 @Model.Model 注解值，若为空则取 @Fun 注解值，若皆为空则取全限定类名。

3）模型元信息

模型的 priority，当展示模型定义列表时，使用 priority 配置来对模型进行排序。

模型的 ordering，使用 ordering 属性来配置该模型的数据列表的默认排序。

模型元信息继承形式：

● 不继承（N）；

● 同编码以子模型为准（C）；

● 同编码以父模型为准（P）；

● 父子需保持一致，子模型可缺省（P=C）。

注：模型上配置的索引和唯一索引不会继承，所以需要在子模型重新定义。数据表的表名、表备注和表编码最终以父模型配置为准；扩展继承父子模型字段编码一致时，数据表字段定义以父模型配置为准。

表 4-9 模型元数据

| 名称 | 描述 | 抽象继承 | 同表继承 | 代理继承 | 多表继承 |
| --- | --- | --- | --- | --- | --- |
| 基本信息 | | | | | |
| displayName | 显示名称 | N | N | N | N |
| summary | 描述摘要 | N | N | N | N |
| label | 数据标题 | N | N | N | N |
| check | 模型校验方法 | N | N | N | N |
| rule | 模型校验表达式 | N | N | N | N |
| 模型编码 | | | | | |
| Model | 模型编码 | N | N | N | N |
| 高级特性 | | | | | |
| name | 技术名称 | N | N | N | N |
| table | 逻辑数据表名 | N | P=C | P=C | N |
| type | 模型类型 | N | N | N | N |
| chain | 是否是链式模型 | N | N | N | N |
| index | 索引 | N | N | N | N |

续表

| 名称 | 描述 | 抽象继承 | 同表继承 | 代理继承 | 多表继承 |
|---|---|---|---|---|---|
| unique | 唯一索引 | N | N | N | N |
| managed | 需要数据管理器 | N | N | N | N |
| priority | 优先级，默认100 | N | N | N | N |
| ordering | 模型查询数据排序 | N | N | N | N |
| relationship | 是否多对多关系模型 | N | N | N | N |
| inherited | 多重继承 | N | N | N | N |
| unInheritedFields | 不从父类继承的字段 | N | N | N | N |
| unInheritedFunctions | 不从父类继承的函数 | N | N | N | N |
| 高级特性——数据源 | | | | | |
| dsKey | 数据源 | N | P=C | P=C | N |
| 高级特性——持久化 | | | | | |
| logicDelete | 是否逻辑删除 | P | P | P | N |
| logicDeleteColumn | 逻辑删除字段 | P | P | P | N |
| logicDeleteValue | 逻辑删除状态值 | P | P | P | N |
| logicNotDeleteValue | 非逻辑删除状态值 | P | P | P | N |
| underCamel | 字段是否驼峰下划线映射 | P | P | P | N |
| capitalMode | 字段是否大小写映射 | P | P | P | N |
| 高级特性——序列生成配置 | | | | | |
| sequence | 配置编码 | C | C | C | N |
| prefix | 前缀 | C | C | C | N |
| suffix | 后缀 | C | C | C | N |
| separator | 分隔符 | C | C | C | N |
| size | 序列长度 | C | C | C | N |
| step | 序列步长 | C | C | C | N |
| initial | 初始值 | C | C | C | N |
| format | 序列格式化 | C | C | C | N |
| 高级特性——关联关系（或逻辑外键） | | | | | |
| unique | 外键值是否唯一 | C | C | C | N |
| foreignKey | 外键名称 | C | C | C | N |
| relationFields | 关系字段列表 | C | C | C | N |
| references | 关联模型 | C | C | C | N |
| referenceFields | 关联字段列表 | C | C | C | N |
| limit | 关系数量限制 | C | C | C | N |
| pageSize | 查询每页个数 | C | C | C | N |
| domainSize | 模型筛选可选项每页个数 | C | C | C | N |
| domain | 模型筛选，前端可选项 | C | C | C | N |
| onUpdate | 更新关联操作 | C | C | C | N |
| onDelete | 删除关联操作 | C | C | C | N |
| 静态配置 | | | | | |
| Static | 静态元数据模型 | N | N | N | N |

字段定义继承形式见表 4-10。

表 4-10　字段定义继承形式

| 名称 | 描述 | 抽象继承 | 同表继承 | 代理继承 | 多表继承 |
|---|---|---|---|---|---|
| 字段定义 | 字段定义 | C | C | C | C |

4）模型约束

（1）主键约束。

每个模型都可以配置自身的主键列表，也可以不配置主键。主键值不可缺省，可以索引到模型所对应数据表中唯一的一条记录。

（2）外键约束。

模型与模型之间的关联关系可以配置外键约束，以此约束关联关系之间数据的变更行为。

（3）校验约束。

模型可以配置校验函数对该模型的数据进行校验，存储数据时，校验数据是否合法合规。

**2. 字段元数据**

模型字段描述的是实体的特征属性。模型与字段之间的关联关系由 Model 的 Model 与 Field 的 Model 进行关联。ModelField 继承关系抽象类 Relation。

使用 @Field 注解来描述模型的字段。如果未配置字段类型，系统会根据 Java 代码的字段声明类型自动获取业务类型。建议配置 displayName 属性来描述字段在前端的显示名称。可以使用 defaultValue 配置字段的默认值。

1）元数据注解说明

元数据注解说明如图 4-13 所示。

2）安装与更新

使用 @Field.field 配置字段的不可变更编码。字段一旦安装，无法再对该字段编码值进行修改，之后的字段配置更新会依据该编码进行查找并更新；如果仍然修改该注解的配置值，则系统会将该字段识别为新字段，存储模型会创建新的数据库表字段，而原字段将会 rename 为废弃字段。

3）基础配置

（1）不可变更字段。

使用 immutable 属性描述该字段前后端都无法进行更新操作，系统会忽略不可变更字段的更新操作。

（2）自动生成编码的字段。

可以使用 @Field.Sequence 注解在字段上配置编码生成规则，为编码为空的字段自动生成编码。详见 3.3.5 节。

（3）字段的序列化与反序列化。

使用 @Field 注解的 serialize 属性配置非字符串类型属性的序列化与反序列化方式，最终会以序列化后的字符串持久化到存储中。详见 3.3.7 节。

（4）前端默认配置。

可以使用 @Field 注解中的以下属性来配置前端的默认视觉与交互规则，也可以在前端设置覆盖以下配置。

- required，是否必填。
- invisible，是否不可见。
- priority，字段优先级，列表的列使用该属性进行排序。

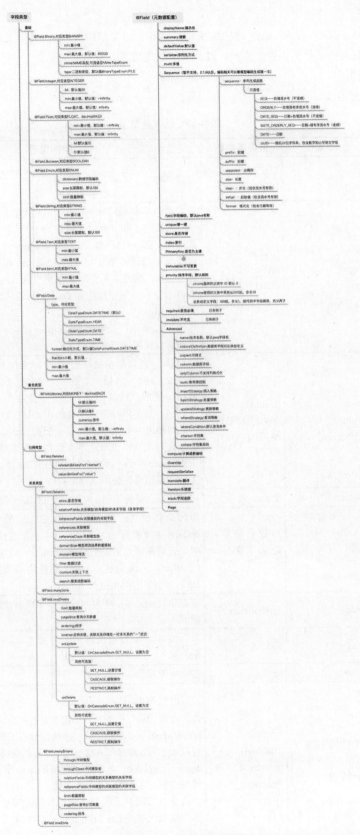

图 4-13 元数据注解说明

4）字段类型

类型系统由基本类型、复合（组件）类型、引用类型和关系类型构成。通过类型系统来决定应用程序、数据库和前端视觉视图是如何进行交互的，以及如何处理数据及数据间关系。

（1）基本类型。

基本类型见表 4-11。

表 4-11　基本类型

| 业务类型 | Java 类型 | 数据库类型 | 规则说明 |
| --- | --- | --- | --- |
| BINARY | Byte<br>Byte[] | TINYINT<br>BLOB | 二进制类型，不推荐使用 |
| INTEGER | Short<br>Integer<br>Long<br>BigInteger | smallint<br>int<br>bigint<br>decimal (size,0) | 整数，包括整数（10～11 位有效数字）、长整数（19～20 位有效数字）和大整数（超过 19 位）。<br>数据库规则：默认使用 int；如果 size 小于 6 则使用 smallint；如果 size 超过 6 则使用 int；如果 size 超过 10 位数字（包含符号位），则使用长整数 bigint；如果 size 超过 19 位数字（包含符号位），则使用大数 decimal。若未配置 size，则按 Java 类型推测。<br>前端交互规则：整数使用 Number 类型，长整数和大整数前后端协议使用字符串类型 |
| FLOAT | Float<br>Double<br>BigDecimal | float (M,D)<br>double (M,D)<br>decimal (M,D) | 浮点数，包括单精度浮点数（7～8 位有效数字）、双精度浮点数（15～16 位有效数字）和大数（超过 15 位）。<br>数据库规则：默认使用单精度浮点数 float；如果 size 超过 7 位数字，则使用双精度浮点数 double；如果 size 超过 15 位数字，则使用大数 decimal。若未配置 size，则按 Java 类型推测。<br>前端交互规则：单精度浮点数 float 和双精度浮点数 double 使用 Number 类型（因为都使用 IEEE754 协议 64 位进行存储），大数前后端协议使用字符串类型 |
| BOOLEAN | Boolean | tinyint (1) | 布尔类型，值为 1,true（真）或 0,false（假） |
| ENUM | Enum | 与数据字典指定基本类型一致 | 前端交互规则：可选项从 ModelField 的 options 字段获取，options 字段值为字段指定数据字典子集的 JSON 序列化字符串。前后端传递的是可选项的 name，数据库存储使用可选项的 value。multi 属性为 true，则使用多选控件；multi 属性为 false，则使用单元控件 |
| STRING | String | varchar (size) | 字符串，size 为长度限制默认值参考，前端可以 view 覆盖该配置 |
| TEXT | String | text | 多行文本，编辑态组件为多行文本框，长度限制为配置项 size 值 |
| HTML | String | text | 富文本编辑器 |
| DATETIME | Java.util.Date<br>Java.sql.Timestamp | datetime (fraction)<br>timestamp (fraction) | 日期时间类型<br>数据库规则：日期和时间的组合，时间格式为 YYYY-MM-DD HH:MM:SS[.fraction]，默认精确到秒，在默认的秒精确度上，可以带小数，最多带 6 位小数，即可以精确到 microseconds（6 digits）precision。可以通过设置 fraction 来设置精确小数位数，最终存储在字段的 decimal 属性上。<br>前端交互规则：前端默认使用日期时间控件，根据日期时间类型格式化格式 format 格式化日期时间 |

| 业务类型 | Java 类型 | 数据库类型 | 规则说明 |
|---|---|---|---|
| YEAR | Java.util.Date | year | 年份类型<br>日期类型<br>数据库规则：默认"YYYY"格式表示的日期值<br>前端交互规则：前端默认使用年份控件，根据日期类型格式化格式 format 格式化日期 |
| DATE | Java.util.Date<br>Java.sql.Date | date<br>date | 日期类型<br>数据库规则：默认"YYYY-MM-DD"格式表示的日期值<br>前端交互规则：前端默认使用日期控件，根据日期类型格式化格式 format 格式化日期 |
| TIME | Java.util.Date<br>Java.sql.Time | time (fraction)<br>time (fraction) | 时间类型<br>数据库规则：默认"HH:MM:SS"格式表示的时间值<br>前端交互规则：前端默认使用时间控件，根据日期类型格式化格式 format 格式化日期 |

（2）复合类型。

复合类型见表 4-12。

表 4-12 复合类型

| 业务类型 | Java 类型 | 数据库类型 | 规则说明 |
|---|---|---|---|
| MONEY | BigDecimal | decimal（M,D） | 金额，前端使用金额控件，可以使用 currency 设置币种字段 |

（3）引用类型。

引用类型见表 4-13。

表 4-13 引用类型

| 业务类型 | Java 类型 | 数据库类型 | 规则说明 |
|---|---|---|---|
| RELATED | 基本类型或关系类型 | 不存储或 varchar、text | 引用字段<br>数据库规则：点表达式最后一级对应字段类型；数据库字段值默认为 Java 字段的序列化值，默认使用 JSON 序列化<br>前端交互规则：点表达式最后一级对应字段控件类型<br>注：可参考 3.3.9 节字段类型关系引用的举例 |

（4）关系类型。

关系类型见表 4-14。

表 4-14 关系类型

| 业务类型 | Java 类型 | 数据库类型 | 规则说明 |
|---|---|---|---|
| O2O | 模型 /DataMap | 不存储或 varchar、text | 一对一关系 |
| M2O | 模型 /DataMap | 不存储或 varchar、text | 多对一关系 |
| O2M | List< 模型 /DataMap> | 不存储或 varchar、text | 一对多关系 |
| M2M | List< 模型 /DataMap> | 不存储或 varchar、text | 多对多关系 |

多值字段或者关系字段需要存储，默认使用 JSON 格式序列化。多值字段数据库字段类型默认为 varchar (1024)；关系字段数据库字段类型默认为 text。

（5）类型默认推断。

M 代表精度，即有效长度（总位数），D 代表标度，即小数点后的位数，fraction 为时间秒以下精度。

multi 表示该字段为多值字段（见表 4-15）。

表 4-15 类型默认推断

| Java 类型 | Field 注解 | 推断 ttype | 推断配置 | 推断数据库配置 |
|---|---|---|---|---|
| Byte | @Field | BINARY | 无 | blob |
| String | @Field | STRING | size=128 | varchar (128) |
| List&lt;primitive type&gt; | @Field | STRING | size=1024,multi=true | varchar (1024) |
| Map | @Field | STRING | size=1024 | varchar (1024) |
| Short | @Field | INTEGER | M=5 | smallint (6) |
| Integer | @Field | INTEGER | M=10 | integer (11) |
| Long | @Field | INTEGER | M=19 | bigint (20) |
| BigInteger | @Field | INTEGER | M=64 | decimal (64,0) |
| Float | @Field | FLOAT | M=7,D=2 | float (7,2) |
| Double | @Field | FLOAT | M=15,D=4 | double (15, 4) |
| BigDecimal | @Field | FLOAT | M=64,D=6 | decimal (64,6) |
| Boolean | @Field | BOOLEAN | 无 | tinyint (1) |
| Java.util.Date | @Field | DATETIME | fraction=0 | datetime |
| Java.util.Date | @Field.Date (type=DateTypeEnum.YEAR) | YEAR | 无 | year |
| Java.util.Date | @Field.Date (type=DateTypeEnum.DATE) | DATE | 无 | date |
| Java.util.Date | @Field.Date (type=DateTypeEnum.TIME) | TIME | fraction=0 | time |
| Java.sql.Timestamp | @Field | DATETIME | fraction=0 | timestamp |
| Java.sql.Date | @Field | DATE | 无 | date |
| Java.sql.Time | @Field | TIME | fraction=0 | time |
| Long | @Field.Date | DATETIME | fraction=0 | datetime |
| enum implements IEnum | @Field | ENUM | 无 | 根据枚举 value 类型 |
| primitive type | @Field.Enum (dictionary= 数据字典编码 ) | ENUM | 无 | 根据枚举 value 类型 |
| List&lt;primitive type&gt; | @Field.Enum (dictionary= 数据字典编码 ) | ENUM | multi=true | varchar (512) |
| 模型类 | @Field.Relation | M2O | 无 | text |
| DataMap | @Field.Relation | M2O | 无 | text |
| List&lt; 模型类 &gt; | @Field.Relation | O2M | multi=true | text |
| List&lt;DataMap&gt; | @Field.Relation | O2M | multi=true | text |

5）字段约束

（1）主键。

可以使用 Yaml 或者 @Model.Advanced 的 keyGenerator 属性来配置模型主键的自动生成规则，AUTO_INCREMENT 或者分布式 ID。如果不配置，将不会自动生成主键值。

（2）逻辑外键约束。

在创建关联关系字段的时候，可以使用 @Field.Relation 注解的 onUpdate 和 onDelete 属性指定在删除模型或更新模型关系字段值时,对关联模型进行的相应操作。操作包括 RESTRICT、NO ACTION、SET NULL 和 CASCADE,默认值为 RESTRICT。

● RESTRICT 是指模型与关联模型有关联记录的情况下,引擎会阻止模型关系字段的更新或删除模型记录。

● NO ACTION 是指不做约束(这里与数据库约束的定义不相同)。

● CASCADE 表示在更新模型关系字段或者删除模型时,级联更新关联模型对应记录的关联字段值或者级联删除关联模型对应记录。

● SET NULL 表示在更新模型关系字段或者删除模型时,关联模型的对应关联字段将被设置为 null(该字段值允许为 null 的情况下,设置为 null,若不允许为 null,则引擎阻止对模型的操作)。

(3)通用校验约束。

通用校验约束见表 4-16~表 4-18。

表 4-16 通用校验约束(表一)

| 字段业务类型 | size | limit | decimal | mime | min | max |
|---|---|---|---|---|---|---|
| BINARY | | | | 文件类型 | 最小比特位 | 最大比特位 |
| INTEGER | 有效数字 | | | | 最小值 | 最大值 |
| FLOAT | 有效数字 | | 小数位数 | | 最小值 | 最大值 |
| BOOLEAN | | | | | | |
| ENUM | 存储字符数 | 多选最多数量 | | | | |
| STRING | 存储字符数 | | | | 字符数 | 字符数 |
| TEXT | | | | | 字符数 | 字符数 |
| HTML | | | | | 字符数 | 字符数 |
| MONEY | 有效数字 | | 小数位数 | | 最小值 | 最大值 |
| RELATED | | | | | | |

表 4-17 通用校验约束(表二)

| 字段业务类型 | fraction | format | min | max |
|---|---|---|---|---|
| DATETIME | 时间精度 | 时间格式 | 最早日期时间 | 最晚日期时间 |
| YEAR | | 时间格式 | 最早年份 | 最晚年份 |
| DATE | | 时间格式 | 最早日期 | 最晚日期 |
| TIME | 时间精度 | 时间格式 | 最早时间 | 最晚时间 |

表 4-18 通用校验约束(表三)

| 字段业务类型 | size | domainSize | limit | pageSize |
|---|---|---|---|---|
| RELATED | 存储字符数(若序列化存储) | | | |
| O2O | 存储字符数(若序列化存储) | 可选项每页个数 | | |
| M2O | 存储字符数(若序列化存储) | 可选项每页个数 | | |
| O2M | 存储字符数(若序列化存储) | 可选项每页个数 | 关系数量限制 | 查询每页个数 |
| M2M | 存储字符数(若序列化存储) | 可选项每页个数 | 关系数量限制 | 查询每页个数 |

在模型或字段上配置 check 函数,则处理前端请求时会进行校验约束。也可以调用模型上的 check 函数进行编程式校验。

(4)默认值约束。

字段默认值 defaultValue 可以是基本类型或者关系类型的序列化值。时间类型可以使用 format 来格式化时间表达式或者使用长整数来设置默认值。枚举类型使用枚举项值 value 来设置默认值。如果需要进行复杂的计算，请使用模型的 construct 构造函数来配置解决。

（5）唯一约束。

将字段或者模型配置 unique（唯一索引），可以为模型或字段添加唯一约束。

（6）可选项约束。

使用枚举定义字段的可选项值，可以为字段提供可选项约束功能。

6）关系字段

关联关系用于描述模型间的关联方式（如图 4-14 所示）：

● 多对一关系，主要用于明确从属关系；
● 一对多关系，主要用于明确从属关系；
● 多对多关系，主要用于弱依赖关系的处理，提供中间模型进行关联关系的操作；
● 一对一关系，主要用于多表继承和行内合并数据。

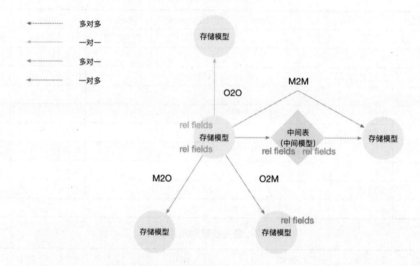

图 4-14 关系字段

（1）名词解释。

关联关系比较重要的名词解释如下：

● 关联关系：使用 relation 表示，包括关联关系类型、关联关系双边的模型和关联关系的读写。
● 关联关系字段：业务类型 ttype 为 o2o、o2m、m2o 或 m2m 的字段。
● 关联模型：使用 references 表示，自身模型关联的模型。
● 关联字段：使用 referenceFields 表示，关联模型的字段，表示关联模型的哪些字段与自身模型的哪些字段建立关系。
● 关系模型：自身模型。
● 关系字段：使用 relationFields 表示，自身模型的字段，表示自身模型的哪些字段与关联模型的哪些字段建立关系。
● 中间模型，使用 through 表示，只有多对多存在中间模型，模型的 relationship=true。

（2）关联关系的默认视图。

● 一对多默认视图，编辑态在行内是下拉多选，在详情态是选项卡表格；展示态在行内是折叠面板表格，在详情态是选项卡表格。

● 多对一默认视图，编辑态在行内是下拉单选，在详情态是下拉单选；展示态在行内是文字，在详情是文字。

● 多对多默认视图，编辑态在行内是下拉多选，在详情态是选项卡表格；展示态在行内是折叠面板表格，在详情是选项卡表格。

● 一对一默认视图，编辑态在行内是平铺，在详情态是分组；展示态在行内是平铺，在详情态是分组。

后端研发时所有关联关系和引用的处理都限制在本模型，即平台至多处理到当前模型的字段，不再继续依赖关联关系和引用处理关联模型，但是可以手动调用模型上的链式方法 fieldQuery、fieldCreate 和 fieldUpdate 来完成关联关系的查询与更新操作。

使用 o2m 或者 m2m 关联关系关联的临时模型没有分页查询操作。

（3）关联关系的配置。

可以使用 @Field.Relation 注解的 relationFields、referenceFields、references 和 through 来配置关联关系。relationFields 与 referenceFields 为存储关联关系的一一映射字段列表。

如果 relationFields 缺省，一对多或者多对多关系的 relationFields 默认为模型主键；一对一或者多对一关系的 relationFields 默认为关联关系字段名加上首字母大写的主键名拼接而成的字符串。如果有多个主键，则 relationFields 和 referenceFields 也对应有多个字段。

如果配置了 relationFields，但 referenceFields 缺省，则 referenceFields 与 relationFields 字段名一致。

一对多关系的 referenceFields 必填。如果 referenceFields 缺省，多对多、多对一或者一对一关系的 referenceFields 默认为主键。

表 4-19 关联关系的配置

| 关系类型 | 缺省关系字段默认值 | 缺省关联字段默认值 |
| --- | --- | --- |
| 一对多 | 默认为关系模型的 pk | 默认为"关系模型名+关系模型"的 pk；如果关系另一端的多对一字段名不是关系模型名，则需明确指定，使两端关系字段与关联字段对应 |
| 一对一 | 默认为"关联关系字段名+关系模型"的 pk | 默认为关联模型的 pk |
| 多对一 | 默认为"关联关系字段名+关系模型"的 pk | 默认为关联模型的 pk |
| 多对多 | 默认为关系模型的 pk | 默认为关联模型的 pk |

注：pk 对应数据库中的 pk，即数据库表的主键字段。

多对多使用 through 来指定中间模型的模型编码，如果指定模型编码的中间模型不存在，系统会根据 through 自动生成中间模型，中间模型的默认字段为与两端模型关联的关系字段。与关系模型关联的关系字段名称为关系模型名称加上关系模型的主键拼接而成的字符串；与关联模型关联的多对一字段名称为关联模型名称加上关联模型主键拼接而成的字符串。如果与关系模型关联的关系字段名称和与关联模型关联的多对一字段名称冲突，需要使用 throughRelationFields 和 throughReferenceFields 明确配置指定字段名称解决冲突。

系统根据模块的依赖关系，自动生成的中间模型将生成在先加载的建立多对多关系的关系模型所在模块。

（4）读写关联关系字段。

默认关联关系字段的 store 属性为 false，relationStore 为 true。若设置关联关系字段 relationStore 属性为 true，则会为关联关系字段生成关系字段用于存储关联关系。若设置关联关系字段 store 属性为 true，则存储时序列化字段值存储到数据库中，查询时从数据库中反序列化得到字段值。字段类型为 varchar 且长度为 128，如果需要改变字段长度，可以使用 @Field.Advanced 的 columnDefinition 设置。

当 store 属性为 false 时，则字段值为关联关系查询得到的结果。如果 store 为 false 且 relationStore 为 false，则只能对字段进行赋值来设置字段值。

（5）关联操作。

调用数据管理器的 API 不会触发关联操作，需要调用 fieldQuery 和 fieldSave 方法进行关联模型的关联操作。

前端的查询接口会根据 GraphQL 协议进行关联查询。

前端的新增和更新接口默认会存储当前模型的关联关系字段、新增的递归并更新一对多关系的关联关系字段。更新接口会检查当前模型的逻辑外键约束，可以调用模型的 ignore 方法或设置模型数据的 ignore 属性来改变递归深度，避免循环操作。

前端的删除接口会默认删除当前模型数据和根据级联配置进行当前模型的关联关系字段的关联操作。删除接口会检查当前模型的逻辑外键约束。可以调用模型的 ignore 方法或设置模型数据的 ignore 属性来改变递归深度，避免循环操作。

（6）关联数据分页。

可以使用关系字段配置中分页数量 pageSize 来限定关联查询的返回结果数量。可选项可以使用 domainSize 来限定可选项返回结果数量，由前端从字段元数据中获取并设置为可选项查询分页数量限制。

（7）反转关系。

一对多关联关系可以设置 inverse 为 true 反转关系，反转关系后关联关系存储在一对多关系中"一"这一端。

（8）引用字段。

引用字段可以通过与其他字段建立引用关系来获取数据。

当引用字段的 store 属性为 true 时，则字段值为存储的字段值，数据存储时将被引用字段值存储到数据存储中（unset 掉被引用字段，则直接存储引用字段值）；当 store 属性为 false 时，则数据为被引用字段的字段值且不会存储。

### 4.1.7 函数之元数据详解

如不了解 Function 的定义，可以参阅 2.3 节中对 Function 的描述，本节主要介绍 Function 元数据构成，说明 Oinone 从哪些维度来描述 Function。

**1. 元数据说明**

1）FunctionDefinition

FunctionDefinition 的详细信息见表 4-20。

表 4-20 FunctionDefinition 的详细信息

| 元素数据构成 | 含义 | 对应注解 | 备注 |
| --- | --- | --- | --- |
| namespace | 函数命名空间 | @Fun (" ") | @Fun 或 @Model.Model |
| name | 技术名称 | @Function (<br>name=" ",<br>scene={},<br>summary=" ",<br>openLevel=FunctionOpenEnum.<br>REMOTE<br>) | |
| scene | 可用场景 | | 见：FunctionSceneEnum |
| description | 描述 | | |
| openLevel | 开放级别 | | 见：FunctionOpenEnum |
| fun | 编码 | @Function.fun (" ") | |

| 元素数据构成 | 含义 | 对应注解 | 备注 |
|---|---|---|---|
| displayName | 显示名称 | @Function.Advanced (<br>displayName=" ",<br>type=FunctionTypeEnum.UPDATE,<br>dataManager=false,<br>language=FunctionLanguageEnum.JAVA,<br>isBuiltin=false,<br>category=FunctionCategoryEnum.OTHER,<br>group="pamirs",<br>version="1.0.0",<br>timeout=5000,<br>retries=0,<br>isLongPolling=false,<br>longPollingKey="userId"<br>longPollingTimeout=1) | |
| type | 函数类型<br>默认：4（改） | | 见 |
| dataManager | 数据管理器函数<br>默认：false | | |
| language | 函数语言<br>默认：DSL | | |
| isBuiltin | 是否内置函数 | | |
| category | 分类 | | 见 FunctionCategoryEnum |
| group | 系统分组<br>pamirs | | |
| version | 系统版本<br>1.0.0 | | |
| timeout | 超时时间 | | |
| retries | 重试次数 | | |
| isLongPolling | 是否支持 long polling，默认 false | | |
| longPollingKey | 支持从上下文中获取字段作为 key | | |
| longPollingTimeout | long polling 超时时间 | | |
| transactionConfig | 事务配置<br>JSON 存储 | | 见 TransactionConfig<br>@PamirsTransactional |
| source | 来源 | | 系统推断值，见<br>FunctionSourceEnum |
| extPointList | 函数包含扩展点 | | 系统推断值 |
| module | 所属模块 | | 系统推断值 |
| bitOptions | 位 | | 系统推断值 |
| attributes | 属性 | | 系统推断值 |
| imports | 上下文引用 | | 系统推断值 |
| context | 上下文变量 | | 系统推断值 |
| codes | 函数内容 | | 系统推断值 |
| beanName | bean 名称 | | 系统推断值 |
| rule | 前端规则 | | 系统推断值，一般通过 Action.rule 传递下来 |
| clazz | 函数位置 | | 系统推断值 |
| method | 函数方法 | | 系统推断值 |
| argumentList | 函数参数 | | 系统推断值，List<Argument> |
| returnType | 返回值类型 | | 系统推断值 |

2）TransactionConfig

TransactionConfig 是函数事务管理中的配置项事务，见表 4-21，具体事务使用请见 4.18 节。

表 4-21 TransactionConfig

| 元素数据构成 | 含义 | 对应注解 | 备注 |
| --- | --- | --- | --- |
| transactionManager | 事务管理器 | @PamirsTransactional (transactionManager=" ", enableXa=false, isolation=Isolation.DEFAULT, propagation=Propagation.REQUIRED, timeout=-1, readOnly=false, rollbackFor={}, rollbackForClassName={}, noRollbackFor={}, noRollbackForClassName={}, rollbackForExpCode={}, noRollbackForExpCode={}) | |
| enableXa | 分布式事务 默认为 false | | |
| isolation | 事务隔离级别 | | |
| propagation | 事务传递类型 | | |
| timeout | 过期时间 | | |
| readOnly | 只读 false | | |
| rollbackForExpCode | 回滚异常编码 | | |
| rollbackForExpCode | 忽略异常编码 | | |
| namespace | 函数命名空间 | | 系统推断值 |
| fun | 函数编码 | | 系统推断值 |
| active | 生效 | | 系统推断值 |

**2. 元数据，在 Java 中对应的注解类说明**

1）命名空间注解

（1）Fun。

@Fun 函数申明

└── value 命名空间

（2）Model。

@Model

└── Model 命名空间

2）函数信息注解

@Function

├── name 技术名称

├── scene 可用场景

├── summary 描述摘要

├── openLevel 开放级别

├── Advanced 更多配置

│  ├── displayName 显示名称

│  ├── type 函数类型，默认 FunctionTypeEnum.UPDATE

│  ├── managed 数据管理器函数，默认 false

│  ├── language 语言，默认 FunctionLanguageEnum.JAVA

│  ├── builtin 是否内置函数，默认否

│  ├── category 分类，FunctionCategoryEnum.OTHER

```
│   ├── group 系统分组，默认 pamirs
│   ├── version 系统版本
│   ├── timeout 超时时间
│   ├── retries 重试次数
│   ├── isLongPolling 是否支持 long polling，默认 false
│   ├── longPollingKey 支持从上下文中获取字段作为 key，默认 userId
│   └── longPollingTimeout long polling 超时时间
├── fun 函数编码
└── value 生效
```

3）函数事务管理注解

@PamirsTransactional
value @AliasFor ("transactionManager")
```
├── transactionManager @AliasFor ("value") 默认值：" "
├── propagation 事务传递类型，默认值为 Propagation.REQUIRED
├── isolation 事务隔离级别，默认值为 Isolation.DEFAULT
├── timeout 过期时间
├── readOnly 只读，默认为 false
├── rollbackFor 回滚异常类
├── rollbackForClassName 回滚异常类名
├── rollbackForExpCode 回滚异常编码
├── noRollbackFor 忽略异常类
├── noRollbackForClassName 忽略异常类名
├── noRollbackForExpCode 忽略异常编码
└── enableXa 分布式事务，默认为 false
```

### 4.1.8 函数之事务管理

**1. 事务管理介绍**

函数 Function 中有标识是否支持事务的字段，为 isTransaction（默认为 false），事务传播行为 propagationBehavior（默认 PROPAGATION_SUPPORTS），事务隔离级别 isolationLevel（默认使用数据库默认的事务隔离级别），所以不会默认为函数添加事务。另外事务配置提供全局配置。

平台事务管理兼容 Spring 声明式与编程式事务，支持多数据源事务管理。事务管理中多数据源嵌套独立事务，不会造成死锁风险。使用多数据源或分表操作，不会导致脏读。如果需要多数据源分布式事务，请使用 PamirsTransational 分布式事务管理方案 [@PamirsTransational (enableXa=true)]。分布式事务一般用于量小的跨模块配置管理场景。

1）使用方式

声明式事务，使用 @PamirsTransactional 注解在需要事务管理的类或方法上标注。在非无代码场景下，与 @Transactional 注解功能一致。

编程式事务，使用 PamirsTransactionTemplate 即可。在非无代码场景下，与 TransactionTemplate 功能一致。

配置式事务，使用 TxConfig 模型在模块安装时初始化存储事务配置数据。

2）事务特性

原子性（atomicity）：强调事务的不可分割。

一致性（consistency）：事务执行的前后数据的完整性保持一致。

隔离性（isolation）：一个事务执行的过程中，不应该受到其他事务的干扰。

持久性（durability）：事务一旦结束，数据就真实保存到数据库。

3）事务隔离级别

事务隔离级别指的是一个事务对数据的修改与另一个并行的事务的隔离程度，当多个事务同时访问相同数据时，如果没有采取必要的隔离机制，就可能发生如表 4-22 所示问题。

表 4-22 事务隔离级别

| 问题 | 描述 |
| --- | --- |
| 脏读 | 所谓脏读，就是一个事务读到另一个事务未提交更新的数据，就是指事务 A 读到了事务 B 还没有提交的数据，比如银行取钱，事务 A 开启事务，此时切换到事务 B，事务 B 开启事务取走 100 元，此时切换回事务 A，事务 A 读取的肯定是数据库里面的原始数据，因为事务 B 取走了 100 块钱，并没有提交，数据库里面的账务余额肯定还是原始余额 |
| 不可重复读 | 在一个事务里面的操作中发现了未被操作的数据，比方在同一个事务中先后执行两条一模一样的 select 语句，其间在此次事务中没有执行过任何 DDL 语句，但先后得到的结果不一致 |
| 幻读 | 是指当事务不是独立执行时发生的一种现象，例如第一个事务对一个表中的数据进行了修改，这种修改涉及表中的全部数据行。同时，第二个事务也修改这个表中的数据，这种修改是向表中插入一行新数据。那么，以后就会发生操作第一个事务的用户发现表中还有没有修改的数据行，就好像发生了幻觉一样 |

4）Pamirs (Spring) 支持的隔离级别

Pamirs (Spring) 支持的隔离级别见表 4-23 和表 4-24。

表 4-23 隔离级别与描述

| 隔离级别 | 描述 |
| --- | --- |
| DEFAULT | 使用数据库本身使用的隔离级别 ORACLE（读已提交） MySQL（可重复读） |
| READ_UNCOMMITTED | 读未提交（脏读）最低的隔离级别，一切皆有可能 |
| READ_COMMITED | 读已提交，ORACLE 默认隔离级别，有不可重复读以及幻读风险 |
| REPEATABLE_READ | 可重复读，解决不可重复读的隔离级别，但还是有幻读风险 |
| SERLALIZABLE | 串行化，最高的事务隔离级别，不管多少事务，挨个运行完一个事务的所有子事务之后才可以执行另外一个事务里面的所有子事务，这样就解决了脏读、不可重复读和幻读的问题 |

表 4-24 隔离级别说明表

| 隔离级别 | 脏读可能性 | 不可重复读可能性 | 幻读可能性 | 加锁度 |
| --- | --- | --- | --- | --- |
| READ_UNCOMMITTED | 是 | 是 | 是 | 否 |
| READ_COMMITED | 否 | 是 | 是 | 否 |
| REPEATABLE_READ | 否 | 否 | 是 | 否 |
| SERLALIZABLE | 否 | 否 | 否 | 是 |

5）事务的传播行为

（1）以下事务的传播行为不会有新事务产生，都会在一个事务内完成。

● PROPAGATION_REQUIRED 支持当前事务，如果不存在就新建一个（默认）。

- PROPAGATION_SUPPORTS 支持当前事务，如果不存在，就不使用事务。
- PROPAGATION_MANDATORY 支持当前事务，如果不存在，抛出异常。

（2）以下的事务传播行为会产生新的事务。
- PROPAGATION_REQUIRES_NEW 如果有事务存在，挂起当前事务，创建一个新的事务。
- PROPAGATION_NOT_SUPPORTED 以非事务方式运行，如果有事务存在，挂起当前事务。
- PROPAGATION_NEVER 以非事务方式运行，如果有事务存在，抛出异常。
- PROPAGATION_NESTED 如果当前事务存在，则嵌套事务执行。

事务 A 中嵌套事务 B，嵌套 PROPAGATION_REQUIRES_NEW 方法不应与事务 A 处在同类中。

表 4-25　事务传播行为

| 异常状态 | PROPAGATION_REQUIRES_NEW（两个独立事务） | PROPAGATION_NESTED（事务 B 嵌套事务 A 中） | PROPAGATION_REQUIRED（同一个事务） |
| --- | --- | --- | --- |
| 事务 A 抛异常<br>事务 B 正常 | 事务 A 回滚，事务 B 正常提交 | 事务 A 与事务 B 一起回滚 | 事务 A 与事务 B 一起回滚 |
| 事务 A 正常<br>事务 B 抛异常 | 1. 如果事务 A 中捕获事务 B 的异常，并没有继续向上抛异常，则事务 B 先回滚，事务 A 再正常提交；<br>2. 如果事务 A 未捕获事务 B 的异常，默认会将事务 B 的异常向上抛，则事务 B 先回滚，事务 A 再回滚 | 事务 B 先回滚，事务 A 再正常提交 | 事务 A 与事务 B 一起回滚 |
| 事务 A 抛异常<br>事务 B 抛异常 | 事务 B 先回滚，事务 A 再回滚 | 事务 A 与事务 B 一起回滚 | 事务 A 与事务 B 一起回滚 |
| 事务 A 正常<br>事务 B 正常 | 事务 B 先提交，事务 A 再提交 | 事务 A 与事务 B 一起提交 | 事务 A 与事务 B 一起提交 |

**2. 声明式事务（举例）**

Step1. 修改 PetShopBatchUpdateAction。

（1）不损害任何性能。

（2）事务保障率超过 4 个 9。

（3）经过阿里等大厂验证，特别是在阿里的结算平台中得到了很好的验证。

@PamirsTransactional 更多配置项请参考 4.14 节。@PamirsTransactional 百分之百兼容 @Transactional。

代码如下：

```
1   @Action(displayName = "确定",bindingType = ViewTypeEnum.FORM,contextType = ActionContextTypeEnum.SINGLE)
2   @PamirsTransactional
3   //@Transactional
4   public PetShopBatchUpdate conform(PetShopBatchUpdate data){
5       if(data.getPetShopList() == null || data.getPetShopList().size()==0){
6           throw PamirsException.construct(DemoExpEnumerate.PET_SHOP_BATCH_UPDATE_SHOPLIST_IS_NULL).errThrow();
7       }
8       List<PetShopProxy> proxyList = data.getPetShopList();
9       for(PetShopProxy petShopProxy:proxyList){
10          petShopProxy.setDataStatus(data.getDataStatus());
11      }
12      new PetShopProxy().updateBatch(proxyList);
13      throw PamirsException.construct(DemoExpEnumerate.SYSTEM_ERROR).errThrow();
14      //      return data;
15  }
```

## Step2. 重启看效果。

进入店铺管理列表页,选择"记录"单击"批量更新数据状态"按钮,修改记录的数据状态为"未启用",提交看效果如图 4-15、图 4-16 所示。期望效果为提示系统异常,数据修改失败,如图 4-17 所示。

图 4-15 数据状态显示已启用

图 4-16 批量更新数据状态

图 4-17 提示系统异常

### 3. 编程式事务(举例)

为了提升性能,特别是在高并发场景,编程式事务开发模式有利于精细化控制事务开启长度,尽可能地在事务开启前,把费时的查询工作、数据准备做完。基本套路如下。

```
1  Tx.build(new TxConfig().setPropagation(Propagation.REQUIRED.value())).executeWitho
   utResult(status -> {
2          //执行逻辑
3  });
```

Step1. 修改 PetShopBatchUpdateAction,代码如下。

```
1   @Action(displayName = "确定",bindingType = ViewTypeEnum.FORM,contextType = ActionC
    ontextTypeEnum.SINGLE)
2   public PetShopBatchUpdate conform(PetShopBatchUpdate data){
3       if(data.getPetShopList() == null || data.getPetShopList().size()==0){
4           throw PamirsException.construct(DemoExpEnumerate.PET_SHOP_BATCH_UPDATE_S
    HOPLIST_IS_NULL).errThrow();
5       }
6       List<PetShopProxy> proxyList = data.getPetShopList();
7       for(PetShopProxy petShopProxy:proxyList){
8           petShopProxy.setDataStatus(data.getDataStatus());
9       }
10      Tx.build(new TxConfig().setPropagation(Propagation.REQUIRED.value())).execute
    WithoutResult(status -> {
11          new PetShopProxy().updateBatch(proxyList);
12          throw PamirsException.construct(DemoExpEnumerate.SYSTEM_ERROR).errThrow
    ();
13      });
14      return data;
15  }
```

Step2. 重启看效果。

跟声明式事务效果一致。

**4. 配置式事务**

该模式一般用于平台内部使用以及无代码编辑器管理事务时使用,就不举例了。

**5. 分布式事务(不建议使用)**

如果要严格意义上的分布式事务,需要配置 enableXa 为 true。同时引入依赖包,代码如下。

```
1   <groupId>pro.shushi.pamirs.framework</groupId>
2   <artifactId>pamirs-connectors-data-xa</artifactId>
```

注:该版本还不支持远程 RPC 后的分布式事务,因该模式有很大的弊端,也就是把原本无状态的服务变成有状态,导致性能和耦合度都极差。所以一般使用事务性消息、异步任务等最终一致性方案去替代。

### 4.1.9 函数之元位指令

**1. 元位指令介绍**

元位指令系统给请求上下文的指令位字段做按位与标记,来对函数进行下发对应指令结构复杂。

元位指令系统分为数据指令和请求上下文指令两种。

1)数据指令

数据指令基本都是系统内核指令,业务开发时用不到,这里不再赘述。

2)请求上下文指令

请求上下文指令见表 4-26。

表 4-26 请求上下文指令

| 位 | 指令 | 指令名 | 前端默认值 | 后端默认值 | 描述 |
|---|---|---|---|---|---|
| 20 | builtAction | 内建动作 | 否 | 否 | 是否平台内置定义的服务器动作对应操作：<br>PamirsSession.directive().disableBuiltAction();<br>PamirsSession.directive().enableBuiltAction(); |
| 21 | unlock | 失效乐观锁 | 否 | 否 | 系统对带有乐观锁模型默认使用乐观锁对应操作：<br>PamirsSession.directive().enableOptimisticLocker();<br>PamirsSession.directive().disableOptimisticLocker(); |
| 22 | check | 数据校验 | 是 | 否 | 系统后端操作默认不进行数据校验，标记后生效数据校验对应操作：<br>PamirsSession.directive().enableCheck();<br>PamirsSession.directive().disableCheck(); |
| 23 | defaultValue | 默认值计算 | 是 | 否 | 是否自动填充默认值对应操作：<br>PamirsSession.directive().enableDefaultValue();<br>PamirsSession.directive().disableDefaultValue(); |
| 24 | extPoint | 执行扩展点 | 是 | 否 | 前端请求默认执行扩展点，可以标记忽略扩展点后端编程式调用数据管理器默认不执行扩展点对应操作：<br>PamirsSession.directive().enableExtPoint();<br>PamirsSession.directive().disableExtPoint(); |
| 25 | hook | 拦截 | 是 | 否 | 是否进行函数调用拦截对应操作：<br>PamirsSession.directive().enableHook();<br>PamirsSession.directive().disableHook(); |
| 26 | authenticate | 鉴权 | 是 | 否 | 系统默认进行权限校验与过滤，标记后使用权限校验对应操作：<br>PamirsSession.directive().sudo();<br>PamirsSession.directive().disableSudo(); |
| 27 | ormColumn | ORM 字段别名 | 否 | 否 | 系统指令，请勿设置 |
| 28 | usePkStrategy | 使用 PK 策略 | 是 | 否 | 使用 PK 是否空作为采用新增还是更新的持久化策略对应操作：<br>PamirsSession.directive().enableUsePkStrategy();<br>PamirsSession.directive().disableUsePkStrategy(); |
| 29 | fromClient | 是否客户端调用 | 是 | 否 | 是否客户端（前端）调用对应操作：<br>PamirsSession.directive().enableFromClient();<br>PamirsSession.directive().disableFromClient(); |
| 30 | sync | 同步执行函数 | 否 | 否 | 异步执行函数强制使用同步方式执行（仅对 Spring Bean 有效） |

| 位 | 指令 | 指令名 | 前端默认值 | 后端默认值 | 描述 |
|---|---|---|---|---|---|
| 31 | ignore FunManagement | 忽略函数管理 | 否 | 否 | 忽略函数管理器处理，防止 Spring 调用重复拦截对应操作：<br>PamirsSession.directive().enableIgnoreFunManagement();<br>PamirsSession.directive().disableIgnoreFunManagement(); |

请求上下文指令：使用 session 上下文中非持久化 META_BIT 属性设置指令。

### 2. 使用指令

（1）普通模式代码如下所示。

```
1  PamirsSession.directive().disableOptimisticLocker();
2  try{
3      更新逻辑
4  } finally {
5      PamirsSession.directive().enableOptimisticLocker();
6  }
```

（2）批量设置模式代码如下所示。

```
1  Models.directive().run(() -> {此处添加逻辑}, SystemDirectiveEnum.AUTHENTICATE)
```

### 3. 使用举例

我们在 4.1.5 节中提到过失效乐观锁，在这里就尝试一下吧。

Step1. 修改 PetItemInventroyAction。

手动失效乐观锁代码如下所示。

```
1  package pro.shushi.pamirs.demo.core.action;
2
3  import org.springframework.stereotype.Component;
4  import pro.shushi.pamirs.demo.api.model.PetItemInventroy;
5  import pro.shushi.pamirs.meta.annotation.Function;
6  import pro.shushi.pamirs.meta.annotation.Model;
7  import pro.shushi.pamirs.meta.api.session.PamirsSession;
8  import pro.shushi.pamirs.meta.constant.FunctionConstants;
9  import pro.shushi.pamirs.meta.enmu.FunctionOpenEnum;
10 import pro.shushi.pamirs.meta.enmu.FunctionTypeEnum;
11
12 import java.util.ArrayList;
13 import java.util.List;
14
15 @Model.model(PetItemInventroy.MODEL_MODEL)
16 @Component
17 public class PetItemInventroyAction {
18
19     @Function.Advanced(type= FunctionTypeEnum.UPDATE)
20     @Function.fun(FunctionConstants.update)
21     @Function(openLevel = {FunctionOpenEnum.API})
22     public PetItemInventroy update(PetItemInventroy data){
23         List<PetItemInventroy> inventroys = new ArrayList<>();
24         inventroys.add(data);
```

```
25          PamirsSession.directive().disableOptimisticLocker();
26          try{
27              //批量更新，会自动抛售
28              int i = data.updateBatch(inventroys);
29              //单记录更新，不自动抛售需要自行判断
30 //           int i = data.updateById();
31 //           if(i!=1){
32 //               throw PamirsException.construct(DemoExpEnumerate.PET_ITEM_INVENTROY_
33 //           }
34          } finally {
35              PamirsSession.directive().enableOptimisticLocker();
36          }
37          return data;
38      }
39
40 }
```

**Step2.** 重启看效果。

体验一：在页面上修改记录，数据库字段中的 opt_version 不再自动加一。

体验二：同时打开两个页面，依次单击，会发现两次都成功。数据库字段中的 opt_version 不再自动加一，如图 4-18 ~ 图 4-20 所示。

图 4-18　编辑宠物商品库存（1）

图 4-19　编辑宠物商品库存（2）

图 4-20　宠物商品库存的数据库记录变化

### 4.1.10　函数之触发与定时

函数的触发与定时在很多场景中会用到，也是 Oinone 的一项基础能力。比如在流程产品中定义流程触发时就会让用户选择模型触发还是时间触发，就是用到了函数的触发与定时能力。

整体链路示意如图 4-21 所示，本节只讲 trigger 里的两类任务，一个是触发任务，一个是定时任务，异步任务放在 4.1.11 节单独介绍。

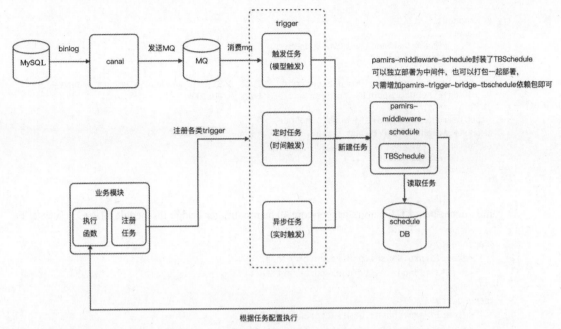

图 4-21　整体链路示意图

**1. 触发任务 TriggerTaskAction（举例）**

触发任务的创建，使用 pamirs-middleware-canal 监听 MySQL 的 binlog 事件，通过 rocketmq 发送变更数据消息，收到 MQ 消息后，创建 TriggerAutoTask。

触发任务的执行，使用 TBSchedule 拉取触发任务后，执行相应函数。

注意：pamirs-middleware-canal 监听的数据库表必须包含触发模型的数据库表。

Step1. 下载 canal 中间件。

下载，去 .txt 后缀为 pamirs-middleware-canal-deployer-3.0.1.zip，解压文件如图 4-22 所示。

图 4-22　下载 canal 中间件

Step2. 引入依赖 pamirs-core-trigger 模块。

（1）pamirs-demo-api 增加 pamirs-trigger-api，代码如下。

```
1  <dependency>
2      <groupId>pro.shushi.pamirs.core</groupId>
3      <artifactId>pamirs-trigger-api</artifactId>
4  </dependency>
```

（2）DemoModule 在模块依赖定义中增加 @Module(dependencies={TriggerModule.MODULE_MODULE})，代码如下。

```
1   @Component
2   @Module(
3           name = DemoModule.MODULE_NAME,
4           displayName = "oinoneDemo工程",
5           version = "1.0.0",
6           dependencies = {ModuleConstants.MODULE_BASE, CommonModule.MODULE_MODULE,
    UserModule.MODULE_MODULE, TriggerModule.MODULE_MODULE}
7   )
8   @Module.module(DemoModule.MODULE_MODULE)
9   @Module.Advanced(selfBuilt = true, application = true)
10  @UxHomepage(PetShopProxy.MODEL_MODEL)
11  public class DemoModule implements PamirsModule {
12      ……其他代码
13  }
```

（3）pamirs-demo-boot 增加 pamirs-trigger-core 和 pamirs-trigger-bridge-tbschedule 的依赖，代码如下。

```
1   <dependency>
2       <groupId>pro.shushi.pamirs.core</groupId>
3       <artifactId>pamirs-trigger-core</artifactId>
4   </dependency>
5   <dependency>
6       <groupId>pro.shushi.pamirs.core</groupId>
7       <artifactId>pamirs-trigger-bridge-tbschedule</artifactId>
8   </dependency>
```

（4）修改 pamirs-demo-boot 的 applcation-dev.yml，代码如下。

① 修改 pamris.event.enabled 和 pamris.event.schedule.enabled 为 true。

② pamirs_boot_modules 增加启动模块：trigger。

```
1   pamirs:
2     event:
3       enabled: true
4       schedule:
5         enabled: true
6     rocket-mq:
7       namesrv-addr: 127.0.0.1:9876
8     boot:
9       init: true
10      sync: true
11      modules:
12        - base
13        - common
14        - sequence
15        - resource
16        - user
17        - auth
18        - message
19        - international
20        - business
21        - trigger
22        - demo_core
23
```

Step3. 启动 canal 中间件。

（1）canal 的库表需要手工建立。

canal 的建库代码如下。

```sql
create schema canal_tsdb collate utf8mb4_bin
```

canal 的建表代码如下。

```sql
CREATE TABLE IF NOT EXISTS `meta_snapshot` (
  `id` bigint(20) unsigned NOT NULL AUTO_INCREMENT COMMENT '主键',
  `gmt_create` datetime NOT NULL COMMENT '创建时间',
  `gmt_modified` datetime NOT NULL COMMENT '修改时间',
  `destination` varchar(128) DEFAULT NULL COMMENT '通道名称',
  `binlog_file` varchar(64) DEFAULT NULL COMMENT 'binlog文件名',
  `binlog_offest` bigint(20) DEFAULT NULL COMMENT 'binlog偏移量',
  `binlog_master_id` varchar(64) DEFAULT NULL COMMENT 'binlog节点id',
  `binlog_timestamp` bigint(20) DEFAULT NULL COMMENT 'binlog应用的时间戳',
  `data` longtext DEFAULT NULL COMMENT '表结构数据',
  `extra` text DEFAULT NULL COMMENT '额外的扩展信息',
  PRIMARY KEY (`id`),
  UNIQUE KEY binlog_file_offest(`destination`,`binlog_master_id`,`binlog_file`,`binlog_offest`),
  KEY `destination` (`destination`),
  KEY `destination_timestamp` (`destination`,`binlog_timestamp`),
  KEY `gmt_modified` (`gmt_modified`)
) ENGINE=InnoDB AUTO_INCREMENT=1 DEFAULT CHARSET=utf8mb4 COMMENT='表结构记录表快照表';

CREATE TABLE IF NOT EXISTS `meta_history` (
  `id` bigint(20) unsigned NOT NULL AUTO_INCREMENT COMMENT '主键',
  `gmt_create` datetime NOT NULL COMMENT '创建时间',
  `gmt_modified` datetime NOT NULL COMMENT '修改时间',
  `destination` varchar(128) DEFAULT NULL COMMENT '通道名称',
  `binlog_file` varchar(64) DEFAULT NULL COMMENT 'binlog文件名',
  `binlog_offest` bigint(20) DEFAULT NULL COMMENT 'binlog偏移量',
  `binlog_master_id` varchar(64) DEFAULT NULL COMMENT 'binlog节点id',
  `binlog_timestamp` bigint(20) DEFAULT NULL COMMENT 'binlog应用的时间戳',
  `use_schema` varchar(1024) DEFAULT NULL COMMENT '执行sql时对应的schema',
  `sql_schema` varchar(1024) DEFAULT NULL COMMENT '对应的schema',
  `sql_table` varchar(1024) DEFAULT NULL COMMENT '对应的table',
  `sql_text` longtext DEFAULT NULL COMMENT '执行的sql',
  `sql_type` varchar(256) DEFAULT NULL COMMENT 'sql类型',
  `extra` text DEFAULT NULL COMMENT '额外的扩展信息',
  PRIMARY KEY (`id`),
  UNIQUE KEY binlog_file_offest(`destination`,`binlog_master_id`,`binlog_file`,`binlog_offest`),
  KEY `destination` (`destination`),
  KEY `destination_timestamp` (`destination`,`binlog_timestamp`),
  KEY `gmt_modified` (`gmt_modified`)
) ENGINE=InnoDB AUTO_INCREMENT=1 DEFAULT CHARSET=utf8mb4 COMMENT='表结构变化明细表';

CREATE TABLE IF NOT EXISTS `canal_filter` (
  `id` bigint(20) unsigned NOT NULL AUTO_INCREMENT COMMENT '主键',
  `destination` varchar(128) NOT NULL COMMENT '通道名称',
  `filter` text NOT NULL COMMENT '过滤表达式',
  `create_date` timestamp NOT NULL DEFAULT CURRENT_TIMESTAMP COMMENT '创建时间',
  `write_date` timestamp NOT NULL DEFAULT CURRENT_TIMESTAMP ON UPDATE CURRENT_TIMESTAMP COMMENT '修改时间',
  PRIMARY KEY (`id`),
  UNIQUE KEY destination(`destination`)
) ENGINE=InnoDB AUTO_INCREMENT=1 DEFAULT CHARSET=utf8mb4 COMMENT='通道过滤';
```

```sql
51  CREATE TABLE IF NOT EXISTS `canal_destination` (
52    `id` bigint unsigned NOT NULL AUTO_INCREMENT COMMENT '主键',
53    `destination` varchar(128) NOT NULL COMMENT '通道名称',
54    `content` text NOT NULL COMMENT '通道配置内容',
55    `create_date` timestamp NOT NULL DEFAULT CURRENT_TIMESTAMP COMMENT '创建时间',
56    `write_date` timestamp NOT NULL DEFAULT CURRENT_TIMESTAMP ON UPDATE CURRENT_TIMESTAMP COMMENT '修改时间',
57    PRIMARY KEY (`id`),
58    UNIQUE KEY `destination` (`destination`)
59  ) ENGINE=InnoDB AUTO_INCREMENT=2 DEFAULT CHARSET=utf8mb4 COLLATE=utf8mb4_0900_ai_ci COMMENT='通道配置'
```

（2）修改 canal 的启动配置，代码如下。

① canal_tsdb 就是通过上面 sql 语句建库，在下面的配置文件中它对应的用户名和密码需要换成本机的；

② 配置 filter: demo\.demo_core_pet_talent，监听 PetTalent 模型数据变化，filter 为 canal 第一次启动默认监听的表。如果数据库中 canal_filter 表有数据这个修改无效；

③ filter 目前需要手工配置，在下个版本中已经去掉了手工配置，而且文中的 canal 中间件已经是 2.2.2 版本了，兼容当前教程的版本。

```yaml
1   pamirs:
2     middleware:
3       data-source:
4         jdbc-url: jdbc:mysql://localhost:3306/canal_tsdb?useUnicode=true&characterEncoding=utf-8&verifyServerCertificate=false&useSSL=false&requireSSL=false
5         driver-class-name: com.mysql.cj.jdbc.Driver
6         username: root
7         password: oinone
8       canal:
9         ip: 127.0.0.1
10        port: 1111
11        metricsPort: 1112
12        zkClusters:
13          - 127.0.0.1:2181
14        destinations:
15          - destinaion: pamirschangedata
16            name: pamirschangedata
17            desc: pamirschangedata
18            slaveId: 1235
19            filter: demo\.demo_core_pet_talent
20            dbUserName: root
21            dbPassword: oinone
22            memoryStorageBufferSize: 65536
23            topic: CHANGE_DATA_EVENT_TOPIC
24            dynamicTopic: false
25            dbs:
26              - { address: 127.0.0.1, port: 3306 }
27            tsdb:
28              enable: true
29              jdbcUrl: "jdbc:mysql://127.0.0.1:3306/canal_tsdb"
30              userName: root
31              password: oinone
32            mq: rocketmq
33            rocketmq:
34              namesrv: 127.0.0.1:9876
35              retryTimesWhenSendFailed: 5
36        dubbo:
37          application:
38            name: canal-server
39            version: 1.0.0
```

```yaml
40      registry:
41        address: zookeeper://127.0.0.1:2181
42      protocol:
43        name: dubbo
44        port: 20881
45      scan:
46        base-packages: pro.shushi
47    server:
48      address: 0.0.0.0
49      port: 10010
50      sessionTimeout: 3600
```

**pamirs.canal 相关配置说明：**

canal.ip：运行时 IP。

canal.port：运行时端口。

canal.zkClusters：连接 ZooKeeper 集群地址。

canal.destinations：运行时 `Canal` 的所有实例。

canal.destinations.destinaion: `Canal` 的单个实例配置值为所有实例（唯一）。

canal.destinations.destinaion.id：与 `canal.destinations.destinaion.slaveId` 配置一致。

canal.destinations.destinaion.slaveId：为 Canal 的实例 ID。

canal.destinations.destinaion.filter：为 Canal 监听过滤正则。

canal.destinations.destinaion.dbUserName：为 Canal 监听的 MySQL 的用户名。

canal.destinations.destinaion.dbPassword：为 Canal 监听的 MySQL 的用户密码。

canal.destinations.destinaion.topic：为监听到数据后往 RocketMQ 的指定 Topic 发送消息。

canal.destinations.destinaion.dbs：为连接的 MySQL 实例的 IP 和端口，如果为主从配置且主从都开启了 Binlog 同步功能则可以配置两个地址。

canal.destinations.destinaion.memoryStorageBufferSiz 与 canal.destinations.destinaion.dynamicTopic：为固定值不需要更改。

canal.tsdb：用来保存 canal 运行时的元数据，配置为 canal_tsdb 库相关的连接信息。

（3）启动 canal 中间件，代码如下。

① 进入 canal 中间件解压目录执行下面命令就可以启动。

② --spring.config.location 请配置绝对路径，换成本机的绝对路径就可以。

```
1  java -jar pamirs-middleware-canal-deployer-3.0.1-SNAPSHOT.jar --spring.config.location=/Users/oinone/Documents/oinone/canel/pamirs-middleware-canal-deployer-3.0.1/canal-config/application-dev.yml --spring.profiles.active=dev
```

**Step4. 新建触发任务，代码如下。**

新建 PetTalentTrigger 类，当 PetTalent 模型的数据记录被新建时触发系统做一些事情。

```
10  @Fun(PetTalent.MODEL_MODEL)
11  @Slf4j
12  public class PetTalentTrigger {
13      @Function
14      @Trigger(displayName = "PetTalent创建时触发",name = "PetTalent#Trigger#onCreate",condition = TriggerConditionEnum.ON_CREATE)
15      public PetTalent onCreate(PetTalent data){
16          log.info(data.getName() + ", 被创建");
```

```
17            //可以增加逻辑
18            return data;
19        }
20    }
```

**Step5.** 重启应用看效果。

（1）解决启动时 dubbo 报错。

启动过程中会报如图 4-23 所示错误，虽然不影响结果，但还是要把它消灭掉。修改 bootstarp.yml 文件，关闭 SpringCloud 的自动注册就好了。

图 4-23　解决启动时 dubbo 报错

解决启动时 dubbo 报错的代码如下。

```
 1    spring:
 2      profiles:
 3        active: dev
 4      application:
 5        name: pamirs-demo
 6      cloud:
 7        service-registry:
 8          auto-registration:
 9            enabled: false
10    pamirs:
11      default:
12        environment-check: true
13        tenant-check: true
14
15    ---
16    spring:
17      profiles: dev
18      cloud:
19        config:
20          enabled: false
21          uri: http://127.0.0.1:7001
22          label: master
23          profile: dev
24        nacos:
25          server-addr: http://127.0.0.1:8848
26          discovery:
27            enabled: false
28            namespace:
29            prefix: application
30            file-extension: yml
31          config:
32            enabled: false
33            namespace:
34            prefix: application
35            file-extension: yml
```

```
36
37    dubbo:
38      application:
39        name: pamirs-demo
40        version: 1.0.0
41      registry:
42        address: zookeeper://127.0.0.1:2181
43      protocol:
44        name: dubbo
45        port: -1
46    #     serialization: pamirs
47      scan:
48        base-packages: pro.shushi
49      cloud:
50        subscribed-services:
51
52    ---
53    spring:
54      profiles: test
55      cloud:
56        config:
57          enabled: false
58          uri: http://127.0.0.1:7001
59          label: master
60          profile: test
61        nacos:
62          server-addr: http://127.0.0.1:8848
63          discovery:
64            enabled: false
65            namespace:
66            prefix: application
67            file-extension: yml
68          config:
69            enabled: false
70            namespace:
71            prefix: application
72            file-extension: yml
73
74    dubbo:
75      application:
76        name: pamirs-demo
77        version: 1.0.0
78      registry:
79        address: zookeeper://127.0.0.1:2181
80      protocol:
81        name: dubbo
82        port: -1
83    #     serialization: pamirs
84      scan:
85        base-packages: pro.shushi
86      cloud:
87        subscribed-services:
88
```

（2）再次重启查看效果。

前端新增一个宠物达人，在后台 Console 搜索"被创建"，我们可以看到对应由触发器打印出来的日志，如图 4-24 所示。

(a)

(b)

图 4-24 再次重启看效果

Step6. 修改 canal 的 topic，代码如下。

在分布式环境下可以通过修改 canal 的 topic（canal.destinations.destinaion.topic）来隔离多个应用的触发消息。

（1）新建 DemoNotifyTopicEdit 利用 topic 修改 api，增加后缀 ""_"+ DemoModule.MODULE_MODULE"。

```
 8    @Component
 9    public class DemoNotifyTopicEdit   implements NotifyTopicEditorApi {
10        @Override
11        public String handlerTopic(String topic) {
12            if(NotifyConstant.AUTO_TRIGGER_TOPIC.equals(topic)){
13                return NotifyConstant.AUTO_TRIGGER_TOPIC +"_"+ DemoModule.MODULE_MODU
    LE;
14            }
15            return topic;
16        }
17    }
18
```

（2）canal 的配置 canal.destinations.destinaion.topic 改为：CHANGE_DATA_EVENT_TOPIC_demo_core，代码如下。

```
 1  pamirs:
 2    middleware:
 3      data-source:
 4        jdbc-url: jdbc:mysql://localhost:3306/canal_tsdb?useUnicode=true&characterEn
 5        driver-class-name: com.mysql.cj.jdbc.Driver
 6        username: root
 7        password: oinone
 8      canal:
 9        ip: 127.0.0.1
10        port: 1111
11        metricsPort: 1112
12        zkClusters:
13          - 127.0.0.1:2181
14        destinations:
15          - destinaion: pamirschangedata
16            name: pamirschangedata
```

```
17          desc: pamirschangedata
18          slaveId: 1235
19          filter: demo\.demo_core_pet_talent
20          dbUserName: root
21          dbPassword: oinone
22          memoryStorageBufferSize: 65536
23          topic: CHANGE_DATA_EVENT_TOPIC_demo_core
24          dynamicTopic: false
25          dbs:
26            - { address: 127.0.0.1, port: 3306 }
27      tsdb:
28        enable: true
29        jdbcUrl: "jdbc:mysql://127.0.0.1:3306/canal_tsdb"
30        userName: root
31        password: oinone
32      mq: rocketmq
33      rocketmq:
34        namesrv: 127.0.0.1:9876
35        retryTimesWhenSendFailed: 5
```

（3）小伙伴自行测试。

**2. 定时任务**

定时任务是一种常见的模式，这里就不介绍概念了，直接进入示例环节。

Step1. 新建 PetTalentAutoTask 实现 ScheduleAction，代码如下。

注：

（1）getInterfaceName() 需要跟 taskAction.setExecuteNamespace 定义保持一致，都是函数的命名空间；

（2）taskAction.setExecuteFun("execute") 的参数跟执行函数名"execute"一致；

（3）TaskType 需配置为 CYCLE_SCHEDULE_NO_TRANSACTION_TASK，把定时任务的 schedule 线程分开，否则有一个时间长的任务会导致普通异步或触发任务全部延时。

```
21      @Fun(PetTalent.MODEL_MODEL)
22      public class PetTalentAutoTask implements ScheduleAction {
23
24          @Autowired
25          private ScheduleTaskActionService scheduleTaskActionService;
26
27          public void initTask(){
28              ScheduleTaskAction taskAction = new ScheduleTaskAction();
29              taskAction.setDisplayName("定时任务测试");   //定时任务描述
30              taskAction.setDescription("定时任务测试");
31              taskAction.setTechnicalName(PetTalent.MODEL_MODEL+"#"+PetTalentAutoTask.c
        lass.getSimpleName()+"#"+"testAutoTask");       //设置定时任务技术名
32              taskAction.setLimitExecuteNumber(-1);    //设置执行次数
33              taskAction.setPeriodTimeValue(1);         //设置执行周期规则
34              taskAction.setPeriodTimeUnit(TimeUnitEnum.MINUTE);
35              taskAction.setPeriodTimeAnchor(TriggerTimeAnchorEnum.START);
36              taskAction.setLimitRetryNumber(1);        //设置失败重试规则
37              taskAction.setNextRetryTimeValue(1);
38              taskAction.setNextRetryTimeUnit(TimeUnitEnum.MINUTE);
39              taskAction.setExecuteNamespace(PetTalent.MODEL_MODEL);
40              taskAction.setExecuteFun("execute");
41              taskAction.setExecuteFunction(new FunctionDefinition().setTimeout(5000));
42              taskAction.setTaskType(TaskType.CYCLE_SCHEDULE_NO_TRANSACTION_TASK.getVal
        ue());  //设置定时任务,执行任务类型
43              taskAction.setContext(null);              //用户传递上下文参数
44              taskAction.setActive(true);                //定时任务是否生效
45              taskAction.setFirstExecuteTime(System.currentTimeMillis());
46              scheduleTaskActionService.submit(taskAction);//初始化任务,幂等可重复执行
47          }
```

```
48
49      @Override
50      public String getInterfaceName() {return PetTalent.MODEL_MODEL;}
51
52      @Override
53      @Function
54      public Result<Void> execute(ScheduleItem item) {
55          log.info("testAutoTask,上次执行时间"+item.getLastExecuteTime());
56          return new Result<>();
57      }
58  }
```

**Step2.** 修改 DemoModuleBizInit，进行定时任务初始化

模块更新的时候调用 petTalentAutoTask.initTask()，initTask 本身是幂等的所以多调几次没有关系。在"模块之生命周期"一节介绍过 InstallDataInit、UpgradeDataInit、ReloadDataInit，如果有兴趣可以去回顾下。

```
17  @Component
18  public class DemoModuleBizInit implements InstallDataInit,
19      UpgradeDataInit, ReloadDataInit {
20
21      @Autowired
22      private PetTalentAutoTask petTalentAutoTask;
23
24      @Override
25      public boolean upgrade(AppLifecycleCommand command, String version,
26                             String existVersion) {
27          petTalentAutoTask.initTask(); //初始化petTalent的定时任务
28          return Boolean.TRUE;
29      }
30
31      @Override
32      public List<String> modules() {
33          return Collections.singletonList(DemoModule.MODULE_MODULE);
34      }
35
36      @Override
37      public int priority() {return 0;}
38  }
```

**Step3.** 重启看效果，如图 4-25 所示。

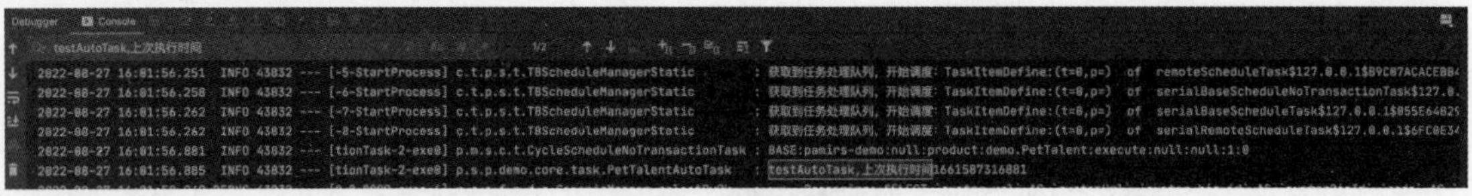

图 4-25　重启看效果

### 4.1.11　函数之异步执行

异步任务是一种常见的开发模式，它在分布式开发模式中有很多应用场景。

（1）高并发场景中，把长流程切短，用异步方式去掉可以异步的非关键功能，缩短主流程响应时间，提升用户体验。

（2）异构系统的集成调用，通过异步任务完成解耦与自动重试。

（3）作为分布式系统最终一致性的可选方案。

现在我们来了解 Oinone 如何结合"Spring+TbSchedule"来完成异步任务。

**1. TbSchedule 介绍**

TbSchedule 是一个支持分布式的调度框架，让批量任务或者不断变化的任务能够被动态地分配到多个主机的 JVM 中，在不同的线程组中并行执行，所有的任务能够被不重复、不遗漏地快速处理。基于 ZooKeeper 的纯 Java 实现，由 Alibaba 开源。在互联网和电商领域 TbSchedule 的使用非常广泛，目前被应用于阿里巴巴、淘宝、支付宝、京东、聚美、汽车之家、国美等互联网企业的流程调度系统，也是笔者早期在阿里参与设计的一款产品。

如图 4-26 所示为 Oinone 的异步任务执行原理，先进行一个大致介绍。

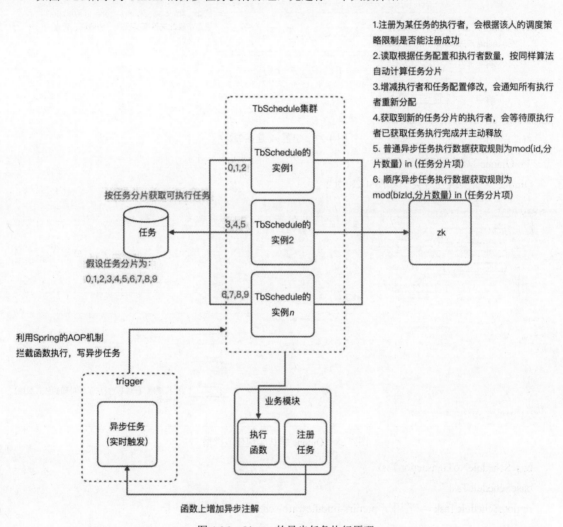

图 4-26　Oinone 的异步任务执行原理

1）基础管理工具

（1）下载 TbSchedule 的控制台 jar 包去除文件后缀 .txt（详见本书末"附录"）。

（2）启动控制台，代码如下。

```
java -jar pamirs-middleware-schedule-console-3.0.1.jar
```

（3）访问地址，代码如下。

```
http://127.0.0.1:10014/schedule/index.jsp?manager=true
```

（4）配置 zk 连接参数，如图 4-27 所示。

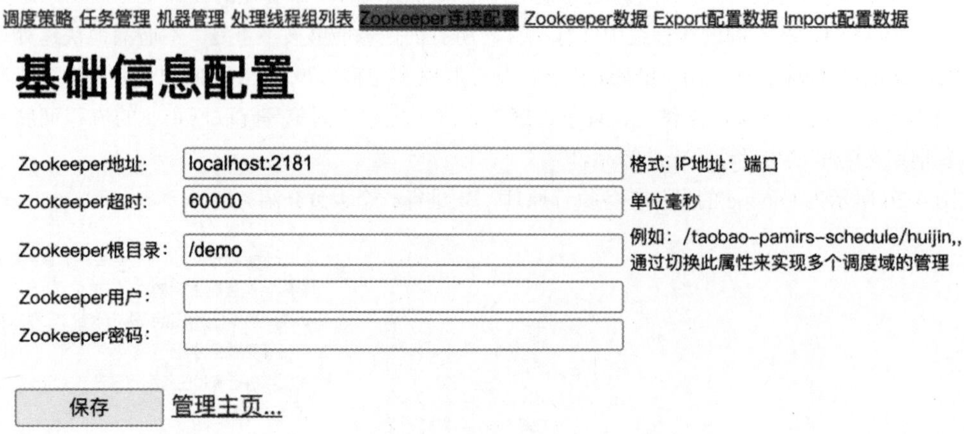

图 4-27 配置 zk 连接参数

2）Oinone 默认实现任务类型

Oinone 默认实现任务类型如图 4-28 所示。

图 4-28 Oinone 默认实现任务类型

baseScheduleNoTransactionTask

baseScheduleTask

remoteScheduleTask --- 适用于 pamirs-middleware-schedule 独立部署场景

serialBaseScheduleNoTransactionTask

serialBaseScheduleTask

serialRemoteScheduleTask --- 适用于 pamirs-middleware-schedule 独立部署场景

cycleScheduleNoTransactionTask

delayMsgTransferScheduleTask

deleteTransferScheduleTask

注：

（1）默认情况下，所有任务的任务项都只配置了一个任务项，只有一台机器能分配任务。修改配置有两种方法。

① 如果要修改配置可以在启动项目中放置 schedule.json，来修改配置。

② 人工进入控制修改任务对应任务项的配置。

（2）如果想为某一个核心任务配置的独立调度器，不受其他任务执行影响，可以参见"独立调度的异步任务"。

（3）任务表相关说明。

任务表相关说明如图 4-29 所示。

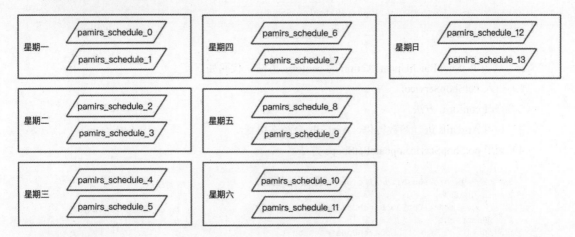

图 4-29　任务表相关说明

### 2. 构建第一个异步任务（举例）

Step1. 新建 PetShopService 和 PetShopServiceImpl。

（1）新建 PetShopService，定义 updatePetShops 方法，代码如下。

```
 9   @Fun(PetShopService.FUN_NAMESPACE)
10 ▾ public interface PetShopService {
11       String FUN_NAMESPACE = "demo.PetShop.PetShopService";
12       @Function
13       void updatePetShops(List<PetShop> petShops);
14
15   }
```

（2）PetShopServiceImpl 实现 PetShopService 接口并在 updatePetShops 增加 @XAsync 注解，代码如下。

① displayName = "异步批量更新宠物商店"，定义异步任务展示名称。

② limitRetryNumber = 3，定义任务失败重试次数，默认：-1 不断重试。

③ nextRetryTimeValue = 60，定义任务失败重试的时间数，默认：3。

④ nextRetryTimeUnit，定义任务失败重试的时间单位，默认：TimeUnitEnum.SECOND。

⑤ delayTime，定义任务延迟执行的时间数，默认：0。

⑥ delayTimeUnit，定义任务延迟执行的时间单位，默认：TimeUnitEnum.SECOND。

```
12      @Fun(PetShopService.FUN_NAMESPACE)
13      @Component
14      public class PetShopServiceImpl implements PetShopService {
15
16          @Override
17          @Function
18          @XAsync(displayName = "异步批量更新宠物商店",limitRetryNumber = 3,nextRetryTimeV
    alue = 60)
19          public void updatePetShops(List<PetShop> petShops) {
20              new PetShop().updateBatch(petShops);
21          }
22      }
```

Step2. 修改 PetShopBatchUpdateAction 的 conform 方法，代码如下。

（1）引入 PetShopService。

（2）修改 conform 方法。

（3）利用 ArgUtils 进行参数转化，ArgUtils 会经常用到。

（4）调用 petShopService.updatePetShops 方法。

```
1   package pro.shushi.pamirs.demo.core.action;
2   …… 引依赖类
3   @Model.model(PetShopBatchUpdate.MODEL_MODEL)
4   @Component
5   public class PetShopBatchUpdateAction {
6
7       @Autowired
8       private PetShopService petShopService;
9       ……其他代码
10      @Action(displayName = "确定",bindingType = ViewTypeEnum.FORM,contextType = Act
    ionContextTypeEnum.SINGLE)
11      public PetShopBatchUpdate conform(PetShopBatchUpdate data){
12          if(data.getPetShopList() == null || data.getPetShopList().size()==0){
13              throw PamirsException.construct(DemoExpEnumerate.PET_SHOP_BATCH_UPDA
    TE_SHOPLIST_IS_NULL).errThrow();
14          }
15          List<PetShopProxy> proxyList = data.getPetShopList();
16          for(PetShopProxy petShopProxy:proxyList){
17              petShopProxy.setDataStatus(data.getDataStatus());
18          }
19          Tx.build(new TxConfig().setPropagation(Propagation.REQUIRED.value())).exe
    cuteWithoutResult(status -> {
20              new PetShopProxy().updateBatch(proxyList);
21              //利用ArgUtils进行参数转化
22              List<PetShop> shops = ArgUtils.convert(PetShopProxy.MODEL_MODEL, PetS
    hop.MODEL_MODEL,proxyList);
23              petShopService.updatePetShops(shops);
24  //          throw PamirsException.construct(DemoExpEnumerate.SYSTEM_ERROR).errT
    hrow();
25          });
26          return data;
27      }
28
29  }
30
```

Step3. 重启看效果。

异步会有一定的延迟，我们按以下步骤测试异步执行效果。

（1）进入商店管理列表页，选择其中一行数据单击"批量更新数据状态"按钮，进入批量修改宠物商店数据状态页面，如图 4-30 所示。

图 4-30　批量修改宠物商店数据状态页面

（2）在批量修改宠物商店数据状态页面，数据状态设置为"未启用"，单击"组合动作"按钮回到商店管理列表页，如图 4-31 所示。

图 4-31　单击"组合动作"按钮回到商店管理列表页

（3）查看商店管理列表页的数据记录的数据状态字段是否修改成功，此时可能未修改成功，也可能已经修改成功，因为本身就是毫秒级的速度，单击"搜索"刷新数据，发现数据记录的数据状态字段修改成功，如图 4-32 所示。

图 4-32　发现数据记录的数据状态字段修改成功

（4）查看任务表，根据任务表与日期的对照关系查询指定表，如图 4-33 所示。

图 4-33　根据任务表与日期的对照关系查询指定表

### 3. 异步任务高级玩法

1）顺序异步任务（举例）

这里的顺序任务是指把任务按一定规则分组以后按时间顺序串行执行，不同分组间的任务不相互影响。有点类似 mq 的顺序消息。

eg：订单的状态变更的异步任务需要根据任务产生时间顺序执行。那么分组规则是按订单 id 分组，执行顺序按任务产生顺序执行。

Step1. PetShopService 和 PetShopServiceImpl。

（1）修改 PetShopService 新增定义 asyncSerialUpdatePetShops 方法，代码如下。

```
 9    @Fun(PetShopService.FUN_NAMESPACE)
10    public interface PetShopService {
11        String FUN_NAMESPACE = "demo.PetShop.PetShopService";
12        @Function
13        void updatePetShops(List<PetShop> petShops);
14        @Function
15        void asyncSerialUpdatePetShops(List<PetShop> petShops);
16
17    }
```

（2）修改 PetShopServiceImpl 实现 ScheduleAction 接口，并增加 asyncSerialUpdatePetShops 方法。

① 引入 executeTaskActionService 用于提交异步串行任务 ExecuteTaskAction。

● setExecuteNamespace (getInterfaceName ())，确保跟 getInterfaceName () 一致。

● setExecuteFun ("execute")；跟执行函数名"execute"一致。

● setTaskType (TaskType.SERIAL_BASE_SCHEDULE_NO_TRANSACTION_TASK.getValue())，必须用 SERIAL_BASE_SCHEDULE_NO_TRANSACTION_TASK，其为顺序执行任务类型。

● setBizId (petShop.getCreateUid ())，根据创建人 id 分组，根据实际业务情况决定。

② getInterfaceName () 跟函数的命名空间保持一致，代码如下。

```
22    @Fun(PetShopService.FUN_NAMESPACE)
23    @Component
24    public class PetShopServiceImpl implements PetShopService, ScheduleAction {
25
26        @Autowired
27        private ExecuteTaskActionService executeTaskActionService;
28
29        @Override
30        @Function
31        @XAsync(displayName = "异步批量更新宠物商店",limitRetryNumber = 3,nextRetryTimeV
      alue = 60)
32        public void updatePetShops(List<PetShop> petShops) {
33            new PetShop().updateBatch(petShops);
34        }
35
36        @PamirsTransactional
37        @Override
38        @Function
```

```
39    public void asyncSerialUpdatePetShops(List<PetShop> petShops){
40        for(PetShop petShop:petShops) {
41            executeTaskActionService.submit((ExecuteTaskAction) new ExecuteTaskAction()
42                    .setBizId(petShop.getCreateUid()))//根据创建人id分组,根据实际业务情况决定
43                    .setTaskType(TaskType.SERIAL_BASE_SCHEDULE_NO_TRANSACTION_TASK.getValue())
44                    .setNextRetryTimeUnit(TimeUnitEnum.SECOND)//失败重试时间单位
45                    .setNextRetryTimeValue(10)//失败重试时间数
46                    .setLimitRetryNumber(6)//最多重试次数
47                    .setDisplayName("异步顺序任务-更新宠物商店,以createUid分组")
48                    .setExecuteNamespace(getInterfaceName())
49                    .setExecuteFun("execute")
50                    .setContext(JsonUtils.toJSONString(petShop)));
51        }
52    }
53
54    @Override
55    public String getInterfaceName() {
56        return PetShopService.FUN_NAMESPACE;
57    }
58
59    @Override
60    @Function
61    public Result<Void> execute(ScheduleItem scheduleItem) {
62        Result<Void> result = new Result<>();
63        PetShop petShop = JsonUtils.parseObject(scheduleItem.getContext(),PetShop.class);
64        petShop.updateById();
65        return result;
66    }
67 }
68
```

Step2. 修改 PetShopBatchUpdateAction 的 conform 方法，代码如下。
改调用异步顺序方法，petShopService.asyncSerialUpdatePetShops(shops)。

```
1  package pro.shushi.pamirs.demo.core.action;
2  …… 引入依赖类
3  @Model.model(PetShopBatchUpdate.MODEL_MODEL)
4  @Component
5  public class PetShopBatchUpdateAction {
6
7      @Autowired
8      private PetShopService petShopService;
9      ……其他代码
10     @Action(displayName = "确定",bindingType = ViewTypeEnum.FORM,contextType = ActionContextTypeEnum.SINGLE)
11     public PetShopBatchUpdate conform(PetShopBatchUpdate data){
12         if(data.getDataStatus() == null){
13             throw PamirsException.construct(DemoExpEnumerate.PET_SHOP_BATCH_UPDATE_DATASTATUS_IS_NULL).errThrow();
14         }
15         List<PetShopProxy> proxyList = data.getPetShopList();
16         for(PetShopProxy petShopProxy:proxyList){
17             petShopProxy.setDataStatus(data.getDataStatus());
18         }
19         Tx.build(new TxConfig().setPropagation(Propagation.REQUIRED.value())).executeWithoutResult(status -> {
20             //利用ArgUtils进行参数转化
```

```
21            List<PetShop> shops = ArgUtils.convert(PetShopProxy.MODEL_MODEL, PetS
   hop.MODEL_MODEL,proxyList);
22 //           petShopService.updatePetShops(shops);
23              petShopService.asyncSerialUpdatePetShops(shops);
24 //           throw PamirsException.construct(DemoExpEnumerate.SYSTEM_ERROR).errT
   hrow();
25       });
26       return data;
27   }
28
29 }
30
```

Step3. 重启看效果。

页面效果跟构建第一个异步任务一样，但任务生产和执行逻辑不一样。会根据 biz_id 分配任务项与分组确保执行顺序。

（1）分配任务项，相同任务项一定会分配给同一个 schedule 的执行者。

（2）分组，任务在同一个 schedule 的执行者，相同分组 id 一定会分配给同一个线程执行，页面操作完以后查看数据任务表，如图 4-34 所示。

图 4-34 根据 biz_id 分配任务项与分组确保执行顺序

2）独立调度的异步任务（举例）

如果把所有任务都放在同一个任务类型下，复用同一套任务策略、任务配置、任务执行器，那么当某些不重要的异步任务大量失败会影响其他任务的执行，所以我们在一些高并发大任务量的场景下会独立给一些核心异步任务配置独立调度策略。

Step1. 修改 pamirs-demo-core 的 pom。

增加对 pamirs-middleware-schedule-core 依赖，代码如下，为了复用 Oinone 默认实现任务类型的基础逻辑，在例子中我们自定义的异步任务继承 SerialBaseScheduleNoTransactionTask 的基础逻辑。

```
1    <dependency>
2        <groupId>pro.shushi.pamirs.middleware.schedule</groupId>
3        <artifactId>pamirs-middleware-schedule-core</artifactId>
4    </dependency>
```

Step2. 新建 PetShopUpdateCustomAsyncTask，代码如下。

```
6  @Component
7  public class PetShopUpdateCustomAsyncTask extends SerialBaseScheduleNoTransaction
   Task {
8
9      public static final String TASK_TYPE = PetShopUpdateCustomAsyncTask.class.get
   SimpleName();
10
11     @Override
12     public String getTaskType() {
13         return TASK_TYPE;
```

```
14        }
15
16    }
```

Step3. 修改 PetShopServiceImpl 的 asyncSerialUpdatePetShops 方法。

修改 TaskType 为 PetShopUpdateCustomAsyncTask.TASK_TYPE，代码如下。

```
1     @PamirsTransactional
2     @Override
3     @Function
4     public void asyncSerialUpdatePetShops(List<PetShop> petShops){
5         for(PetShop petShop:petShops) {
6             executeTaskActionService.submit((ExecuteTaskAction) new ExecuteTaskAction()
7                     .setBizId(petShop.getCreateUid()))//根据创建人id分组，根据实际业务情况决定
8     //              .setTaskType(TaskType.SERIAL_BASE_SCHEDULE_NO_TRANSACTION_TASK.getValue())
9                     .setTaskType(PetShopUpdateCustomAsyncTask.TASK_TYPE)
10                    .setNextRetryTimeUnit(TimeUnitEnum.SECOND)//失败重试时间单位
11                    .setNextRetryTimeValue(10)//失败重试时间数
12                    .setLimitRetryNumber(6)//最多重试次数
13                    .setDisplayName("异步顺序任务-更新宠物商店，以createUid分组")
14                    .setExecuteNamespace(getInterfaceName())
15                    .setExecuteFun("execute")
16                    .setContext(JsonUtils.toJSONString(petShop)));
17        }
18    }
```

Step4. 初始化数据。

下载文件 schedule.json，放在 pamirs-demo-boot 的 src/main/resources/init 目录下（详见本书末"附录"）。

我们在系统原有提供的 schedule.json 中增加任务类型为 petShopUpdateCustomAsyncTask 的配置，代码如下，配置项 "taskType": "CUSTOM"，标志为客户自定义。实际注册到 TbSchedule 会按 beanNames 转化为 taskType，其他参数含义见 TbSchedule 的管理控制台，有对应中文说明。

```
1
2     {
3         "taskType": "CUSTOM",
4         "beanNames": "petShopUpdateCustomAsyncTask",
5         "values": {
6             "heartBeatRate": 1000,
7             "judgeDeadInterval": 10000,
8             "sleepTimeNoData": 500,
9             "sleepTimeInterval": 500,
10            "fetchDataNumber": 500,
11            "executeNumber": 1,
12            "threadNumber": 8,
13            "processorType": "SLEEP",
14            "expireOwnSignInterval": 1.0,
15            "taskParameter": "",
16            "taskKind": "static",
17            "taskItems": [
18                0,1,2,3,4,5,6,7
19            ],
20            "maxTaskItemsOfOneThreadGroup": 0,
21            "version": 0,
22            "sts": "resume",
```

```
23              "fetchDataCountEachSchedule": -1
24          },
25          "strategy": {
26              "IPList": [
27                  "127.0.0.1"
28              ],
29              "numOfSingleServer": 1,
30              "assignNum": 4,
31              "kind": "Schedule",
32              "taskParameter": "",
33              "sts": "resume"
34          }
35      }
```

**Step5.** 重启看效果。

页面效果跟构建第一个异步任务一样，但任务生产和执行逻辑不一样。会根据 biz_id 分配任务项与分组确保执行顺序，同时会有独立的调度器以及规则配置。

（1）在 tbSchedule 的管理控制台，可以看见多了一个"petShopUpdateCustomAsyncTask"的任务类型，单击"编辑"就可以看到我们配置任务类型对应的参数，如图 4-35 所示。

图 4-35 tbSchedule 管理控制台

（2）页面操作完以后查看对应数据任务表，如图 4-36 所示。

图 4-36 查看对应数据任务表

### 4. 不同应用如何隔离执行单元

schedule 和模块部署在一起的时候，多模块独立 boot 的情况下，需要做必要的配置。如果 schedule 独立部署则没有必要，因为全部走远程，不存在类找不到的问题。

（1）通过如下代码配置 pamirs.zookeeper.rootPath，确保两组机器都能覆盖所有任务分片，这样不会漏数据。

（2）通过 pamirs.event.schedule.ownSign 来隔离。确保两组机器只取各自产生的数据，这样不会重复执行数据。

```
1   pamirs:
2     zookeeper:
3       zkConnectString: 127.0.0.1:2181
4       zkSessionTimeout: 60000
5       rootPath: /demo
6     event:
7       enabled: true
8       schedule:
9         enabled: true
10        ownSign: demo
11      rocket-mq:
12        namesrv-addr: 127.0.0.1:9876
```

### 4.1.12 函数之内置函数与表达式

本文意在列全所有内置函数与表达式，方便大家查阅。

**1. 内置函数**

内置函数是系统预先定义好的函数，并且提供表达式调用支持。

1）通用函数

（1）数学函数。

数字函数见表 4-27。

表 4-27　数学函数

| 表达式 | 名称 | 说明 |
| --- | --- | --- |
| ABS | 绝对值 | 函数场景：表达式<br>函数示例：ABS (number)<br>函数说明：获取 number 的绝对值 |
| FLOOR | 向下取整 | 函数场景：表达式<br>函数示例：FLOOR (number)<br>函数说明：对 number 向下取整 |
| CEIL | 向上取整 | 函数场景：表达式<br>函数示例：CEIL (number)<br>函数说明：对 number 向上取整 |
| ROUND | 四舍五入 | 函数场景：表达式<br>函数示例：ROUND (number)<br>函数说明：对 number 四舍五入 |
| MOD | 取余 | 函数场景：表达式<br>函数示例：MOD (A,B)<br>函数说明：A 对 B 取余 |
| SQRT | 平方根 | 函数场景：表达式<br>函数示例：SQRT (number)<br>函数说明：对 number 取平方根 |
| SIN | 正弦 | 函数场景：表达式<br>函数示例：SIN (number)<br>函数说明：对 number 取正弦 |
| COS | 余弦 | 函数场景：表达式<br>函数示例：COS (number)<br>函数说明：对 number 取余弦 |

续表

| 表达式 | 名称 | 说明 |
| --- | --- | --- |
| PI | 圆周率 | 函数场景：表达式<br>函数示例：PI ()<br>函数说明：圆周率 |
| ADD | 相加 | 函数场景：表达式<br>函数示例：ADD (A,B)<br>函数说明：A 与 B 相加 |
| SUBTRACT | 相减 | 函数场景：表达式<br>函数示例：SUBTRACT (A,B)<br>函数说明：A 与 B 相减 |
| MULTIPLY | 乘积 | 函数场景：表达式<br>函数示例：MULTIPLY (A,B)<br>函数说明：A 与 B 相乘 |
| DIVIDE | 相除 | 函数场景：表达式<br>函数示例：DIVIDE (A,B)<br>函数说明：A 与 B 相除 |
| MAX | 取最大值 | 函数场景：表达式<br>函数示例：MAX (collection)<br>函数说明：返回集合中的最大值，参数 collection 为集合或数组 |
| MIN | 取最小值 | 函数场景：表达式<br>函数示例：MIN (collection)<br>函数说明：返回集合中的最小值，参数 collection 为集合或数组 |
| SUM | 求和 | 函数场景：表达式<br>函数示例：SUM (collection)<br>函数说明：返回对集合的求和，参数 collection 为集合或数组 |
| AVG | 取平均值 | 函数场景：表达式<br>函数示例：AVG (collection)<br>函数说明：返回集合的平均值，参数 collection 为集合或数组 |
| COUNT | 计数 | 函数场景：表达式<br>函数示例：COUNT (collection)<br>函数说明：返回集合的总数，参数 collection 为集合或数组 |
| UPPER_MONEY | 大写金额 | 函数场景：表达式<br>函数示例：UPPER_MONEY (number)<br>函数说明：返回金额的大写，参数 number 为数值或数值类型的字符串 |

（2）文本函数。

文本函数见表 4-28。

表 4-28　文本函数

| 表达式 | 名称 | 说明 |
| --- | --- | --- |
| TRIM | 空字符串过滤 | 函数场景：表达式<br>函数示例：TRIM (text)<br>函数说明：去掉文本字符串 text 中的首尾空格，文本为空时，返回空字符串 |
| IS_BLANK | 是否为空字符串 | 函数场景：表达式<br>函数示例：IS_BLANK (text)<br>函数说明：判断文本字符串 text 是否为空 |

| 表达式 | 名称 | 说明 |
| --- | --- | --- |
| STARTS_WITH | 是否以指定字符串开始 | 函数场景：表达式<br>函数示例：STARTS_WITH (text,start)<br>函数说明：判断文本字符串 text 是否以文本字符串 start 开始，文本为空时，按照空字符串处理 |
| ENDS_WITH | 是否以指定字符串结束 | 函数场景：表达式<br>函数示例：ENDS_WITH (text,start)<br>函数说明：判断文本字符串 text 是否以文本字符串 end 结束，文本为空时，按照空字符串处理 |
| CONTAINS | 包含 | 函数场景：表达式<br>函数示例：CONTAINS (text,subtext)<br>函数说明：判断文本字符串 text 是否包含文本字符串 subtext，文本 text 为空时，按照空字符串处理 |
| LOWER | 小写 | 函数场景：表达式<br>函数示例：LOWER (text)<br>函数说明：小写文本字符串 text，文本为空时，按照空字符串处理 |
| UPPER | 大写 | 函数场景：表达式<br>函数示例：UPPER (text)<br>函数说明：大写文本字符串 text，文本为空时，按照空字符串处理 |
| REPLACE | 替换字符串 | 函数场景：表达式<br>函数示例：REPLACE (text,oldtext,newtext)<br>函数说明：使用文本字符串 newtext 替换文本字符串 text 中的文本字符串 oldtext |
| LEN | 获取字符串长度 | 函数场景：表达式<br>函数示例：LEN (text)<br>函数说明：获取文本字符串 text 的长度，文本为空时，按照空字符串处理 |
| JOIN | 连接字符串 | 函数场景：表达式<br>函数示例：JOIN (text,join)<br>函数说明：将文本字符串 text 连接文本字符串 join，文本为空时，按照空字符串处理 |
| PARSE | 反序列化 JSON 字符串 | 函数场景：表达式<br>函数示例：PARSE (text)<br>函数说明：将 JSON 文本字符串 text 反序列化为集合或者 map |
| JSON | 序列化为 JSON 字符串 | 函数场景：表达式<br>函数示例：JSON (object)<br>函数说明：将记录 object 序列化为 JSON 字符串 |

（3）正则函数。

正则函数见表 4-29。

表 4-29 正则函数

| 表达式 | 名称 | 说明 |
| --- | --- | --- |
| MATCHES | 正则匹配 | 函数场景：表达式<br>函数示例：MATCHES (text,regex)<br>函数说明：校验字符串是否满足正则匹配，例如 regex 为 [a-zA-Z][a-zA-Z0-9]*$，来校验 text 是否匹配 |

续表

| 表达式 | 名称 | 说明 |
| --- | --- | --- |
| CHECK_PHONE | 手机号校验 | 函数场景：表达式<br>函数示例：CHECK_PHONE (text)<br>函数说明：校验手机号是否正确 |
| CHECK_EMAIL | 邮箱校验 | 函数场景：表达式<br>函数示例：CHECK_EMAIL (text)<br>函数说明：校验邮箱是否正确 |
| CHECK_USER_NAME | 用户名校验 | 函数场景：表达式<br>函数示例：CHECK_USER_NAME (text)<br>函数说明：校验用户名是否正确 |
| CHECK_PWD | 密码强弱校验 | 函数场景：表达式<br>函数示例：CHECK_PWD (text)<br>函数说明：判断密码是否满足强弱校验 |
| CHECK_INTEGER | 整数校验 | 函数场景：表达式<br>函数示例：CHECK_INTEGER (text)<br>函数说明：校验是否为整数 |
| CHECK_ID_CARD | 身份证校验 | 函数场景：表达式<br>函数示例：CHECK_ID_CARD (text)<br>函数说明：校验身份证是否正确 |
| CHECK_URL | 合法 URL 校验 | 函数场景：表达式<br>函数示例：CHECK_URL (text)<br>函数说明：校验 URL 是否正确 |
| CHECK_CHINESE | 中文校验 | 函数场景：表达式<br>函数示例：CHECK_CHINESE (text)<br>函数说明：校验是否为中文文本 |
| CHECK_NUMBER | 纯数字校验 | 函数场景：表达式<br>函数示例：CHECK_NUMBER (text)<br>函数说明：校验是否为纯数字 |
| CHECK_TWO_DIG | 验证是否两位小数 | 函数场景：表达式<br>函数示例：CHECK_TWO_DIG (text)<br>函数说明：校验是否两位小数 |
| CHECK_IP | IP 地址校验 | 函数场景：表达式<br>函数示例：CHECK_IP (text)<br>函数说明：校验 IP 地址是否正确 |
| CHECK_CONTAINS_CHINESE | 包含中文校验 | 函数场景：表达式<br>函数示例：CHECK_CONTAINS_CHINESE (text)<br>函数说明：校验是否包含中文 |
| CHECK_SIZE_MAX | 只能输入 n 个字符 | 函数场景：表达式<br>函数示例：CHECK_SIZE_MAX (text,n)<br>函数说明：只能输入 n 个字符 |
| CHECK_SIZE_MIN | 至少输入 n 个字符 | 函数场景：表达式<br>函数示例：CHECK_SIZE_MIN (text,n)<br>函数说明：至少输入 n 个字符 |
| CHECK_SIZE | 输入 m-n 个字符 | 函数场景：表达式<br>函数示例：CHECK_SIZE (text,m,n)<br>函数说明：输入 m-n 个字符 |

| 表达式 | 名称 | 说明 |
| --- | --- | --- |
| CHECK_CODE | 只能由英文、数字、下画线组成 | 函数场景：表达式<br>函数示例：CHECK_CODE (text)<br>函数说明：只能由英文、数字、下画线组成 |
| CHECK_ENG_NUM | 只能包含英文和数字 | 函数场景：表达式<br>函数示例：CHECK_ENG_NUM (text)<br>函数说明：只能包含英文和数字 |

（4）时间函数。

时间函数见表 4-30。

表 4-30　时间函数

| 表达式 | 名称 | 说明 |
| --- | --- | --- |
| NOW | 返回当前时间 | 函数场景：表达式<br>函数示例：NOW ()<br>函数说明：返回当前时间 |
| NOW_STR | 返回当前时间字符串 | 函数场景：表达式<br>函数示例：NOW_STR ()<br>函数说明：返回当前时间字符串，精确到时分秒，格式为 yyyy-MM-dd hh:mm:ss |
| TODAY_STR | 返回今天的日期字符串 | 函数场景：表达式<br>函数示例：TODAY_STR ()<br>函数说明：返回今天的日期字符串，精确到天，格式为 yyyy-MM-dd |
| ADD_DAY | 加减指定天数 | 函数场景：表达式<br>函数示例：ADD_DAY (date,days)<br>函数说明：将指定日期加 / 减指定天数，date 为指定日期，days 为指定天数，当为负数时在 date 上减去此天数 |
| ADD_MONTH | 加减指定月数 | 函数场景：表达式<br>函数示例：ADD_MONTH (date,months)<br>函数说明：将指定日期加 / 减指定月数，date 为指定日期，months 为指定月数，当为负数时在此 date 上减去此月数 |
| ADD_YEAR | 加减指定年数 | 函数场景：表达式<br>函数示例：ADD_YEAR (date,years)<br>函数说明：将指定日期加 / 减指定年数，date 为指定日期，years 为指定年数，当为负数时在此 date 上减去此年数 |
| TO_DATE | 转换为时间 | 函数示例：TO_DATE (date,pattern)<br>函数说明：将 date 字符串按格式转换为时间 |
| ADD_WORK_DAY | 工作日加减天数（跳过周末） | 函数示例：ADD_WORK_DAY (date,days) |

（5）集合函数。

集合函数见表4-31。

表4-31 集合函数

| 表达式 | 名称 | 说明 |
|---|---|---|
| LIST_GET | 获取集合（或数组）中的元素 | 函数场景：表达式<br>函数示例：LIST_GET (list,index)<br>函数说明：获取集合list中索引为数字index的元素 |
| LIST_IS_EMPTY | 判断集合（或数组）是否为空 | 函数场景：表达式<br>函数示例：LIST_IS_EMPTY (list)<br>函数说明：传入一个对象集合，判断是否为空 |
| LIST_CONTAINS | 判断集合（或数组）是否包含元素 | 函数场景：表达式<br>函数示例：LIST_CONTAINS (list,item)<br>函数说明：判断集合list是否包含元素item |
| LIST_ADD | 将元素添加到集合（或数组） | 函数场景：表达式<br>函数示例：LIST_ADD (list,item)<br>函数说明：将元素item添加到集合list |
| LIST_ADD_BY_INDEX | 将元素添加到集合（或数组）的指定位置 | 函数场景：表达式<br>函数示例：LIST_ADD_BY_INDEX (list,index,item)<br>函数说明：将元素item添加到集合list的索引index处 |
| LIST_REMOVE | 移除集合（或数组）中的元素 | 函数场景：表达式<br>函数示例：LIST_REMOVE (list,item)<br>函数说明：从集合list中移除元素item |
| LIST_COUNT | 获取集合（或数组）元素数量 | 函数场景：表达式<br>函数示例：LIST_COUNT (list)<br>函数说明：传入一个对象集合，获取集合元素数量 |
| LIST_IDS | 获取集合（或数组）中的所有id | 函数场景：表达式<br>函数示例：LIST_IDS (list)<br>函数说明：传入一个对象集合，获取集合中的所有id组成的列表 |
| LIST_FIELD_VALUES | 将对象集合（或数组）转化为属性集合 | 函数场景：表达式<br>函数示例：LIST_FIELD_VALUES (list,Model,field)<br>函数说明：传入一个对象集合，该对象的模型和属性字段，返回属性值集合 |
| LIST_FIELD_EQUALS | 判断对象集合（或数组）中属性值匹配情况 | 函数场景：表达式<br>函数示例：LIST_FIELD_EQUALS (list,Model,field,value)<br>函数说明：判断对象集合（或数组）中属性值匹配情况，返回布尔集合 |
| LIST_FIELD_NOT_EQUALS | 判断对象集合（或数组）中属性值不匹配情况 | 函数场景：表达式<br>函数示例：LIST_FIELD_NOT_EQUALS (list,Model,field,value)<br>函数说明：判断对象集合（或数组）中属性值不匹配情况，返回布尔集合 |
| LIST_FIELD_IN | 判断对象集合（或数组）中属性值是否在指定集合（或数组）中 | 函数场景：表达式<br>函数示例：LIST_FIELD_IN (list,Model,field,list)<br>函数说明：判断对象集合（或数组）中属性值是否在指定集合（或数组）中，返回布尔集合 |

续表

| 表达式 | 名称 | 说明 |
|---|---|---|
| LIST_FIELD_NOT_IN | 判断对象集合（或数组）中属性值是否不在指定集合（或数组）中 | 函数场景：表达式<br>函数示例：<br>LIST_FIELD_NOT_IN (list,Model,field,list)<br>函数说明：判断对象集合（或数组）中属性值是否不在指定集合（或数组）中，返回布尔集合 |
| LIST_AND | 将一个布尔集合进行逻辑与运算 | 函数场景：表达式<br>函数示例：LIST_AND (list)<br>函数说明：将一个布尔集合进行逻辑与运算，返回布尔值 |
| LIST_OR | 将一个布尔集合进行逻辑或运算 | 函数场景：表达式<br>函数示例：LIST_OR (list)<br>函数说明：将一个布尔集合进行逻辑或运算，返回布尔值 |
| STRING_LIST_TO_NUMBER_LIST | 将一个字符集合转换为数值集合 | 函数场景：表达式<br>函数示例：<br>STRING_LIST_TO_NUMBER_LIST (list)<br>函数说明：将一个字符集合转换为数值集合，如果转换成功，返回一个数值集合；转换失败，返回集合本身 |
| COMMA | 将集合里面的值用逗号拼接 | 函数场景：表达式<br>函数示例：COMMA (list)<br>函数说明：将集合里面的值用逗号拼接"集合里面的值只能是 Number 或者 String 类型"，返回一个字符串 |
| CONCAT | 将集合里面的值用指定的符号拼接 | 函数场景：表达式<br>函数示例：CONCAT (list, split)<br>函数说明：将集合里面的值用 split 拼接（集合里面的值只能是 Number 或者 String 类型），返回一个字符串 |

（6）键值对函数。

键值对函数见表 4-32。

表 4-32　键值对函数

| 表达式 | 名称 | 说明 |
|---|---|---|
| MAP_GET | 从键值对中获取指定键的值 | 函数场景：表达式<br>函数示例：MAP_GET (map,key)<br>函数说明：从键值对中获取键为 key 的值 |
| MAP_IS_EMPTY | 判断键值对是否为空 | 函数场景：表达式<br>函数示例：MAP_IS_EMPTY (map)<br>函数说明：判断键值对 map 是否为空 |
| MAP_PUT | 向键值对中添加键值 | 函数场景：表达式<br>函数示例：MAP_PUT (map,key,value)<br>函数说明：将键为 key 的值为 value 添加到键值对 map 中 |
| MAP_REMOVE | 移除键值对中的元素 | 函数场景：表达式<br>函数示例：MAP_REMOVE (map,key)<br>函数说明：从键值对 map 中移除键 key |

续表

| 表达式 | 名称 | 说明 |
| --- | --- | --- |
| MAP_COUNT | 获取键值数量 | 函数场景：表达式<br>函数示例：MAP_COUNT (map)<br>函数说明：获取键值对 map 的键值数量 |

（7）上下文函数。

上下文函数见表 4-33。

表 4-33 上下文函数

| 表达式 | 名称 | 说明 |
| --- | --- | --- |
| CURRENT_UID | 获取当前用户 id | 函数场景：表达式<br>函数示例：CURRENT_UID ()<br>函数说明：获取当前用户 id |
| CURRENT_USER_NAME | 获取当前用户名 | 函数场景：表达式<br>函数示例：CURRENT_USER_NAME ()<br>函数说明：获取当前用户的用户名 |
| CURRENT_USER | 获取当前用户 | 函数场景：表达式<br>函数示例：CURRENT_USER ()<br>函数说明：获取当前用户 |
| CURRENT_ROLE_IDS | 获取当前用户的角色 id 列表 | 函数场景：表达式<br>函数示例：CURRENT_ROLE_IDS ()<br>函数说明：获取当前用户的角色 id 列表 |
| CURRENT_ROLES | 获取当前用户的角色列表 | 函数场景：表达式<br>函数示例：CURRENT_ROLES ()<br>函数说明：获取当前用户的角色列表 |
| CURRENT_PARTNER_ID | 获取当前用户的合作伙伴 id | 函数场景：表达式<br>函数示例：CURRENT_PARTNER_ID ()<br>函数说明：获取当前用户的合作伙伴 id |
| CURRENT_PARTNER | 获取当前用户的合作伙伴 | 函数场景：表达式<br>函数示例：CURRENT_PARTNER ()<br>函数说明：获取当前用户的合作伙伴 |

（8）对象函数。

对象函数见表 4-34。

表 4-34 对象函数

| 表达式 | 名称 | 说明 |
| --- | --- | --- |
| IS_NULL | 判断是否为空 | 函数场景：表达式<br>函数示例：IS_NULL ( 文本或控件 )<br>函数说明：判断对象是否为空，为空则返回 true，不为空则返回 false，可用于判断具体值或者控件 |
| EQUALS | 判断是否相等 | 函数场景：表达式<br>函数示例：EQUALS (A,B)<br>函数说明：判断 A 和 B 是否相等 |

续表

| 表达式 | 名称 | 说明 |
|---|---|---|
| FIELD_GET | 获取对象属性值 | 函数场景：表达式<br>函数示例：FIELD_GET (obj,dotExpression)<br>函数说明：从对象中根据点表达式获取属性值 |

（9）逻辑函数。

逻辑函数见表 4-35。

表 4-35　逻辑函数

| 表达式 | 名称 | 说明 |
|---|---|---|
| IF | 条件表达式 | 函数场景：表达式<br>函数示例：IF (A,B,C)<br>函数说明：如果 F 满足条件 A，则返回 B，否则返回 C，支持多层嵌套 IF 函数 |
| AND | 逻辑与 | 函数场景：表达式<br>函数示例：AND (A,B)<br>函数说明：返回条件 A 逻辑与条件 B 的值 |
| OR | 逻辑或 | 函数场景：表达式<br>函数示例：OR (A,B)<br>函数说明：返回条件 A 逻辑或条件 B 的值 |
| NOT | 逻辑非 | 函数场景：表达式<br>函数示例：NOT (A)<br>函数说明：返回 逻辑非 条件 A 的值 |

2）特定场景函数

（1）商业函数。

商业函数见表 4-36。

表 4-36　商业函数

| 表达式 | 名称 | 说明 |
|---|---|---|
| CURRENT_CORP_ID | 获取当前用户的公司 id | 函数场景：商业公司<br>函数示例：CURRENT_CORP_ID ()<br>函数说明：获取当前用户的公司 id |
| CURRENT_CORP | 获取当前用户的公司 | 函数场景：商业公司<br>函数示例：CURRENT_CORP ()<br>函数说明：获取当前用户的公司 |
| CURRENT_SHOP_ID | 获取当前用户的店铺 id | 函数场景：商业店铺<br>函数示例：CURRENT_SHOP_ID ()<br>函数说明：获取当前用户的店铺 id |
| CURRENT_SHOP | 获取当前用户的店铺 | 函数场景：商业店铺<br>函数示例：CURRENT_SHOP ()<br>函数说明：获取当前用户的店铺 |

（2）逻辑 DSL 函数。

逻辑 DSL 函数见表 4-37。

表 4-37　逻辑 DSL 函数

| 表达式 | 名称 | 说明 |
| --- | --- | --- |
| FOR_INDEX | 获取循环的 index | 函数场景：逻辑 DSL<br>函数示例：FOR_INDEX（context）<br>函数说明：获取循环的 index |

**2. 表达式**

表达式，是由数字、算符、函数、数字分组符号（括号）、自由变量和约束变量等以能求得数值的有意义排列方法所得的组合。约束变量在表达式中已被指定数值，而自由变量则可以在表达式之外另行指定数值。

表达式可以使用运算符（+、-、*、/、&&、||、!、==、!=）、点表达式（例如：模型 A. 字段 C. 关联模型字段 A）和内置函数。表达式格式如：IF（ISNULL（模型 A. 字段 x），模型 A. 字段 y. 关联模型字段 z，模型 A. 字段 m）。

表达式中模型字段的前端展现使用展示名称 displayName，表达式原始内容使用技术名称 name。

1）点表达式

点表达式是表达式的子集，由变量名与点组成。点前的变量与点后的变量为从属关系，点后的变量从属于点前的变量。可以使用点表达式获取由全表达式确定的最后一个点后变量的值。

2）正则表达式

正则表达式见表 4-38。

表 4-38　正则表达式

| 对应内置函数 | 说明 | 正则表达式 |
| --- | --- | --- |
| CHECK_PHONE | 手机号校验 | ^（1[3-9]）\\d{9}$ |
| CHECK_EMAIL | 邮箱校验 | ^[a-z0-9A-Z]+[-\|a-z0-9A-Z._]+@（[a-z0-9A-Z]+（-[a-z0-9A-Z]+）?\\.）+[a-z]{2,}$ |
| CHECK_USER_NAME | 用户名校验 | 非空校验 |
| CHECK_PWD | 密码强弱校验<br>（强密码校验） | ^（?=.*[a-z]）（?=.*[A-Z]）[a-zA-Z0-9~!@&%#_（.）]{8,16}$ |
| CHECK_INTEGER | 整数校验 | ^-{0,1}[1-9]\d*$ |
| CHECK_ID_CARD | 身份证校验 | ^\d{15}$）\|（^\d{18}$）\|（^\d{17}（\d\|X\|x）$ |
| CHECK_URL | 合法 URL 校验 | ^（?:（?:https?）://）（?:（?:1\d{2}\|2[0-4]\d\|25[0-5]\|[1-9]\d\|[1-9]）（?:\.（?:1\d{2}\|2[0-4]\d\|25[0-5]\|[1-9]\d\|\d））{2}（?:\.（?:1\d{2}\|2[0-4]\d\|25[0-5]\|[1-9]\d\|\d））\|（?:（?:[a-z\u00a1-\uffff0-9]-*）*[a-z\u00a1-\uffff0-9]+）（?:\.（?:[a-z\u00a1-\uffff0-9]-*）*[a-z\u00a1-\uffff0-9]+）*）（?::（[1-9][1-9]\d\|[1-9]\d{2}\|[1-9]\d{3}\|[1-5]\d{4}\|6[0-4]\d{3}\|65[0-4]\d{2}\|655[0-2]\d\|6553[0-5]））?（?:/\S*）?$ |
| CHECK_CHINESE | 中文校验 | ^[\u4e00-\u9fa5]{0,}$ |
| CHECK_NUMBER | 纯数字校验 | ^[0-9]*$ |
| CHECK_TWO_DIG | 验证是否两位小数 | ^[0-9]+（\.[0-9]{2}）?$ |
| CHECK_IP | IP 地址校验 | ^（（2[0-4]\d\|25[0-5]\|[01]?\d\d?）\.）{3}（2[0-4]\d\|25[0-5]\|[01]?\d\d?）$ |
| CHECK_CONTAINS_CHINESE | 包含中文校验 | ^.?[\u4e00-\u9fa5]{0,}.?$ |

| 对应内置函数 | 说明 | 正则表达式 |
| --- | --- | --- |
| CHECK_SIZE | 只能输入 n 个字符 | ^.{n}$ |
| CHECK_SIZE_MIN | 至少输入 n 个字符 | ^.{n,}$ |
| CHECK_SIZE_MAX | 最多输入 n 个字符 | ^.{0,n}$ |
| CHECK_SIZE_RANGE | 输入 m-n 个字符 | ^.{m,n}$ |
| CHECK_CODE | 只能由英文、数字、下划线组成 | ^[a-z0-9A-Z_]*$ |
| CHECK_ENG_NUM | 只能包含英文和数字 | ^[a-z0-9A-Z]*$ |

3）内置变量

在表达式中可以使用点表达式来获取内置变量的属性及子属性的属性。例如，使用 activeRecord 来获取当前记录，activeRecord.id 来获取当前选中行记录的 id。

（1）数据变量。

数据变量见表 4-39。

表 4-39　数据变量

| 变量 | 名称 | 说明 |
| --- | --- | --- |
| activeRecord | 当前选中值 | 选中单行记录跳转视图初始化时，值为单条当前选中记录；选中多行记录跳转视图初始化时，值为当前选中记录列表；整表单校验时，值为当前表单提交记录；单字段校验时，值为当前字段值。作为动作筛选条件时，值为动作模型定义数据 |

（2）上下文变量。

上下文变量见表 4-40。

表 4-40　上下文变量

| 变量 | 名称 | 说明 |
| --- | --- | --- |
| module | 模块 | 使用示例：context.module<br>示例说明：请求上下文中的模块 |
| tenant | 租户 | 使用示例：context.tenant<br>示例说明：请求上下文中的租户 |
| lang | 语言 | 使用示例：context.lang<br>示例说明：请求上下文中的语言 |
| country | 国家 | 使用示例：context.country<br>示例说明：请求上下文中的国家 |
| env | 环境 | 使用示例：context.env<br>示例说明：请求上下文中的环境 |
| extend | 扩展信息 | 使用示例：context.extend.扩展变量名<br>示例说明：请求上下文中的扩展信息 |

4）内置函数

内置函数章节介绍的内置函数可以在表达式中使用。例如，使用 ABS (ActiveValue.amount) 来获取当前选中记录金额的绝对值。

## 4.1.13　Action 之校验

在 3.5.3 节中有涉及 ServerAction 的校验，本节介绍一个特殊的写法，当内置函数和表达式不够用的时候，怎么扩展。还是拿 PetShopProxyAction 举例，修改如下。

```
1   package pro.shushi.pamirs.demo.core.action;
2
3   ……引依赖类
4
5   @Model.model(PetShopProxy.MODEL_MODEL)
6   @Component
7   public class PetShopProxyAction extends DataStatusBehavior<PetShopProxy> {
8
9       ……其他代码
10
11      //     @Validation(ruleWithTips = {
12      //             @Validation.Rule(value = "!IS_BLANK(data.code)", error = "编码为必填项"),
13      //             @Validation.Rule(value = "LEN(data.shopName) < 128", error = "名称过长，不能超过128位"),
14      //     })
15      @Validation(check = "checkName")
16      @Action(displayName = "启用")
17      @Action.Advanced(rule="activeRecord.code !== undefined && !IS_BLANK(activeRecord.code)")
18      public PetShopProxy dataStatusEnable(PetShopProxy data){
19          data = super.dataStatusEnable(data);
20          data.updateById();
21          return data;
22      }
23
24      @Function
25      public Boolean checkName(PetShopProxy data) {
26          String field = "name";
27          String name = data.getShopName();
28          boolean success = true;
29          if (StringUtils.isBlank(name)) {
30              PamirsSession.getMessageHub()
31                      .msg(Message.init()
32                              .setLevel(InformationLevelEnum.ERROR)
33                              .setField(field)
34                              .setMessage("名称为必填项"));
35              success = false;
36          }
37          if (name.length() > 128) {
38              PamirsSession.getMessageHub()
39                      .msg(Message.init()
40                              .setLevel(InformationLevelEnum.ERROR)
41                              .setField(field)
42                              .setMessage("名称过长，不能超过128位"));
43              success = false;
44          }
45          return success;
46      }
47
48      ……其他代码
49
50  }
51
```

注：

（1）check 属性指定了校验函数名称，命名空间必须与服务器动作一致。

（2）校验函数的入参必须与服务器动作一致。

（3）使用 PamirsSession#getMessageHub 方法可通知前端错误的属性及需要展示的提示信息，允许多个。

### 4.1.14　Search 之非存储字段条件

Search 默认查询的是模型的 queryPage 函数，但我们有时候需要替换调用的函数，下个版本支持这一特性。其核心场景为当搜索条件中有非存储字段时，如果直接用 queryPage 函数的 RSQL 拼接就会报错。在此介绍一个比较友好的临时替代方案。

**1. 非存储字段条件（举例）**

Step1. 为 PetTalent 新增一个非存储字段 unStore，代码如下。

```
1    @Field(displayName = "非存储字段测试",store = NullableBoolEnum.FALSE)
2    private String unStore;
```

Step2. 修改 PetTalent 的 Table 视图的 Template，代码如下。

在 <template slot="search" cols="4"></template> 标签内增加一个查询条件：

```
1    <field data="unStore" />
```

Step3. 重启看效果。

进入宠物达人列表页，在搜索框"非存储字段测试"输入查询内容，单击"搜索"报错。

Step4. 修改 PetTalentAction 的 queryPage 方法，代码如下。

```
1    package pro.shushi.pamirs.demo.core.action;
2
3    ……引入依赖类
4
5    @Model.model(PetTalent.MODEL_MODEL)
6    @Component
7    public class PetTalentAction {
8
9        ……其他代码
10
11       @Function.Advanced(type= FunctionTypeEnum.QUERY)
12       @Function.fun(FunctionConstants.queryPage)
13       @Function(openLevel = {FunctionOpenEnum.API})
14       public Pagination<PetTalent> queryPage(Pagination<PetTalent> page, IWrapper<PetTalent> queryWrapper){
15           String rsql = RSQLHelper.toTargetString(RSQLHelper.parse(PetTalent.MODEL_MODEL,queryWrapper.getOriginRsql()), new RSQLNodeConnector() {
16               @Override
17               public String comparisonConnector(RSQLNodeInfo nodeInfo) {
18                   //判断字段为unStore,则进行替换
19                   if ("unStore".equals(nodeInfo.getField())) {
20                       //获取前端传过的搜索值
21                       String unStoreSearchValue = nodeInfo.getArguments().get(0);
22                       //新建一个查询RSQL节点，并设置RSQLNodeInfoType: AND、OR、COMPARISON，这里nodeInfo.getType()是COMPARISON
23                       //利用AND、OR可以组合成更复杂的RSQL
24                       RSQLNodeInfo newNode = new RSQLNodeInfo(nodeInfo.getType());
25                       //设置查询字段为name
26                       newNode.setField("name");
27                       //设置RSQL操作符
28                       newNode.setOperator(RsqlSearchOperation.LIKE.getOperator());
29                       //设置RSQL的查询值集合
30                       newNode.setArguments(Lists.newArrayList(unStoreSearchValue));
31                       //返回
```

```
32                return super.comparisonConnector(newNode);
33            }
34            return super.comparisonConnector(nodeInfo);
35        }
36    });
37    //新建查询类
38    QueryWrapper queryWrapper1 = Pops.<PetTalent>f(Pops.query().from(PetTalen
      t.MODEL_MODEL)).get();
39    //把RSQL转换成SQL
40    String sql = RsqlParseHelper.parseRsql2Sql(PetTalent.MODEL_MODEL,rsql);
41    //设置queryWrapper1的SQL查询语句
42    queryWrapper1.apply(sql);
43    return new PetTalent().queryPage(page,queryWrapper1);
44  }
45  ……其他代码
46
47 }
48
```

Step5. 重启看效果。

在搜索框非存储字段测试输入查询内容，跟搜索达人是一样的效果，如图 4-37 所示。

图 4-37　重启看效果

### 4.1.15　框架之网关协议

**1. 多端协议**

协议内容格式如下。

1）请求头

请求头由头信息、请求地址和 HTTP 参数键值对组成。

（1）头信息 headerMap。

头信息 headerMap 代码如下。

```
1   "sec-fetch-mode" -> "cors"
2   "content-length" -> "482"
3   "sec-fetch-site" -> "none"
4   "accept-language" -> "zh-CN,zh;q=0.9"
5   "cookie" -> "pamirs_uc_session_id=241af6a1dbba41a4b35afc96ddf15915"
6   "origin" -> "chrome-extension://flnheeellpciglgpaodhkhmapeljopja"
7   "accept" -> "application/json"
8   "host" -> "127.0.0.1:8090"
9   "connection" -> "keep-alive"
10  "content-type" -> "application/json"
11  "accept-encoding" -> "gzip, deflate, br"
12  "user-agent" -> "Mozilla/5.0 (Macintosh; Intel Mac OS X 10_15_7) AppleWebKit/537.
    36 (KHTML, like Gecko) Chrome/88.0.4324.192 Safari/537.36"
13  "sec-fetch-dest" -> "empty"
```

（2）请求地址 requestUrl。

例如：http://127.0.0.1:8090/pamirs/DemoCore?scene=redirectListPage。

（3）HTTP 参数键值对 parameterMap。

URL 中 queryString 在服务端最终会转化为参数键值对。

2）请求体格式

请求体格式采用 GraphQL 协议。请求体格式分为 API 请求和上下文变量。以商品的 test 接口为例，请求格式如下。

（1）API 请求格式。

API 请求格式如下所示。

```
1   query{
2     petShopProxyQuery {
3       queryPage(page: {currentPage: 1, size: 1}, queryWrapper: {rsql: "(1==1)"}) {
4         content {
5           income
6           id
7           code
8           creater {
9             id
10            nickname
11          }
12          relatedShopName
13          shopName
14          petTalents {
15            id
16            name
17          }
18          items {
19            id
20            itemName
21          }
22        }
23        size
24        totalPages
25        totalElements
26      }
27    }
28  }
```

（2）上下文变量 variables。

请求策略 requestStrategy 见表 4-41。

表 4-41　请求策略 requestStrategy

| 名称 | 类型 | 说明 |
| --- | --- | --- |
| checkStrategy | CheckStrategyEnum | 校验策略：<br>RETURN_WHEN_COMPLETED：全部校验完成再返回结果<br>RETURN_WHEN_ERROR：校验错误即返回结果 |
| msgLevel | InformationLevelEnum | 消息级别：<br>DEBUG ("debug"," 调试 "," 调试 ")，<br>INFO ("info"," 信息 "," 信息 ")，<br>WARN ("warn"," 警告 "," 警告 ")，<br>SUCCESS ("success"," 成功 "," 成功 ")，<br>ERROR ("error"," 错误 "," 错误 ")<br>不设置，则只返回错误消息；<br>上方消息级别清单，越往下级别越高。<br>只有消息的级别高于或等于该设定级别才返回，否则会被过滤 |
| onlyValidate | Boolean | 只校验不提交数据 |

上下文变量示例如下。

```
{
  "requestStrategy": {
    "checkStrategy": "RETURN_WHEN_COMPLETED",
    "msgLevel":"INFO"
  }
}
```

3）响应体格式

协议响应内容包括 data、extensions 和 errors 三部分，extensions 和 errors 是可缺省的，data 部分为业务数据返回值。应用业务层可以在 extensions 中添加 API 返回值之外的扩展信息。extensions 中包含 success、messages 和 extra 三部分，success 标识请求是否成功，如果业务正确处理并返回，则 errors 部分为空；如果业务处理返回失败，则将错误信息添加到 errors 中。

正确响应格式示例如下。

```
{
  "data": {
    "petShopProxyQuery": {
      "queryPage": {
        "content": [
          {
            "id": "246675081504233477",
            "creater": {
              "id": "10001"
            },
            "relatedShopName": "oinone宠物店铺001",
            "shopName": "oinone宠物店铺001",
            "petTalents": [
              {
                "id": "248149320438706183",
                "name": "老邓头"
              }
            ],
            "items": [
              {
                "id": "246675081504233480",
                "itemName": "萌猫商品001"
              }
            ]
          }
        ],
        "size": "1",
        "totalPages": 2,
        "totalElements": "2"
      }
    }
  },
  "errors": [],
  "extensions": {
    "success": true,
    "dataloader": {
      "overall-statistics": {
```

错误响应格式示例如下。

```
{
  "data": { "itemMutation": {}},
  "errors": [{
    "message": "校验失败, 数据错误",
    "extensions": {
      "errorCode": "10050009",
      "errorType": "SYSTEM_ERROR",
      "level": "ERROR",
      "messages": [{
        "code": "10080016",
        "errorType": "DATA_ERROR",
        "level": "ERROR",
        "message": "长度必须大于或等于4",
```

```
14         "path": [
15            "itemMutation",
16            "test",
17            "item",
18            "description"]},
19       {
20         "code": "10080016",
21         "errorType": "DATA_ERROR",
22         "level": "ERROR",
23         "message": "字段值必须大于或等于2",
24         "path": [
25            "itemMutation",
26            "test",
27            "item",
28            "view"]
29       }, {
30         "errorType": "BIZ_ERROR",
31         "level": "ERROR",
32         "message": "输入错误,请输入正确的重量",
33         "path": [
34            "itemMutation",
35            "test"]
36       },{
37         "data": "IS_NULL(activeValue[0].price)",
38         "errorType": "BIZ_ERROR",
39         "level": "ERROR",
40         "message": "输入错误,必填字段",
41         "path": [
42            "itemMutation",
43            "test"]
44       }],
45       "classification": "DataFetchingException"
46      }
47   }],
48   "extensions": {
49      "success": false,
50      "extra": { "variableA": "" }
51   }
52 }
```

**2. Pamirs API DSL**

Pamirs API DSL 采用 GraphQL 协议。

GraphQL 是一种用于查询和操作数据的查询语言,同时也是一个由 Facebook 开发和开源的运行时系统。与传统的 RESTful API 不同,GraphQL 允许客户端明确指定需要获取的数据,从而避免了过度获取或获取不足的数据问题。它有一个丰富的类型系统,可以自定义数据类型。这个类型系统有助于明确数据的结构和关系,从而提供了更好的文档和可理解性。

Pamirs aPaaS 在 GraphQL 的基础上支持了 BigDecimal、BigInteger、Date、Double、Html、Money、Void、Map、Obj。

**3. Pamirs Query DSL**

Pamirs Query DSL 采用 RSQL 协议。

协议原文:RSQL is a query language for parametrized filtering of entries in RESTful APIs. It's based on FIQL (Feed Item Query Language) – an URI-friendly syntax for expressing filters across the entries in an Atom Feed. FIQL is great for use in URI; there are no unsafe characters, so URL encoding is not required. On the other side, FIQL's syntax is not very intuitive and URL encoding isn't always that big deal, so RSQL also provides a friendlier syntax for logical operators and some of the comparison operators.

中译说明,RSQL 是一种查询语言,用于对 RESTful API 中的条目进行参数化过滤。它基于 FIQL(Feed Item Query Language)——一种 URI 友好的语法,用于跨 Atom Feed 中的条目表达过滤器。FIQL 非常适合在 URI 中使用;没有不安全的字符,因此不需要 URL 编码。但是,FIQL 的语法不太直观,URL 编码也不总是那么重要,因此 RSQL 还为逻辑运算符和一些比较运算符提供了更友好的语法。

## 4.1.16 框架之网关协议——RSQL 及扩展

**1. RSQL / FIQL 解析器**

例如，您可以像这样查询资源：/movies?query=name=="Kill Bill";year=gt=2003 or /movies?query=director.lastName==Nolan and year>=2000。参见以下示例：

这是一个用 JavaCC 和 Java 编写的完整且经过彻底测试的 RSQL 解析器。因为 RSQL 是 FIQL 的超集，所以它也可以用于解析 FIQL。

1）语法和语义

以下语法规范采用 EBNF 表示法（ISO 14977）编写。

RSQL 表达式由一个或多个比较组成，通过逻辑运算符相互关联：

```
Logical AND : ; or and
Logical OR  : , or or
```

默认情况下，AND 运算符优先级更高（即，在任何 OR 运算符之前对其求值）。但是，可以使用带括号的表达式来更改优先级，从而产生所包含表达式产生的任何结果。

```
input = or, EOF;
or = and, { "," , and };
and = constraint, { ";" , constraint };
constraint= ( group | comparison );
group = "(", or, ")";
```

比较由选择器、运算符和参数组成。

```
comparison= 选择器、比较运算、参数；
```

选择器标识要筛选的资源表示形式的字段（或属性、元素…）。它可以是任何不包含保留字符的非空 Unicode 字符串（见下文）或空格。选择器的特定语法不由此解析器强制执行。

```
selector= 未保留 str;
```

比较运算符采用 FIQL 表示法，其中一些运算符还具有另一种语法：

- Equal to : ==
- Not equal to : !=
- Less than : =lt= or <
- Less than or equal to : =le= or <=
- Greater than operator : =gt= or >
- Greater than or equal to : =ge= or >=
- In : =in=
- Not in : =out=

您还可以使用自己的运算符简单地扩展此解析器（请参阅下一节）。

```
comparison-op = comp-fiql | comp-alt;
comp-fiql = ( ( "=", { ALPHA } ) | "!" ), "=";
comp-alt = ( ">" | "<" ), [ "=" ];
```

参数可以是单个值，也可以是用逗号分隔的括号中的多个值。不包含任何保留字符或空格的值可以不加引号，其他参数必须用单引号或双引号括起来。

```
arguments = ( "(", value, { "," , value }, ")" ) | value;
value = unreserved-str | double-quoted | single-quoted;
unreserved-str = unreserved, { unreserved }
single-quoted= "'", { ( escaped | all-chars - ( "'" | "\" ) ) }, "'";
double-quoted= '"', { ( escaped | all-chars - ( '"' | "\" ) ) }, '"';
reserved= '"' | "'" | "(" | ")" | ";" | "," | "=" | "!" | "~" | "<" | ">";
unreserved= all-chars - reserved - " ";
escaped= "\", all-chars;
all-chars= ? all unicode characters ?;
```

如果需要在带引号的参数中同时使用单引号和双引号，则必须使用\（反斜杠）转义其中一个引号。如果要按字面意思使用\，请将其加倍为\\。反斜杠只有在引用的参数中才有特殊含义，而不是在未引用的参数。

2）示例

```
- name=="Kill Bill";year=gt=2003
- name=="Kill Bill" and year>2003
- genres=in=(sci-fi,action);(director=='Christopher Nolan',actor==*Bale);year=ge=2000
- genres=in=(sci-fi,action) and (director=='Christopher Nolan' or actor==*Bale) and year>=2000
- director.lastName==Nolan;year=ge=2000;year=lt=2010
- director.lastName==Nolan and year>=2000 and year<2010
- genres=in=(sci-fi,action);genres=out=(romance,animated,horror),director==Que*Tarantino
- genres=in=(sci-fi,action) and genres=out=(romance,animated,horror) or director==Que*Tarantino
```

**2. Oinone 拓展协议**

（1）正常类型。

- Is null : =isnull=
- Not null : =notnull=
- Like to : =like=
- Not like to : =notlike=
- Column equal to : =cole=
- Not column equal to : =colnot=
- like 'xxxx%' : =starts=
- not like 'xxxx%' :=notstarts=
- like '%xxxx' : =ends=
- not like '%xxxx' : =notends=

（2）二进制枚举。

- Intersect : =has=
- Not Intersect : =hasnt=

- Contain：=contain=
- Not Contain：=notcontain=

### 4.1.17　框架之网关协议——GraphQL 协议

一个 GraphQL 服务是通过定义类型和类型上的字段来创建的，然后给每个类型上的每个字段提供解析函数。例如，一个 GraphQL 服务告诉我们当前登录用户是 me，这个用户的名称可能如下所示。

```
type Query {
    me: User
}
type User {
    id: ID
    name: String
}
```

一并的还有每个类型上字段的解析函数。

```
function Query_me(request) {
    return request.auth.user;
}
function User_name(user) {
    return user.getName();
}
```

一旦一个 GraphQL 服务运行起来（通常在 Web 服务的一个 URL 上），它就能接收 GraphQL 查询，并验证和执行。接收到的查询首先会被检查确保它只引用了已定义的类型和字段，然后运行指定的解析函数来生成结果。

例如如下查询。

```
{
    me {
        name
    }
}
```

会产生如下的 JSON 结果。

```
{
    "me": {
        "name": "Luke Skywalker"
    }
}
```

了解更多登录网址 https://graphql.cn/learn/。

### 4.1.18　框架之网关协议——Variables 变量

我们在应用开发过程中有一种特殊情况：在后端逻辑编写的时候需要知道请求的发起入口，平台利用 GQL 协议中的 variables 属性来传递信息，本节就介绍如何获取。

**1. 前端附带额外变量**

前端附带额外变量见表 4-42。

表 4-42　前端附带额外变量

| 属性名 | 类型 | 说明 |
| --- | --- | --- |
| scene | String | 菜单入口 |

variables 信息中 scene 如图 4-38 所示。

图 4-38　variables 信息中的 scene

### 2. 后端如何接收 variables 信息

通过 PamirsSession.getRequestVariables() 可以得到 PamirsRequestVariables 对象。

### 3. 第一个 variable（举例）

Step1. 修改 PetTalentAction，获取前端传递的 variables，代码如下。

```
package pro.shushi.pamirs.demo.core.action;

……类引用

@Model.model(PetTalent.MODEL_MODEL)
@Component
public class PetTalentAction {
    ……其他代码
    @Function.Advanced(type= FunctionTypeEnum.QUERY)
    @Function.fun(FunctionConstants.queryPage)
    @Function(openLevel = {FunctionOpenEnum.API})
    public Pagination<PetTalent> queryPage(Pagination<PetTalent> page, IWrapper<PetTalent> queryWrapper){
        String scene = (String)PamirsSession.getRequestVariables().getVariables().get("scene");
        System.out.println("scene: "+ scene);
        ……其他代码
    }
    ……其他代码
}
```

Step2. 重启验证。

单击宠物达人不同菜单入口，查看效果，如图 4-39 和图 4-40 所示。

图 4-39　效果（1）

图 4-40　效果（2）

## 4.1.19 框架之网关协议——后端占位符

在我们日常开发中会碰到一些特殊场景，需要由前端来传一些如"当前用户 id""当前用户 code"诸如此类只有后端才知道值的参数，那么后端占位符就是来解决类似问题的。如前端传 ${currentUserId}，后端会自动替换为当前用户 id。

Step1. 后端定义占位符。

我们新建一个 UserPlaceHolder 继承 AbstractPlaceHolderParser，用 namespace 来定义一个"currentUserId"的占位符，其对应值由 value () 决定为"PamirsSession.getUserId ().toString ()"，active 要为真才有效，priority 为优先级。

```java
    @Component
    public class UserPlaceHolder extends AbstractPlaceHolderParser {
        @Override
        protected String value() {
            return PamirsSession.getUserId().toString();
        }

        @Override
        public Integer priority() {
            return 10;
        }

        @Override
        public Boolean active() {
            return Boolean.TRUE;
        }

        @Override
        public String namespace() {
            return "currentUserId";
        }
    }
```

Step2. 前端使用后端占位符，代码如下。

我们经常在 o2m 和 m2m 中设置 domain 来过滤数据，这里案例就是在 field 中设置来过滤条件，domain="createUid == $#{currentUserId}"，注意这里用的是 $#{currentUserId} 而不是 ${currentUserId}，这是前端为了区分真正变量和后端占位符，提交的时候会把 # 过滤掉提交。修改宠物达人表格视图 Template 中的 search 部分。

```xml
<template slot="search"  cols="4">
    <field data="name" label="达人"/>
    <field data="petTalentSex" multi="true" label="达人性别"/>
    <field data="creater" />
<!--        <field data="petShops" label="宠物商店" domain="createUid == ${activeRecord.creater.id}"/>-->
    <field data="petShops" label="宠物商店" domain="createUid == $#{currentUserId}"/>
    <field data="dataStatus" label="数据状态" multi="true">
        <options>
            <option name="DRAFT" displayName="草稿" value="DRAFT" state="ACTIVE"/>
            <option name="NOT_ENABLED" displayName="未启用" value="NOT_ENABLED" state="ACTIVE"/>
            <option name="ENABLED" displayName="已启用" value="ENABLED" state="ACTIVE"/>
            <option name="DISABLED" displayName="已禁用" value="DISABLED" state="ACTIVE"/>
        </options>
    </field>
    <field data="createDate" label="创建时间"/>
    <field data="unStore" />
</template>
```

Step3. 重启看效果。

请求上都带上了 createUid==${currentUserId}，如图 4-41 所示。

图 4-41　请求上都带上了 createUid==${currentUserId}

## 4.1.20　框架之 Session

在日常开发中，我们经常需要把一些通用的信息放入程序执行的上下文中，以便业务开发人员快速获取。那么 Oinone 的 PamirsSession 就是来解决此类问题的。

**1. PamirsSession 介绍**

在 Oinone 的体系中 PamirsSession 是执行上下文的承载，能从中获取业务基础信息、指令信息、元数据信息、环境信息、请求参数以及前后端 MessageHub 等。在前面的学习过程中我们已经多次接触到了如何使用 PamirsSession：

在 4.1.19 一节中，使用 PamirsSession.getUserId() 来获取当前登录用户 id，诸如此类的业务基础信息。

在 4.1.18 一节中，使用 PamirsSession.getRequestVariables() 得到 PamirsRequestVariables 对象，进而获取前端请求的相关信息。

在 4.1.5 节中，使用 PamirsSession.directive() 来操作元位指令系统，进而影响执行策略。

在 4.1.15、3.4.1 等节中，都用到 PamirsSession.getMessageHub() 来设置返回消息。

**2. 构建模块自身 Session（举例）**

不同的应用场景对 PamirsSession 的诉求是不一样的，这个时候我们就可以去扩展 PamirsSession 来达到我们的目的。

1）构建模块自身 Session 的步骤

（1）构建自身特有的数据结构 XSessionData。

（2）对 XSessionData 进行线程级缓存封装。

（3）利用 Hook 机制初始化 XSessionData 并放到 ThreadLocal 中。

（4）定义自身 XSessionApi。

（5）实现 XSessionApi 接口、SessionClearApi。在请求结束时会调用 SessionClearApi 的 clear 方法。

（6）定义 XSession 继承 PamirsSession。

扩展 PamirsSession 的经典案例设计图如图 4-42 所示。

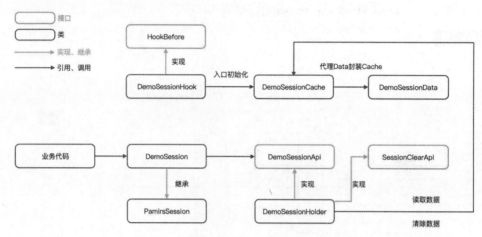

图 4-42　扩展 PamirsSession 的经典案例设计图

2）构建 Demo 应用自身 Session

下面的例子为给 Session 放入当前登录用户。

Step1. 新建 DemoSessionData 类，代码如下。

构建自身特有的数据结构 DemoSessionData，增加一个模型为 PamirsUser 的字段 user，DemoSessionData 用 Data 注解，注意要用 Oinone 平台提供的 @Data。

```
 6    @Data
 7    public class DemoSessionData {
 8
 9        private PamirsUser user;
10
11    }
```

Step2. 新建 DemoSessionCache。

对 DemoSessionData 进行线程级缓存封装，代码如下。

```
 8    public class DemoSessionCache {
 9        private static final ThreadLocal<DemoSessionData> BIZ_DATA_THREAD_LOCAL = new ThreadLocal<>();
10        public static PamirsUser getUser(){
11            return BIZ_DATA_THREAD_LOCAL.get()==null?null:BIZ_DATA_THREAD_LOCAL.get().getUser();
12        }
13        public static void init(){
14            if(getUser()!=null){
15                return ;
16            }
17            Long uid = PamirsSession.getUserId();
18            if(uid == null){
19                return;
20            }
21            PamirsUser user = CommonApiFactory.getApi(UserService.class).queryById(uid);
22            if(user!=null){
23                DemoSessionData demoSessionData = new DemoSessionData();
24                demoSessionData.setUser(user);
25                BIZ_DATA_THREAD_LOCAL.set(demoSessionData);
26            }
27        }
28        public static void clear(){
29            BIZ_DATA_THREAD_LOCAL.remove();
30        }
31    }
```

Step3. 新建 DemoSessionHook，代码如下。

利用 Hook 机制，调用 DemoSessionCache 的 init 方法初始化 DemoSessionData 并放到 ThreadLocal 中。

@Hook (module= DemoModule.MODULE_MODULE)，规定只有增加 DemoModule 模块访问的请求该拦截器才会生效，否则其他模块的请求都会被 DemoSessionHook 拦截。

```java
10    @Component
11    public class DemoSessionHook implements HookBefore {
12        @Override
13        @Hook(priority = 1,module = DemoModule.MODULE_MODULE)
14        public Object run(Function function, Object... args) {
15            DemoSessionCache.init();
16            return function;
17        }
18    }
19
```

Step4. 新建 DemoSessionApi，代码如下。

```java
6    public interface DemoSessionApi extends  CommonApi {
7        PamirsUser getUser();
8    }
9
```

Step5. 新建 DemoSessionHolder，代码如下。

（1）实现 DemoSessionApi 接口。

（2）实现 SessionClearApi 接口，在请求结束时会调用 SessionClearApi 的 clear 方法。

```java
7     @Component
8     public class DemoSessionHolder implements DemoSessionApi, SessionClearApi {
9         @Override
10        public PamirsUser getUser() {
11            return DemoSessionCache.getUser();
12        }
13
14        @Override
15        public void clear() {
16            DemoSessionCache.clear();
17        }
18    }
19
```

Step6. 新建 DemoSession，代码如下。

```java
7     public class DemoSession extends PamirsSession {
8         public static PamirsUser getUser(){
9             return CommonApiFactory.getApi(DemoSessionApi.class).getUser();
10        }
11    }
12
```

Step7. 修改 UserPlaceHolder，代码如下。

使用 DemoSession.getUser ().getId () 替代 PamirsSession.getUserId ()。

```java
7     @Component
8     public class UserPlaceHolder extends AbstractPlaceHolderParser {
9         @Override
10        protected String value() {
11            return DemoSession.getUser().getId().toString();
12        }
13
14        @Override
15        public Integer priority() {
16            return 10;
17        }
18
19        @Override
20        public Boolean active() {
21            return Boolean.TRUE;
```

```
22          }
23
24          @Override
25  ▾       public String namespace() {
26              return "currentUserId";
27          }
28      }
29
```

**Step8.** 重启看效果。

与 4.1.19 节的例子效果一样。

### 4.1.21　框架之分布式消息

消息中间件是在分布式开发中常见的一种技术手段，用于模块间的解耦、异步处理、数据最终一致等场景。

**1. 介绍**

Oinone 对开源的 RocketMQ 进行了封装，是平台提供的一种较为简单的使用方式，并非是对 RocketMQ 进行的功能扩展。同时也伴随着以下两个非常至关重要的目的。

适配不同企业对 RocketMQ 的不同版本选择，不致改上层业务代码。目前已经适配的有 RocketMQ 的开源版本和阿里云版本。下个版本会对 API 进行升级支持不同类型 MQ，以适配不同企业对 MQ 的不同要求，应对一些企业客户已经对 MQ 进行技术选择。

对协议头进行扩展，如多租户的封装，SaaS 模式中为了共用 MQ 基础资源，需要在消息头中加入必要租户信息。

**2. 使用准备**

demo 工程默认已经依赖消息，这里只是做介绍，无须大家额外操作，大家可以用 maven 依赖树命令查看引用关系。

1）依赖包

（1）核心 POM。

核心 POM 如下所示。

```
1  ▾ <dependency>
2        <groupId>pro.shushi.pamirs.framework</groupId>
3        <artifactId>pamirs-connectors-event</artifactId>
4    </dependency>
```

（2）多租户 POM。

只要在 boot 工程中引入就可以，不影响开发，代码如下。

```
1    <!-- RocketMQ 多租户支持 -->
2  ▾ <dependency>
3        <groupId>pro.shushi.pamirs.tenant</groupId>
4        <artifactId>pamirs-connectors-event-tenant</artifactId>
5    </dependency>
```

（3）相关功能引入。

增强模型、触发器都依赖 MQ，代码如下。

```
1    <!-- 增强模型 -->
2  ▾ <dependency>
3        <groupId>pro.shushi.pamirs.core</groupId>
4        <artifactId>pamirs-channel</artifactId>
5    </dependency>
```

```xml
6      <!-- 触发器 API -->
7      <dependency>
8          <groupId>pro.shushi.pamirs.core</groupId>
9          <artifactId>pamirs-trigger-api</artifactId>
10     </dependency>
11     <!-- 触发器 core -->
12     <dependency>
13         <groupId>pro.shushi.pamirs.core</groupId>
14         <artifactId>pamirs-trigger-core</artifactId>
15     </dependency>
```

2）yml 配置文件参考

请见 4.1.1 节"pamirs.event"部分。

**3. 使用说明**

1）发送消息（NotifyProducer）

NotifyProducer 是 Pamirs Event 中所有生产者的基本 API，它仅仅定义了消息发送的基本行为，例如生产者自身的属性、启动和停止、当前状态以及消息发送方法。它本身并不决定消息如何发送，而是根据具体的实现确定其功能。

目前仅实现了 RocketMQProducer，你可以使用下面介绍的方法轻松使用这些功能：

（1）Notify 注解方式。

① 使用示例如下所示。

```java
1   @Component
2   public class DemoProducer {
3   
4       @Notify(topic = "test", tags = "model")
5       public DemoModel sendModel() {
6           return new DemoModel();
7       }
8   
9       @Notify(topic = "test", tags = "dto")
10      public DemoDTO sendDTO() {
11          return new DemoDTO();
12      }
13  }
```

② 解释说明。

● 使用 Component 注解方式注册 Spring Bean。

● Notify 注解指定 topic 和 tags。

● topic 和 tags 对应 NotifyEvent 中的 topic 和 tags。

（2）RocketMQProducer 方法调用。

① 使用示例如下所示。

```java
1   @Component
2   public class SendRocketMQMessage {
3   
4       @Autowired
5       private RocketMQProducer rocketMQProducer;
6   
7       /**
8        * 发送普通消息
9        */
10      public void sendNormalMessage() {
11          rocketMQProducer.send(new NotifyEvent("test", "model", new DemoModel()));
```

```
12            rocketMQProducer.send(new NotifyEvent("test", "dto", new DemoDTO()));
13        }
14
15        /**
16         * 发送有序消息
17         */
18        public void sendOrderlyMessage() {
19            DemoModel data = new DemoModel();
20            data.setAge(10);
21            rocketMQProducer.send(new NotifyEvent("test", "model", data)
22                    .setQueueSelector((queueSize, event) -> {
23                        DemoModel body = (DemoModel) event.getBody();
24                        return body.getAge() % queueSize;
25                    }));
26        }
27
28        /**
29         * 发送事务消息
30         */
31        public void sendTransactionMessage() {
32            rocketMQProducer.send(new NotifyEvent("test", "model", new DemoModel())
33                    .setIsTransaction(true)
34                    .setGroup("demoTransactionListener"));
35        }
36    }
```

② 解释说明。

● 使用 Component 注解方式注册 Spring Bean。

● 使用 Autowired 注解方式装配 RocketMQProducer 实例。

● 使用 send 方法发送指定消息。

● 在"发送普通消息"方法中，实现的效果与 Notify 注解方式完全一致。

● 在"发送有序消息"方法中，队列选择器是必须配置的，queueSize 属性为 MQ 队列的总数量，在 broker 中配置。有序消息必须配合有序消费者才能达到有序消费的目的，否则还是无序的普通消息，消费者需要配置 @NotifyListener (consumerType=ConsumerType.ORDERLY)。

● 在"发送事务消息"方法中指定的 group 为 ProducerGroup，事务消息是通过不同的 Producer 发出的，事务消息监听请参考 TransactionListener 注解的相关使用方法（示例中的 group 与下文介绍中的一致）。

（3）使用 TransactionListener 开启事务消息监听。

① 使用示例如下所示。

```
1    @Component
2    @TransactionListener
3    public class DemoTransactionListener implements NotifyTransactionListener {
4
5        @Override
6        public NotifyTransactionState checkLocalTransaction(NotifyEvent event) {
7            return NotifyTransactionState.COMMIT;
8        }
9
10   }
```

② 解释说明。

● 实现 NotifyTransactionListener 接口。

● 使用 Component 注解方式注册 Spring Bean。

● 添加 TransactionListener 注解注册事务监听，生成对应的生产者。

● 当前 ProducerGroup 将使用这个类的 BeanName，即 "demoTransactionListener"。如果你想自定义 ProducerGroup，可以使用 TransactionListener 的 value 属性进行设置。

2）消费消息（NotifyConsumer）

NotifyConsumer 是 Pamirs Evnet 中所有消费者的基本 API，与 NotifyProducer 类似，它仅仅定义了消息消费的基本行为，例如消费者自身的属性、启动和停止、当前状态以及消息的监听和订阅方法。它本身并不决定消息如何消费以及如何被订阅，而是根据具体的实现确定其功能。

目前仅实现了 RocketMQEventPushConsumer，你可以使用下面介绍的方法轻松使用这些功能：

（1）在类上使用 NotifyListener 注解。

① 使用示例如下所示。

```
@Component
@NotifyListener(topic = "test", tags = "model")
public class DemoConsumerClass implements NotifyEventListener {

    @Override
    public void consumer(NotifyEvent event) {
        DemoModel data = (DemoModel) event.getBody();
        // do some things.
    }
}
```

② 解释说明。

● 实现 NotifyEventListener 接口。

● 使用 Component 注解方式注册 Spring Bean。

● 当前 ConsumerGroup 将使用这个类的 BeanName，即 "demoConsumerClass"。如果你想自定义 ConsumerGroup，可以使用 Component 的 value 属性进行设置。

● 从 body 中将可以获取生产者发送的数据对象，并且已经做好了类型处理，可以直接使用。

● 使用原生的 RocketMQ 发送的消息，类型可能是无法识别的，你可以使用 NotifyListener 中提供的 bodyClass 来指定类型。

● topic 和 tags 对应 NotifyEvent 中的 topic 和 tags。

（2）在方法上使用 NotifyListener 注解。

① 使用示例如下所示。

```
@Component
public class DemoConsumerMethod {

    @Bean
    @NotifyListener(topic = "test", tags = "model")
    public NotifyEventListener modelConsumer() {
        return event -> {
            DemoModel data = (DemoModel) event.getBody();
            // do some things.
        };
    }

    @Bean
    @NotifyListener(topic = "test", tags = "dto")
    public NotifyEventListener dtoConsumer() {
        return event -> {
            DemoDTO data = (DemoDTO) event.getBody();
            // do some things.
        };
    }
}
```

② 解释说明。
● 使用 Bean 注解方式注册 Spring Bean。
● 方法返回值为 NotifyEventListener 类型。
● 当前 ConsumerGroup 将使用对应方法生成的 BeanName，即 "ModelConsumer" 和 "dtoConsumer"。如果你想自定义 ConsumerGroup，可以使用 Bean 的 value 属性进行设置。

3）实战

约定：每个模块下的 Topic 和 Tags 必须定义常量池进行统一管理，主要是为了方便维护与管理，技术没有限制。

（1）常规消息（举例）。

Step1. 新建 PetNotifyEnum，代码如下。

用 PetNotifyEnum 来管理模块的所有 Topic 和 Tags 的常量定义。

```java
package pro.shushi.pamirs.demo.api.mq;

public class PetNotifyEnum {

    public static class PetItemInventroy{

        public interface Topic{
            String PET_ITEM_INVENTROY_CHANGE ="petItemInventroyChange";
        }
        public interface Tag{
            String CREATE ="create";
            String UPDATE ="update";
            String DELETE ="delete";
        }
    }
}
```

Step2. 新建 PetItemInventoryMqProducer，代码如下。

新建 PetItemInventroy 模型对应消息生产者，用于发送 PetItemInventroy 模型变动的相关消息。

```java
@Component
public class PetItemInventoryMqProducer {

    @Autowired
    private RocketMQProducer rocketMQProducer;

    /**
     * 发送普通消息
     */
    public void sendNormalMessage(PetItemInventroy data,String tag) {
        rocketMQProducer.send(new NotifyEvent(PetNotifyEnum.PetItemInventroy.Topic.PET_ITEM_INVENTROY_CHANGE, tag, data));
    }

}
```

Step3. 新建 PetItemInventoryMqConsumer，代码如下。

新建 PetItemInventoryMqConsumer，订阅 PetItemInventroy 模型变动的相关消息，并进行相关处理。

```java
@Component
@NotifyListener(topic = PetNotifyEnum.PetItemInventroy.Topic.PET_ITEM_INVENTROY_CHANGE, tags = "*")
@Slf4j
public class PetItemInventoryMqConsumer implements NotifyEventListener {
    @Override
    public void consumer(NotifyEvent event) {
        PetItemInventroy petItemInventroy = (PetItemInventroy)event.getBody();
        log.info("petItemInventroy的消息，库存数量为: " + petItemInventroy.getQuantity());
    }
}
```

Step4. 修改 PetItemInventroyAction，代码如下。

在修改 PetItemInventroy 完成之后，发送 topic 为"PetNotifyEnum.PetItemInventroyMq.Topic.PET_ITEM_INVENTROY_CHANGE"，tag 为"PetNotifyEnum.PetItemInventroyMq.Tag.UPDATE"的消息出去。

```
 1  package pro.shushi.pamirs.demo.core.action;
 2  ……引入依赖类
 3  @Model.model(PetItemInventroy.MODEL_MODEL)
 4  @Component
 5  public class PetItemInventroyAction {
 6
 7      @Autowired
 8      private PetItemInventoryMqProducer petItemInventoryMqProducer;
 9
10      @Function.Advanced(type= FunctionTypeEnum.UPDATE)
11      @Function.fun(FunctionConstants.update)
12      @Function(openLevel = {FunctionOpenEnum.API})
13      public PetItemInventroy update(PetItemInventroy data){
14          ……其他代码
15          petItemInventoryMqProducer.sendNormalMessage(data, PetNotifyEnum.PetItemI
    nventroy.Tag.UPDATE);
16          return data;
17      }
18  }
```

Step5. 重启看效果。

① 编辑商品库存记录，如图 4-43 所示。

图 4-43  编辑商品库存记录

② 查看后端日志是否打印，如图 4-44 所示。

图 4-44  查看后端日志是否打印

（2）顺序消息（举例）。

Step1. 修改 PetItemInventoryMqProducer，代码如下。

增加一个顺序消息发送的方法，发送时根据商品库存的 id 分队列，相同队列顺序消费。

```
10  @Component
11  public class PetItemInventoryMqProducer {
12
13      @Autowired
14      private RocketMQProducer rocketMQProducer;
15
16      /**
```

```
17          * 发送普通消息
18          */
19         public void sendNormalMessage(PetItemInventroy data,String tag) {
20             rocketMQProducer.send(new NotifyEvent(PetNotifyEnum.PetItemInventroy.Topic.PET_ITEM_INVENTROY_CHANGE, tag, data));
21         }
22
23         /**
24          * 发送有序消息
25          */
26         public void sendOrderlyMessage(PetItemInventroy data,String tag) {
27
28             rocketMQProducer.send(new NotifyEvent(PetNotifyEnum.PetItemInventroy.Topic.PET_ITEM_INVENTROY_CHANGE, tag, data)
29                     .setQueueSelector((queueSize, event) -> {
30                         PetItemInventroy body = (PetItemInventroy) event.getBody();
31                         Long queue = body.getId().longValue() % Long.valueOf(queueSize);
32                         return queue.intValue();
33                     }));
34
35         }
36     }
```

Step2. 修改 PetItemInventoryMqConsumer，代码如下。

增加 @NotifyListener (consumerType= ConsumerType.ORDERLY) 的注解。

```
12     @Component
13     //@NotifyListener(topic = PetNotifyEnum.PetItemInventroy.Topic.PET_ITEM_INVENTROY_CHANGE, tags = "*")
14     @NotifyListener(topic = PetNotifyEnum.PetItemInventroy.Topic.PET_ITEM_INVENTROY_CHANGE, tags = "*",consumerType= ConsumerType.ORDERLY)
15     @Slf4j
16     public class PetItemInventoryMqConsumer implements NotifyEventListener {
17         @Override
18         public void consumer(NotifyEvent event) {
19             PetItemInventroy petItemInventroy = (PetItemInventroy)event.getBody();
20             log.info("petItemInventroy的消息, 库存数量为:" + petItemInventroy.getQuantity());
21         }
22     }
```

Step3. 修改 PetItemInventroyAction，代码如下。

调用 petItemInventoryMqProducer 的顺序消息发送接口。

```
1      package pro.shushi.pamirs.demo.core.action;
2      ……引入依赖类
3      @Model.model(PetItemInventroy.MODEL_MODEL)
4      @Component
5      public class PetItemInventroyAction {
6
7          @Autowired
8          private PetItemInventoryMqProducer petItemInventoryMqProducer;
9
10         @Function.Advanced(type= FunctionTypeEnum.UPDATE)
11         @Function.fun(FunctionConstants.update)
12         @Function(openLevel = {FunctionOpenEnum.API})
13         public PetItemInventroy update(PetItemInventroy data){
14             ……其他代码
15     //        petItemInventoryMqProducer.sendNormalMessage(data, PetNotifyEnum.PetItemInventroy.Tag.UPDATE);
16             petItemInventoryMqProducer.sendOrderlyMessage(data, PetNotifyEnum.PetItemInventroy.Tag.UPDATE);
17             return data;
18         }
19     }
```

Step4. 重启看效果。

同常规消息。

（3）事务消息（举例）。

注：这种写法默认第一次是 UNKNOWN，然后通过 MQ 回调二次确认，在消息时效性要求高的场景下是不符合要求的。对实效性要求高的，请见"事务消息——优化（举例）"章节。

Step1. 新建 NotifyTransactionListener，代码如下。

```
@Component
@Slf4j
@TransactionListener
public class PetItemInventoryMqTransactionListener implements NotifyTransactionListener {

    @Override
    public NotifyTransactionState checkLocalTransaction(NotifyEvent event) {
        log.info("消息回滚");
        return NotifyTransactionState.ROLLBACK;
    }
}
```

Step2. 修改 PetItemInventoryMqProducer，代码如下。

增加发送事务消息的方法，通过 .setIsTransaction (true) 来显示设置该消息是事务消息，通过 setGroup ("petItemInventoryMqTransactionListener") 来匹配事务监听处理器。

```
/**
 * 发送事务消息
 */
public void sendTransactionMessage(PetItemInventroy data,String tag) {
    rocketMQProducer.send(new NotifyEvent(PetNotifyEnum.PetItemInventroy.Topic.PET_ITEM_INVENTROY_CHANGE, tag, data)
            .setQueueSelector((queueSize, event) -> {
                PetItemInventroy body = (PetItemInventroy) event.getBody();
                Long queue = body.getId().longValue() % Long.valueOf(queueSize);
                return queue.intValue();
            })
            .setIsTransaction(true)
            .setGroup("petItemInventoryMqTransactionListener")
    );
}
```

Step3. 修改 PetItemInventroyAction，代码如下。

调用 petItemInventoryMqProducer 的事务性消息发送接口。

```
package pro.shushi.pamirs.demo.core.action;
......引入依赖类
@Model.model(PetItemInventroy.MODEL_MODEL)
@Component
public class PetItemInventroyAction {

    @Autowired
    private PetItemInventoryMqProducer petItemInventoryMqProducer;

    @Function.Advanced(type= FunctionTypeEnum.UPDATE)
    @Function.fun(FunctionConstants.update)
    @Function(openLevel = {FunctionOpenEnum.API})
    public PetItemInventroy update(PetItemInventroy data){
```

```
14            ……其他代码
15            //petItemInventoryMqProducer.sendNormalMessage(data, PetNotifyEnum.PetIte
   mInventroy.Tag.UPDATE);
16            //petItemInventoryMqProducer.sendOrderlyMessage(data, PetNotifyEnum.PetIt
   emInventroy.Tag.UPDATE);
17            petItemInventoryMqProducer.sendTransactionMessage(data, PetNotifyEnum.Pet
   ItemInventroy.Tag.UPDATE);
18            return data;
19        }
20    }
```

Step4. 重启看效果。

① 编辑商品库存记录，如图 4-45 所示。

图 4-45　编辑商品库存记录

② 查看后端日志是否打印。

消息回滚，没有消费消费的日志，从日志打印时间上看 MQ 回调会有延迟，如图 4-46、图 4-47 所示。

图 4-46　MQ 日志打印（1）

图 4-47　MQ 日志打印（2）

Step5. 修改 NotifyTransactionListener，代码如下。

```
10    @Component
11    @Slf4j
12    @TransactionListener
13    public class PetItemInventoryMqTransactionListener implements NotifyTransactionLi
      stener {
14
15        @Override
16        public NotifyTransactionState checkLocalTransaction(NotifyEvent event) {
17            log.info("消息提交");
18            //正常业务情况下，这里要增加自己的逻辑判断，来确定返回状态值
19            return NotifyTransactionState.COMMIT;
20        }
21    }
22
```

Step6. 重启看效果。

消息提交，并看到消息消费的日志，从日志打印时间上看 MQ 回调会有延迟，如图 4-48、图 4-49 所示。

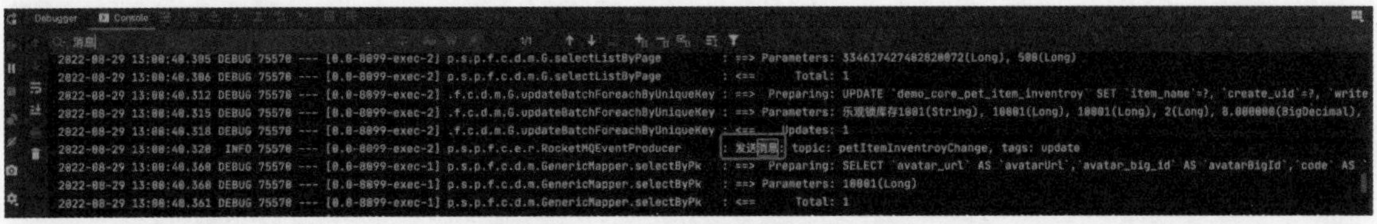

图 4-48 MQ 日志打印（3）

图 4-49 MQ 日志打印（4）

（4）事务消息——优化（举例）。

Step1. 修改 PetItemInventoryMqProducer，代码如下。

修改 PetItemInventoryMqProducer 事务性消息发送方法 sendTransactionMessage，增加 NotifyExecute LocalTransactionCallback 入参。

```
/**
 * 发送事务消息
 */
public void sendTransactionMessage(PetItemInventory data, String tag, NotifyExecu
teLocalTransactionCallback callback) {

    rocketMQProducer.send(new NotifyEvent(PetNotifyEnum.PetItemInventory.Topic.PE
T_ITEM_INVENTROY_CHANGE, tag, data)
            .setQueueSelector((queueSize, event) -> {
                PetItemInventory body = (PetItemInventory) event.getBody();
                Long queue = body.getId().longValue() % Long.valueOf(queueSize);
                return queue.intValue();
            })
            .setIsTransaction(true)
            .setGroup("petItemInventoryMqTransactionListener")
            .setExecuteLocalTransactionCallback(callback)
    );
}
```

Step2. 修改 PetItemInventoryAction，代码如下。

修改 PetItemInventoryAction 的 update 方法。

```
package pro.shushi.pamirs.demo.core.action;
......引入依赖类
@Model.model(PetItemInventroy.MODEL_MODEL)
@Component
public class PetItemInventroyAction {

    @Autowired
    private PetItemInventoryMqProducer petItemInventoryMqProducer;

    @Function.Advanced(type= FunctionTypeEnum.UPDATE)
    @Function.fun(FunctionConstants.update)
    @Function(openLevel = {FunctionOpenEnum.API})
    public PetItemInventroy update(PetItemInventroy data){
        //petItemInventoryMqProducer.sendNormalMessage(data, PetNotifyEnum.PetIte
mInventroy.Tag.UPDATE);
        //petItemInventoryMqProducer.sendOrderlyMessage(data, PetNotifyEnum.PetIt
emInventroy.Tag.UPDATE);
        //     petItemInventoryMqProducer.sendTransactionMessage(data, PetNotifyEnum.P
etItemInventroy.Tag.UPDATE);
```

```
17          petItemInventoryMqProducer.sendTransactionMessage(data, PetNotifyEnum.Pet
   ItemInventroy.Tag.UPDATE,new NotifyExecuteLocalTransactionCallback(){
18              @Override
19              public NotifyTransactionState callback(NotifyTransactionState execute
   State, NotifyEvent event) {
20
21                  List<PetItemInventroy> inventroys = new ArrayList<>();
22                  inventroys.add(data);
23                  PamirsSession.directive().disableOptimisticLocker();
24                  try{
25                      int i = data.updateBatch(inventroys);
26                  } finally {
27                      PamirsSession.directive().enableOptimisticLocker();
28                  }
29                  //业务代码执行完毕，提交消息
30                  return NotifyTransactionState.COMMIT;
31              }
32          });
33          return data;
34      }
35  }
```

Step3. 重启看效果。

编辑商品库存记录，发现消息的处理几乎没有延迟，如图 4-50 所示。

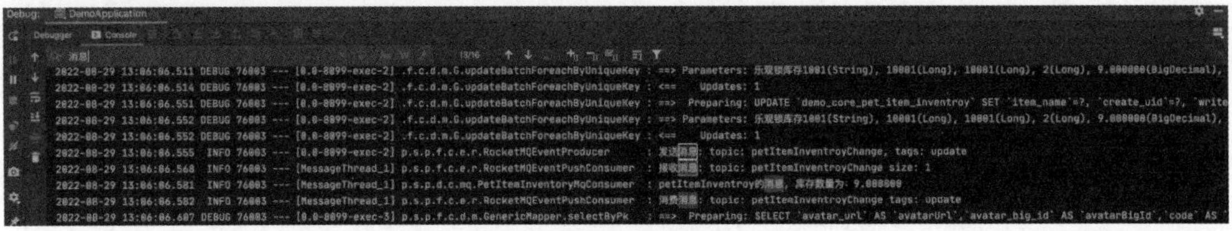

图 4-50　编辑商品库存记录

Step4. 模拟业务成功，消息发送失败。

修改 PetItemInventoryAction 的 update 方法，代码如下。

```
1   @Function.Advanced(type= FunctionTypeEnum.UPDATE)
2   @Function.fun(FunctionConstants.update)
3   @Function(openLevel = {FunctionOpenEnum.API})
4   public PetItemInventory update(PetItemInventory data){
5       //petItemInventoryMqProducer.sendNormalMessage(data, PetNotifyEnum.PetItemInv
    entory.Tag.UPDATE);
6       //petItemInventoryMqProducer.sendOrderlyMessage(data, PetNotifyEnum.PetItemIn
    ventory.Tag.UPDATE);
7       //  petItemInventoryMqProducer.sendTransactionMessage(data, PetNotifyEnum.PetIte
    mInventory.Tag.UPDATE);
8       petItemInventoryMqProducer.sendTransactionMessage(data, PetNotifyEnum.PetItem
    Inventory.Tag.UPDATE,new NotifyExecuteLocalTransactionCallback(){
9
10          @Override
11          public NotifyTransactionState callback(NotifyTransactionState executeStat
    e, NotifyEvent event) {
12
13              List<PetItemInventory> inventroys = new ArrayList<>();
14              inventroys.add(data);
15              PamirsSession.directive().disableOptimisticLocker();
16              try{
17                  int i = data.updateBatch(inventroys);
18              } finally {
19                  PamirsSession.directive().enableOptimisticLocker();
20              }
21              //模拟业务成功，消息发送失败
```

```
22                  throw new RuntimeException();
23                  //业务代码执行完毕，提交消息
24                  //return NotifyTransactionState.COMMIT;
25              }
26          });
27          return data;
28      }
```

Step5. 重启看效果。

这个效果跟第一种事务消息一样，消息提交延迟严重，如图 4-51、图 4-52 所示。

图 4-51　MQ 日志打印（5）

图 4-52　MQ 日志打印（6）

Step6. 总结。

优化后的事务消息，在正常情况下时效性非常高，异常情况下也能通过 NotifyTransactionListener 做保障。

### 4.1.22　框架之分布式缓存

分布式缓存 Oinone 平台主要用到了 Redis，为了让业务研发时可以无感使用 RedisTemplate 和 StringRedisTemplate，已经提前注册好了 RedisTemplate 和 StringRedisTemplate，而且内部会自动处理相关特殊逻辑以应对多租户环境，小伙伴们不能自己重新定义 Redis 的相关 bean。

（1）配置说明。

分布式缓存配置说明如下所示。

```
 1  spring:
 2    redis:
 3      database: 0
 4      host: 127.0.0.1
 5      port: 6379
 6      timeout: 2000
 7  #    cluster:
 8  #      nodes:
 9  #        - 127.0.0.1:6379
10  #      timeout: 2000
11  #      max-redirects: 7
12      jedis:
13        pool:
14          # 连接池中的最大空闲连接  默认8
15          max-idle: 8
16          # 连接池中的最小空闲连接  默认0
17          min-idle: 0
18          # 连接池最大连接数  默认8，负数表示没有限制
19          max-active: 8
20          # 连接池最大阻塞等待时间（使用负值表示没有限制）  默认-1
21          max-wait: -1
```

（2）代码示例。

代码示例如下所示。

```
7   @Component
8   public class Test {
9
10      @Autowired
11      private RedisTemplate redisTemplate;
12      @Autowired
13      private StringRedisTemplate stringRedisTemplate
14
15  }
```

## 4.1.23　框架之信息传递

在 4.1.13、3.4.1、4.2.3、3.3.4 等章节中，都用到了 PamirsSession.getMessageHub () 来设置返回信息，基本上都是在传递后端逻辑判断是异常的信息，而且在系统报错时也会通过它来返回错误信息，前端接收到错误信息则会以提示框的方式进行错误提示。其实后端除了可以返回错误信息以外，还可以返回调试、告警、成功、信息级别等的信息给前端。但是默认情况下前端只提示错误信息，可以通过前端的统一配置放开提示级别，这有点类似后端的日志级别。

**1. 不同信息类型的举例**

Step1. 新建 PetTypeAction。

借用 PetType 模型的表格页作为信息传递的测试入口，为 PetType 模型新增一个 ServerAction，代码如下，在代码中对信息的所有类型进行模拟。

```
14  @Model.model(PetType.MODEL_MODEL)
15  @Component
16  public class PetTypeAction {
17
18      @Action(displayName = "消息",bindingType = ViewTypeEnum.TABLE,contextType = Ac
    tionContextTypeEnum.CONTEXT_FREE)
19      public PetType message(PetType data){
20          PamirsSession.getMessageHub().info("info1");
21          PamirsSession.getMessageHub().info("info2");
22          PamirsSession.getMessageHub().error("error1");
23          PamirsSession.getMessageHub().error("error2");
24          PamirsSession.getMessageHub().msg(new Message().msg("success1").setLevel
    (InformationLevelEnum.SUCCESS));
25          PamirsSession.getMessageHub().msg(new Message().msg("success2").setLevel
    (InformationLevelEnum.SUCCESS));
26          PamirsSession.getMessageHub().msg(new Message().msg("debug1").setLevel(In
    formationLevelEnum.DEBUG));
27          PamirsSession.getMessageHub().msg(new Message().msg("debug2").setLevel(In
    formationLevelEnum.DEBUG));
28          PamirsSession.getMessageHub().msg(new Message().msg("warn1").setLevel(Inf
    ormationLevelEnum.WARN));
29          PamirsSession.getMessageHub().msg(new Message().msg("warn2").setLevel(Inf
    ormationLevelEnum.WARN));
30          return data;
31      }
32  }
33
```

Step2.（前端）修改提示级别。

在项目初始化时使用 CLI 构建初始化前端工程，在 src/middleware 有拦截器的默认情况下，修改信息提示的默认级别为 ILevel.SUCCESS，如图 4-53、图 4-54 所示。

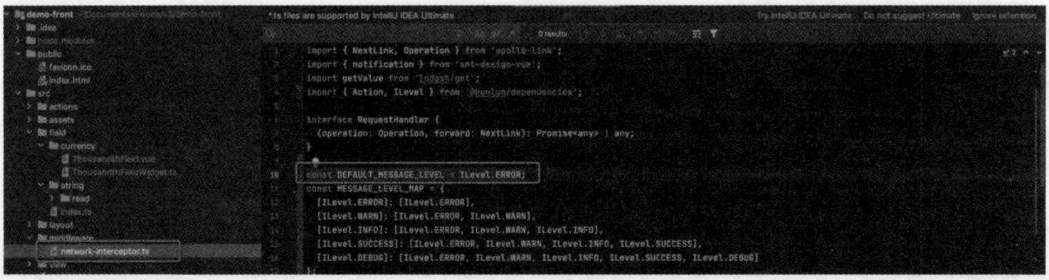

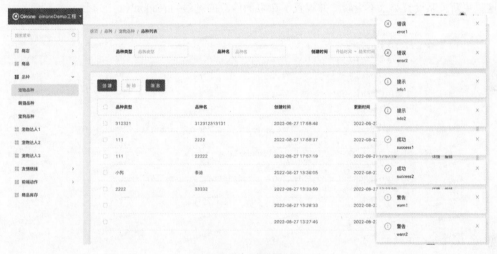

图 4-53　（前端）修改提示级别（1）

```
const DEFAULT_MESSAGE_LEVEL = ILevel.SUCCESS;
```

图 4-54　（前端）修改提示级别（2）

Step3. 重启系统看效果。

从页面效果中看到已经不再只提示错误信息。从协议端看错误级别的信息是在 errors 下，其他级别的信息是在 extensions 下，如图 4-55、图 4-56 所示。

图 4-55　系统提示


```
▾{data: {petTypeMutation: {message: {}}}, errors: [{extensions: {,…}}],…}
 ▸data: {petTypeMutation: {message: {}}}
 ▾errors: [{extensions: {,…}}]
  ▾0: {extensions: {,…}}
   ▾extensions: {,…}
     classification: "DataFetchingException"
    ▾messages: [{errorType: "BIZ_ERROR", level: "ERROR", message: "error1", path: ["petTypeMutation", "message"]},…]
     ▸0: {errorType: "BIZ_ERROR", level: "ERROR", message: "error1", path: ["petTypeMutation", "message"]}
     ▸1: {errorType: "BIZ_ERROR", level: "ERROR", message: "error2", path: ["petTypeMutation", "message"]}
 ▾extensions: {success: false, messages: [{level: "INFO", message: "info1", path: ["petTypeMutation", "message"]},…],…}
  ▸dataloader: {,…}
  ▾messages: [{level: "INFO", message: "info1", path: ["petTypeMutation", "message"]},…]
   ▸0: {level: "INFO", message: "info1", path: ["petTypeMutation", "message"]}
   ▸1: {level: "INFO", message: "info2", path: ["petTypeMutation", "message"]}
   ▸2: {level: "SUCCESS", message: "success1", path: ["petTypeMutation", "message"]}
   ▸3: {level: "SUCCESS", message: "success2", path: ["petTypeMutation", "message"]}
   ▸4: {level: "DEBUG", message: "debug1", path: ["petTypeMutation", "message"]}
   ▸5: {level: "DEBUG", message: "debug2", path: ["petTypeMutation", "message"]}
   ▸6: {level: "WARN", message: "warn1", path: ["petTypeMutation", "message"]}
   ▸7: {level: "WARN", message: "warn2", path: ["petTypeMutation", "message"]}
  success: false
```


图 4-56　系统提示的请求返回结果

**2. MessageHub 的其他说明**

在实现上看 MessageHub 是基于 GQL 协议，前后端都有配套实现。同时前端还提供了订阅 MessageHub 的信息功能，以满足前端更多交互要求，前端 MessageHub 提供的订阅能力使用教程请见前端高级特性之 4.2.2 节。

## 4.1.24 框架之分库分表

随着数据库技术的发展如分区设计、分布式数据库等，业务层的分库分表的技术终将成为老一辈程序员的回忆，谈笑间既羡慕又自吹地说道"现在的研发真简单，连分库分表都不需要考虑了"。既然这样为什么要写这篇文章呢？因为现今的数据库虽能解决大部分场景的数据量问题，但涉及核心业务数据真到过亿数据后性能加速降低，能拿出的方案都还有一定的局限性，或者说性价比不高。相对于性价比比较高的分库分表，也会是现阶段一种不错的补充。言归正传，Oinone 的分库分表方案是基于 Sharding-JDBC 的整合方案，所以大家得先具备一点 Sharding-JDBC 的知识。

**1. 分表（举例）**

做分库分表前，大家要有一个明确注意的点就是分表字段的选择，它是非常重要的，与业务场景非常相关。在明确了分库分表字段以后，甚至在功能上都要做一些妥协。比如分库分表字段在查询管理中作为查询条件是必须带上的，否则效率只会更低。

Step1. 新建 ShardingModel 模型，代码如下。

ShardingModel 模型是用于分表测试的模型，我们选定 userId 作为分表字段。分表字段不允许更新，所以这里更新策略设置类永不更新，并设置了在页面修改的时候为 readonly。

```
10    @Model.model(ShardingModel.MODEL_MODEL)
11    @Model(displayName = "分表模型",summary="分表模型",labelFields ={"name"} )
12    public class ShardingModel extends AbstractDemoIdModel {
13        public static final String MODEL_MODEL="demo.ShardingModel";
14
15        @Field(displayName = "名称")
16        private String name;
17
18        @Field(displayName = "用户id",summary = "分表字段",immutable=true/* 不可修改 *
    */)
19        @UxForm.FieldWidget(@UxWidget(readonly = "scene == 'redirectUpdatePage'"/* 在
    编辑页面只读 **/ ))
20        @Field.Advanced(updateStrategy = FieldStrategyEnum.NEVER)
21        private Long userId;
22
23    }
```

Step2. 配置分表策略。

（1）配置 ShardingModel 模型走分库分表的数据源 pamirsSharding。

（2）为 pamirsSharding 配置数据源以及 sharding 规则。

① pamirs.sharding.define 用于 Oinone 的数据库表创建用。

② pamirs.sharding.rule 用于分表规则配置。

分库分表配置如图 4-57、图 4-58 所示。

```
1    pamirs:
2      load:
3        sessionMode: true
4      framework:
5        system:
6          system-ds-key: base
7          system-models:
8            - base.WorkerNode
9        data:
10         default-ds-key: pamirs
11         ds-map:
12           base: base
13         modelDsMap:
14           "[demo.ShardingModel]": pamirsSharding  #配置模型对应的库
```

图 4-57　分库分表配置（1）

```
 1  pamirs:
 2    sharding:
 3      define:
 4        data-sources:
 5          ds: pamirs
 6          pamirsSharding: pamirs #申明pamirsSharding库对应的pamirs数据源
 7        models:
 8          "[trigger.PamirsSchedule]":
 9            tables: 0..13
10          "[demo.ShardingModel]":
11            tables: 0..7
12            table-separator: _
13      rule:
14        pamirsSharding:  #配置pamirsSharding库的分库分表规则
15          actual-ds:
16            - pamirs   #申明pamirsSharding库对应的pamirs数据源
17          sharding-rules:
18            # Configure sharding rule，以下配置跟sharding-JDBC配置一致
19            - tables:
20                demo_core_sharding_model: #demo_core_sharding_model表规则配置
21                  actualDataNodes: pamirs.demo_core_sharding_model_${...}
22                  tableStrategy:
23                    standard:
24                      shardingColumn: user_id
25                      shardingAlgorithmName: table_inline
26                  shardingAlgorithms:
27                    table_inline:
28                      type: INLINE
29                      props:
30                        algorithm-expression: demo_core_sharding_model_${(Long.valueOf(user_id) % 8)}
31              props:
32                sql.show: true
```

图 4-58　分库分表配置（2）

Step3. 配置测试入口，代码如下。

修改 DemoMenus 类增加一行代码，为测试提供入口。

```
1  @UxMenu("分表模型")@UxRoute(ShardingModel.MODEL_MODEL) class ShardingModelMenu{}
```

Step4. 重启看效果。

（1）自行尝试增删改查，如图 4-59 所示。

图 4-59　自行尝试增删改查

（2）观察数据库表与数据分布，如图 4-60 所示。

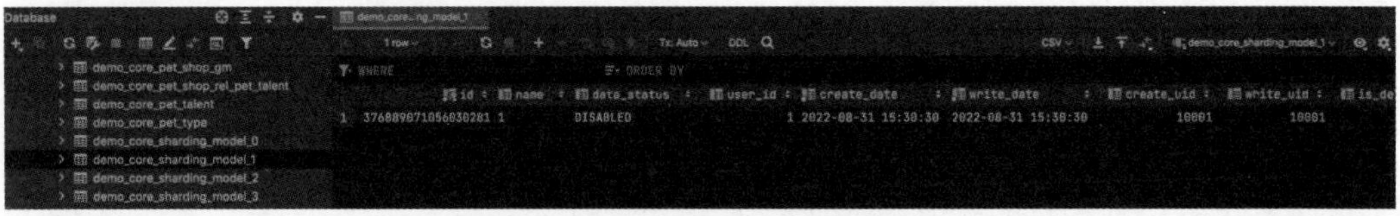

图 4-60　观察数据库表与数据分布

**2. 分库分表（举例）**

Step1. 新建 ShardingModel2 模型，代码如下。

ShardingModel2 模型用于分库分表测试模型，我们选定 userId 作为分表字段。分库分表字段不允许更新，所以这里更新策略设置类永不更新，并设置了在页面修改的时候为 readonly。

```java
10    @Model.model(ShardingModel2.MODEL_MODEL)
11    @Model(displayName = "分库分表模型",summary="分库分表模型",labelFields ={"name"} )
12    public class ShardingModel2 extends AbstractDemoIdModel {
13        public static final String MODEL_MODEL="demo.ShardingModel2";
14
15        @Field(displayName = "名称")
16        private String name;
17
18
19        @Field(displayName = "用户id",summary = "分库分表字段",immutable=true/* 不可修改 **/)
20        @UxForm.FieldWidget(@UxWidget(readonly = "scene == 'redirectUpdatePage'"/* 在编辑页面只读 **/ ))
21        @Field.Advanced(updateStrategy = FieldStrategyEnum.NEVER)
22        private Long userId;
23
24    }
```

Step2. 配置分库分表策略。

（1）配置 ShardingModel2 模型走分库分表的数据源 testShardingDs。

（2）新增两个数据库配置：testShardingDs_0、testShardingDs_1。

（3）为 testShardingDs 配置数据源以及 sharding 规则。

① pamirs.sharding.define 用于 Oinone 的数据库表创建用。

② pamirs.sharding.rule 用于分库分表规则配置。

分库分表配置如图 4-61 ～图 4-63 所示。

```yaml
1   pamirs:
2     load:
3       sessionMode: true
4     framework:
5       system:
6         system-ds-key: base
7         system-models:
8           - base.WorkerNode
9       data:
10        default-ds-key: pamirs
11        ds-map:
12          base: base
13        modelDsMap:
14          "[demo.ShardingModel]": pamirsSharding  #配置模型对应的库
15          "[demo.ShardingModel2]": testShardingDs  #配置模型对应的库
```

图 4-61　分库分表配置（3）

```yaml
pamirs:
  datasource:
    pamirs:
      driverClassName: com.mysql.cj.jdbc.Driver
      type: com.alibaba.druid.pool.DruidDataSource
      url: jdbc:mysql://127.0.0.1:3306/demo2?useSSL=false&allowPublicKeyRetrieval=true&useServerPrepStmts=true&cachePrepStmts=true&useUnicode=true&characterEncoding=utf8&serverTimezone=Asia/Shanghai&autoReconnect=true&allowMultiQueries=true
      username: root
      password: oinone
      initialSize: 5
      maxActive: 200
      minIdle: 5
      maxWait: 60000
      timeBetweenEvictionRunsMillis: 60000
      testWhileIdle: true
      testOnBorrow: false
      testOnReturn: false
      poolPreparedStatements: true
      asyncInit: true
    base:
      driverClassName: com.mysql.cj.jdbc.Driver
      type: com.alibaba.druid.pool.DruidDataSource
      url: jdbc:mysql://127.0.0.1:3306/demo2_base?useSSL=false&allowPublicKeyRetrieval=true&useServerPrepStmts=true&cachePrepStmts=true&useUnicode=true&characterEncoding=utf8&serverTimezone=Asia/Shanghai&autoReconnect=true&allowMultiQueries=true
      username: root
      password: oinone
      initialSize: 5
      maxActive: 200
      minIdle: 5
      maxWait: 60000
      timeBetweenEvictionRunsMillis: 60000
      testWhileIdle: true
      testOnBorrow: false
      testOnReturn: false
      poolPreparedStatements: true
      asyncInit: true
    testShardingDs_0:
      driverClassName: com.mysql.cj.jdbc.Driver
      type: com.alibaba.druid.pool.DruidDataSource
      url: jdbc:mysql://127.0.0.1:3306/demo2_testShardingDs_0?useSSL=false&allowPublicKeyRetrieval=true&useServerPrepStmts=true&cachePrepStmts=true&useUnicode=true&characterEncoding=utf8&serverTimezone=Asia/Shanghai&autoReconnect=true&allowMultiQueries=true
      username: root
      password: oinone
      initialSize: 5
      maxActive: 200
      minIdle: 5
      maxWait: 60000
      timeBetweenEvictionRunsMillis: 60000
      testWhileIdle: true
      testOnBorrow: false
      testOnReturn: false
      poolPreparedStatements: true
      asyncInit: true
    testShardingDs_1:
      driverClassName: com.mysql.cj.jdbc.Driver
      type: com.alibaba.druid.pool.DruidDataSource
      url: jdbc:mysql://127.0.0.1:3306/demo2_testShardingDs_1?useSSL=false&allowPublicKeyRetrieval=true&useServerPrepStmts=true&cachePrepStmts=true&useUnicode=tru
```

图 4-62 分库分表配置（4）

```
e&characterEncoding=utf8&serverTimezone=Asia/Shanghai&autoReconnect=true&allowMul
tiQueries=true
55          username: root
56          password: oinone
57          initialSize: 5
58          maxActive: 200
59          minIdle: 5
60          maxWait: 60000
61          timeBetweenEvictionRunsMillis: 60000
62          testWhileIdle: true
63          testOnBorrow: false
64          testOnReturn: false
65          poolPreparedStatements: true
66          asyncInit: true
```

图 4-62 （续）

```
1   pamirs:
2     sharding:
3       define:
4         data-sources:
5           ds: pamirs
6           pamirsSharding: pamirs    #申明pamirsSharding库对应的pamirs数据源
7           testShardingDs:           #申明testShardingDs库对应的testShardingDs_0\1数据源
8             - testShardingDs_0
9             - testShardingDs_1
10        models:
11          "[trigger.PamirsSchedule]":
12            tables: 0..13
13          "[demo.ShardingModel]":
14            tables: 0..7
15            table-separator: _
16          "[demo.ShardingModel2]":
17            ds-nodes: 0..1          #申明testShardingDs库对应的建库规则
18            ds-separator: _
19            tables: 0..7
20            table-separator: _
21      rule:
22        pamirsSharding:  #配置pamirsSharding库的分库分表规则
23          actual-ds:
24            - pamirs    #申明pamirsSharding库对应的pamirs数据源
25          sharding-rules:
26            # Configure sharding rule, 以下配置跟sharding-JDBC配置一致
27            - tables:
28                demo_core_sharding_model:
29                  actualDataNodes: pamirs.demo_core_sharding_model_${0..7}
30                  tableStrategy:
31                    standard:
32                      shardingColumn: user_id
33                      shardingAlgorithmName: table_inline
34                shardingAlgorithms:
35                  table_inline:
36                    type: INLINE
37                    props:
38                      algorithm-expression: demo_core_sharding_model_${(Long.valueOf
    (user_id) % 8)}
39              props:
40                sql.show: true
41        testShardingDs: #配置testShardingDs库的分库分表规则
42          actual-ds: #申明testShardingDs库对应的pamirs数据源
43            - testShardingDs_0
```

图 4-63 分库分表配置（5）

```yaml
        - testShardingDs_1
    sharding-rules:
      # Configure sharding rule, 以下配置跟sharding-JDBC配置一致
      - tables:
          demo_core_sharding_model2:
            actualDataNodes: testShardingDs_${0..1}.demo_core_sharding_model2_${0..7}
            databaseStrategy:
              standard:
                shardingColumn: user_id
                shardingAlgorithmName: ds_inline
            tableStrategy:
              standard:
                shardingColumn: user_id
                shardingAlgorithmName: table_inline
        shardingAlgorithms:
          table_inline:
            type: INLINE
            props:
              algorithm-expression: demo_core_sharding_model2_${(Long.valueOf(user_id) % 8)}
          ds_inline:
            type: INLINE
            props:
              algorithm-expression: testShardingDs_${(Long.valueOf(user_id) % 2)}
    props:
      sql.show: true
```

图 4-63 （续）

Step3. 配置测试入口，代码如下。

修改 DemoMenus 类增加一行代码，为测试提供入口。

```
@UxMenu("分库分表模型")@UxRoute(ShardingModel2.MODEL_MODEL) class ShardingModel2Menu {}
```

Step4. 重启看效果。

（1）自行尝试增删改查，如图 4-64 所示。

图 4-64　自行尝试增删改查

（2）观察数据库表与数据分布，如图 4-65 所示。

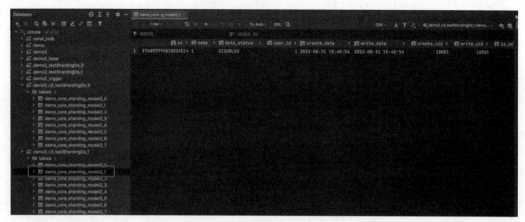

图 4-65　观察数据库表与数据分布

### 4.1.25　框架之搜索引擎

**1. 使用场景**

在碰到大数据量并且需要全文检索的场景时，我们在分布式架构中基本会架设 ElasticSearch 来作为一个常规解决方案。在 Oinone 体系中增强模型就是应对这类场景，其背后也是整合了 ElasticSearch。

**2. 整体介绍**

Oinone 与 es 整合设计如图 4-66 所示。

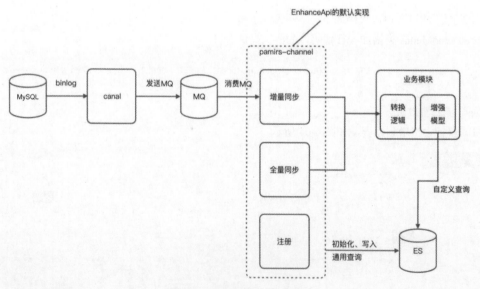

图 4-66　Oinone 与 es 整合设计

1）基础环境安装

（1）Canal 安装。

参见 4.1.10 节。

（2）修改 Canal 配置并重启，代码如下。

新增 Canal 的实例 destinaion: pamirs，监听分表模型的 binlog-filter: demo\.demo_core_sharding_Model……用于增量同步。

```yaml
pamirs:
  middleware:
    data-source:
      jdbc-url: jdbc:mysql://localhost:3306/canal_tsdb?useUnicode=true&characterEncoding=utf-8&verifyServerCertificate=false&useSSL=false&requireSSL=false
      driver-class-name: com.mysql.cj.jdbc.Driver
      username: root
      password: oinone
    canal:
      ip: 127.0.0.1
      port: 1111
      metricsPort: 1112
      zkClusters:
        - 127.0.0.1:2181
      destinations:
        - destinaion: pamirschangedata
          name: pamirschangedata
          desc: pamirschangedata
          slaveId: 1235
          filter: demo\.demo_core_pet_talent
          dbUserName: root
          dbPassword: oinone
          memoryStorageBufferSize: 65536
          topic: CHANGE_DATA_EVENT_TOPIC
          dynamicTopic: false
          dbs:
            - { address: 127.0.0.1, port: 3306 }
        - destinaion: pamirs
          id: 1234
          name: pamirs
          desc: pamirs
          slaveId: 1234
          filter: demo\.demo_core_sharding_model_0,demo\.demo_core_sharding_model_1,demo\.demo_core_sharding_model_2,demo\.demo_core_sharding_model_3,demo\.demo_core_sharding_model_4,demo\.demo_core_sharding_model_5,demo\.demo_core_sharding_model_6,demo\.demo_core_sharding_model_7
          dbUserName: root
          dbPassword: oinone
          memoryStorageBufferSize: 65536
          topic: BINLOG_EVENT_TOPIC
          dynamicTopic: false
          dbs:
            - { address: 127.0.0.1, port: 3306 }
      tsdb:
        enable: true
        jdbcUrl: "jdbc:mysql://127.0.0.1:3306/canal_tsdb"
        userName: root
        password: oinone
      mq: rocketmq
      rocketmq:
        namesrv: 127.0.0.1:9876
        retryTimesWhenSendFailed: 5
dubbo:
  application:
    name: canal-server
    version: 1.0.0
  registry:
    address: zookeeper://127.0.0.1:2181
  protocol:
    name: dubbo
    port: 20881
  scan:
    base-packages: pro.shushi
server:
  address: 0.0.0.0
  port: 10010
  sessionTimeout: 3600
```

(3) ES 安装。

① 下载安装包或直接下载,下载后去除后缀 .txt,然后解压文件(下载详情见本书末"附录")。

② 替换安装目录 /config 下的 elasticsearch.yml,主要是文件中追加了三个配置,代码如下。

```
1  xpack.security.enabled: false
2  xpack.security.http.ssl.enabled: false
3  xpack.security.transport.ssl.enabled: false
```

③ 启动。

导入环境变量,代码如下(ES 运行时需要 JDK18 及以上版本 JDK 运行环境,ES 安装包中包含了一个 JDK18 版本)。

```
1  # export JAVA_HOME=/Users/oinone/Documents/oinone/es/elasticsearch-8.4.1/jdk.app/Contents/Home/
2  export JAVA_HOME=ES解压安装目录/jdk.app/Contents/Home/
```

运行 ES,代码如下。

```
1  ## nohup /Users/oinone/Documents/oinone/es/elasticsearch-8.4.1/bin/elasticsearch >> $TMPDIR/elastic.log 2>&1 &
2  nohup ES安装目录/bin/elasticsearch >> $TMPDIR/elastic.log 2>&1 &
```

④ 停止 ES,代码如下。

```
1  lsof -nP -iTCP:9300 -sTCP:LISTEN | grep java | awk '{print $2;}' | xargs kill
```

(4) ES 控制台安装。

① 下载文件,下载后去除后缀 .txt,然后解压文件,解压之后执行 bin/cerebro 或者 bin/cerebro.bat(Windows)(下载详情见本书末"附录")。

② 进入 cerebro-0.9.4/bin 目录下执行以下命令。

```
1  ./cerebro
```

③ 启动后访问地址:http://localhost:9000,如图 4-67 所示。

图 4-67　启动后访问地址:http://localhost:9000

④ Node address 填入 ES 服务地址，如图 4-68 所示。

图 4-68　Cerebro 连接 ES 成功

恭喜 ES 环境搭建完毕，我们开始进行学习。

**3. 第一个增强模型（举例）**

Step1. 相关依赖包引入 boot 工程。

（1）boot 工程需要指定 ES 客户端包版本，不指定版本会隐性依赖顶层 spring-boot 依赖管理指定的低版本。

（2）boot 工程加入 pamris-channel 的工程依赖，代码如下。

```xml
<dependency>
    <groupId>org.elasticsearch.client</groupId>
    <artifactId>elasticsearch-rest-client</artifactId>
    <version>8.4.1</version>
</dependency>
<dependency>
    <groupId>jakarta.json</groupId>
    <artifactId>jakarta.json-api</artifactId>
    <version>2.1.1</version>
</dependency>

<dependency>
    <groupId>pro.shushi.pamirs.core</groupId>
    <artifactId>pamirs-channel-core</artifactId>
</dependency>
```

Step2. 在 pamirs-demo-api 中增加 pamirs-channel-api 的依赖，代码如下。

```xml
<dependency>
    <groupId>pro.shushi.pamirs.core</groupId>
    <artifactId>pamirs-channel-api</artifactId>
</dependency>
```

Step3. 修改 application-dev.yml，代码如下。

在 pamirs-demo-boot 的 application-dev.yml 文件中增加配置 pamirs.boot.modules 增加 channel，即在启动模块中增加 channel 模块。同时注意 ES 的配置，是否跟 ES 的服务一致。

```yaml
pamirs:
  boot:
    modules:
      - channel
    elastic:
      url: 127.0.0.1:9200
```

Step4. DemoModule 增加对 FileModule 的依赖，代码如下。

```java
@Module(dependencies = {ChannelModule.MODULE_MODULE})
```

Step5. 为 ShardingModel 新建一个增强模型，代码如下。

在大数据量的情况下，我们经常会通过 ES 来提高查询速度。

```
10    @Model(displayName = "测试EnhanceModel")
11    @Model.model(ShardingModelEnhance.MODEL_MODEL)
12    @Model.Advanced(type = ModelTypeEnum.PROXY, inherited = {EnhanceModel.MODEL_MODE
      L})
13    @Enhance(shards = "3", replicas = "1", reAlias = true,increment= IncrementEnum.OP
      EN)
14  ▸ public class ShardingModelEnhance extends ShardingModel {
15        public static final String MODEL_MODEL="demo.ShardingModelEnhance";
16
17
18    }
19
```

Step6. 为增强模型增加菜单入口，代码如下。

```
1   @UxMenu("增强模型")@UxRoute(ShardingModelEnhance.MODEL_MODEL) class ShardingModelEn
    hanceMenu{}
```

Step7. 重启系统看效果。

（1）第一次访问页面会报错，因为针对该增强模型的相关初始化工作还未完成如图 4-69 所示。

图 4-69　首次访问页面报错

（2）进入传输增强模型应用，访问增强模型列表我们会发现一条记录，并单击"全量同步"初始化 ES，并全量 dump 数据，如图 4-70 所示。

图 4-70　全量 dump 数据

（3）再次回到 Demo 应用，进入增强模型页面，可以正常访问并进行增删改查操作，如图 4-71 所示。

图 4-71　再次回到 Demo 应用

对数据进行修改,我们可以看到如图 4-72 所示日志。

图 4-72　数据修改后的日志

**4. 个性化 dump 逻辑**

有时候我们 dump 逻辑是有个性化需求,那么我们可以重写模型的 synchronize 方法,函数重写特性我们在 3.4.3 节中已经有详细介绍。

Step1. 重写 ShardingModelEnhance 模型的 synchronize 方法,代码如下。

下面例子中我们给 ShardingModelEnhance 模型增加一个 nick 字段,在同步搜索引擎的时候把 name 赋值给 nick 字段。

```java
@Model(displayName = "测试EnhanceModel")
@Model.model(ShardingModelEnhance.MODEL_MODEL)
@Model.Advanced(type = ModelTypeEnum.PROXY, inherited = {EnhanceModel.MODEL_MODEL})
@Enhance(shards = "3", replicas = "1", reAlias = true,increment= IncrementEnum.OPEN)
public class ShardingModelEnhance extends ShardingModel {
    public static final String MODEL_MODEL="demo.ShardingModelEnhance";

    @Field(displayName = "nick")
    private String nick;

    @Function.Advanced(displayName = "同步数据", type = FunctionTypeEnum.UPDATE)
    @Function(summary = "数据同步函数")
    public List<ShardingModelEnhance> synchronize(List<ShardingModelEnhance> data) {
        for(ShardingModelEnhance shardingModelEnhance:data){
            shardingModelEnhance.setNick(shardingModelEnhance.getName());
        }
        return data;
    }
}
```

Step2. 重启应用看效果。

我们修改记录数据,可以看到 nick 字段也自动赋值了。如果针对老数据记录,我们需要把新增的字段都自动填充，可以进入传输增强模型应用，访问增强模型列表，找到对应的记录并单击"全量同步"，如图 4-73 所示。

图 4-73　重启应用看效果

**5. 给搜索增加个性化逻辑**

如果我们需要在查询方法中增加逻辑，在前面的教程中一般是重写 queryPage 函数，但对于增强模型我们需要重写的是 search 函数。

Step1. 重写 ShardingModelEnhance 模型的 search 方法。

```
 1  @Function(
 2          summary = "搜索函数",
 3          openLevel = {FunctionOpenEnum.LOCAL, FunctionOpenEnum.REMOTE, FunctionOpenEnum.API}
 4  )
 5  @pro.shushi.pamirs.meta.annotation.Function.Advanced(
 6          type = {FunctionTypeEnum.QUERY},
 7          category = FunctionCategoryEnum.QUERY_PAGE,
 8          managed = true
 9  )
10  public Pagination<ShardingModelEnhance> search(Pagination<ShardingModelEnhance> page, IWrapper<ShardingModelEnhance> queryWrapper) {
11      System.out.println("您的个性化搜索逻辑");
12      return ((IElasticRetrieve) CommonApiFactory.getApi(IElasticRetrieve.class)).search(page, queryWrapper);
13  }
```

Step2. 重启应用看效果，如图 4-74 所示。

图 4-74　重启应用看效果

## 4.2 前端高级特性

### 4.2.1 组件之生命周期

组件生命周期的意义所在：比如动态创建了视图、字段，等它们初始化完成或者发生了修改后要执行业务逻辑，这个时候只能去自定义当前字段或者视图，体验极差，平台应该提供一系列的生命周期，允许其他人调用生命周期的 API 去执行对应的逻辑。

**1. 实现原理**

当用户通过内部 API 去监听某个生命周期的时候，内部会动态地去创建该生命周期，每个生命周期都有唯一标识，内部会根据该唯一标识去创建对应的 Effect，Effect 会根据生命周期的唯一标识实例化一个 lifeCycle，lifeCycle 创建完成后，会被存储到 Heart 中，Heart 是整个生命周期的心脏，当心脏每次跳动的时候（生命周期被监听触发）都会触发对应的生命周期，如图 4-75 所示。

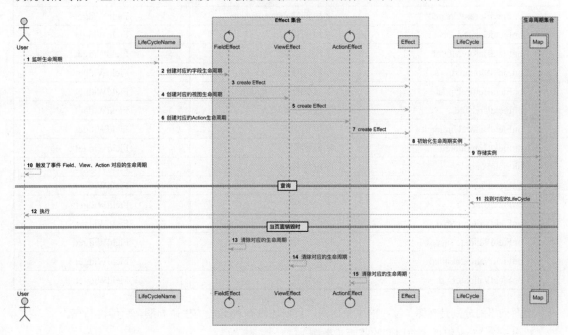

图 4-75 实现原理

**2. 生命周期 API**

生命周期 API 见表 4-43。

表 4-43 生命周期 API

| API | 描述 | 返回值 |
| --- | --- | --- |
| View LifeCycle | | |
| onViewBeforeCreated | 视图创建前 | ViewWidget |
| onViewCreated | 视图创建后 | ViewWidget |
| onViewBeforeMount | 视图挂载前 | ViewWidget |
| onViewMounted | 视图挂载后 | ViewWidget |
| onViewBeforeUpdate | 视图数据发生修改前 | ViewWidget |
| onViewUpdated | 视图数据修改后 | ViewWidget |
| onViewBeforeUnmount | 视图销毁前 | ViewWidget |

续表

| API | 描述 | 返回值 |
|---|---|---|
| onViewUnmounted | 视图销毁 | ViewWidget |
| onViewSubmit | 提交数据 | ViewWidget |
| onViewSubmitStart | 数据开始提交 | ViewWidget |
| onViewSubmitSuccess | 数据提交成功 | ViewWidget |
| onViewSubmitFailed | 数据提交失败 | ViewWidget |
| onViewSubmitEnd | 数据提交结束 | ViewWidget |
| onViewValidateStart | 视图字段校验 | ViewWidget |
| onViewValidateSuccess | 校验成功 | ViewWidget |
| onViewValidateFailed | 校验失败 | ViewWidget |
| onViewValidateEnd | 校验结束 | ViewWidget |
| Field LifeCycle | | |
| onFieldBeforeCreated | 字段创建前 | FieldWidget |
| onFieldCreated | 字段创建后 | FieldWidget |
| onFieldBeforeMount | 字段挂载前 | FieldWidget |
| onFieldMounted | 字段挂载后 | FieldWidget |
| onFieldBeforeUpdate | 字段数据发生修改前 | FieldWidget |
| onFieldUpdated | 字段数据修改后 | FieldWidget |
| onFieldBeforeUnmount | 字段销毁前 | FieldWidget |
| onFieldUnmounted | 字段销毁 | FieldWidget |
| onFieldFocus | 字段聚焦 | FieldWidget |
| onFieldChange | 字段的值发生了变化 | FieldWidget |
| onFieldBlur | 字段失焦 | FieldWidget |
| onFieldValidateStart | 字段开始校验 | FieldWidget |
| onFieldValidateSuccess | 校验成功 | FieldWidget |
| onFieldValidateFailed | 校验失败 | FieldWidget |
| onFieldValidateEnd | 校验结束 | FieldWidget |

上面列出的分别是视图、字段的生命周期，目前还没有 Action 的生命周期，后续再补充。

**3. 第一个 View 组件生命周期的监听（举例）**

Step1. 新建 registryLifeCycle.ts，代码如下。

新建 registryLifeCycle.ts，监听宠物达人的列表页。'宠物达人 table_demo_core' 为视图名，您需要找后端配合。

```typescript
import { onViewCreated } from '@kunlun/dependencies'

function registryLifeCycle(){

    onViewCreated('宠物达人table_demo_core', (viewWidget) => {
        console.log('宠物达人table_demo_core');
        console.log(viewWidget);
    });

}

export {registryLifeCycle}
```

Step2. 修改 main.ts，代码如下。

全局注册 lifeCycle。

```
1  import { registryLifeCycle } from './registryLifeCycle';
2
3  registryLifeCycle();
```

Step3. 看效果，如图 4-76 所示。

图 4-76　实际效果

**4. 第一个 Filed 组件生命周期的监听（举例）**

Step1. 修改 registryLifeCycle.ts，代码如下。

通过 onFieldValueChange 增加宠物达人搜索视图的 name（达人）字段的值变化进行监听。

宠物达人 search:name 代表，视图名：字段名。

```
1  import { onViewCreated , onFieldValueChange} from '@kunlun/dependencies'
2
3  function registryLifeCycle(){
4
5      onViewCreated('宠物达人table_demo_core', (viewWidget) => {
6          console.log('宠物达人table_demo_core');
7          console.log(viewWidget);
8      });
9      onFieldValueChange('宠物达人search:name', (filedWidget) => {
10         console.log('宠物达人search:name');
11         console.log(filedWidget);
12     });
13
14 }
15
16 export {registryLifeCycle}
```

Step2. 看效果。

输入三个 1，执行三次，如图 4-77 所示。

图 4-77　输入三个 1，执行三次

### 4.2.2 框架之 MessageHub

**1. MessageHub**

请求出现异常时,提供"点对点"的通信能力。

**2. 何时使用**

错误提示是用户体验中特别重要的组成部分,大部分的错误体现在整页级别、字段级别、按钮级别。友好的错误提示应该是怎么样的呢?我们假设它是这样的:

(1)与用户操作精密契合。

① 当字段输入异常时,错误展示在错误框底部。

② 按钮触发服务时异常,错误展示在按钮底部。

(2)区分不同的类型。

① 错误;

② 成功;

③ 警告;

④ 提示;

⑤ 调试。

(3)简洁易懂的错误信息。

在 Oinone 平台中,我们怎么做到友好地错误提示呢?接下来介绍我们的 MessageHub,它为自定义错误提示提供无限的可能。

**3. 如何使用**

1)订阅

订阅代码如下所示。

```
1  import { useMessageHub, ILevel } from "@kunlun/dependencies"
2  const messageHub = useMessageHub('当前视图的唯一标识');
3  /* 订阅错误信息 */
4  messageHub.subscribe((errorResult) => {
5    console.log(errorResult)
6  })
7  /* 订阅成功信息 */
8  messageHub.subscribe((errorResult) => {
9    console.log(errorResult)
10 }, ILevel.SUCCESS)
```

2)销毁

销毁代码如下所示。

```
1  /**
2   * 在适当的时机销毁它
3   * 如果页面逻辑运行时都不需要销毁,在页面destroyed是一定要销毁,重要!!!
4   */
5  messageHub.unsubscribe()
```

**4. 实战**

让我们在前文中自定义表单,加入我们的 MessageHub,模拟在表单提交时,后端报错信息在字段下方给予提示。

Step1.(后端)重写 PetType 的创建函数,代码如下。

重写 PetType 的创建函数,在创建逻辑中通过 MessageHub 返回错误信息,返回错误信息的同时要设置 paths 信息方便前端处理。

```java
@Action.Advanced(name = FunctionConstants.create, managed = true)
    @Action(displayName = "确定", summary = "创建", bindingType = ViewTypeEnum.FORM)
    @Function(name = FunctionConstants.create)
    @Function.fun(FunctionConstants.create)
    public PetType create(PetType data){
        List<Object> paths = new ArrayList<>();
        paths.add("demo.PetType");
        paths.add("kind");
        PamirsSession.getMessageHub().msg(new Message().msg("kind error").setPath(paths).setLevel(InformationLevelEnum.ERROR).setErrorType(ErrorTypeEnum.BIZ_ERROR));

        List<Object> paths2 = new ArrayList<>();
        paths2.add("demo.PetType");
        paths2.add("name");
        PamirsSession.getMessageHub().msg(new Message().msg("name error").setPath(paths2).setLevel(InformationLevelEnum.ERROR).setErrorType(ErrorTypeEnum.BIZ_ERROR));
//        data.create();
        return data;
    }
```

Step 2. 修改 PetForm.vue，代码如下。

```
<template>
  <div class="petFormWrapper">
    <form :model="formState" @finish="onFinish">
      <a-form-item label="品种种类" id="name" name="kind" :rules="[{ required: true, message: '请输入品种种类!', trigger: 'focus' }]">
        <a-input v-model:value="formState.kind" @input="(e) => onNameChange(e, 'kind')" />
        <span style="color: red">{{ getServiceError('kind') }}</span>
      </a-form-item>

      <a-form-item label="品种名" id="name" name="name" :rules="[{ required: true, message: '请输入品种名!', trigger: 'focus' }]">
        <a-input v-model:value="formState.name" @input="(e) => onNameChange(e, 'name')" />
        <span style="color: red">{{ getServiceError('name') }}</span>
      </a-form-item>
    </form>
  </div>
</template>

<script lang="ts">
import { defineComponent, reactive } from 'vue';
import { Form } from 'ant-design-vue';

export default defineComponent({
  props: ['onChange', 'reloadData', 'serviceErrors'],
  components: { Form },
  setup(props) {
    const formState = reactive({
      kind: '',
      name: '',
    });

    const onFinish = () => {
      console.log(formState);
```

```
32      };
33
34      const onNameChange = (event, name) => {
35        props.onChange(name, event.target.value);
36      };
37
38      const reloadData = async () => {
39        await props.reloadData();
40      };
41      // 提示服务器异常消息
42      const getServiceError = (name: string) => {
43        const error = props.serviceErrors.find(error => error.name === name);
44        return error ? error.error : '';
45      }
46
47      return {
48        formState,
49        reloadData,
50        onNameChange,
51        onFinish,
52        getServiceError
53      };
54    }
55  });
56 </script>
```

Step3. PetFormViewWidget.ts，代码如下。

```
1  import { constructOne, FormWidget, queryOne, SPI, ViewWidget, Widget, IModel, getModelByUrl, getModel, getIdByUrl, FormWidgetV3, CustomWidget, MessageHub, useMessageHub, ILevel, CallChaining } from '@kunlun/dependencies';
2  import PetFormView from './PetForm.vue';
3
4  @SPI.ClassFactory(CustomWidget.Token({ widget: 'PetForm' }))
5  export class PetFormViewWidget extends FormWidgetV3 {
6    public initialize(props) {
7      super.initialize(props);
8      this.setComponent(PetFormView);
9      return this;
10   }
11
12   /**
13    * 数据提交
14    * @protected
15    */
16   @Widget.Reactive()
17   @Widget.Inject()
18   protected callChaining: CallChaining | undefined;
19
20   private modelInstance!: IModel;
21
22   // MessageHub相关逻辑
23   private messageHub!: MessageHub;
24
25   @Widget.Reactive()
26   private serviceErrors: Record<string, unknown>[] = [];
27
28   /**
29    * 重要！！！！
30    * 当字段改变时修改formData
31    */
32   @Widget.Method()
```

```ts
33    public onFieldChange(fieldName: string, value) {
34      this.setDataByKey(fieldName, value);
35    }
36
37    /**
38     * 表单编辑时查询数据
39     * */
40    public async fetchData(content: Record<string, unknown>[] = [], options: Record<string, unknown> = {}, variables: Record<string, unknown> = {}) {
41      this.setBusy(true);
42      const context: typeof options = { sourceModel: this.modelInstance.model, ...options };
43      const fields = this.modelInstance?.modelFields;
44      try {
45        const id = getIdByUrl();
46        const data = (await queryOne(this.modelInstance.model, (content[0] || { id }) as Record<string, string>, fields, variables, context)) as Record<string, unknown>;
47
48        this.loadData(data);
49        this.setBusy(false);
50        return data;
51      } catch (e) {
52        console.error(e);
53      } finally {
54        this.setBusy(false);
55      }
56    }
57
58    /**
59     * 新增数据时获取表单默认值
60     * */
61    @Widget.Method()
62    public async constructData(content: Record<string, unknown>[] = [], options: Record<string, unknown> = {}, variables: Record<string, unknown> = {}) {
63      this.setBusy(true);
64      const context: typeof options = { sourceModel: this.modelInstance.model, ...options };
65      const fields = this.modelInstance.modelFields;
66      const reqData = content[0] || {};
67      const data = await constructOne(this.modelInstance!.model, reqData, fields, variables, context);
68      return data as Record<string, unknown>;
69    }
70
71    @Widget.Method()
72    private async reloadData() {
73      const data = await this.constructData();
74      // 覆盖formData
75      this.setData(data);
76    }
77
78    @Widget.Method()
79    public onChange(name, value) {
80      this.formData[name] = value;
81    }
82
83    protected async mounted() {
84      super.mounted();
85      const modelModel = getModelByUrl();
86      this.modelInstance = await getModel(modelModel);
```

```
 87          this.fetchData();
 88          this.messageHub = useMessageHub('messageHubCode');
 89          this.messageHub.subscribe((result) => {
 90            this.serviceErrors = [];
 91            // 收集错误信息
 92            if (Array.isArray(result)) {
 93              const errors = result.map((res) => {
 94                const path = res.path[1],
 95                  error = res.message;
 96                return {
 97                  name: path,
 98                  error
 99                };
100              });
101              this.serviceErrors = errors;
102            }
103          });
104          this.messageHub.subscribe((result) => {
105            console.log(result);
106          }, ILevel.SUCCESS);
107          // 数据提交钩子函数！！！
108          this.callChaining?.callBefore(() => {
109            return this.formData;
110          });
111        }
112      }
113
```

**Step4.** 刷新页面看效果，如图 4-78 所示。

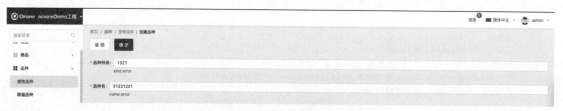

图 4-78　刷新页面看效果

### 4.2.3　框架之 SPI 机制

SPI（Service Provider Interface，服务提供接口）是一套用来被第三方实现或者扩展的 API，它可以用来启用框架扩展和替换组件，简单来说就是用来解耦，实现组件的自由插拔，这样我们就能在平台提供的基础组件外扩展新组件或覆盖平台组件，目前定义的 SPI 组件见表 4-44。

表 4-44　目前定义 SPI 组件

| 目前定义的 SPI 组件 | |
| --- | --- |
| ViewWidget | 视图组件 |
| FieldWidget | 字段组件 |
| ActionWidget | 动作组件 |

使用 TypeScript 装饰器（注解）装饰你的代码：https://juejin.cn/post/6844903876605280269。

（1）通过注解定义一种 SPI 接口（Interface），代码如下。

```
1  @SPI.Base<IViewFilterOptions, IView>('View', ['id', 'name', 'type', 'model', 'widg
   et'])
2  export abstract class ViewWidget<ViewData = any> extends DslNodeWidget {
3
4  }
```

（2）通过注解注册提供 View 类型接口的一个或多个实现，代码如下。

```
1  @SPI.ClassFactory(ViewWidget.Token({ type: [ViewType.Form] }))
2  export class FormWidget extends ViewWidget<any> {
3
4  }
```

（3）视图的 xml 内通过配置来调用已定义的一种 SPI 组件，代码如下。

```
1  <view widget="form" model="demo.shop">
2    <field name="id" />
3  </view>
```

组件集成示意图如图 4-79 所示。

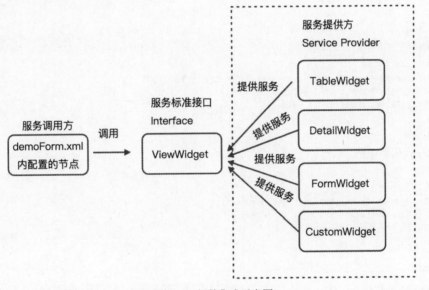

图 4-79　组件集成示意图

当有多个服务提供方时，按以下规则匹配出最符合条件的服务提供方。

SPI 组件没有严格地按匹配选项属性限定，而是一个匹配规则。

（1）按 widget 最优先匹配，配置了 widget 等于是指定了需要调用哪个 widget，此时其他属性直接忽略。

（2）按最大匹配原则（匹配到的属性越多优先级越高）。

（3）按后注册优先原则。

### 4.2.4　框架之网络请求——HttpClient

Oinone 提供统一的网络请求底座，基于 graphql 二次封装。

**1. 初始化**

初始化代码如下所示。

```
1  import { HttpClient } from '@kunlun/dependencies';
2  const http = HttpClient.getInstance();
3  http.setMiddleware() // 必须设置，请求回调。具体查看文章https://shushi.yuque.com/yqitv
   f/oinone/vwo80g
4  http.setBaseURL() // 必须设置，后端请求的路径
```

**2. HttpClient 详细介绍**

1）获取实例

获取实例代码如下所示。

```
1  import { HttpClient } from '@kunlun/dependencies';
2  const http = HttpClient.getInstance();
```

2）接口地址

接口地址代码如下所示。

```
1  import { HttpClient } from '@kunlun/dependencies';
2  const http = HttpClient.getInstance();
3  http.setBaseURL('接口地址');
4  http.getBaseURL(); // 获取接口地址
```

3）请求头

请求头代码如下所示。

```
1  import { HttpClient } from '@kunlun/dependencies';
2  const http = HttpClient.getInstance();
3  http.setHeader({key: value});
```

4）variables

variables 代码如下所示。

```
1  import { HttpClient } from '@kunlun/dependencies';
2  const http = HttpClient.getInstance();
3  http.setExtendVariables((moduleName: string) => {
4    return customFuntion();
5  });
```

5）回调

回调代码如下所示。

```
1  import { HttpClient } from '@kunlun/dependencies';
2  const http = HttpClient.getInstance();
3  http.setMiddleware([middleware]);
```

6）业务使用 -query

业务使用 -query 如下所示。

```
1  private http = HttpClient.getInstance();
2
3  private getTestQuery = async () => {
4    const query = `gql str`;
5
6    const result = await this.http.query('module name', query);
7    console.log(result)
8    return result.data[`xx`][`xx`]; // 返回的接口，打印出result对象层次返回
9  };
```

7）业务使用 -mutate

业务使用 -mutate 如下所示。

```
1   private http = HttpClient.getInstance();
2
3   private getTestMutate = async () => {
4     const mutation = `gql str`;
5
6     const result = await this.http.mutate('module name', mutation);
7     console.log(result)
8     return result.data[`xx`]['xx']; // 返回的接口，打印出result对象层次返回
9   };
```

**3. 如何使用 HttpClient**

1）初始化

在项目目录 src/main.ts 下初始化 httpClient。

初始化必须要做的事：

（1）设置服务接口链接。

（2）设置接口请求回调。

2）业务实战

前文说到自定义新增宠物表单，让我们在这个基础上加入我们的 httpClient。

Step1. 新增 service.ts，如图 4-80 所示。

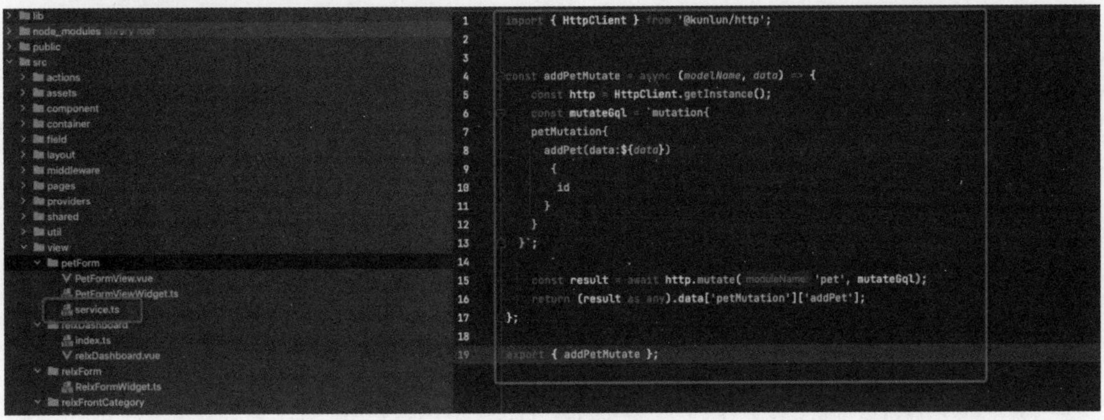

图 4-80　新增 service.ts

service.ts，代码如下。

```
1   import { HttpClient } from '@kunlun/dependencies';
2
3
4   const addPetMutate = async (modelName, data) => {
5       const http = HttpClient.getInstance();
6       const mutateGql = `mutation{
7       petMutation{
8         addPet(data:${data})
9         {
10          id
11        }
12      }
13    }`;
14
15      const result = await http.mutate('pet', mutateGql);
16      return (result as any).data['petMutation']['addPet'];
```

```
17    };
18
19    export { addPetMutate };
```

Step2. 业务中使用，代码如下。

```
1  import { constructOne, queryOne, SPI, ViewWidget, Widget, IModel, getModelByUrl,
   getModel, getIdByUrl, FormWidgetV3, CustomWidget, CallChaining } from '@kunlun/de
   pendencies';
2  import PetFormView from './PetFormView.vue';
3  // 引入
4  import { addPetMutate } from './service';
5
6  @SPI.ClassFactory(CustomWidget.Token({ widget: 'PetForm' }))
7  export class PetFormViewWidget extends FormWidgetV3 {
8    public initialize(props) {
9      super.initialize(props);
10     this.setComponent(PetFormView);
11     return this;
12   }
13
14   /**
15    * 数据提交
16    * @protected
17    */
18   @Widget.Reactive()
19   @Widget.Inject()
20   protected callChaining: CallChaining | undefined;
21
22   private modelInstance!: IModel;
23
24   /**
25    * 重要！！！！
26    * 当字段改变时修改formData
27    * */
28   @Widget.Method()
29   public onFieldChange(fieldName: string, value) {
30     this.setDataByKey(fieldName, value);
31   }
32
33   /**
34    * 表单编辑时查询数据
35    * */
36   public async fetchData(content: Record<string, unknown>[] = [], options: Record
   <string, unknown> = {}, variables: Record<string, unknown> = {}) {
37     this.setBusy(true);
38     const context: typeof options = { sourceModel: this.modelInstance.model, ...o
   ptions };
39     const fields = this.modelInstance?.modelFields;
40     try {
41       const id = getIdByUrl();
42       const data = (await queryOne(this.modelInstance.model, (content[0] || { id
   }) as Record<string, string>, fields, variables, context)) as Record<string, unkn
   own>;
43
44       this.loadData(data);
45       this.setBusy(false);
46       return data;
47     } catch (e) {
48       console.error(e);
49     } finally {
50       this.setBusy(false);
51     }
52   }
```

```
53
54     /**
55      * 新增数据时获取表单默认值
56      * */
57     @Widget.Method()
58     public async constructData(content: Record<string, unknown>[] = [], options: Record<string, unknown> = {}, variables: Record<string, unknown> = {}) {
59       this.setBusy(true);
60       const context: typeof options = { sourceModel: this.modelInstance.model, ...options };
61       const fields = this.modelInstance.modelFields;
62       const reqData = content[0] || {};
63       const data = await constructOne(this.modelInstance!.model, reqData, fields, variables, context);
64       return data as Record<string, unknown>;
65     }
66
67     @Widget.Method()
68     private async reloadData() {
69       const data = await this.constructData();
70       // 覆盖formData
71       this.setData(data);
72     }
73
74     @Widget.Method()
75     public onChange(name, value) {
76       this.formData[name] = value;
77     }
78
79     protected async mounted() {
80       super.mounted();
81       const modelModel = getModelByUrl();
82       this.modelInstance = await getModel(modelModel);
83       this.fetchData();
84       // 数据提交钩子函数！！！
85       this.callChaining?.callBefore(() => {
86         return this.formData;
87       });
88     }
89
90     @Widget.Method()
91     private async addPet() {
92       // 调用
93       const data = await addPetMutate('petModule', this.getData());
94     }
95   }
96
```

## 4.2.5 框架之网络请求——Request

在中后台业务场景中，大部分的请求是可以被枚举的，比如创建、删除、更新、查询。在上文中，我们讲了 HttpClient 如何自定义请求，来实现自己的业务诉求。本节中讲到的 Request 是离业务更进一步的封装，它提供了开箱即用的 API，比如 insertOne、updateOne，它是基于 HttpClient 做的二次封装，当你熟悉 Request 时，在中后台的业务场景中，所有的业务接口自定义将事半功倍。

**1. Request 详细介绍**

1）元数据——Model

（1）获取模型实例代码如下所示。

```
1   import { getModel } from '@kunlun/dependencies'
2   getModel('modelName');
```

(2) 清除所有缓存的模型，代码如下。

```
1  import { cleanModelCache } from '@kunlun/dependencies'
2  cleanModelCache();
```

2) 元数据——module

(1) 获取应用实例，包含应用入口和菜单，代码如下。

```
1  import { queryModuleByName } from '@kunlun/dependencies'
2  queryModuleByName('moduleName')
```

(2) 查询当前用户所有的应用，代码如下。

```
1  import { loadModules } from '@kunlun/dependencies'
2  loadModules()
```

3) query

(1) 分页查询，代码如下。

```
1  import { queryPage } from '@kunlun/dependencies'
2
3  queryPage(modelName, {
4      pageSize: 15,        // 一次查询几条
5      currentPage, 1,      // 当前页码
6      condition?: '',      // 查询条件
7      maxDepth?: 1,        // 查几层模型出来，如果有2，会把所有查询字段的关系字段都查出来
8      sort?: [];           // 排序规则
9  }, fields, variables, context)
```

(2) 自定义分页查询——可自定义后端接口查询数据，代码如下。

```
1  import { customQueryPage } from '@kunlun/dependencies'
2
3  customQueryPage(modelName, methodName, {
4      pageSize: 15,        // 一次查询几条
5      currentPage, 1,      // 当前页码
6      condition?: '',      // 查询条件
7      maxDepth?: 1,        // 查几层模型出来，如果有2，会把所有查询字段的关系字段都查出来
8      sort?: [];           // 排序规则
9  }, fields, variables, context)
```

(3) 查询一条——根据 params 匹配出一条数据，代码如下。

```
1  import { queryOne } from '@kunlun/dependencies'
2
3  customQueryPage(modelName, params, fields, variables, context)
```

(4) 自定义查询，代码如下。

```
1  import { customQuery } from '@kunlun/dependencies'
2
3  customQuery(methodName, modelName, record, fields, variables, context)
```

4) update

update 代码如下所示。

```
1  import { updateOne } from '@kunlun/dependencies'
2
3  updateOne(modelName, record, fields, variables, context)
```

5) insert

insert 代码如下所示。

```
1  import { insertOne } from '@kunlun/dependencies'
2
3  insertOne(modelName, record, fields, variables, context)
```

6) delete

delete 代码如下所示。

```
1  import { deleteOne } from '@kunlun/dependencies'
2
3  deleteOne(modelName, record, variables, context)
```

**2. construct**

（1）构造一条数据——获取初始化的值，一个页面一般只会调一次，代码如下。

```
1  import { constructOne } from '@kunlun/dependencies'
2
3  constructOne(modelName, record, fields, variables, context)
```

（2）构造一条数据——当需要重复获取初始化值时，第一次使用 constructOne，后面的调用使用 constructMirror，代码如下。

```
1  import { constructMirror } from '@kunlun/dependencies'
2
3  constructMirror(modelName, record, fields, variables, context)
```

直接调用后端的 function，一般在特殊的业务场景中使用，比如导入 / 导出等，代码如下。

```
1  import { callFunction } from '@kunlun/dependencies'
2  // action 后端定义的serverAction
3  callFunction(modelName, action, params, fields, variables, context)
```

**3. 如何使用**

让我们用 Request 里的函数改造 PetForm/service.ts 里的 addPet 方法。

（1）request insertOne，如图 4-81 所示。

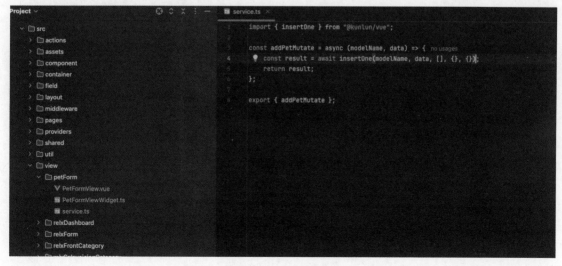

图 4-81　request insertOne

（2）httpClient insertOne，如图 4-82 所示。

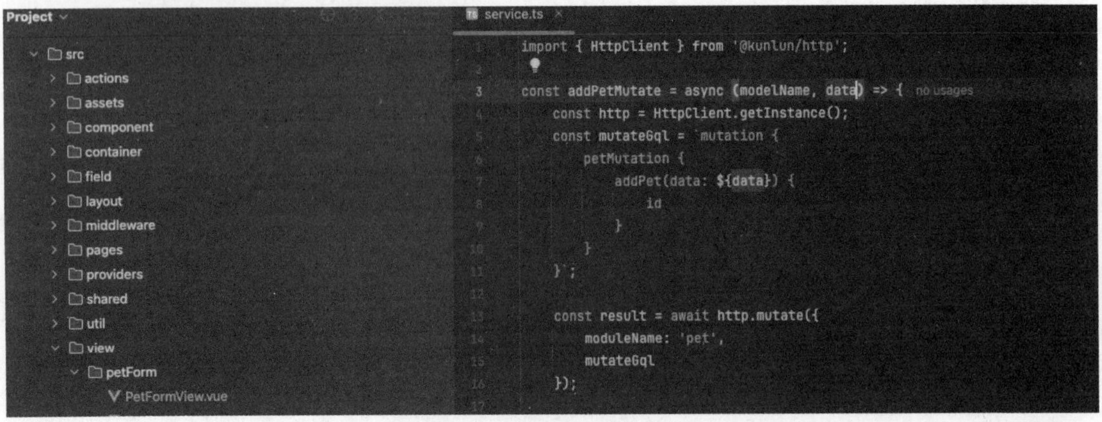

图 4-82　httpClient insertOne

insertOne 使用示例如下所示。

```
1   import { insertOne } from "@kunlun/dependencies";
2
3
4   const addPetMutate = async (modelName, data) => {
5       const result = await insertOne(modelName, data, [], {}, {})
6       return result
7   };
8
9   export { addPetMutate };
```

可以看到，在使用 request 的 insertOne 时，我们的代码量大大地减少。为了满足业务的多变性和便捷性，我们提供了两种方式，大家根据自己的场景自由选择。

### 4.2.6　框架之网络请求——拦截器

在整个 http 的链路中，异常错误对前端来说尤为重要，它作用在很多不同的场景，通用的比如 500、502 等。一个好的软件通常需要在不同的错误场景中做不同的事情。当用户 cookie 失效时，希望能自动跳转到登录页；当用户权限发生变更时，希望能跳转到一个友好的提示页；那么如何满足这些个性化的诉求呢？接下来让我们一起了解 Oinone 前端网络请求——拦截器。

**1. 入口**

在 src 目录下 main.ts 中可以看到 VueOioProvider，这是系统功能提供者的注册入口，如图 4-83 所示。

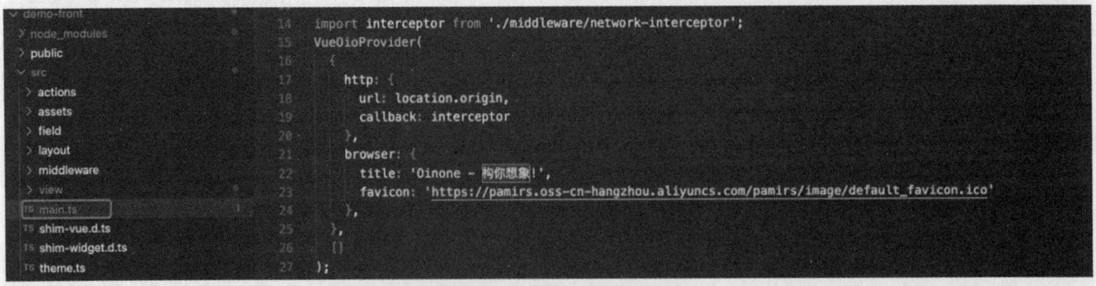

图 4-83　VueOioProvider

拦截器的申明入口如下所示。

```
1  import interceptor from './middleware/network-interceptor';
2  VueOioProvider(
3    {
4      http: {
5        callback: interceptor
6      }
7    },
8    []
9  );
```

### 2. middleware

在项目初始化时使用 CLI 构建初始化前端工程，在 src/middleware 有拦截器的默认实现，如图 4-84 所示。

图 4-84　在 src/middleware 有拦截器的默认实现

### 3. interceptor

interceptor 在请求返回后触发，interceptor 有两个回调函数：error 和 next。

error 参数：

● graphQLErrors 处理业务异常；

● networkError 处理网络异常。

next 参数：

● extensions 后端返回扩展参数，代码如下。

```
1  const interceptor: RequestHandler = (operation, forward) => {
2    return forward(operation).subscribe({
3      error: ({ graphQLErrors, networkError }) => {
4        console.log(graphQLErrors, networkError);
5        // 默认实现 => interceptor error
6      },
7      next: ({ extensions }) => {
8        console.log(extensions);
9        // 默认实现 => interceptor next
10     },
11   });
12 };
```

### 4. interceptor error

interceptor error 如下所示。

```
// 定义错误提示等级
const DEFAULT_MESSAGE_LEVEL = ILevel.ERROR;
// 错误提示等级 对应提示的报错
const MESSAGE_LEVEL_MAP = {
  [ILevel.ERROR]: [ILevel.ERROR],
  [ILevel.WARN]: [ILevel.ERROR, ILevel.WARN],
  [ILevel.INFO]: [ILevel.ERROR, ILevel.WARN, ILevel.INFO],
  [ILevel.SUCCESS]: [ILevel.ERROR, ILevel.WARN, ILevel.INFO, ILevel.SUCCESS],
  [ILevel.DEBUG]: [ILevel.ERROR, ILevel.WARN, ILevel.INFO, ILevel.SUCCESS, ILevel.DEBUG]
};
// 错误提示通用函数
const notificationMsg = (type: string = 'error', tip: string = '错误', desc: string = '') => {
  notification[type]({
    message: tip,
    description: desc
  });
};
// 根据错误等级 返回错误提示和类型
const getMsgInfoByLevel = (level: ILevel) => {
  let notificationType = 'info';
  let notificationText = translate('kunlun.common.info');
  switch (level) {
    case ILevel.DEBUG:
      notificationType = 'info';
      notificationText = translate('kunlun.common.debug');
      break;
    case ILevel.INFO:
      notificationType = 'info';
      notificationText = translate('kunlun.common.info');
      break;
    case ILevel.SUCCESS:
      notificationType = 'success';
      notificationText = translate('kunlun.common.success');
      break;
    case ILevel.WARN:
      notificationType = 'warning';
      notificationText = translate('kunlun.common.warning');
      break;
  }
  return {
    notificationType,
    notificationText
  };
};

error: ({ graphQLErrors, networkError }) => {
    if (graphQLErrors) {
        graphQLErrors.forEach(async ({ message, locations, path, extensions }) => {
            let { errorCode, errorMessage, messages } = extensions || {};
            // FIXME: extensions.errorCode
            if (errorCode == null) {
                const codeArr = /code: (\d+),/.exec(message);
                if (codeArr) {
                    errorCode = Number(codeArr[1]);
                }
            }
```

```
56            }
57            if (errorMessage == null) {
58              const messageArr = /msg: (.*),/.exec(message);
59              if (messageArr) {
60                errorMessage = messageArr[1];
61              }
62            }
63            // 错误通用提示
64            if (messages && messages.length) {
65              messages.forEach((m) => {
66                notificationMsg('error', translate('kunlun.common.error'), m.message || '');
67              });
68            } else {
69              notificationMsg('error', translate('kunlun.common.error'), errorMessage || message);
70            }
71            // 提示扩展信息 根据错误等级来提示对应级别的报错
72            const extMessage = getValue(response, 'extensions.messages');
73            if (extMessage && extMessage.length) {
74              const messageLevelArr = MESSAGE_LEVEL_MAP[DEFAULT_MESSAGE_LEVEL];
75              extMessage.forEach((m) => {
76                if (messageLevelArr.includes(m.level)) {
77                  const { notificationType, notificationText } = getMsgInfoByLevel(m.level);
78                  notificationMsg(notificationType, notificationText, m.message || '');
79                }
80              });
81            }
82
83            // 消息模块的用户未登录错误码
84            const MAIL_USER_NOT_LOGIN = 20080002;
85            // 基础模块的用户未登录错误码
86            const BASE_USER_NOT_LOGIN_ERROR = 11500001;
87            if (
88              [MAIL_USER_NOT_LOGIN, BASE_USER_NOT_LOGIN_ERROR].includes(Number(errorCode)) &&
89              location.pathname !== '/auth/login'
90            ) {
91              const redirect_url = location.pathname;
92              location.href = `/login?redirect_url=${redirect_url}`;
93            }
94            /**
95             * 应用配置异常跳转至通用的教程页面
96             */
97            // 模块参数配置未完成
98            const BASE_SYSTEM_CONFIG_IS_NOT_COMPLETED_ERROR = 11500004;
99            if ([BASE_SYSTEM_CONFIG_IS_NOT_COMPLETED_ERROR].includes(Number(errorCode))) {
100             const action = getValue(response, 'extensions.extra.action');
101             if (action) {
102               Action.registerAction(action.model, action);
103               const searchParams: string[] = [];
104               searchParams.push(`module=${action.module}`);
105               searchParams.push(`model=${action.model}`);
106               searchParams.push(`viewType=${action.viewType}`);
107               searchParams.push(`actionId=${action.id}`);
108               const href = `${origin}/page;${searchParams.join(';')}`;
109               location.href = href;
110             }
```

```
111            }
112          });
113        }
114        if (networkError) {
115          const { name, result } = networkError;
116          const errMsg = (result && result.message) || `${networkError}`;
117          if (name && result && result.message) {
118            notification.error({
119              message: translate('kunlun.common.error'),
120              description: `[${name}]: ${errMsg}`,
121            });
122          }
123        }
124      }
```

## 5. interceptor next

interceptor next 如下所示。

```
1  next: ({ extensions }) => {
2    if (extensions) {
3      const messages = extensions.messages as {
4        level: 'SUCCESS';
5        message: string;
6      }[];
7      if (messages)
8        messages.forEach((msg) => {
9          notification.success({
10            message: '操作成功',
11            description: msg.message,
12          });
13        });
14    }
15  }
```

## 6. 完整代码

完整代码如下所示。

```
1   import { NextLink, Operation } from 'apollo-link';
2   import { notification } from 'ant-design-vue';
3   import getValue from 'lodash/get';
4   import { Action, ILevel, translate } from '@kunlun/dependencies';
5
6   interface RequestHandler {
7     (operation: Operation, forward: NextLink): Promise<any> | any;
8   }
9
10  const DEFAULT_MESSAGE_LEVEL = ILevel.ERROR;
11  const MESSAGE_LEVEL_MAP = {
12    [ILevel.ERROR]: [ILevel.ERROR],
13    [ILevel.WARN]: [ILevel.ERROR, ILevel.WARN],
14    [ILevel.INFO]: [ILevel.ERROR, ILevel.WARN, ILevel.INFO],
15    [ILevel.SUCCESS]: [ILevel.ERROR, ILevel.WARN, ILevel.INFO, ILevel.SUCCESS],
16    [ILevel.DEBUG]: [ILevel.ERROR, ILevel.WARN, ILevel.INFO, ILevel.SUCCESS, ILevel.DEBUG]
17  };
18
19  const notificationMsg = (type: string = 'error', tip: string = '错误', desc: string = '') => {
20    notification[type]({
21      message: tip,
22      description: desc
23    });
24  };
```

```
25
26  const getMsgInfoByLevel = (level: ILevel) => {
27    let notificationType = 'info';
28    let notificationText = translate('kunlun.common.info');
29    switch (level) {
30      case ILevel.DEBUG:
31        notificationType = 'info';
32        notificationText = translate('kunlun.common.debug');
33        break;
34      case ILevel.INFO:
35        notificationType = 'info';
36        notificationText = translate('kunlun.common.info');
37        break;
38      case ILevel.SUCCESS:
39        notificationType = 'success';
40        notificationText = translate('kunlun.common.success');
41        break;
42      case ILevel.WARN:
43        notificationType = 'warning';
44        notificationText = translate('kunlun.common.warning');
45        break;
46    }
47    return {
48      notificationType,
49      notificationText
50    };
51  };
52
53  const interceptor: RequestHandler = (operation, forward) => {
54    return forward(operation).subscribe({
55      error: ({ graphQLErrors, networkError, response }) => {
56        if (graphQLErrors) {
57          graphQLErrors.forEach(async ({ message, locations, path, extensions }) => {
58            let { errorCode, errorMessage, messages } = extensions || {};
59            // FIXME: extensions.errorCode
60            if (errorCode == null) {
61              const codeArr = /code: (\d+),/.exec(message);
62              if (codeArr) {
63                errorCode = Number(codeArr[1]);
64              }
65            }
66            if (errorMessage == null) {
67              const messageArr = /msg: (.*),/.exec(message);
68              if (messageArr) {
69                errorMessage = messageArr[1];
70              }
71            }
72            if (messages && messages.length) {
73              messages.forEach((m) => {
74                notificationMsg('error', translate('kunlun.common.error') || '', m.message || '');
75              });
76            } else {
77              notificationMsg('error', translate('kunlun.common.error') || '', errorMessage || message);
78            }
79
80            const extMessage = getValue(response, 'extensions.messages');
81            if (extMessage && extMessage.length) {
82              const messageLevelArr = MESSAGE_LEVEL_MAP[DEFAULT_MESSAGE_LEVEL];
```

```
 83            extMessage.forEach((m) => {
 84              if (messageLevelArr.includes(m.level)) {
 85                const { notificationType, notificationText } = getMsgInfoByLevel(m.level);
 86                notificationMsg(notificationType, notificationText, m.message || '');
 87              }
 88            });
 89          }
 90          console.log(extMessage);
 91
 92          // 消息模块的用户未登录错误码
 93          const MAIL_USER_NOT_LOGIN = 20080002;
 94          // 基础模块的用户未登录错误码
 95          const BASE_USER_NOT_LOGIN_ERROR = 11500001;
 96          if (
 97            [MAIL_USER_NOT_LOGIN, BASE_USER_NOT_LOGIN_ERROR].includes(Number(errorCode)) &&
 98            location.pathname !== '/auth/login'
 99          ) {
100            const redirect_url = location.pathname;
101            location.href = `/login?redirect_url=${redirect_url}`;
102          }
103          /**
104           * 应用配置异常跳转至通用的教程页面
105           */
106          // 模块参数配置未完成
107          const BASE_SYSTEM_CONFIG_IS_NOT_COMPLETED_ERROR = 11500004;
108          if ([BASE_SYSTEM_CONFIG_IS_NOT_COMPLETED_ERROR].includes(Number(errorCode))) {
109            const action = getValue(response, 'extensions.extra.action');
110            if (action) {
111              Action.registerAction(action.model, action);
112              const searchParams: string[] = [];
113              searchParams.push(`module=${action.module}`);
114              searchParams.push(`model=${action.model}`);
115              searchParams.push(`viewType=${action.viewType}`);
116              searchParams.push(`actionId=${action.id}`);
117              const href = `${origin}/page;${searchParams.join(';')}`;
118              location.href = href;
119            }
120          }
121        });
122      }
123      if (networkError) {
124        const { name, result } = networkError;
125        const errMsg = (result && result.message) || `${networkError}`;
126        if (name && result && result.message) {
127          notification.error({
128            message: translate('kunlun.common.error') || '',
129            description: `[${name}]: ${errMsg}`
130          });
131        }
132      }
133    }
134  });
135 };
136
137 export default interceptor;
138
```

## 4.2.7 框架之翻译工具

**1. 翻译**

Oinone 目前的默认文案是中文，如果需要使用其他语言，Oinone 也提供一系列的翻译能力。

**2. 定义**

（1）首先定义文件，如图 4-85 所示。

图 4-85　定义文件

（2）zh_cn.ts 如下所示。

```
const zhCN = {
    demo: {
        test: '这是测试'
    }
}

export default zhCN
```

（3）main.ts 注册如下所示。

```
import { LanguageType, registryLanguage } from '@kunlun/dependencies';
import Zh_cn from './local/zh_cn'
registryLanguage(LanguageType['zh-CN'], Zh_cn);
```

**3. Vue 模板使用**

Vue 模板使用如下所示。

```
<template>
  <div class="petFormWrapper">
    <form :model="formState" @finish="onFinish">
      <a-form-item :label="translate('demo.test')" id="name" name="kind" :rules="[{ required: true, message: '请输入品种种类!', trigger: 'focus' }]">
        <a-input v-model:value="formState.kind" @input="(e) => onNameChange(e, 'kind')" />
        <span style="color: red">{{ getServiceError('kind') }}</span>
      </a-form-item>

      <a-form-item label="品种名" id="name" name="name" :rules="[{ required: true, message: '请输入品种名!', trigger: 'focus' }]">
        <a-input v-model:value="formState.name" @input="(e) => onNameChange(e, 'name')" />
```

```
11              <span style="color: red">{{ getServiceError('name') }}</span>
12          </a-form-item>
13        </form>
14      </div>
15    </template>
16
17    <script lang="ts">
18    import { defineComponent, reactive } from 'vue';
19    import { Form } from 'ant-design-vue';
20
21  ▼ export default defineComponent({
22      // 引入translate
23      props: ['onChange', 'reloadData', 'serviceErrors', 'translate'],
24      components: { Form },
25  ▼   setup(props) {
26  ▼     const formState = reactive({
27          kind: '',
28          name: '',
29        });
30
31  ▼     const onFinish = () => {
32          console.log(formState);
33        };
34
35  ▼     const onNameChange = (event, name) => {
36          props.onChange(name, event.target.value);
37        };
38
39  ▼     const reloadData = async () => {
40          await props.reloadData();
41        };
42
43  ▼     const getServiceError = (name: string) => {
44          const error = props.serviceErrors.find(error => error.name === name);
45          return error ? error.error : '';
46        }
47
48  ▼     return {
49          formState,
50          reloadData,
51          onNameChange,
52          onFinish,
53          getServiceError
54        };
55      }
56    });
57    </script>
```

实际效果如图 4-86 所示。

图 4-86　实际效果

**4. JS 使用——translate**

JS 使用——translate 如下所示。

```vue
<template>
  <div class="petFormWrapper">
    <form :model="formState" @finish="onFinish">
      <a-form-item :label="translate('demo.test')" id="name" name="kind" :rules="[{ required: true, message: '请输入品种种类!', trigger: 'focus' }]">
        <a-input v-model:value="formState.kind" @input="(e) => onNameChange(e, 'kind')" />
        <span style="color: red">{{ getServiceError('kind') }}</span>
      </a-form-item>

      <a-form-item label="品种名" id="name" name="name" :rules="[{ required: true, message: '请输入品种名!', trigger: 'focus' }]">
        <a-input v-model:value="formState.name" @input="(e) => onNameChange(e, 'name')" />
        <span style="color: red">{{ getServiceError('name') }}</span>
      </a-form-item>
    </form>
  </div>
</template>

<script lang="ts">
import { defineComponent, reactive, onMounted } from 'vue';
// 引入
import { translate } from '@kunlun/dependencies';
import { Form } from 'ant-design-vue';

export default defineComponent({
  props: ['onChange', 'reloadData', 'serviceErrors', 'translate'],
  components: { Form },
  setup(props) {
    const formState = reactive({
      kind: '',
      name: '',
    });

    const onFinish = () => {
      console.log(formState);
    };

    const onNameChange = (event, name) => {
      props.onChange(name, event.target.value);
    };

    const reloadData = async () => {
      await props.reloadData();
    };

    const getServiceError = (name: string) => {
      const error = props.serviceErrors.find(error => error.name === name);
      return error ? error.error : '';
    }

    onMounted(() => {
      // 使用
      const test = translate('demo.test');
      console.log(test);
    });

    return {
      formState,
      reloadData,
      onNameChange,
```

```
59        onFinish,
60        getServiceError
61    };
62   }
63 });
64 </script>
```

实际效果如图 4-87 所示。

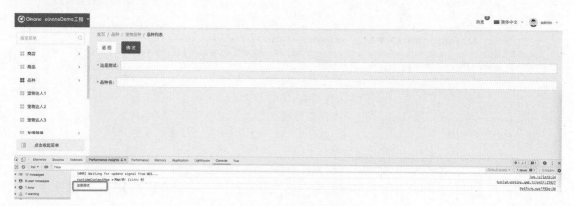

图 4-87　实际效果

**5. JS 使用——translateValueByKey**

translateValueByKey 和 translate 的区别是，获取资源的路径不同，translate 从前端定义中获取对应的翻译，translateValueByKey 是从平台翻译功能中配置得来，建议使用，代码如下。

```
1 import { translateValueByKey } from '@kunlun/dependencies';
2 translateValueByKey("中文名称") || "默认值"
```

**6. 翻译 dslNode**

处理 xml 的国际化，代码如下。

```
1 import { translateNode } from '@kunlun/translateValueByKey';
2 translateNode(dslNode);
```

## 4.3　Oinone 的分布式体验

在 Oinone 的体系中分布式比较独特，boot 工程中启动模块包含被调用函数所属模块就走本地调用，不包含就走远程调用，在此带您体验分布式部署以及分布式部署需要注意点。

看下面例子之前先把说法统一一下：启动或请求 SecondModule 代表启动或请求 pamirs-second-boot 工程，启动或请求 DemoModule 代表启动或请求 pamirs-demo-boot 工程，并没有严格意义上启动哪个模块之说，只有启动工程包含哪个模块。

**1. 构建 SecondModule 模块**

Step1. 构建模块工程。

参考 3.2.1 节，利用脚手架工具构建一个 SecondModule，记住需要修改脚本。

脚本修改如下所示。

```bash
#!/bin/bash

# 脚手架使用目录
# 本地 local
# 本地脚手架信息存储路径 ~/.m2/repository/archetype-catalog.xml
archetypeCatalog=local
# 以下参数以pamirs-second为例
# 新项目的groupId
groupId=pro.shushi.pamirs.second
# 新项目的artifactId
artifactId=pamirs-second
# 新项目的version
version=1.0.0-SNAPSHOT
# Java包名前缀
packagePrefix=pro.shushi
# Java包名后缀
packageSuffix=pamirs.second
# 新项目的pamirs platform version
pamirsVersion=3.0.1-SNAPSHOT
# Java类名称前缀
javaClassNamePrefix=Second
# 项目名称 module.displayName
projectName=oinoneSecond工程
# 模块 MODULE_MODULE 常量
moduleModule=second_core
# 模块 MODULE_NAME 常量
moduleName=SecondCore
# spring.application.name
applicationName=pamirs-second
```

生产工程如图 4-88 所示。

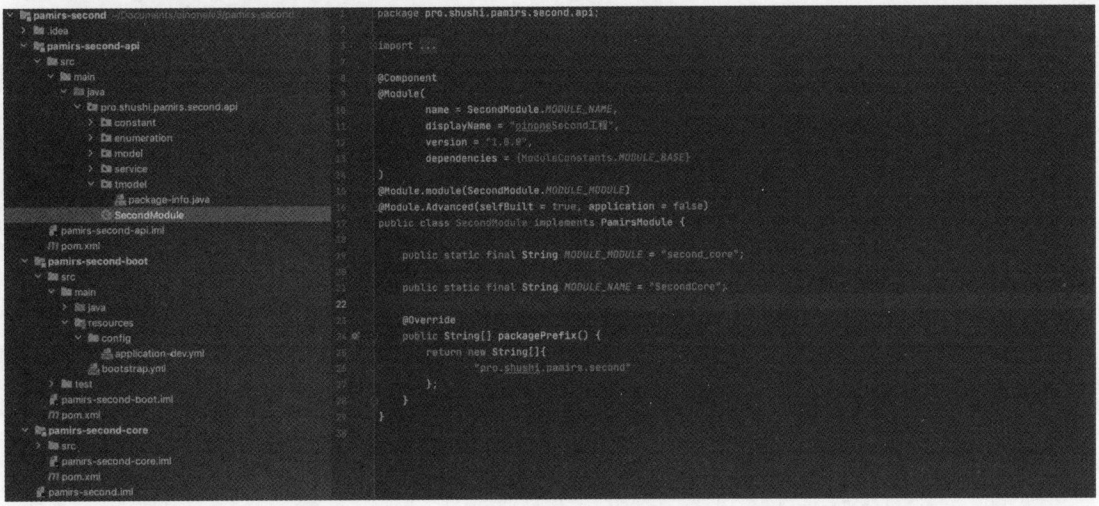

图 4-88　生产工程

**Step2. 调整配置。**

修改 application-dev.yml 文件。

修改 SecondModule 的 application-dev.yml 的内容。

（1）base 库换成与 DemoModule 一样的配置，配置项为 pamirs.datasource.base，代码如下。

```yaml
1  pamirs:
2    datasource:
3      base:
4        driverClassName: com.mysql.cj.jdbc.Driver
5        type: com.alibaba.druid.pool.DruidDataSource
6        url: jdbc:mysql://127.0.0.1:3306/demo_base?useSSL=false&allowPublicKeyRetrieval=true&useServerPrepStmts=true&cachePrepStmts=true&useUnicode=true&characterEncoding=utf8&serverTimezone=Asia/Shanghai&autoReconnect=true&allowMultiQueries=true
7        username: root # 数据库用户
8        password: oinone # 数据库用户对应的密码
9        initialSize: 5
10       maxActive: 200
11       minIdle: 5
12       maxWait: 60000
13       timeBetweenEvictionRunsMillis: 60000
14       testWhileIdle: true
15       testOnBorrow: false
16       testOnReturn: false
17       poolPreparedStatements: true
18       asyncInit: true
```

（2）修改后端 Server 的启动端口号为 9091，代码如下。

```yaml
1  server:
2    address: 0.0.0.0
3    port: 9091
4    sessionTimeout: 3600
```

（3）修改 bootstrap.yml 文件，代码如下。

设置 dubbo 序列化方式为 pamirs，记得 DemoModule 也要改。

```yaml
1  spring:
2    profiles:
3      active: dev
4    application:
5      name: pamirs-second
6
7  pamirs:
8    default:
9      environment-check: true
10     tenant-check: true
11
12 ---
13 spring:
14   profiles: dev
15   cloud:
16     config:
17       enabled: false
18       uri: http://127.0.0.1:7001
19       label: master
20       profile: dev
21     nacos:
22       server-addr: http://127.0.0.1:8848
23       discovery:
24         enabled: false
25         namespace:
26         prefix: application
27         file-extension: yml
28       config:
29         enabled: false
```

```yaml
30        namespace:
31        prefix: application
32        file-extension: yml
33
34  dubbo:
35    application:
36      name: pamirs-second
37      version: 1.0.0
38    registry:
39      address: zookeeper://127.0.0.1:2181
40    protocol:
41      name: dubbo
42      port: -1
43      serialization: pamirs
44    scan:
45      base-packages: pro.shushi
46    cloud:
47      subscribed-services:
48
49  ---
50  spring:
51    profiles: test
52    cloud:
53      config:
54        enabled: false
55        uri: http://127.0.0.1:7001
56        label: master
57        profile: test
58      nacos:
59        server-addr: http://127.0.0.1:8848
60        discovery:
61          enabled: false
62          namespace:
63          prefix: application
64          file-extension: yml
65        config:
66          enabled: false
67          namespace:
68          prefix: application
69          file-extension: yml
70
71  dubbo:
72    application:
73      name: pamirs-second
74      version: 1.0.0
75    registry:
76      address: zookeeper://127.0.0.1:2181
77    protocol:
78      name: dubbo
79      port: -1
80      serialization: pamirs
81    scan:
82      base-packages: pro.shushi
83    cloud:
84      subscribed-services:
```

Step3. 构建一个 RemoteTestModel，代码如下。

新建 RemoteTestModel 模型，用于远程调用体验。

```
 8        @Model.model(RemoteTestModel.MODEL_MODEL)
 9        @Model(displayName = "远程测试模型",labelFields = "name")
10      public class RemoteTestModel extends CodeModel {
11
12            public static final String MODEL_MODEL="second.RemoteTestModel";
13
14            @Field(displayName = "name")
15            private String name;
16
17
18        }
```

Step4. 新增 SecondSessionHook，代码如下。

@Hook (module= SecondModule.MODULE_MODULE)，规定只有增加对 SecondModule 模块访问的请求该拦截器才会生效，否则其他模块的请求都会被 SecondSessionHook 拦截。

```
10      @Component
11      @Slf4j
12    public class SecondSessionHook implements HookBefore {
13          @Override
14          @Hook(priority = 1,module= SecondModule.MODULE_MODULE)
15          public Object run(Function function, Object... args) {
16              log.info("second hook");
17              return function;
18          }
19      }
```

Step5. 新建 RemoteTestModelAction，代码如下。

在 SecondModule 中新建 RemoteTestModelAction，自定义 RemoteTestModel 模型的 queryPage 函数。方便 debug，看看效果。

```
12        @Model.model(RemoteTestModel.MODEL_MODEL)
13      public class RemoteTestModelAction {
14
15            @Function.Advanced(type= FunctionTypeEnum.QUERY)
16            @Function.fun(FunctionConstants.queryPage)
17            @Function(openLevel = {FunctionOpenEnum.API})
18          public Pagination<RemoteTestModel> queryPage(Pagination<RemoteTestModel> pag
      e, IWrapper<RemoteTestModel> queryWrapper){
19              return new RemoteTestModel().queryPage(page,queryWrapper);
20          }
21
22        }
```

Step6. boot 工程需要引入 Oinone 的 RPC 包。

（1）父 pom 的依赖管理中先加 pamirs-distribution-faas 的依赖，代码如下。

```
1    <dependency>
2        <groupId>pro.shushi.pamirs.distribution</groupId>
3        <artifactId>pamirs-distribution-faas</artifactId>
4        <version>${pamirs.boot.version}</version>
5    </dependency>
```

（2）在 pamirs-second-boot 中增加 pamirs-distribution-faas 的依赖，代码如下。

```
1    <dependency>
2        <groupId>pro.shushi.pamirs.distribution</groupId>
3        <artifactId>pamirs-distribution-faas</artifactId>
4    </dependency>
```

（3）为 SecondApplication 类增加类注解 @EnableDubbo，代码如下。

```
1  @EnableDubbo
2  public class SecondApplication {
3  }
```

Step7. 启动 SecondModule。

别忘了启动指令要为 INSTALL。参考"DemoModule 的启动说明"一节。

Step8. Second 工程本地 mvn install，方便 DemoModule 包依赖。

**2. DemoModule 模块准备**

Step1. DemoModule 引入 Oinone 的 RPC 包和 Second 的 API 包。

（1）父 pom 的依赖管理中先加入 pamirs-second-api 和 pamirs-distribution-faas 的依赖，代码如下。

```
 1  <dependency>
 2      <groupId>pro.shushi.pamirs.second</groupId>
 3      <artifactId>pamirs-second-api</artifactId>
 4      <version>1.0.0-SNAPSHOT</version>
 5  </dependency>
 6
 7  <dependency>
 8      <groupId>pro.shushi.pamirs.distribution</groupId>
 9      <artifactId>pamirs-distribution-faas</artifactId>
10      <version>${pamirs.boot.version}</version>
11  </dependency>
```

（2）在 pamirs-demo-api 中增加 pamirs-second-api 的依赖，代码如下。

```
1  <dependency>
2      <groupId>pro.shushi.pamirs.second</groupId>
3      <artifactId>pamirs-second-api</artifactId>
4  </dependency>
```

（3）在 pamirs-demo-boot 中增加 pamirs-distribution-faas 的依赖，代码如下。

```
1  <dependency>
2      <groupId>pro.shushi.pamirs.distribution</groupId>
3      <artifactId>pamirs-distribution-faas</artifactId>
4  </dependency>
```

（4）为 DemoApplication 类增加类注解 @EnableDubbo，代码如下。

```
1  @EnableDubbo
2  public class DemoApplication {
3  }
```

Step2. 修改 DemoModule 定义，代码如下。

修改 DemoModule 的依赖注解，增加 SecondModule.MODULE_MODULE。

```
1  @Module(dependencies = {SecondModule.MODULE_MODULE})
```

Step3. 修改 pamirs-demo-boot 的 bootstrap.yml 文件。

参考 SecondModule 修改 dubbo 的序列化方式为 pamirs。

Step4. 修改 pamirs-demo-boot 的 DemoApplication，如图 4-89 所示。

为依赖模块配置扫描包路径，修改 DemoApplication 增加 second 的扫描包，在日常开发中小伙伴的应用肯定不是以 pamirs 开头，所以大家别忘了 SpringBoot 的基本配置。在"构建第一个 Module"一节也提到我们在启动工程中需要配置启动模块和依赖模块的扫描路径。

```
@ComponentScan(
        basePackages = {"pro.shushi.pamirs.meta",
                "pro.shushi.pamirs.framework.connectors.event",
                "pro.shushi.pamirs.framework",
                "pro.shushi.pamirs",
                "pro.shushi.himalaya",
                "pro.shushi.tanggula",
                "pro.shushi.pamirs.demo",
                "pro.shushi.pamirs.second"
        },
        excludeFilters = {
                @ComponentScan.Filter(
                        type = FilterType.ASSIGNABLE_TYPE,
                        value = {RedisAutoConfiguration.class, RedisRepositoriesAutoConfiguration.class, RedisClusterConfig.class}
                )
        })
@Slf4j
@EnableTransactionManagement
@EnableAsync
@EnableDubbo
@MapperScan(value = "pro.shushi.pamirs", annotationClass = Mapper.class)
@SpringBootApplication(exclude = {DataSourceAutoConfiguration.class, FreeMarkerAutoConfiguration.class})
public class DemoApplication {
```

图 4-89　修改 pamirs-demo-boot 的 DemoApplication

**3. 分布式部署体验**

以上工作准备好以后，我们就可以通过 DemoModule 来远程调用 SecondModule。

**第一个案例（前端把请求分别打到 DemoModule、SecondModule）**

在日常研发中，不同模块的菜单整合也是非常常见的，比如把 B 模块的菜单挂载在 A 模块中。这个案例就是讲解如何做到跨模块的菜单整合。

Step1. 修改前端工程的 vue.config.js，代码如下。

利用 node 的 proxy，分别把 DemoCore 转发到 8090 端口，把 SecondCore 转发到 8091 端口，把默认其他模块转发到 8090 端口。localhost 还是 127.0.0.1 跟浏览器地址框输入保持一致。

```
const WidgetLoaderPlugin = require('@kunlun/widget-loader/dist/plugin.js').default;
const Dotenv = require('dotenv-webpack');

module.exports = {
  lintOnSave: false,
  configureWebpack: {
    module: {
      rules: [
        {
          test: /\.widget$/,
          loader: '@kunlun/widget-loader',
        },
      ],
    },
    plugins: [new WidgetLoaderPlugin(), new Dotenv()],
    resolveLoader: {
      alias: {
        '@kunlun/widget-loader': require.resolve('@kunlun/widget-loader'),
      },
    },
  },
  devServer: {
    port: 8080,
    disableHostCheck: true,
    progress: false,
    proxy: {
```

```
27         'pamirs/DemoCore': {
28             // 支持跨域
29             changeOrigin: true,
30             target: 'http://localhost:8090',
31         },
32         'pamirs/SecondCore': {
33             // 支持跨域
34             changeOrigin: true,
35             target: 'http://localhost:8091',
36         },
37         pamirs: {
38             // 支持跨域
39             changeOrigin: true,
40             target: 'http://localhost:8090',
41         },
42     },
43   },
44 };
```

Step2. 修改 DemoMenus，代码如下。

增加一个 RemoteTestModel 的管理入口，这里需要指定菜单的 module 为 DemoModule，否则应用会切换到 SecondModule

```
1  @UxMenu("远程模型")@UxRoute(model = RemoteTestModel.MODEL_MODEL,module= DemoModule.
   MODULE_MODULE) class RemoteTestModelMenu{}
```

Step3. 重启前端应用看效果。

（1）单击"远程模型"菜单，可以正常进行增、删、改、查操作，如图4-90所示。

图 4-90　单击"远程模型"菜单

（2）SecondSessionHook 起作用。

"second hook"并打印出来，前端请求没有再进过 DemoModule，远程调用 SecondModule，而是直接打到了 SecondModule 上，如图4-91所示。

图 4-91　SecondSessionHook 起作用

## 第二个案例（跨模块代理，自动走远程）

Step1. 新建 RemoteTestModelProxy，代码如下。

在 DemoModule 中新建 RemoteTestModelProxy 代理继承 RemoteTestModel。

```
 7  @Model.model(RemoteTestModelProxy.MODEL_MODEL)
 8  @Model.Advanced(type = ModelTypeEnum.PROXY)
 9  @Model(displayName = "远程模型的代理",summary="远程模型的代理")
10  public class RemoteTestModelProxy extends RemoteTestModel {
11
12      public static final String MODEL_MODEL="demo.RemoteTestModelProxy";
13
14  }
```

Step2. 修改 DemoMenus。

增加一个 RemoteTestModelProxy 的管理入口，代码如下。

```
1  @UxMenu("远程代理")@UxMenu.route(RemoteTestModelProxy.MODEL_MODEL) class RemoteTestModelProxyMenu{}
```

Step3. 重启看效果。

（1）单击"远程代理"菜单，如图 4-92 所示。

图 4-92 单击"远程代理"菜单

（2）在 SecondModule 的 RemoteTestModelAction 中 debug，会发现 DemoModule 会自动调用 RemoteTestModel 模型的 queryPage，如图 4-93 所示。

图 4-93 DemoModule 会自动调用 RemoteTestModel 模型的 queryPage

（3）SecondSessionHook 不会生效。

"second hook"并没有打印出来，为什么？在函数相关特性系列文章"面向切面——拦截器"一节中介绍道"不是前端直接发起的请求不会生效"，可能小伙伴有疑问，我不是通过前端单击的吗？是的，但经过 DemoModule 再调用 SecondModule，系统会判定为后端调用。

### 第三个案例（远程调用自定义函数）

Step1. 新增函数 RemoteTestModelService 和 RemoteTestModelServiceImpl。

SecondModule 定义 RemoteTestModelService，用于 DemoModule 的调用。SecondModule 记得再次 mvn install。

新增函数 RemoteTestModelService，代码如下。

```java
 9  @Fun(RemoteTestModelService.FUN_NAMESPACE)
10  public interface RemoteTestModelService {
11
12      String FUN_NAMESPACE ="second.RemoteTestModelService";
13      @Function
14      Pagination<RemoteTestModel> queryPage(Pagination<RemoteTestModel> page, IWrapper<RemoteTestModel> queryWrapper);
15      @Function
16      String hello(String name);
17      @Function
18      String remoteTest(RemoteTestModel remoteTest);
19  }
20
```

新增函数 RemoteTestModelServiceImpl，代码如下。

```java
 1  package pro.shushi.pamirs.second.core.service;
 2
 3  import org.springframework.stereotype.Component;
 4  import pro.shushi.pamirs.meta.annotation.Fun;
 5  import pro.shushi.pamirs.meta.annotation.Function;
 6  import pro.shushi.pamirs.meta.api.dto.condition.Pagination;
 7  import pro.shushi.pamirs.meta.api.dto.wrapper.IWrapper;
 8  import pro.shushi.pamirs.second.api.model.RemoteTestModel;
 9  import pro.shushi.pamirs.second.api.service.RemoteTestModelService;
10
11  @Fun(RemoteTestModelService.FUN_NAMESPACE)
12  @Component
13  public class RemoteTestModelServiceImpl implements RemoteTestModelService {
14
15      @Override
16      @Function
17      public Pagination<RemoteTestModel> queryPage(Pagination<RemoteTestModel> page, IWrapper<RemoteTestModel> queryWrapper) {
18          return new RemoteTestModel().queryPage(page,queryWrapper);
19      }
20      @Override
21      @Function
22      public String hello(String name) {
23          return "hello" + name ;
24      }
25
26      @Override
27      @Function
28      public String remoteTest(RemoteTestModel remoteTest) {
29          return "remoteTest" + remoteTest.getName() ;
30      }
31  }
```

Step2. 新增 RemoteTestModelProxyAction，代码如下。

（1）新增 RemoteTestModelProxyAction，并定义 queryPage、hello、remoteTest 等 Action，方便前端单击测试。

（2）引入 RemoteTestModelService，分别调用 queryPage、hello、remoteTest 方法。

```java
19    @Model.model(RemoteTestModelProxy.MODEL_MODEL)
20    @Component
21    @Slf4j
22    public class RemoteTestModelProxyAction {
23        @Autowired
24        private RemoteTestModelService remoteTestModelService;
25
26        @Function.Advanced(type= FunctionTypeEnum.QUERY)
27        @Function.fun(FunctionConstants.queryPage)
28        @Function(openLevel = {FunctionOpenEnum.API})
29        public Pagination<RemoteTestModelProxy> queryPage(Pagination<RemoteTestModelProxy> page, IWrapper<RemoteTestModelProxy> queryWrapper){
30
31            Pagination<RemoteTestModel> pageConvert = ArgUtils.convert(RemoteTestModelProxy.MODEL_MODEL,RemoteTestModel.MODEL_MODEL,page);
32            IWrapper<RemoteTestModel> queryWrapperConvert = ArgUtils.convert(RemoteTestModelProxy.MODEL_MODEL,RemoteTestModel.MODEL_MODEL,queryWrapper);
33            Pagination<RemoteTestModel> resultTmp= remoteTestModelService.queryPage(pageConvert,queryWrapperConvert);
34            Pagination<RemoteTestModelProxy> result = ArgUtils.convert(RemoteTestModel.MODEL_MODEL,RemoteTestModelProxy.MODEL_MODEL,resultTmp);
35            return result;
36        }
37
38        @Action(displayName = "hello")
39        public RemoteTestModelProxy hello(RemoteTestModelProxy data){
40            String hello = remoteTestModelService.hello(data.getName());
41            log.info(hello);
42            return data;
43        }
44        @Action(displayName = "remoteTest")
45        public RemoteTestModelProxy remoteTest(RemoteTestModelProxy data){
46            RemoteTestModel s = ArgUtils.convert(RemoteTestModelProxy.MODEL_MODEL,RemoteTestModel.MODEL_MODEL,data);
47            String testRemote = remoteTestModelService.remoteTest(s);
48            log.info(testRemote);
49            return data;
50        }
51    }
```

Step3. 重启看效果。

单击 remoteTest 按钮查看效果，如图 4-94 所示。

图 4-94　单击 remoteTest 按钮查看效果

总结案例一、二、三。

把 SecondModule 模块停掉，则上面一、二、三 Case 都会报错。如果 DemoModule 的 pamirs-demo-boot 中依赖加上 pamirs-second-core，同时 application-dev.yml 文件中 pamirs.boot.modules 加上 second_core 的配置，则 DemoModule 和 SecondModule 部署在一起，再把前端 node 请求代理都转发到 DemoModule 模块上，则只需要启动 pamirs-demo-boot 就可以了。这个留给小伙伴们自己试验。

分布式或不分布式只是部署方式的差异，代码层面没有差异，有了体感以后再去看 2.4.1 一节可能会有更深的体会。

注：

（1）编码远程调用服务时要确保入参和服务端定义是一样的，可以用平台提供 ArgUtils.convert 进行转换。

（2）部署分布式情况下要用 ng 或其他方式进行转发，针对前端发起的请求，根据请求 url 中带的模块信息转发到对应有启动该模块的 boot 应用中。

## 4.4  Oinone 的分布式体验进阶

在分布式开发中，每个人基本只负责自己相关的模块开发。所以每个研发就都需要一个环境，比如一般公司会有 N 个项目环境、1 个日常环境、1 个预发环境、1 个线上环境。在配置项目环境的时候就特别麻烦，Oinone 的好处在于每个研发可以通过 boot 工程把涉及的模块都启动在一个 jvm 中进行开发，并不依赖任何环境，在项目开发中，特别方便。但当公司系统膨胀到一定规模，大到很多人都不知道有哪些模块，或者公司出于安全策略考虑，或者为了提高启动速度（毕竟模块多了启动的速度也会降下来），就需要将 Oinone 与经典分布式组织模式融合。本文就给大家介绍 Oinone 与经典分布式组织模式的兼容性。

**1. 模块启动的最小集**

我们来改造 SecondModule 模块，让该模块的用户权限相关都远程走 DemoModule。

Step1. 修改 SecondModule 的启动工程 application-dev.yml 文件。

除了 base、second_core 两个模块保留，其他模块都去除了。

```yaml
pamirs:
  boot:
    init: true
    sync: true
    modules:
      - base
      - second_core
```

Step2. 去除 boot 工程的依赖，代码如下。

去除 SecondModule 启动工程的 pom 依赖。

```xml
<!-- <dependency>
    <groupId>pro.shushi.pamirs.core</groupId>
    <artifactId>pamirs-resource-core</artifactId>
</dependency>
<dependency>
    <groupId>pro.shushi.pamirs.core</groupId>
    <artifactId>pamirs-user-core</artifactId>
</dependency>
<dependency>
    <groupId>pro.shushi.pamirs.core</groupId>
    <artifactId>pamirs-auth-core</artifactId>
</dependency>
<dependency>
    <groupId>pro.shushi.pamirs.core</groupId>
```

```
15      <artifactId>pamirs-message-core</artifactId>
16    </dependency>
17    <dependency>
18      <groupId>pro.shushi.pamirs.core</groupId>
19      <artifactId>pamirs-international</artifactId>
20    </dependency>
21    <dependency>
22      <groupId>pro.shushi.pamirs.core</groupId>
23      <artifactId>pamirs-business-core</artifactId>
24    </dependency>
25    <dependency>
26      <groupId>pro.shushi.pamirs.core</groupId>
27      <artifactId>pamirs-apps-core</artifactId>
28    </dependency> -->
```

Step3. 重启 SecondModule。

远程模型和远程代理菜单均能正常访问，如图 4-95 所示。

图 4-95　远程模型和远程代理菜单均能正常访问

Step4. SecondModule 增加对模块依赖。

我们让 SecondModule 增加用户和权限模块的依赖，期待效果是：SecondModule 对用户和权限的访问都会走 Dome 应用，因为 Demo 模块的启动工程中包含了 user、auth 模块。

（1）修改 pamirs-second-api 的 pom 文件，增加对 user 和 auth 的 API 包依赖，代码如下。

```
1  <dependency>
2    <groupId>pro.shushi.pamirs.core</groupId>
3    <artifactId>pamirs-user-api</artifactId>
4  </dependency>
5  <dependency>
6    <groupId>pro.shushi.pamirs.core</groupId>
7    <artifactId>pamirs-auth-api</artifactId>
8  </dependency>
```

（2）修改 SecondModule 类，增加依赖定义，代码如下。

```
1  @Module(
2      dependencies = {ModuleConstants.MODULE_BASE, AuthModule.MODULE_MODULE, UserModule.MODULE_MODULE}
3  )
```

Step5. 修改 RemoteTestModel 模型。

为 RemoteTestModel 模型增加 user 字段，代码如下。

```
1   @Field.many2one
2   @Field(displayName = "用户")
3   private PamirsUser user;
```

Step6. 重启系统看效果。

（1）mvn install pamirs-second 工程，因为需要让 pamirs-demo 工程能依赖到最新的 pamirs-second-API 包。

（2）重启 pamirs-second 和 pamirs-demo。

（3）两个页面都正常，如图 4-96、图 4-97 所示。

图 4-96　远程模型的代理列表

图 4-97　页面显示正常

**2. PmetaOnline 的 NEVER 指令（开发时环境共享）**

我们在 4.1.2 节中介绍过 "-PmetaOnline 指令"，该参数用于设置元数据在线的方式，如果不使用该参数，则 profile 属性的默认值请参考 "服务启动可选项"。-PmetaOnline 参数可选项为：

（1）NEVER——不持久化元数据，会将 pamirs.boot.options 中的 updateModule、reloadMeta 和 updateMeta 属性设置为 false。

（2）MODULE——只注册模块信息，会将 pamirs.boot.options 中的 updateModule 属性设置为 true，reloadMeta 和 updateMeta 属性设置为 false。

（3）ALL——注册持久化所有元数据，会将 pamirs.boot.options 中的 updateModule、reloadMeta 和 updateMeta 属性设置为 true。

Oinone 的默认模式下元数据都是注册持久化到 DB 的，但当我们在分布式场景下新开发模块或者对已有模块进行本地化开发时，作为开发阶段我们肯定是希望复用原有环境，但不对原有环境造成影响。那么 -PmetaOnline 就很有意义。让我们还没有经过开发自测的代码产生的元数据仅限于开发本地环境，而不是直接影响整个大的项目环境。

**PmetaOnline 指令设置为 NEVER（举例）**

Step1. 为 DemoCore 新增一个 DevModel 模型，代码如下。

```
 6    @Model.model(DevModel.MODEL_MODEL)
 7    @Model(displayName = "开发阶段模型",summary="开发阶段模型，当PmetaOnline指令设置为NEVER
      时，本地正常启动但元数据不落库",labelFields={"name"})
 8  public class DevModel  extends AbstractDemoCodeModel{
 9
10      public static final String MODEL_MODEL="demo.DevModel";
11
12      @Field(displayName = "名称")
13      private String name;
14
15  }
```

Step2. 为 DevModel 模型配置菜单，代码如下。

```
 1    @UxMenu("开发模型")@UxRoute(DevModel.MODEL_MODEL) class DevModelProxyMenu{}
```

Step3. 启动 Demo 应用时指定 -PmetaOnline，如图 4-98 所示。

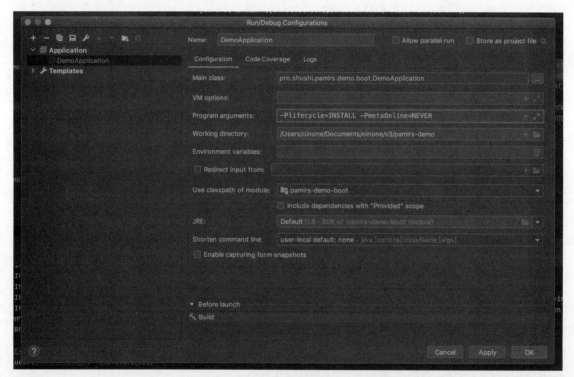

图 4-98　启动 Demo 应用时指定 -PmetaOnline

Step4. 重启系统看效果。

（1）查看元数据，如图 4-99 所示。

图 4-99　查看元数据

（2）菜单与详情页面能正常操作，如图 4-100、图 4-101 所示。

图 4-100　开发模型菜单可正常操作

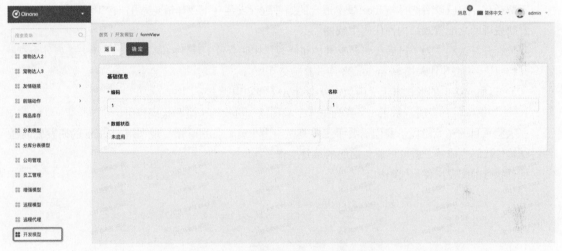

图 4-101　开发模型详情页面可正常操作

Step5. NEVER 模式需注意的事项。

（1）业务库需设定为本地开发库，这样才不会影响公共环境，因为对库表结构的修改还是会正常进行的。

（2）如果不小心影响了公共环境，需要对公共环境进行重启恢复。

（3）系统新产生的元数据（如：例子中的"开发模式"菜单）不受权限管控。

### 3. 分布式开发约定

1）设计约定

（1）对于跨模块的存储模型间继承，在部署时需要跟依赖模块配置相同数据源。这个涉及模块规划问题，比如业务上的 user 扩展模块，需要跟 user 模块一起部署。

（2）对于跨模块的多对多关系，配置中间表模型，3.3.9 节中的 m2m 字段有相关介绍。

（3）基础依赖：base。

（4）canal 相关参考：4.1.10 节中提到的修改 topic 方案。

2）编码约定

（1）跨模块间用 service 来调用，避免直接调用模型的数据管理器（如 new 非本模块 Model().queryPage 等）。结构性代码可以用 Oinone 提供的研发辅助工具进行提效率。

（2）在 3.4.1 节中提到，需要在 API 包中声明注解了 @Fun 注解的函数接口，并在接口的方法上加上 @Function 注解。

## 4.5 研发辅助

如下都是一些提升研发效率的小工具。

### 4.5.1 研发辅助之插件——结构性代码

研发辅助目的：

（1）消灭研发过程中的重复性工作，提升研发效率，如使用结构性代码；

（2）提供生产示例性代码，如根据模型生成导入/导出、view 自定义配置等经常性开发功能。

**1. 插件安装**

下载插件，解压以后选择与 Idea 版本适合的插件进行安装（下载详情见本书末"附录"）。

**2. 研发辅助之配置式结构性代码生成器**

我们在开发过程中为了日后代码易于维护和修改，往往会做工程性的职责划分。除去模型外会有：

（1）代理模型和代理模型 Action 来负责前端交互；

（2）以面向接口的形式来定义函数，就会有 API 和实现类之分；

（3）如果项目有多端，Action 又要为每一个端构建一份代理模型和代理模型。

在大型项目的初始阶段，我们需要手工重复做很多事情，特别麻烦。现在用 Oinone 的研发辅助插件的结构性代码生成器，就可以避免前面的重复工作。

插件执行的配置文件，代码如下。

```xml
<?xml version="1.0" encoding="utf-8" ?>
<oinone>
    <makers>
        <!-- 根据模型生成代理类、代理类的Action、Service、ServiceImpl -->
        <maker>
            <!-- 选择模型所在位置 -->
            <modelPath>/Users/oinone/Documents/oinone/demo/pamirs-second/pamirs-second-api/src/main/java/pro/shushi/pamirs/second/api/model</modelPath>
            <!-- 代理模型、代理模型Action生成相关配置信息 -->
            <proxyModules>
                <module>
                    <!-- 代理模型和代理模型Action的生成位置信息 -->
                    <generatePath>/Users/oinone/Documents/oinone/demo/pamirs-second/pamirs-second-api/src/main/java/pro/shushi/pamirs/second/api</generatePath>
                    <!-- 代理模型和代理模型Action的模块前缀 -->
                    <modulePrefix>second</modulePrefix>
                    <!-- 代理模型和代理模型Action的模块名，代理模型和代理模型Action类名为moduleName+模型名+"Proxy"+"Action" -->
                    <moduleName>second</moduleName>
                    <!-- 代理模型和代理模型Action的包名,实际包名为 packageName+".proxy"或packageName+".action"-->
                    <packageName>pro.shushi.pamirs.second.api</packageName>
                </module>
            </proxyModules>
```

```xml
21        <!-- 根据模型生成API，包括service（写方法）和queryService（读方法） -->
22        <apiModule>
23            <!-- service和queryService的生成位置信息 -->
24            <generatePath>/Users/oinone/Documents/oinone/demo/pamirs-second/pamirs-second-api/src/main/java/pro/shushi/pamirs/second/api</generatePath>
25            <!-- service和queryService的模块前缀 -->
26            <modulePrefix>second</modulePrefix>
27            <!-- service和queryService的模块名 -->
28            <moduleName>second</moduleName>
29            <!-- service和queryService的包名,实际包名为 packageName+".service" -->
30            <packageName>pro.shushi.pamirs.second.api</packageName>
31        </apiModule>
32        <!-- 根据模型生成API实现类，包括serviceImpl（写方法）和queryServiceImpl（读方法） -->
33        <coreModule>
34            <!-- serviceImpl和queryServiceImpl的生成位置信息 -->
35            <generatePath>/Users/oinone/Documents/oinone/demo/pamirs-second/pamirs-second-core/src/main/java/pro/shushi/pamirs/second/core</generatePath>
36            <!-- serviceImpl和queryServiceImpl的模块前缀 -->
37            <modulePrefix>second</modulePrefix>
38            <!-- serviceImpl和queryServiceImpl的模块名 -->
39            <moduleName>second</moduleName>
40            <!-- serviceImpl和queryServiceImpl的包名,实际包名为 packageName+".service" -->
41            <packageName>pro.shushi.pamirs.second.core</packageName>
42        </coreModule>
43    </maker>
44   </makers>
45 </oinone>
46
```

**3. 研发辅助之多模型结构性代码生成器**

多模型结构性代码生成器是配置式结构性代码生成器的补充，应对开发后期维护中新增模型的场景。它与配置式结构性代码生成器的不同点在于只要选择模型文件就可以，不需要专门编写 xml 文件。生成的文件默认就在模型所在路径下。

Step1. 在菜单栏上选择"Oinone"选项，并在下拉菜单中选择"多模型结构性代码生成器"选项，如图 4-102 所示。

图 4-102　并单击子菜单：多模型结构性代码生成器

Step2. 设置必要的信息，如图 4-103 所示。
（1）模型前缀。
（2）模型的所属模块。
（3）代理模型的模块。
这三个信息分别用于构建：
（1）代理模型的 MODEL_MODEL = 模型前缀 . 代理模型的模块 . 代理模型类名；
（2）服务的 FUN_NAMESPACE = 模型前缀 . 代理模型的模块 . 服务类名。

图 4-103　设置必要的信息

Step3. 选择为哪些模型生成对应的结构性代码，如图 4-104 所示。

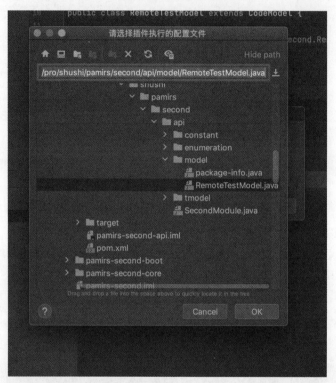

图 4-104　选择为哪些模型生成对应的结构性代码

Step4. 选择代码在模型所在目录，如图 4-105 所示。

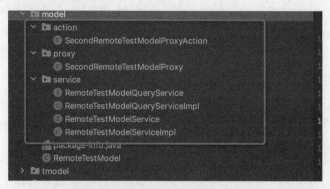

图 4-105　选择代码在模型所在目录

生成的文件默认就在模型所在路径下，您可以手动拖动到对应的包路径中。

## 4.5.2 研发辅助之 SQL 优化

在 Oinone 体系中不需要针对模型写 SQL 语句，默认提供了通用的数据管理器。在带来便利的情况下，也导致传统的 SQL 审查就没办法开展。但是我们可以使用技术的手段收集慢 SQL 并限制问题 SQL 执行。

（1）慢 SQL 收集目的：去发现非原则性问题的慢 SQL，并进行整改。

（2）限制问题 SQL 执行：对一些不规范的 SQL 在系统上直接做限制，如果有特殊情况手动放开。

**1. 发现慢 SQL**

这个功能并没有直接加入到 Oinone 的版本中，需要业务自行写插件，插件代码如下。大家可以根据实际情况进行改造。

（1）堆栈入口，例子中只是使用了 pamirs，可以根据实际情况改成业务包路径。

（2）对慢 SQL 的定义是运行时间为 5s 还是 3s，根据实际情况设置。

```java
@Intercepts({
        @Signature(type = Executor.class,method = "query",args = {MappedStatement.class,Object.class, RowBounds.class, ResultHandler.class})
})
@Component
@Slf4j
public class SlowSQLAnalysisInterceptor implements Interceptor {

    @Override
    public Object intercept(Invocation invocation) throws Throwable {
        long start = System.currentTimeMillis();
        Object result = invocation.proceed();
        long end = System.currentTimeMillis();
        if (end - start > 10000) {//大于10秒
            try {
                StackTraceElement[] stackTraceElements = Thread.currentThread().getStackTrace();
                StringBuffer slowLog = new StringBuffer();
                slowLog.append(System.lineSeparator());
                for (StackTraceElement element : stackTraceElements) {
                    if (element.getClassName().indexOf("pamirs") > 0) {
                        slowLog.append(element.getClassName()).append(":").append(element.getMethodName()).append(":").append(element.getLineNumber()).append(System.lineSeparator());
                    }
                }
                Object parameter = null;
                if (invocation.getArgs().length > 1) {
                    parameter = invocation.getArgs()[1];
                }
                MappedStatement mappedStatement = (MappedStatement) invocation.getArgs()[0];
                BoundSql boundSql = mappedStatement.getBoundSql(parameter);
                Configuration configuration = mappedStatement.getConfiguration();
                String originalSql = showSql(configuration, boundSql);
                originalSql = originalSql.replaceAll("\'", "").replace("\"", "");
                log.warn("检测到的慢SQL为：" + originalSql);
                log.warn("业务慢SQL入口为：" + slowLog.toString());
            } catch (Throwable e1) {
                //忽略
            }
```

```java
48              }
49              return result;
50          }
51  
52      public String showSql(Configuration configuration, BoundSql boundSql) {
53          Object parameterObject = boundSql.getParameterObject();
54          List<ParameterMapping> parameterMappings = boundSql.getParameterMappings();
55          String sql = boundSql.getSql().replaceAll("[\\s]+", " ");
56          if (parameterMappings.size() > 0 && parameterObject != null) {
57              TypeHandlerRegistry typeHandlerRegistry = configuration.getTypeHandlerRegistry();
58              if (typeHandlerRegistry.hasTypeHandler(parameterObject.getClass())) {
59                  sql = sql.replaceFirst("\\?", getParameterValue(parameterObject));
60  
61              } else {
62                  MetaObject metaObject = configuration.newMetaObject(parameterObject);
63                  for (ParameterMapping parameterMapping : parameterMappings) {
64                      String propertyName = parameterMapping.getProperty();
65                      if (metaObject.hasGetter(propertyName)) {
66                          Object obj = metaObject.getValue(propertyName);
67                          sql = sql.replaceFirst("\\?", getParameterValue(obj));
68                      } else if (boundSql.hasAdditionalParameter(propertyName)) {
69                          Object obj = boundSql.getAdditionalParameter(propertyName);
70                          sql = sql.replaceFirst("\\?", getParameterValue(obj));
71                      }
72                  }
73              }
74          }
75          return sql;
76      }
77  
78      private String getParameterValue(Object obj) {
79          String value = null;
80          if (obj instanceof String) {
81              value = "'" + obj.toString() + "'";
82          } else if (obj instanceof Date) {
83              DateFormat formatter = DateFormat.getDateTimeInstance(DateFormat.DEFAULT, DateFormat.DEFAULT, Locale.CHINA);
84              value = "'" + formatter.format(obj) + "'";
85          } else {
86              if (obj != null) {
87                  value = obj.toString();
88              } else {
89                  value = "";
90              }
91          }
92          return value;
93      }
94  
95  }
```

**2. 限制问题 SQL**

Oinone 的非法 SQL 校验插件：IllegalSQLInterceptor。

目前版本的 Oinone 并没有生效该非法 SQL 校验的拦截器，该拦截器会在后续版本中开放，让伙伴自行决定是否生效。

# 第 5 章　Oinone 的 CDM

Oinone 的 CDM 是商业领域的通用模型，更是结合 Oinone 的技术特性提出的新工程建议。本章重点针对 CDM 的设计背景、原理以及商业领域的基础支撑模型展开介绍。

1. CDM 的背景介绍
2. CDM 的设计原理
3. 基础支撑之用户与客户域
4. 基础支撑之商业关系域
5. 基础支撑之结算域
6. 商业支撑之商品域
7. 商业支撑之库存域
8. 商业支撑之执行域

## 5.1　CDM 的背景介绍

如果说低代码开发框架输出技术标准，CDM 则是结合 Oinone 技术特性和软件工程设计，让输出数据标准变成可能。

**1. 背景介绍**

1）无法照搬的最佳实践

要了解引入 CDM 的初衷，得从互联网架构的演进开始，了解其过程，就知道为什么说 Oinone 的 CDM 是中台架构的最佳技术实践的核心！我们在 2.2 节中介绍过互联网技术发展的四个阶段，特别是平台化到中台化的阶段，目的是在一套规范下让听得见炮火声音的团队自行决定业务系统发展，适用多业务线（或多场景应用）独立发展。

互联网架构在演进过程中碰到的问题跟企业数字化转型过程中碰到的问题是非常类似的：

（1）随着企业业务在线化后对系统的性能和稳定性都提出了更高的要求，而且大部分企业的内部很多系统相互割裂，导致很多重复建设，所以我们需要服务化、平台化。

（2）没有一个供应商能同时解决企业的所有商业场景问题，而是需要多个供应商共同参与，所以把供应商类比成各个业务线，在一套规范下让供应商自行决定业务系统发展。

既然跟阿里当初在架构演进过程中碰到的问题非常类似，那么，是不是照搬阿里中台架构方案到企业就好了？当然不是，因为历史原因，阿里的中台架构采用的是平台共建模式："让业务线研发以平台设计好的规范进来共同开发。"其本质还是平台主导模式，它有非常大的历史包袱。我们想象各个供应商共建一个交易平台或商品平台，那是多么荒唐的事情，平台化已经足够复杂了，还让不同背景、不同企业的研发一起共建，最后往往导致企业架构负载过重，这时对企业来说便不再是赋能而是"内耗"。

那么，如果没有历史包袱，我们重新设计，站在上帝视角去看看有没有更好的方式呢？当然有。

2）借鉴微软的 CDM

这里我们借鉴微软的 CDM 理念。CDM 这个概念最早是于 2016 年由微软宣布"以 Dynamics 365 的形式改造其 CRM 和 ERP"战略时提出的。微软给它的定义是：用于存储和管理业务实体的业务数据库，而且是"开箱即用"的。CDM 不仅提供标准实体，它还允许用户建立个性化的实体，用户既可以扩展标准实体也可以增加和标准实体相关的新实体。

CDM 可能并不性感，但绝对是非常必要的。它成为了微软很多产品的基础，构建了无数业务领域的原型。同时微软也期望它能快速成为数据交换和迁移的标准，这个有点像菜鸟网络推出的奇门，让所有的 TMS、OMS、WMS 都基于一套数据接口 API 进行互通，一套标准是为了解决一个行业问题，而不是具体到某一个企业、某一个集团的问题。

我们发现 CDM 的理念跟我们想要的"企业级的数据标准"非常吻合。但是我们也不能照搬照抄。虽然微软的 CDM 很好地解决了数据割裂问题，但对于模型方面来说，模型库非常庞大且复杂，学习成本巨高。

3）数字化时代软件会产生新的技术流派

我们知道传统软件的设计理念：侧重在模型对业务支撑的全面性上。优点体现为配置丰富，缺点为模型设计过于复杂，刚开始有前瞻性，但理解、维护都非常困难，随着业务发展系统原先的设计逐渐腐化，会变得异常笨重。

而 Oinone 的 CDM 设计理念：侧重在简单、灵活、统一上，体现为在上层应用开发时，每一业务领域保持独立，模型简单易懂，并结合 Oinone 的低代码开发机制进行快速开发，灵活应对业务变化。

所以我更想说 Oinone 的 CDM 是微软的 CDM 在原有基础上，与互联网架构结合，利用 Oinone 低代码开发平台特性形成新的工程化建议。Oinone 的 CDM 不以把模型抽象到极致、支撑"所有业务可能性"为目标，而是抽象 80% 通用的设计，保持模型简单可复用，来解决数据割裂问题，并保持业务线独立自主性、快速创新的能力，如图 5-1 所示。

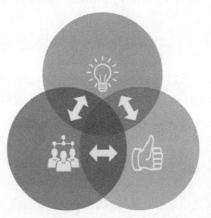

图 5-1　Oinone 的 CDM 特性

**2. Oinone 的 CDM 本质是创新的工程化建议**

引入 CDM 以后系统工程结构会有什么变化，跟大家认知的互联网架构有什么区别。

原本上层的业务线系统，需要调用各个业务平台提供的功能、增加 CDM 以后如图 5-2（b）所示，每个业务线都是独立的。看上去复杂了，其实对业务线来说更加简单了。

互联网整体平台化带来的问题：

（1）业务线每次业务调整都需要给各个平台提需求；

（2）业务平台研发需要了解所有业务线的知识再做设计，对研发要求非常高；

（3）各个业务域的不同需求相互影响，包括系统稳定性、研发对需求响应的及时性。

结合 Oinone 特性提出的新工程建议：

（1）一些通用性模块继续以平台化的方式存在，能力完全复用；

（2）业务线自建业务平台，保持业务线的独立性和敏捷性；

（3）业务线以 CDM 为原型，保证核心数据不割裂，形成一致的数据规范。

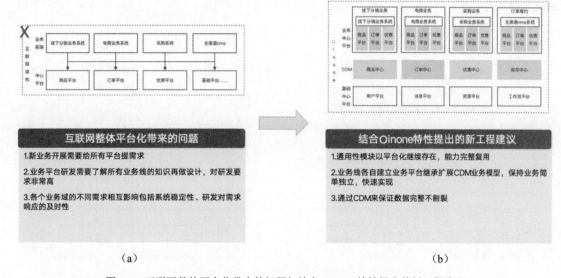

图 5-2　互联网整体平台化带来的问题与结合 Oinone 特性提出的新工程建议

**3. CDM 思路示意图**

该示例中 Oinone 的 CDM 商品域不仅仅提供标准实体，还保证各个业务系统对商品的通用需求、简单易懂。在我们的星空系列业务产品中如全渠道运营、B2B 交易等系统以此为基础建立属于自身个性化的实体，既可以扩展标准实体也可以增加和标准实体相关的新实体，如图 5-3 所示。

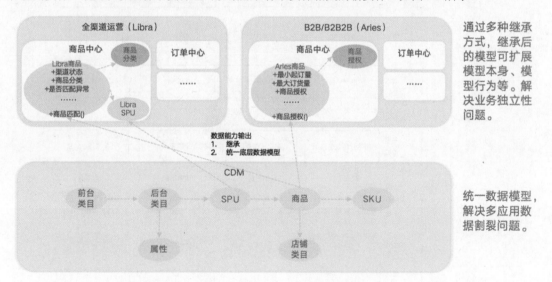

图 5-3　Oinone 的 CDM 思路示意图

带来的好处：

（1）通过多种继承方式，继承后的模型可扩展模型本身、模型行为等，从而解决业务独立性问题；

（2）通过 CDM 层统一数据模型，从而解决多应用数据割裂问题。

## 5.2　CDM 之工程模式

**1. 两种工程模式介绍**

Oinone 推荐的两种 CDM 工程模式都保留互联网特性，如设置跟业务无关的基础平台以及采用平台化思路建设，如图 5-4 所示。

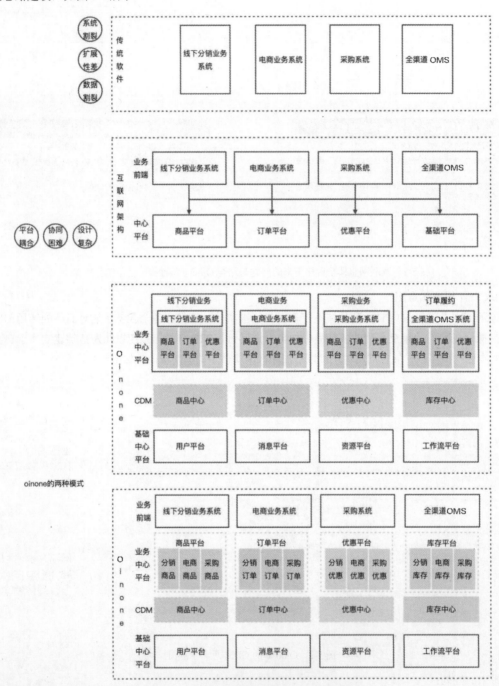

图 5-4　Oinone 推荐的两种 CDM 工程模式

两种侧重点差异如下：

第一种：比较适合企业采用多供应商联合开发场景，先以业务区分，各个业务线有独立的领域平台，最大限度保持不同业务线的独立性，有利于各个业务线独立发展（目前 Oinone 上层星空系列产品采用这种工程模式，因为我们期望帮助企业构建软件生态，必然要考虑不同供应商联合开发场景）。

第二种：比较接近传统互联网架构，先按平台领域区分。如在商品领域，商品平台做总工程，但里面按业务区分模块分子工程来保持业务相互独立，相对于第一种模式基础上把领域的代码放一起，带来的好处是强化大家思考模型的通用性，但不适用于跨公司主体间配合。

注意事项：

（1）Oinone 兼容传统互联网架构。

（2）不管哪种模式，都需要解决 CDM 的维护问题。

**2. CDM 维护常见问题**

Q1：CDM 层缺少模型怎么办？

A1：CDM 层模型是逐步完善和丰富的。如果是特定业务需要的模型，这类模型无通用性，则自己加到工程中；如果是通用的，则通过架构组确定是否需要纳入 CDM。

Q2：CDM 层已有的模型缺少字段怎么办？

A2：CDM 层模型的字段也是逐步完善和丰富的，通用的字段在架构组确定后也会被吸收进来。

Q3：CDM 层不同业务线相互影响怎么办？

A3：扩展字段最好带上自有前缀标志，如果觉得通用，则提交至架构组通过模型缺少字段加入。

Q4：CDM 层某模型新增加了的字段，但原先业务线已经加了相同含义字段怎么办？

A4：业务线可以把自己的字段关联到 CDM 增加的新字段，并做数据迁移。

## 5.3 基础支撑之用户与客户域

**1. 三户模型概念**

1）三户模型的由来

经典的"三户模型"是电信运营支持系统的基础。三户模型即客户、用户和账户，来源于 etom 模型。三者应该是一个相互关联但又彼此独立的三个实体，这种关联只是一个归属和映射的关系，分别体现了完全不同的几个域的信息，客户体现了社会域的信息，用户体现了业务域的信息，账户体现的是资金域的信息。

客户是个社会化概念，指一个自然人或一个法人。

用户是客户使用运营商开发的一个产品以及基于该产品之上的增值业务时，产生的一个实体。如果说一个客户使用了多个产品，那么一个客户就会对应好几个用户（即产品）。

账户的概念起源于金融业，是一个客户在运营商存放资金的实体，目的是为选择的产品付费。

2）Oinone 的三户模式

在原三户模型中"用户"是购买关系产生的产品与客户关系的服务实例，在互联网发展中用户的概念发生了非常大的变化，"用户"概念变成了"使用者"，是指使用电脑或网络服务的人，通常拥有一个用户账号，并以用户名识别。而且新概念在互联网强调用户数的大背景下已经被普遍使用，再去强调电信行业的用户概念就会吃力不讨好。而且不管是企业应用领域还是互联网领域，原用户概念都显得过于复杂和没有必要。这就有了 Oinone 特色的三户模型：

客户：它是个社会化的概念，一个自然人或一个法人。

用户：使用者，是指使用电脑或网络服务的人，通常拥有一个用户账号，并以用户名识别。

除此之外账户的定义与经典三户模型相同。

**2. Oinone 的客户与用户**

三户模型是构建上层应用的基础支撑能力，任何业务行为都跟这里的两个实体脱不了干系。以客户为中心建立商业关系与商业行为主体，以用户为中心构建一致体验与操作行为主体。在底层设计上二者相互独立并无关联，由上层应用自行做关联绑定，往往在登录时在 Session 的处理逻辑中会根据用户去找到对应一个或多个商业（主体）客户，如图 5.5 所示。Session 的实现可以参考 4.1.20 节。

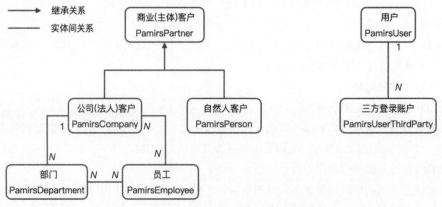

图 5-5 Oinone 的客户与用户

1）客户设计说明

（1）PamirsPartner 作为商业关系与商业行为的主体，派生了两个子类，PamirsCompany 与 PamirsPerson，分别对应公司（法人）客户与自然人客户。

（2）公司（法人）客户 PamirsCompany 对应多个组织部门 PamirsDepartment，公司（法人）客户 PamirsCompany 对应多个员工 PamirsEmployee。

（3）部门 PamirsDepartment 对应一个公司（法人）客户 PamirsCompany，对应多个员工 PamirsEmployee。

（4）员工 PamirsEmployee 对应多个部门 PamirsDepartment，对应一个或多个公司（法人）客户 PamirsCompany，其中有一个主要的公司。

2）用户设计说明

PamirsUser 作为一致体验与操作行为主体，本身绑定登录账号，并且可以关联多个三方登录账户 PamirsUserThirdParty。

3）客户与用户如何关联（举例）

（1）新建 Demo 系统的 PetCompany 和 PetEmployee，用 PetEmployee 去关联用户。

（2）当用户登录时，根据用户 ID 找到 PetEmployee，再根据 PetEmployee 找到 PetComany，把 PetComany 放到 Session 中去。

（3）修改 PetShop 模型关联一个 PamirsPartner，PamirsPartner 的信息从 Session 获取。

Step1. pamirs-demo-api 工程增加依赖，并且 DemoModule 中增加对 BusinessModule 的依赖，代码如下。

```
1  <dependency>
2      <groupId>pro.shushi.pamirs.core</groupId>
3      <artifactId>pamirs-business-api</artifactId>
4  </dependency>
```

@Module.dependencies 中增加 BusinessModule.MODULE_MODULE，代码如下。

```
1  @Module(
2      dependencies = { BusinessModule.MODULE_MODULE}
3  )
4
```

Step2. 新建 PetCompany 和 PetEmployee，以及对应的服务。

新建 PetEmployee，代码如下。

```java
 8      @Model.model(PetEmployee.MODEL_MODEL)
 9      @Model(displayName = "宠物公司员工",labelFields = "name")
10      public class PetEmployee extends PamirsEmployee {
11          public static final String MODEL_MODEL="demo.PetEmployee";
12
13          @Field(displayName = "用户")
14          private PamirsUser user;
15
16      }
17
```

新建 PetComany，代码如下。

```java
 7      @Model.model(PetCompany.MODEL_MODEL)
 8      @Model(displayName = "宠物公司",labelFields = "name")
 9      public class PetCompany extends PamirsCompany {
10
11          public static final String MODEL_MODEL="demo.PetCompany";
12
13          @Field.Text
14          @Field(displayName = "简介")
15          private String introductoin;
16
17      }
```

新建 PetEmployee 对应服务，代码如下。

```java
 7      @Fun(PetEmployeeQueryService.FUN_NAMESPACE)
 8      public interface PetEmployeeQueryService {
 9          String FUN_NAMESPACE ="demo.PetEmployeeQueryService";
10          @Function
11          PetEmployee queryByUserId(Long userId);
12
13      }
```

新建 PetEmployee 对应服务，代码如下。

```java
10      @Fun(PetEmployeeQueryService.FUN_NAMESPACE)
11      @Component
12      public class PetEmployeeQueryServiceImpl implements PetEmployeeQueryService {
13
14          @Override
15          @Function
16          public PetEmployee queryByUserId(Long userId) {
17              if(userId==null){
18                  return null;
19              }
20              QueryWrapper<PetEmployee> queryWrapper = new QueryWrapper<PetEmployee>().from(PetEmployee.MODEL_MODEL).eq("user_id", userId);
21              return new PetEmployee().queryOneByWrapper(queryWrapper);
22          }
23
24      }
25
```

新建 PetComany 对应服务，代码如下。

```
1  package pro.shushi.pamirs.demo.api.service;
2
3  import pro.shushi.pamirs.demo.api.model.PetCompany;
4  import pro.shushi.pamirs.meta.annotation.Fun;
5  import pro.shushi.pamirs.meta.annotation.Function;
6
7  @Fun(PetCompanyQueryService.FUN_NAMESPACE)
8  public interface PetCompanyQueryService {
9      String FUN_NAMESPACE ="demo.PetCompanyQueryService";
10     @Function
11     PetCompany queryByCode(String code);
12 }
```

新建 PetComany 对应服务，代码如下。

```
10  @Fun(PetCompanyQueryService.FUN_NAMESPACE)
11  @Component
12  public class PetCompanyQueryServiceImpl implements PetCompanyQueryService {
13
14      @Override
15      @Function
16      public PetCompany queryByCode(String code) {
17          if(StringUtils.isBlank(code)){
18              return null;
19          }
20          return new PetCompany().queryByCode(code);
21      }
22
23  }
24
```

Step3. Session 中增加 PamirsPartner，代码如下。

对 DemoSession\DemoSessionApi\DemoSessionData\DemoSessionHolder 进行修改，增加 PetCompany getCompany () 相关方法。可以参考 4.1.20 节。

```
12  public class DemoSessionCache {
13      private static final ThreadLocal<DemoSessionData> BIZ_DATA_THREAD_LOCAL = new ThreadLocal<>();
14      public static PamirsUser getUser(){
15          return BIZ_DATA_THREAD_LOCAL.get()==null?null:BIZ_DATA_THREAD_LOCAL.get().getUser();
16      }
17      public static PetCompany getCompany(){
18          return BIZ_DATA_THREAD_LOCAL.get()==null?null:BIZ_DATA_THREAD_LOCAL.get().getCompany();
19      }
20      public static void init(){
21          if(getUser()!=null){
22              return ;
23          }
24          Long uid = PamirsSession.getUserId();
25          if(uid == null){
26              return;
27          }
28          PamirsUser user = CommonApiFactory.getApi(UserService.class).queryById(uid);
29          if(user!=null){
30              DemoSessionData demoSessionData = new DemoSessionData();
31              demoSessionData.setUser(user);
```

```
32
33                PetEmployee employee = CommonApiFactory.getApi(PetEmployeeQueryServic
    e.class).queryByUserId(uid);
34            if(employee!=null){
35                PetCompany company = CommonApiFactory.getApi(PetCompanyQueryServi
    ce.class).queryByCode(employee.getCompanyCode());
36                demoSessionData.setCompany(company);
37            }
38            BIZ_DATA_THREAD_LOCAL.set(demoSessionData);
39        }
40    }
41    public static void clear(){
42        BIZ_DATA_THREAD_LOCAL.remove();
43    }
44 }
```

Step4. 修改 PetShop 模型，以及重写 PetShop 的默认 create 方法。

PetShop 模型增加 partner 字段，修改 openTime 为 readonly=true 的配置，变成带条件 readonly。scene == 'redirectUpdatePage' 表示只有在编辑页面为只读，代码如下。

```
1  @Field(displayName = "所属主体" )
2  @UxForm.FieldWidget(@UxWidget(readonly = "scene == 'redirectUpdatePage'"/* 在编辑页
   面只读 **/ ))
3  private PamirsPartner partner;
4
```

创建 PetShopAction 类重写 PetShop 模型的 create 方法，代码如下。

```
13  @Component
14  @Model.model(PetShop.MODEL_MODEL)
15  public class PetShopAction {
16
17      @Action.Advanced(name= FunctionConstants.create,managed = true)
18      @Action(displayName = "确定",summary = "确定",bindingType = ViewTypeEnum.FORM)
19      @Function(name=FunctionConstants.create)
20      @Function.fun(FunctionConstants.create)
21      public PetShop create(PetShop data){
22          //从session中获取登陆主体信息
23          PetCompany company =  DemoSession.getCompany();
24          if(company!=null){
25              data.setPartner(company);
26          }
27          data.create();
28          return data;
29      }
30
31  }
32
```

Step5. 增加 PetEmployee 和 PetCompany 的管理入口。

DemoMenus 增加 @UxMenu 注解声明，代码如下。

```
1  @UxMenu("公司管理")@UxRoute(PetCompany.MODEL_MODEL) class PetCompanyMenu{}
2  @UxMenu("员工管理")@UxRoute(PetEmployee.MODEL_MODEL) class PetEmployeeMenu{}
```

Step6. 重启应用看效果。

（1）创建公司与员工，在创建的同时建立公司与员工和员工与用户的关联，如图 5-6、图 5-7 所示。

图 5-6　创建公司与员工

图 5-7　员工与用户关联

（2）创建一个宠物店铺。

新增 Oinone 的"宠物店铺 003"，但不要选择所属主体。单击"确定"按钮后期望效果是：会从 session 中自动获取 admin 关联的 PetCompany，并填充到宠物店铺的所属主体字段中，如图 5-8、图 5-9 所示。

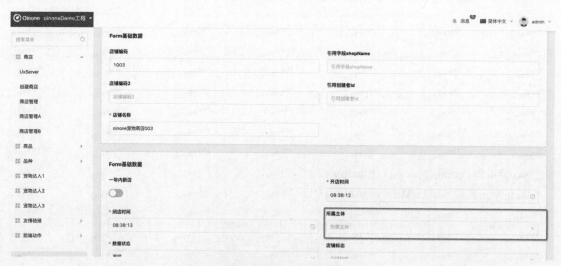

图 5-8　新增 Oinone 的"宠物店铺 003"

图 5-9 自动获取 admin 关联的 PetCompany

Step7. 注意事项。

（1）PetEmployee 的 create 方法应该重写，再调用 PamirsEmployeeService 的 create 方法。

（2）PetEmployeeQueryServiceImpl 的 queryByUserId 方法实现上也没有考虑一个用户绑定多个员工的模式。

（3）employeeType 字段默认是不展示的，但是例子中又没有重写 PetEmployee 的 create 方法，所以这个值是空的。但这个字段为空，会导致报错。解决办法有以下两个（如图 5-10 所示）：

① 重写 create 方法，手动给 employeeType 赋值；

② 自定义员工创建页面，拿到公司列表和部门列表两个字段。

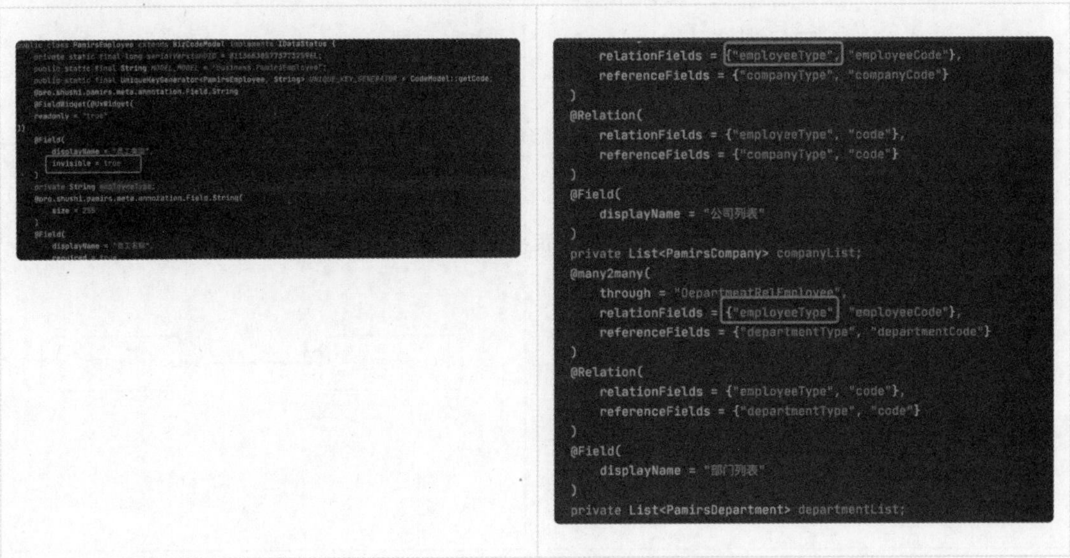

图 5-10 注意事项

### 3. 客户扩展

按 CDM 的设计理念，我们不以把模型抽象到极致，支撑所有业务为目标，而是抽象 80% 通用的设计，保持模型简单可理解。我们只提供了基础模型统一数据存储，用面向对象特性来解决多应用模型复用和数据割裂问题。

Oinone 的 CDM 是商业领域的通用模型，更是结合 Oinone 特性提出的新工程建议。

## 5.4 基础支撑之商业关系域

PamirsPartner 作为商业关系与商业行为的主体，关于 PamirsPartner 之间的关系如何描述，本文将介绍两种常见的设计思路，从思维和实现两个方面进行对比，给出 Oinone 为什么选择关系设计模式的原因。

**1. 两种设计模式对比**

1）设计模式思路介绍

两种设计模式思路分别是角色设计模式思路和关系设计模式思路。

（1）角色设计模式思路介绍。

从产品角度枚举所有商业角色，每个商业角色对应一个派生的商业主体，并对主体间的关系类型进行整理，如图 5-11 所示。

### 角色设计模式

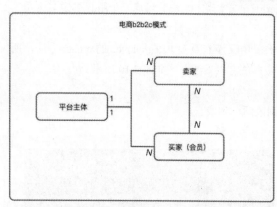
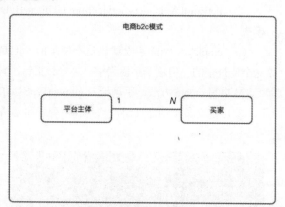
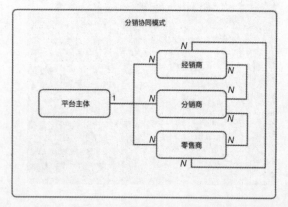
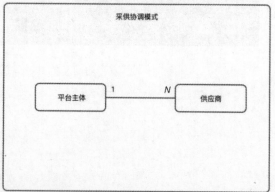

图 5-11　角色设计模式思路介绍

（2）关系设计模式思路介绍。

从产品角度枚举所有商业角色，每个商业角色对应一个派生的主体间商业关系，如图 5-12 所示。

## 关系设计模式

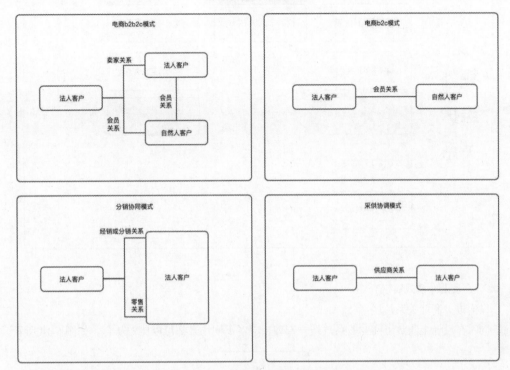

图 5-12 关系设计模式思路介绍

2）设计模式对应实现介绍

（1）角色设计模式实现介绍。

① 不仅商业主体需要扩展，关系也需要额外维护，可以是字段或是关系表。一般 M2O 和 O2M 进行字段维护，M2M 进行关系表维护。

② 创建合同场景中甲方选择"商业主体 A"，乙方必须是与"商业主体 A"有关联的经销商、分销商、零售商、供应商等，在角色设计模式下就非常麻烦，因为关系都是独立维护的，如图 5-13 所示。

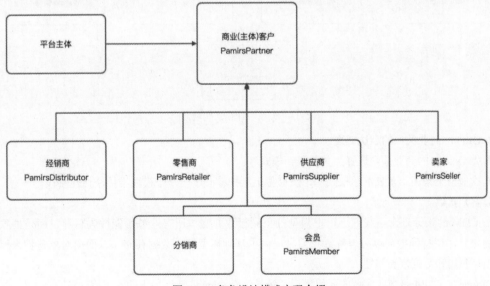

图 5-13 角色设计模式实现介绍

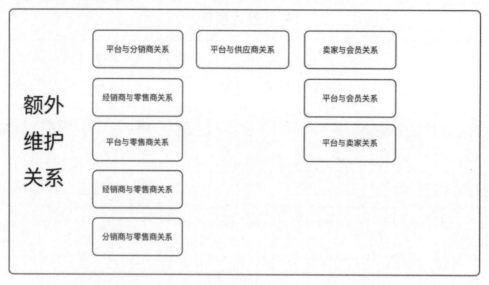

图 5-13 （续）

（2）关系设计模式实现介绍。

① 只需维护商业关系扩展；

② 同时在设计上收敛了商业关系，统一管理，应对不同场景都可以比较从容，如图 5-14 所示。

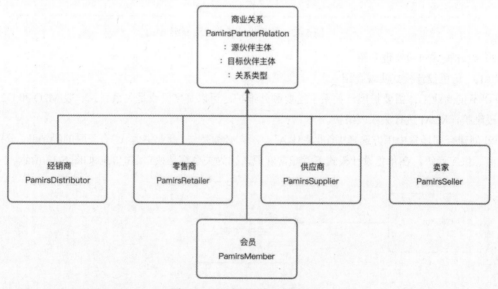

图 5-14　关系设计模式实现介绍

**2. Oinone 商业关系的默认实现**

首先 Oinone 的商业关系选择"关系设计模式"；

其次模型上采用"多表继承模式"，父模型上维护核心字段，子模型维护个性化字段。

**3. 客户扩展**

按 CDM 的设计理念，我们不以把模型抽象到极致并支撑所有业务可能性为目标，而是抽象 80% 通用的设计，保持模型简单可理解。我们只提供了基础模型统一数据存储，以面向对象特性来解决多应用模型复用和数据割裂问题。

Oinone 的 CDM 是商业领域的通用模型，更是结合 Oinone 特性提出的新工程建议。

## 5.5 基础支撑之结算域

**1. 基础介绍**

企业的业务不断进行数字化改造、业务越来越在线化，给企业财务工作带来几个明显的变化和挑战。

（1）变化。

● 业务在线后，不同类型收费、预售、授信模式的创新层出不穷，需要财务不仅只从事单一传统的会计核算工作，还需要积极地参与到业务中去。

● 从事后算账、事后报账，变成财务业务一体化信息的实时处理。

（2）挑战。

● 业务系统与财务系统明显割裂，业务部门与财务部门各自采用一套软件处理其数据，不能及时沟通信息和协同更正信息。

● 财务系统往往都是单体的传统架构，凭证处理能力无法适应今天企业不断爆发的业务发展。

● 财务的严谨性与业务的灵活性中间有巨大的鸿沟，导致业务要采用一种创新的模式，财务可能是最大阻碍。

不论是传统软件公司喜欢讲的业财一体化，还是互联网平台公司喜欢讲的结算平台，都是为了解决以上变化和挑战。业财一体化主要是从财务部门角度出发，在业务支撑上化被动为主动。结算平台往往是结合财务部门和业务运营部门的需求。如果以下面介绍的计费、账务、会计三个领域来说，业财一体化项目往往只包括账务和会计，结算中心往往包括计费、账务、会计。或者说业财一体化弱化了计费，没有将计费纳入企业统一管理，把如何计价交给业务系统自行决定，或者简单处理只产生应收应付单据（计费详单）就好了。

结算域是一个相对比较专业的领域，没有一定背景知识甚至连一些专业名词都很难理解，更不用说模型设计了，这里笔者尽快地简单去描述定位而不是描述细节。而且2.1.9版本的结算领域相对没有那么完善，这里介绍的是下个版本的内容，所以大家看当前版本的时候会有一些对不上。

**2. 子领域职责**

子领域职责如图5-15所示。

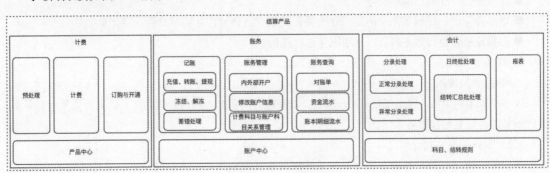

图5-15　子领域职责

1）计费

（1）计费的价值。

随着企业多业务发展以及融合计费需求，我们需要引入计费模型，对灵活计价模式进行支持，快速支撑未来可能的计费方式等。

（2）计费的核心设计理念。

所有的计算器都继承自虚函数计算器 $y=f(x)$。

计费的核心设计理念如图 5-16 所示。

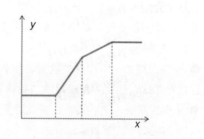

平滑兼容–默认斜率计算器 $y=a+bx$
$y$ — 求值结果(用下标描述结果是什么)
$a$ — 偏移量(计算固定值)
$b$ — 斜率(费率值)
$x$ — 变量(数量)
任何计算都是通过一组斜率组合出来的

利用区间限定定义各种斜率组合出各种算法
交易额
　　　　0~100w： $y=0.03x$
　　　　>100w： $y=0.02x;$
时间
　　　　0:00–6:00： $y=0.02x$
　　　　6:00–24:00： $y=0.03x$
$x$ – 变量，数量

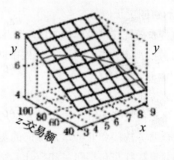

图 5-16　计费的核心设计理念

计费的核心设计理念可实现更灵活、多维区间组合，时间维度、计数器维度、其他属性维度在计数器区间斜率限定，比如交易额、空间、使用月份数等。

（3）计费的核心功能。
● 通过产品定义运营方案。
● 通过订购产品完成商务合同的签订来决定客户计费策略，或者通过系统产品定义通用计费策略。
● 支撑各类产品的模拟计费。
● 以事件驱动，根据事件、产品、订购关系完成产品路由，并实时产生计费详单。
● 根据计费科目与账务科目，打通账务进行核销。

2）账务
（1）账务的价值。
以账户账本为中心，提供记账、账户管理，以及账务的实时监控与持续对账。如果计费是对接业务，那么账务的价值是对接财务系统。

（2）账务的核心设计理念。
不依赖计费，可独立对接，所有业务最终都需要反馈到账户账本的操作上，并通过账本明细记录所有操作。

（3）账务的核心功能。
● 记账：充值、转账、提现、冻结、解冻、差错处理。
● 账务管理：开户、科目维护。
● 账务查询：对账。

3）会计（暂不在计划内）
会计的价值：结算平台的会计模块不是严格意义上的会计系统，它主要是衔接其他的财务系统，做凭证前置处理，目的在于汇总凭证、产出业务账、对接到财务总账系统、缓解财务系统压力。

**3. 模型介绍**

模型如图 5-17 所示。

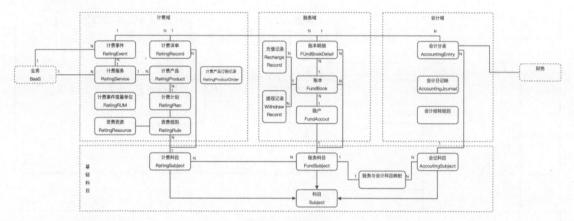

图 5-17　模型介绍

**4. 结算基础流程**

结算基础流程如图 5-18 所示。

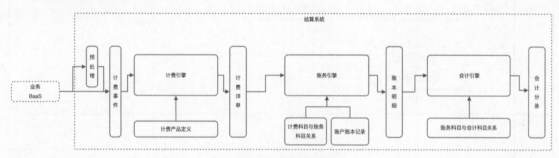

图 5-18　结算基础流程

**5. 客户扩展**

按 CDM 的设计理念，我们不以把模型抽象到极致支撑所有业务可能性为目标，而是抽象 80% 通用的设计，保持模型简单可理解。我们只提供了基础模型统一数据存储，以用面向对象特性来解决多应用模型复用和数据割裂问题。

Oinone 的 CDM 是商业领域的通用模型，更是结合 Oinone 特性提出的新工程建议。

## 5.6　商业支撑之商品域

**1. 基础介绍**

当业务在线化后，用于内部管理的产品主数据，叠加一堆销售属性变成了商品被推到了前台，成为导购链路中最重要的信息载体。看似最基础和最简单的商品模块也有很多注意事项，主要集中在以下几个方面：

（1）商品的属性如何管理、呈现、参与导购（类目、搜索的过滤条件）；

（2）如何解决固定不变的内部管理需求与基于销售特性长期变化的运营需求之间的矛盾；

（3）在多渠道情况下的渠道商品，如何映射到实际 SKU 进行履约。

**2. 模型介绍**

模型如图 5-19 所示。

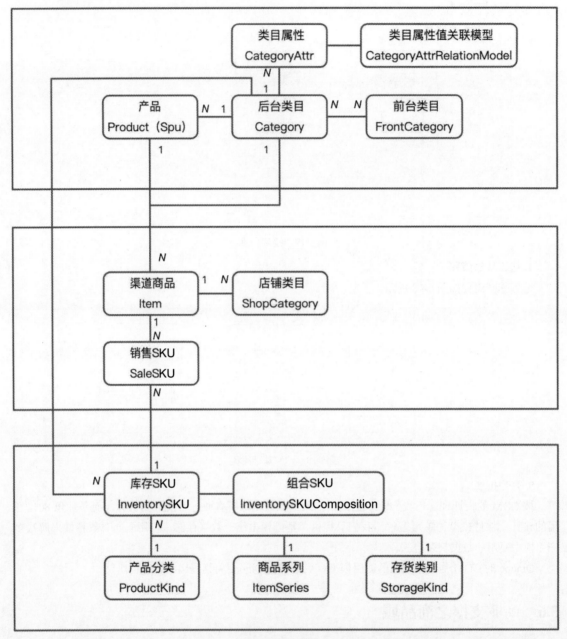

图 5-19　模型介绍

（1）类目属性，解决"商品的属性如何管理、呈现、参与导购（类目、搜索的过滤条件）"；

（2）前后台类目设计，解决"如何解决固定不变的内部管理需求与基于销售特性长期变化的运营需求之间的矛盾"；

（3）销售 SKU 和库存 SKU 设计，解决"在多渠道情况下的渠道商品，如何映射到实际 SKU 进行履约"。

要把这些问题搞清楚，得先把名词讲一下，具体见表 5-1。

表 5-1 各领域定义说明

| 领域 | 名称 | Oinone 的定义 | 说明 | 举例 |
|---|---|---|---|---|
| 平台运营视角 | SPU | Product → SPU<br>2.1.9 → 3.0.0 | SPU (Standard Product Unit)：标准化产品单元。SPU 是商品信息聚合的最小单位，是一组可复用、易检索的标准化信息的集合，该集合描述了一个产品的特性 | iPhone X 可以确定一个产品 |
| | 后台类目 | 后台类目（Category） | 商品分类分级管理，以及规范该类目下公共属性可以分为普通属性、销售属性 | 比如<br>类目：3C 数码 / 手机<br>销售属性：内存大小、颜色等<br>普通属性：分辨率 |
| | 前台类目 | 前台类目（FrontCategory） | 平台导购类目 | 通过前台类目关联后台类目或前台类目属性，用于满足运营需求 |
| 商家销售视角 | 大体上 SPU 处于最上层、Item 属于下一级，而 SKU 属于最低一层<br>SPU 是平台层面，Item 是商家层面，SKU 是商家的 Item 确定销售属性<br>SPU 非必需，在平台类交易中，平台方为了规范商家发布商品信息，进行统一运营时需要 | | | |
| | Item | 渠道商品（Item） | 简单来说：Item 是 SPU 加上归属商家以及商家自有的价格与描述 | 商家 A 的 iPhone X |
| | SKU | 销售 SKU（SaleSKU） | SKU=Stock Keeping Unit（库存保有单位）<br>是对每一个产品和服务的唯一标识符，该系统使用 SKU 的值基于数据管理，使公司能够跟踪系统，如仓库和零售商店或产品的库存情况 | iPhone X 64G 银色则是一个 SKU |
| | 店铺类目 | ShopCategory | 商家店铺导购类目 | 在平台类电商，商家都会有自己独立的店铺主页，商家类目跟前台类目作用类似，影响范围只局限为商家店铺内 |
| 商家管理视角 | 销售 SKU 中会有一个 InvSKUCode 来关联 InventorySKU，比如：品牌上在不同渠道（淘宝、京东、自建电商）中会有不同的销售 SKU，从渠道同步销售 SKU 会根据外部 code | | | |
| | 产品或库存 SKU | InventorySKU | 跟销售领域的 SKU 的定义类似，但销售领域是为了规范购买行为，这里规范企业内部管理 | iPhone X 64G 银色 |
| | 组合 SKU | InventorySKU Composition | | 空调由内外机组合而成，这就是一个组合 SKU |
| | 产品分类 | ProductKind | | |
| | 商品系列 | ItemSeries | 指互相关联或相似的产品，是按照一定的分类标准对企业生产经营的全部产品进行划分的结果。一个产品系列内往往包括多个产品项目。产品系列的划分标准有产品功能、消费上的连带性、面向的顾客群、分销渠道、价格范围等 | 企业内部管理划分 |
| | 存货类别 | StorageKind | 为了反映存货的组成内容，正确计算产品的生产成本以及销售成本，会计上必须对存货进行科学的分类，按存货的不同类别进行核算 | |

**3. 客户扩展**

按 CDM 的设计理念，我们不以把模型抽象到极致支撑所有业务可能性为目标，而是抽象 80% 通用的设计，保持模型简单可理解。我们只提供了基础模型统一数据存储，以用面向对象特性来解决多应用模型复用和数据割裂问题。

Oinone 的 CDM 是商业领域的通用模型，更是结合 Oinone 特性提出的新工程建议。

## 5.7 商业支撑之库存域

库存的差异会反馈到企业的整个价值链上，所以对库存的设计是至关重要的。

**1. 基础介绍**

我们先抛开仓库中对库存的实操管理和整个流通领域的库存，在企业自身一级的采销链路上我们可以从管理和销售两个角度去看。

从管理角度上我们会关心实物库存、在途库存、在产库存、库存批次等，也就是企业有多少库存，分布在哪里、什么环节。

从销售角度上我们会关心可售库存、安全库存等，也就是企业在特定渠道销售中库存的分配规则。

在商业场景中库存管理一头对接仓库、生成、采购，另一头对接多个销售渠道。它的挑战在于不同行业不同特征商品都有比较大的差异，比如家具行业的生产能力、家电区域化销售、生鲜拼车销售、服饰一仓销全国等特点。热销的商品要提升用户体验、防止超卖，滞销的商品要通过活动拉动流量，普通的商品要通过渠道共享最大化可售库存。库存管理的差异会反馈到企业的整个价值链上，所以对库存的设计是至关重要的。

库存设计挑战在于：

（1）技术上：库存设计类似账户、账本的设计，需要能追溯到库存变化的过程，且库存操作都可追溯业务单据，以及要注意热点数据的并发控制。

（2）业务上：在管理角度，上游能跟仓库、采购、生产等进行对接、对账，并为其设置可售规则，下游能为各个销售渠道设置库存分配与同步规则。

**2. 模型介绍**

核心设计逻辑：

（1）单据链路：业务单据（外部业务单据＋库存业务单据）产生库存指令（库存调整入/出库单），再由库存指令操作库存并记录库存流水。

（2）管理链路：基础数据维护仓库、供应商、服务范围与费用。这些数据是订单履约路由和可售库存同步的基础。

（3）库存数据：对外跟商品域同步，通过库存指令进行操作。不同库存各自维护自身库存与流水记录，确保可追溯。

（4）如果跟销售渠道对接，还需要扩展可售库存逻辑规则以及同步规则，比如与 oms 类似的应用。

模型如图 5-20 所示。

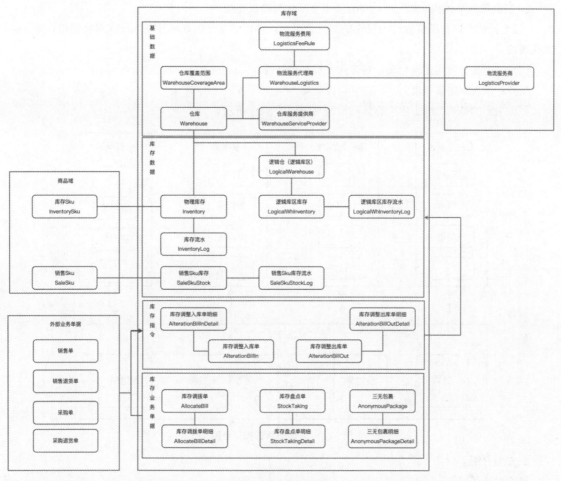

图 5-20　模型介绍

**3. 客户扩展**

按 CDM 的设计理念，我们不以把模型抽象到极致支撑所有业务可能性为目标，而是抽象 80% 通用的设计，保持模型简单可理解。我们只提供了基础模型统一数据存储，以用面向对象特性来解决多应用模型复用和数据割裂问题。

Oinone 的 CDM 是商业领域的通用模型，更是结合 Oinone 特性提出的新工程建议。

## 5.8　商业支撑之执行域

**1. 基础介绍**

执行域包括两个核心：一是订单的产生；二是订单的履约。往往品牌商既有自营渠道（包括 2C、2B）又有第三方渠道。那么有两种设计思路：

（1）把第三方渠道的订单当作自有渠道的订单，产生一种特殊方式，开放订单创建接口，并统一履约。

优点：简单，适合在三方渠道不多且自有渠道单一的情况下使用，并且逻辑相似时系统结构会简单。

缺点：

① 当三方渠道的履约方式、库存分配方式、逆向逻辑等有差异时，会让自有渠道掺杂很多不相干

的逻辑，引入不必要的复杂度；

②自有渠道不够独立和纯粹，自有渠道多样化时难以支撑。

（2）把商家自营渠道假设为特殊的第三方渠道，再建立统一的订单管理系统来对接渠道订单，并完成履约。

优点：交易与履约逻辑分离，对未来发展有扩展性。

缺点：引入一定复杂度。

我们采用的是第二套方案，整体结构简易图如图 5-21 所示。

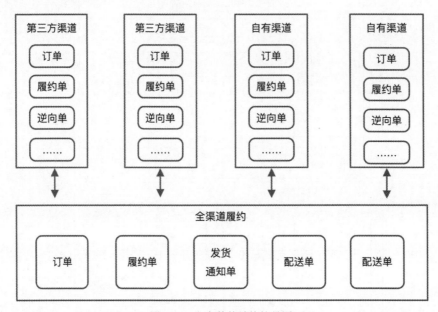

图 5-21　方案整体结构简易图

**2. 模型介绍**

核心设计逻辑：

（1）首先我们看到图 5-21 中交易域和履约域有很多相同父模型的子模型，交易域和履约域的父模型在 CDM 的 himalaya-trade 里。履约域看 oms（libra）对 himalaya-trade 扩展，交易域看 b2c（leo）和 b2b（aries）对 himalaya-trade 扩展。libra、leo、aries 是我们对上层业务产品的命名，取自黄道十二星座。

（2）交易域是从多商家平台视角设计，由自身渠道完成必要的履约相关信息，完成自闭环。

（3）履约域是从单一商家对接多渠道视角设计，由渠道交易订单同步后完成履约发货相关设计，完成自闭环。

（4）在履约域的合单、拆单、发货设计方面，渠道订单只能合单为履约单，不可拆，履约单可以拆单发货，不可合。用 M2O 和 O2M 的组合设计来降低难度，而非采用两个 M2M 的设计。

模型如图 5-22 所示。

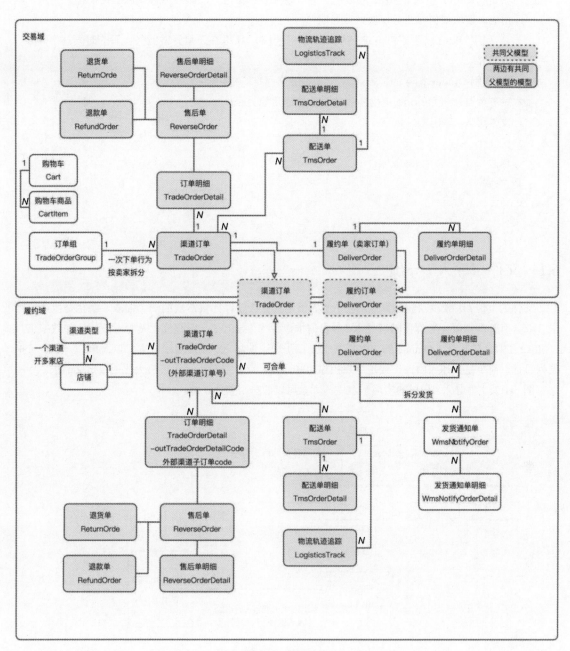

图 5-22　模型介绍

# 第 6 章 Oinone 的通用能力

通用能力是指一些跟具体业务场景无关，但是在业务支撑中必不可少的功能。比如，权限、文件导入/导出、工作流、异构系统集成、国际化等。Oinone 的通用能力本质上也是一个个独立的应用（application=true 的模块）或模块，可以独立部署也可以跟业务模块部署在一起。

**通用能力模块如何与业务模块相互影响？**

（1）利用 Oinone 函数的 hook 特性影响业务模块，所以在正常开发业务代码的时候并不需要关心这一部分；

（2）利用 Oinone 函数的扩展点特性，为业务层扩展提供支持。

**本章节重点介绍 Oinone 提供的一些企业级通用能力：**

1. 文件导入/导出
2. 国际化
3. 集成
4. 工作流
5. 权限

## 6.1 文件与导入/导出

文件的上传/下载以及业务数据的导入/导出是企业级软件比较常规的一个需求，甚至是巨量的需求。业务有管理需要一般都伴随有导入/导出需求，导入/导出在一定程度上是企业级软件和效率工具（office 工具）的桥梁。Oinone 的文件模块就提供了通用的导入/导出实现方案，以简单、一致、可扩展为目标，简单是快速入门，一致是用户操作感知一致，可扩展是满足用户最大化的自定义需求。

图 6-1 为文件导入/导出的实现示意图，大家可以做一个整体了解。

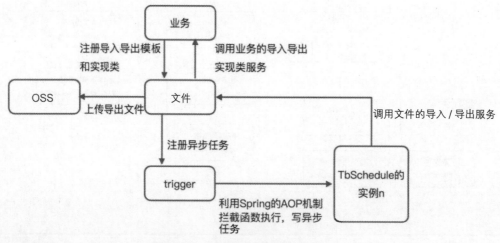

图 6-1 文件导入/导出的实现示意图

**1. 基础能力**

1）准备工作

（1）在 pamirs-demo-api 的 pom 文件中引入 pamirs-file2-api 包依赖，代码如下。

```
1    <dependency>
2        <groupId>pro.shushi.pamirs.core</groupId>
3        <artifactId>pamirs-file2-api</artifactId>
4    </dependency>
```

（2）在 DemoModule 中增加对 FileModule 的依赖，代码如下。

```
1    @Module(dependencies = {FileModule.MODULE_MODULE})
```

（3）pamirs-demo-boot 的 pom 文件中引入 pamirs-file2-core 包依赖，代码如下。

```
1    <dependency>
2        <groupId>pro.shushi.pamirs.core</groupId>
3        <artifactId>pamirs-file2-core</artifactId>
4    </dependency>
```

（4）pamirs-demo-boot 的 application-dev.yml 文件配置 pamirs.boot.modules 中增加 file，即在启动模块中增加 file 模块。

```
1    pamirs:
2      boot:
3        modules:
4          - file
```

（5）pamirs-demo-boot 的 application-dev.yml 文件中增加 OSS 的配置，代码如下。更多有关文件相关配置参见"模块之 yml 文件结构详解"一节。

```
1    cdn:
2      oss:
3        name: 阿里云
4        type: OSS
5        bucket: demo
6        uploadUrl: #换成自己的oss上传服务地址
7        downUrl: #换成自己的oss下载服务地址
8        accessKeyId: #阿里云oss的accessKeyId
9        accessKeySecret: #阿里云oss的accessKeySecret
10       mairDir: upload/demo #换成自己的目录
11       validTime: 360000
12       timeout: 60000
13       active: true
14       referer: www.shushi.pro
```

### 2）文件上传

文件上传/下载是常用功能，也是导入/导出功能的基础。接下来，我们来看看 Oinone 是如何实现文件上传的功能的。

Step1. 修改 PetTalent 模型。

为 PetTalent 模型增加一个字段 imgUrls，用于保存上传后的文件地址。同时定义 Form 视图的时候该字段的前端展示 widget 为 "Upload"。

```
1    @Field.String
2    @Field(displayName = "图片" ,multi = true)
3    @UxForm.FieldWidget(@UxWidget(widget = "Upload"))
4    private List<String> imgUrls;
```

Step2. 修改 PetTalent 的 Form 视图。

因为前面教程中为 PetTalent 的 Form 视图写了自定义视图，所以这里需要手工加上字段配置，不然展示不出来，同时也要加上 widget="Upload" 的配置，因为自定义视图不会再去读字段上定义的 UxWidget，字段上的 UxWidget 只对默认视图生效。

```
1  <field data="imgUrls" widget="Upload"/>
2
```

Step3. 重启看效果如图 6-2 所示。

图 6-2　重启看效果

3）默认导入 / 导出功能

因为前面我们在 Demo 的启动工程中加入对 pamirs-file2-core 的 jar 包依赖，同时启动模块列表中也增加了 file 模块，所以所有的默认表格页面都出现了默认的"导入""导出"按钮，如图 6-3 所示。

图 6-3　所有的默认表格页面都出现了默认的"导入""导出"按钮

但是我们发现宠物达人的表格就没有出现"导入""导出"按钮，首先我们自定义了表格视图，而且在学习自定义 Action 的时候修改了表格视图的 Action 的填充方式为手工指定，如图 6-4 所示。

图 6-4　宠物达人的表格没有出现"导入""导出"按钮

Step1. 修改 PetTalent 的 Table 视图

修改 PetTalent 的自定义 Table 视图，代码如下。

```
1  <template slot="actions" autoFill="true">
2    <!--        <action actionType="client" name="demo.doNothing" label="第一个自定义A
ction" />-->
3    <!--        <action name="delete" label="删除" />-->
4    <!--        <action name="redirectCreatePage" label="创建" />-->
5  </template>
```

把表格视图 Action 的填充方式改为"自动填充"，或者根据平台默认前端动作增加导入、导出两个前端动作，代码如下。

```
1  <template slot="actions" autoFill="true">
2    <action actionType="client" name="demo.doNothing" label="第一个自定义Action" />
3    <action name="delete" label="删除" />
4    <action name="redirectCreatePage" label="创建" />
5    <!--  <action name="$$internal_GotoListExportDialog" label="导出" />
6          <action name="$$internal_GotoListImportDialog" label="导入" /> -->
7    <action name="internalGotoListExportDialog" label="导出" />
8    <action name="internalGotoListImportDialog" label="导入" />
9  </template>
```

Step2. 重启看效果，如图 6-5 所示。

图 6-5　重启看效果

Step3. 注意点。

（1）系统生成的导入模板以默认的 Form 视图为准。

（2）系统生成的导出模板以默认的 Table 视图为准。

（3）有自定义导入导出模板后，默认模板会失效。

**2. 自定义导入 / 导出模板与逻辑**

1）数据的导出（举例一）

通常我们只需定义一个导出模板。默认导出方法是调用模型的 queryPage 函数（即在日常开发中我们经常会自定义模型的 queryPage 函数），为了跟页面查询效果保持一致，默认导出方式实现中手工生效了 hook，所以类似数据权限的 hook 都会生效，与页面查询逻辑一致。

如果不一致可以实现 ExcelExportFetchDataExtPoint 扩展点，在"举例二"中会介绍。

Step1. 新建 PetTalentExportTemplate。

新建 PetTalentExportTemplate 类实现 ExcelTemplateInit 接口，代码如下。

（1）实现 ExcelTemplateInit 接口，系统会自动调用 generator 方法，并注册该方法返回的模板定义列表，例子中定义了一个导出模板"宠物达人导出"。

（2）关联对象为集合，petShops[*].shopName。

（3）关联对象为模型，creater.name。

```
13      @Component
14    public class PetTalentExportTemplate implements ExcelTemplateInit {
15        public static final String TEMPLATE_NAME ="宠物达人导出";
16
17        @Override
18        public List<ExcelWorkbookDefinition> generator() {
19            //可以返回多个模板,导出的时候页面上由用户选择导出模板
20            return Collections.singletonList(
21                    ExcelHelper.fixedHeader(PetTalent.MODEL_MODEL,TEMPLATE_NAME)
22                    .createBlock(TEMPLATE_NAME,PetTalent.MODEL_MODEL)
23                    .setType(ExcelTemplateTypeEnum.EXPORT)
24                    .addColumn("name","名称")
25                    .addColumn("annualIncome","年收入")
26                    .addColumn("petShops[*].shopName","店铺名")
27                    .addColumn("creater.name","创建者")
28                    .build());
29        }
30
31    }
```

Step2. 重启看效果。

（1）单击宠物达人菜单进入宠物达人列表页，选择"记录"选项，单击"导出"按钮，如图6-6所示。

图6-6　宠物达人列表页选择"导出"按钮

导出规则为：

① 如果勾选列表记录行，则只导出勾选记录；

② 如果没有勾选，则根据查询条件进行导出。

（2）在导出弹出框中选择导出模板，单击"导出"按钮提示成功，则表示导出任务被正常创建，如图6-7所示。

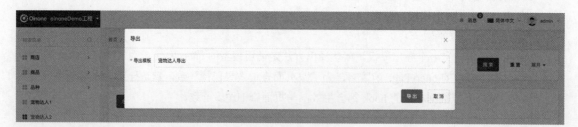

图6-7　选择导出模版

（3）切换模块，进入文件应用，选择"导出任务"菜单查看导出任务情况，如图6-8所示。

导出任务有以下三种情况：

① Table 视图中带来条件但查询结果没有数据，则会报错；

② Table 视图中带来条件或无条件但查询结果有数据，则会成功；

③ Table 视图中勾选了数据行，则只导出勾选数据行对应数据。

图 6-8 查看导出任务情况

2）数据的导出（举例二）

如果有特殊需求，请通过实现 ExcelExportFetchDataExtPoint 扩展点来完成，用这种模式请自行控制数据权限。因为数据权限是通过 hook 机制来完成控制，在"面向切面——拦截器"一节中有说明拦截器只有在前端请求时才会默认生效，如果后端代码自行调用生效 hook 机制，需要根据"函数之元位指令"一节手动去生效。

下面例子中复用了 ExcelExportSameQueryPageTemplate 的逻辑，主要是 hook 的机制。ExcelExportSameQueryPageTemplate 手工生效了 hook，例子中继承了 ExcelExportSameQueryPageTemplate 类，只做返回结果的后置处理。

Step1. 修改 PetTalentExportTemplate，代码如下。

```
@Component
public class PetTalentExportTemplate extends ExcelExportSameQueryPageTemplate implements ExcelTemplateInit , ExcelExportFetchDataExtPoint {
    public static final String TEMPLATE_NAME ="宠物达人导出";

    @Override
    public List<ExcelWorkbookDefinition> generator() {
        //可以返回多个模版，导出的时候页面上由用户选择导出模版
        return Collections.singletonList(
                ExcelHelper.fixedHeader(PetTalent.MODEL_MODEL,TEMPLATE_NAME)
                        .createBlock(TEMPLATE_NAME,PetTalent.MODEL_MODEL)
                        .setType(ExcelTemplateTypeEnum.EXPORT)
                        .addColumn("name","名称")
                        .addColumn("annualIncome","年收入")
                        .addColumn("petShops[*].shopName","店铺名")
                        .addColumn("creater.name","创建者")
                        .build());
    }

    @Override
    @ExtPoint.Implement(expression = "context.model == \"" + PetTalent.MODEL_MODEL+"\" && context.name == \"" +TEMPLATE_NAME+"\" )
    public List<Object> fetchExportData(ExcelExportTask exportTask, ExcelDefinitionContext context) {
        //复用父类的权限控制等
        List<Object> result =  super.fetchExportData(exportTask,context);
        //对petTalents做一些操作
        List<PetTalent>  petTalents =  (List<PetTalent>)result.get(0);
        for(PetTalent petTalent:petTalents){
            petTalent.setName(petTalent.getName()+":被修改过");
        }
        return result;
    }
}
```

Step2. 重启看效果。

相同的操作，发现导出文件内容被特定修改过，如图 6-9 所示。

| 名称 | 年收入 | 店铺名 | 创建者 |
|---|---|---|---|
| 老人:被修改过 |  |  | admin |
| 张学友:被修改过 | 12.00 | oinone宠物商店003 | admin |
| 刘德华:被修改过 |  |  | admin |

图 6-9　导出文件内容被特定修改过

3）数据的导入（举例）

Step1. 创建 PetTalentImportTemplate，代码如下。

```java
30   @Component
31   public class PetTalentImportTemplate extends AbstractExcelImportDataExtPointImpl<PetTalent> implements ExcelTemplateInit , ExcelImportDataExtPoint<PetTalent> {
32       public static final String TEMPLATE_NAME ="宠物达人导入";
33
34       @Override
35       public List<ExcelWorkbookDefinition> generator() {
36           //自定义Excel单元格, 设置单元可选值
37           Map<String,String> sexType = new HashMap<>();
38           sexType.put(String.valueOf(PetTalentSexEnum.FEMAL.value()),PetTalentSexEnum.FEMAL.displayName());
39           sexType.put(String.valueOf(PetTalentSexEnum.MAN.value()),PetTalentSexEnum.MAN.displayName());
40           ExcelCellDefinition sexExcelCellDefinition = new ExcelCellDefinition();
41           sexExcelCellDefinition.setType(ExcelValueTypeEnum.ENUMERATION).setValue("性别").setFormat(JSON.toJSONString(sexType));
42
43           //可以返回多个模板，导出的时候页面上由用户选择导出模板
44           return Collections.singletonList(
45                   ExcelHelper.fixedHeader(PetTalent.MODEL_MODEL,TEMPLATE_NAME)
46                       .createBlock(TEMPLATE_NAME,PetTalent.MODEL_MODEL)
47                       .setType(ExcelTemplateTypeEnum.IMPORT)
48                       //相同key, 相连行会进行合并操作
49                       .addUnique(PetTalent.MODEL_MODEL,"name")
50                       .addColumn("name","名称")
51                       .addColumn("annualIncome","年收入")
52                       .addColumn("petShops[*].shopName","店铺名")
53                       .addColumn("petTalentSex",sexExcelCellDefinition)
54                       .build());
55       }
56
57       @Override
58       @ExtPoint.Implement(expression = "importContext.definitionContext.model == \"" + PetTalent.MODEL_MODEL+"\" && importContext.definitionContext.name == \"" +TEMPLATE_NAME+"\"" )
59       public Boolean importData(ExcelImportContext importContext, PetTalent data) {
60           ExcelImportTask importTask = importContext.getImportTask();
61           if(StringUtils.isBlank(data.getName())){
62               //返回true表示继续读下一行; 返回false表示中断
63               importTask.addTaskMessage(TaskMessageLevelEnum.ERROR,"达人名称不能为空");
64               return Boolean.TRUE;
65           }
```

```
66              if(CollectionUtils.isNotEmpty(data.getPetShops())){
67                  List<String> petShopNams = new ArrayList<>();
68                  for(PetShop petShop:data.getPetShops()){
69                      petShopNams.add(petShop.getShopName());
70                  }
71                  //FIXME 抽到Service中去, 散落的查询逻辑不好维护
72                  IWrapper<PetShop> queryWrapper = Pops.<PetShop>lambdaQuery().from(PetShop.MODEL_MODEL).in(PetShop::getShopName, petShopNams);
73                  List<PetShop> shops = new PetShop().queryList(queryWrapper);
74                  data.setPetShops(shops);
75              }
76              Tx.build(new TxConfig().setPropagation(Propagation.REQUIRED.value())).executeWithoutResult(status -> {
77                  data.create();
78                  data.fieldSave(PetTalent::getPetShops);
79              });
80              return Boolean.TRUE;
81          }
82
83      }
```

新建 PetTalentImportTemplate 类，实现 ExcelImportDataExtPoint 和 ExcelTemplateInit 接口。

（1）实现 ExcelTemplateInit 接口，系统会自动调用 generator 方法，并注册该方法返回的模板定义列表，例子中定义了一个导入模板"宠物达人导入"。注意：

● 通过 ExcelCellDefinition 自定义单元可选值，并通过 addColumn 时绑定，如：.addColumn ("petTalentSex",sexExcelCellDefinition)。

● 通过 addUnique，设置导入时的唯一 key，相同 key 的相连行会进行合并操作。

（2）实现 ExcelImportDataExtPoint 的 importData 扩展点，实现自定义的导入逻辑。注意：

● 声明扩展点的生效规则：@ExtPoint.Implement (expression = "importContext.definitionContext.Model == \"" + PetTalent.MODEL_MODEL+"\" && importContext.definitionContext.name == \"" +TEMPLATE_NAME+"\"" )。可以用表达式的上下文就是方法入参。

● 记录错误：importTask.addTaskMessage (TaskMessageLevelEnum.ERROR," 达人名称不能为空 ")。

● importData 方法返回 true，代表继续下一行的导入；返回 false 则表示中断导入。

Step2. 重启看效果。

（1）单击"宠物达人"菜单进入宠物达人列表页面，单击"导入"按钮，如图 6-10 所示。

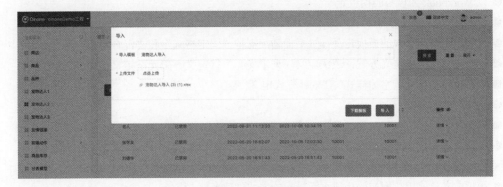

图 6-10　导入数据

（2）在导入弹出框中选择导入模板后单击"下载模板"按钮，并按模板填写数据，如图 6-11 所示。

（3）在导入弹出框中单击"上传"按钮，上传填写好的文件，并单击"导入"按钮。

| | A | B | C | D | E | F |
|---|---|---|---|---|---|---|
| | 名称 | 年收入 | 店铺名 | 性别 | | |
| | 华哥 | 100.00 | oinone宠物商店003 | 男 | | |
| | 华哥 | 100.00 | oinone宠物商店002 | 男 | | |

图 6-11 按照模板填写数据

（4）列表页增加对应数据行，单击"详情"按钮，可以看到所有相关数据，如图 6-12 所示。

图 6-12 单击"详情"按钮，可看到所有相关数据

这个例子中用了 ExcelHelper 这个简单工具类来定义模板，基本能满足日常需求。复杂的需求可以直接用 WorkbookDefinitionBuilder 来完成。

4）导入导出的更多配置项

（1）基础设计。

以 POI 和 EasyExcel 共有的 Excel 处理能力为基础进行设计，有以下功能。

● 以 Workbook 为核心的模板定义。
● 支持多个 Sheet 定义。
● 支持固定表头形式的模板定义。（后面陆续支持"固定格式"和"混合格式"）
● 支持三个维度的定义方式：表头定义（列定义）、行定义、单元格定义。
● 支持单元格样式、字体样式等 Excel 样式定义。
● 支持合并单元格。
● 支持数字、字符串、时间、日期、公式等 Excel 的数据处理能力。
● 支持直接存数据库和获取导入数据两种方式。

（2）Excel 功能。

● 模板定义，使用数据初始化或页面操作的方式创建 Excel 模板。
● 导出任务的创建，页面选择导出模板、导出条件后，单击"确定"按钮，创建导出任务。导出方式为异步，在导出任务页面查看导出任务，并允许多次下载。
● 导入任务的创建，页面选择模板，并单击"下载"按钮，获取对应 Excel 模板。按照指定规则（平台默认规则或业务自定义规则）填写完成后，上传 Excel 文件进行异步导入操作。

Excel 各部分属性如下。

① Excel 工作簿。

Excel 工作簿（ExcelWorkbookDefinition）见表 6-1。

表 6-1  Excel 工作簿

| 属性 | 类型 | 名称 | 必填 | 默认值 | 说明 |
| --- | --- | --- | --- | --- | --- |
| name | String | 名称 | 是 | | Excel 工作簿的定义名称 |
| filename | String | 文件名 | 否 | | 导出时使用的文件名，不指定则默认使用名称作为文件名 |
| version | Enum | Office 版本 | 否 | AUTO | OfficeVersionEnum |
| sheetList | List<ExcelSheetDefinition>（JSONArray） | 工作表定义列表 | 是 | | 对多个 Sheet 进行定义 |

② Excel 工作表。

Excel 工作表（ExcelSheetDefinition）见表 6-2。

表 6-2  Excel 工作表

| 属性 | 类型 | 名称 | 必填 | 默认值 | 说明 |
| --- | --- | --- | --- | --- | --- |
| name | String | 工作表名称 | 否 | | 指定工作表名称，当未指定工作表名称时，默认使用模型的显示名称，未绑定模型生成时，默认使用"Sheet + ${index}"作为工作表名称 |
| blockDefinitionList | List<ExcelBlockDefinition> | 区块列表 | 是 | | 至少一个区块 |
| mergeRangeList | List<ExcelCellRangeDefinition> | 单元格合并范围 | 否 | | 工作表的合并范围相对于 A1 单元格行列索引，绝对定义 |
| uniqueDefinitions | List<ExcelUniqueDefinition> | 唯一定义 | 否 | | 全局模型唯一定义，当区块中未定义时，使用该定义 |

③ Excel 区块。

Excel 区块（ExcelBlockDefinition）见表 6-3。

表 6-3  Excel 区块

| 属性 | 类型 | 名称 | 必填 | 默认值 | 说明 |
| --- | --- | --- | --- | --- | --- |
| designRegion | ExcelCellRangeDefinition | 设计区域 | 是 | | 设计区域是区块的唯一标识，不允许重叠，定义时应示意扩展方向 |
| analysisType | Enum | 解析类型 | 是 | FIXED_HEADER | ExcelAnalysisTypeEnum |

续表

| 属性 | 类型 | 名称 | 必填 | 默认值 | 说明 |
|---|---|---|---|---|---|
| usingCascadingStyle | Boolean | 是否使用层叠样式 | 否 | false | 将样式覆盖变为样式层叠;单个工作表有效;优先级顺序为:列样式 < 行样式 < 单元格样式,三个属性共同定义一个 Sheet 的样式 |
| headerList | List<ExcelHeaderDefinition> | 表头定义 | 否 | | "固定表头"类型下,至少一行,不显示的表头行使用配置行定义 |
| rowList | List<ExcelRowDefinition> | 行定义 | 否 | | "固定格式"类型下,至少一行 |
| bindingModel | String | 绑定模型 | 是 | | 绑定任意模型,允许使用类的全限定名 |
| mergeRangeList | List<ExcelCellRangeDefinition> | 单元格合并范围 | 否 | | 区块内的单元格合并范围相对于区块起始行列索引。当合并范围指定到最后一行的表头定义时,属于相对定义,即扩展时会自动填充合并样式。否则与工作表合并相同 |
| uniqueDefinitions | List<ExcelUniqueDefinition> | 唯一定义 | 否 | | 默认使用模型定义的 unique 属性 |

④ Excel 行。

Excel 行(ExcelRowDefinition)见表 6-4。

表 6-4　Excel 行

| 属性 | 类型 | 名称 | 必填 | 默认值 | 说明 |
|---|---|---|---|---|---|
| cellList | List<ExcelCellDefinition> | 单元格列表 | 否 | | |
| style | ExcelStyleDefinition | 样式 | 否 | | 行样式 |

⑤ Excel 表头。

Excel 表头(ExcelHeaderDefinition 继承 ExcelRowDefinition)见表 6-5。

表 6-5　Excel 表头

| 属性 | 类型 | 名称 | 必填 | 默认值 | 说明 |
|---|---|---|---|---|---|
| cellList | List<ExcelCellDefinition> | 单元格列表 | 否 | | |
| style | ExcelStyleDefinition | 样式 | 否 | | 行样式 |
| direction | Enum | 排列方向 | 否 | HORIZONTAL | ExcelDirectionEnum |
| selectRange | ExcelCellRangeDefinition | 应用范围 | 否 | | 仅起始属性有效 |
| isConfig | Boolean | 是否配置表头 | 否 | | |
| autoSizeColumn | Boolean | 自动列宽 | 否 | true | 多行表头需使用配置表头进行定义 |
| isFrozen | Boolean | 是否冻结 | 否 | | Excel 冻结功能 |

⑥ Excel 单元格范围。

Excel 单元格范围（ExcelCellRangeDefinition）见表 6-6。

表 6-6　Excel 单元格范围

| 属性 | 类型 | 名称 | 必填 | 默认值 | 说明 |
| --- | --- | --- | --- | --- | --- |
| beginRowIndex | Integer | 起始行索引 | 否 | | |
| endRowIndex | Integer | 结束行索引 | 否 | | |
| beginColumnIndex | Integer | 起始列索引 | 否 | | |
| endColumnIndex | Integer | 结束列索引 | 否 | | |
| fixedBeginRowIndex | Boolean | 固定起始行索引 | 否 | | |
| fixedEndRowIndex | Boolean | 固定结束行索引 | 否 | | |
| fixedBeginColumnIndex | Boolean | 固定起始列索引 | 否 | | |
| fixedEndColumnIndex | Boolean | 固定结束列索引 | 否 | | |

⑦ Excel 单元格。

Excel 单元格（ExcelCellDefinition）见表 6-7。

表 6-7　Excel 单元格

| 属性 | 类型 | 名称 | 必填 | 默认值 | 说明 |
| --- | --- | --- | --- | --- | --- |
| field | String | 属性 | 否 | | 当且仅当单元格在表头中有效；多行表头需使用配置表头进行定义；多个配置表头定义属性时，仅首个属性定义生效 |
| value | String | 值 | 否 | | |
| type | Enum | 值类型 | 否 | STRING | ExcelValueTypeEnum |
| formatContextString | String (JSONObject) | 格式化上下文 | 否 | | 针对不同类型的值设置不同的参数内容（类型待定） |
| isStatic | Boolean | 是否静态值 | 否 | false | 静态值在数据解析时将使用配置值，不读取单元格的值 |
| style | ExcelStyleDefinition | 样式 | 否 | | 单元格样式 |

⑧ Excel 单元格样式。

Excel 单元格样式（ExcelStyleDefinition）见表 6-8。

表 6-8　Excel 单元格样式

| 属性 | 类型 | 名称 | 必填 | 默认值 | 说明 |
| --- | --- | --- | --- | --- | --- |
| horizontalAlignment | Enum | 水平对齐 | 否 | GENERAL | ExcelHorizontalAlignmentEnum |
| verticalAlignment | Enum | 垂直对齐 | 否 | | ExcelVerticalAlignmentEnum |
| fillBorderStyle | Enum | 全边框样式 | 否 | NONE | ExcelBorderStyleEnum |
| topBorderStyle | Enum | 上边框样式 | 否 | | ExcelBorderStyleEnum |
| rightBorderStyle | Enum | 右边框样式 | 否 | | ExcelBorderStyleEnum |
| bottomBorderStyle | Enum | 下边框样式 | 否 | | ExcelBorderStyleEnum |
| leftBorderStyle | Enum | 左边框样式 | 否 | | ExcelBorderStyleEnum |
| wrapText | Boolean | 是否自动换行 | 否 | true | |
| typefaceDefinition | ExcelTypefaceDefinition | 字体 | 否 | | |
| width | Integer | 宽 | 否 | | 首列样式中有效 |
| height | Integer | 高 | 否 | | 首列样式中有效 |

⑨ Excel 字体。

Excel 字体（ExcelTypefaceDefinition）见表 6-9。

表 6-9　Excel 字体

| 属性 | 类型 | 名称 | 必填 | 默认值 | 说明 |
| --- | --- | --- | --- | --- | --- |
| typeface | Enum | 字体 | 否 | | ExcelTypefaceEnum |

5）注意事项

如果是在分布式模式下，启动工程中没有加入 file 模块，则需要手工调用 ExcelTemplateInitHelper. init 方法。

## 6.2　集成平台

企业在数字化转型过程中内外部集成是一个必然需求，也是趋势。

集成的诉求主要来自两个方面：一个是企业的数字化改造是由外而内逐步进行的（内部异构集成）；另一个是企业数字化方向是朝越来越开放的方向发展（外部平台、工具集成）。所以我们不能简单地将集成理解为做个 API 对接功能，而是要统一规划，构建成企业的集成门户对 API 定义、安全、控制、记录等做全方位管理。Oinone 在下个版本规则中也纳入了基于集成平台之上做产品化配置的需求。

**1. 概述**

pamirs-eip 为平台提供企业集成门户的相关功能，如请求外部接口使用集成接口和对外开放被其他系统请求调用的开放接口功能。在请求外部接口时，还支持了多个接口调用（路由定义）、分页控制（paging）、增量控制（incremental）等功能。

**2. 准备工作**

Step1. POM 与模块依赖。

在 pamirs-demo-api 和 pamirs-second-api 的 pom 文件中引入 pamirs-eip2-api 包依赖，代码如下。

```xml
<dependency>
    <groupId>pro.shushi.pamirs.core</groupId>
    <artifactId>pamirs-eip2-api</artifactId>
</dependency>
```

DemoModule 和 SecondModule 增加对 EipModule 的依赖，代码如下。

```
@Module(dependencies = {EipModule.MODULE_MODULE})
```

pamirs-demo-boot 和 pamirs-second-boot 工程的 pom 文件中引入 pamirs-eip2-core 包依赖，代码如下。

```xml
<dependency>
    <groupId>pro.shushi.pamirs.core</groupId>
    <artifactId>pamirs-eip2-core</artifactId>
</dependency>
```

Step2. yaml 配置文件参考。

pamirs-demo-boot 和 pamirs-second-boot 工程的 application-dev.yml 文件中配置 pamirs.boot.modules 增加 eip，即在启动模块中增加 eip 模块，代码如下。

```
1  pamirs:
2    boot:
3      modules:
4        - eip
```

pamirs-demo-boot 和 pamirs-second-boot 工程的 application-dev.yml 文件中增加 eip 模块的数据源与路由配置，代码如下。

```
1   pamirs:
2     framework:
3       data:
4         ds-map:
5           eip: eip
6     datasource:
7       eip:
8         driverClassName: com.mysql.cj.jdbc.Driver
9         type: com.alibaba.druid.pool.DruidDataSource
10        url: jdbc:mysql://127.0.0.1:3306/eip_v3?useSSL=false&allowPublicKeyRetrieva
    l=true&useServerPrepStmts=true&cachePrepStmts=true&useUnicode=true&characterEncod
    ing=utf8&serverTimezone=Asia/Shanghai&autoReconnect=true&allowMultiQueries=true
11        username: root
12        password: oinone
13        initialSize: 5
14        maxActive: 200
15        minIdle: 5
16        maxWait: 60000
17        timeBetweenEvictionRunsMillis: 60000
18        testWhileIdle: true
19        testOnBorrow: false
20        testOnReturn: false
21        poolPreparedStatements: true
22        asyncInit: true
```

pamirs-demo-boot 工程的 application-dev.yml 文件中修改 eip 的配置，代码如下。

```
1  pamirs:
2    eip:
3      open-api:
4        enabled: false
```

pamirs-second-boot 工程的 application-dev.yml 文件中修改 eip 的配置，代码如下。

```
1
2   pamirs:
3     eip:
4       open-api:
5         enabled: true
6         standalone:
7           # 开放接口访问IP
8           host: 127.0.0.1
9           # 开放接口访问端口，second为8094
10          port: 8094
11          # 认证Token加密的AES密钥
12          aes-key: NxDZUddmvdu3QQpd5jIww2skNx6U0w0u0AXj3NUCLu8=
13          routes:
14            pamirs:
15              # 开放接口访问IP
16              host: 127.0.0.1
17              # 开放接口访问端口，second为8094
18              port: 8094
19              # 认证Token加密的AES密钥
20              aes-key: NxDZUddmvdu3QQpd5jIww2skNx6U0w0u0AXj3NUCLu8=
```

注：

hosts 配置在远程调用时不能使用 127.0.0.1，可配置为 0.0.0.0 进行自动识别。若自动识别仍无法访问，请准确配置其他已知的可访问 IP 地址。

AES Key 用下面代码生成。

pro.shushi.pamirs.core.common.EncryptHelper 加解密帮助类，默认支持 AES、RSA 类型的数据加解密方法，也可自定义其他类型的加解密方法。

```
1  System.out.println(EncryptHelper.getKey(EncryptHelper.getAESKey()));
```

Step3. 在 pamirs-second-api 新建一个 SessionTenantApi 实现类，代码如下。

只要在我们公共的 jar 包中构建类似 DemoSessionTenant 类就可以了，之所以要构建 SessionTenantApi 实现类，是因为 EIP 是以租户信息做路由的，所以这里写死返回一个"pamirs"租户就好了。

记得要重启 mvn install second 工程，再刷新 demo 工程。

```
 9   @Order(99)
10   @Component
11   @SPI.Service
12   public class DemoSessionTenant implements SessionTenantApi, SessionClearApi {
13       public String getTenant() {
14           return "pamirs";
15       }
16
17       public void setTenant(String tenant) {
18       }
19
20       public void clear() {
21       }
22   }
23
```

### 3. 开放接口（举例）

Step1. 用于演示的模型定义，代码如下。

```
 7   @Model.model(TestOpenApiModel.MODEL_MODEL)
 8   @Model(displayName = "演示开放接口模型")
 9   public class TestOpenApiModel extends IdModel {
10
11       public static final String MODEL_MODEL = "demo.second.TestOpenApiModel";
12
13       @Field.String
14       @Field(displayName = "名称")
15       private String name;
16
17       @Field.Integer
18       @Field(displayName = "年龄")
19       private Integer age;
20   }
21
```

用于演示的返回结果对象，代码如下。

```
 7   @Data
 8   public class TestOpenApiResponse implements Serializable {
 9
10       private Long id;
11       private String name;
12       private Integer age;
13
14   }
```

Step2. 用于演示的接口定义。

（1）TestOpenApiModelService 接口定义了内部访问函数 queryById。

（2）TestOpenApiModelServiceImpl 实现 TestOpenApiModelService 接口，并定义了两个开放接口：

① queryById4Open 函数用 @Open 进行注解，请求路径是：openapi/pamirs/ 方法名 ?tenant=pamirs，返回对象结构为函数返回值类型的 JSON 格式；

② queryById4OpenError 函数用 @Open (config = TestEipConfig.class,path = "error") 进行注解，请求路径为：openapi/pamirs/error?tenant=pamirs，返回对象结构为函数返回值类型的 JSON 格式；

③ config = TestEipConfig.class 设置通用配置类，可以在 TestEipConfig 增加 @Open.Advanced。优先级低于方法上的注解。TestEipConfig 内容详见 6.2 节的"集成接口举例的集成配置模型定义"。配置模型在开放接口定义中是非必需的，在集成接口定义中时是必需的；

④ 更多注解配置见 6.2 节中的"Eip 注解说明"。

定义内部访问函数 queryByI，代码如下。

```
1   package pro.shushi.pamirs.second.api.service;
2
3   import pro.shushi.pamirs.meta.annotation.Fun;
4   import pro.shushi.pamirs.meta.annotation.Function;
5   import pro.shushi.pamirs.second.api.model.TestOpenApiModel;
6
7   @Fun(TestOpenApiModelService.FUN_NAMESPACE)
8   public interface TestOpenApiModelService {
9
10      String FUN_NAMESPACE ="demo.second.TestOpenApiModelService";
11      @Function
12      TestOpenApiModel queryById(Long id);
13
14  }
```

实现 TestOpenApiModelService 接口，代码如下。

```
19  @Fun(TestOpenApiModelService.FUN_NAMESPACE)
20  @Component
21  public class TestOpenApiModelServiceImpl implements TestOpenApiModelService {
22
23      @Override
24      @Function
25      public TestOpenApiModel queryById(Long id) {
26          return new TestOpenApiModel().queryById(id);
27      }
28
29      @Function
30      @Open
31      public OpenEipResult<TestOpenApiResponse> queryById4Open(IEipContext<SuperMap
    > context , ExtendedExchange exchange) {
32          String id = Optional.ofNullable(String.valueOf(context.getInterfaceContex
    t().getIteration("id"))).orElse("");
33          TestOpenApiModel temp  = queryById(Long.valueOf(id));
34          TestOpenApiResponse response = new TestOpenApiResponse();
35          if(temp != null ) {
36              response.setAge(temp.getAge());
37              response.setId(temp.getId());
38              response.setName(temp.getName());
```

```
39          }else{
40              response.setAge(1);
41              response.setId(1L);
42              response.setName("oinone eip test");
43          }
44          OpenEipResult<TestOpenApiResponse> result = new OpenEipResult<TestOpenApiResponse>(response);
45          return result;
46      }
47
48      @Function
49      // @Open(config = TestEipConfig.class,path = "error")
50      @Open(path = "error")
51      @Open.Advanced(
52              httpMethod = "post"
53      )
54      public OpenEipResult<TestOpenApiResponse> queryById4OpenError(IEipContext<SuperMap> context , ExtendedExchange exchange) {
55          throw PamirsException.construct(EipExpEnumerate.SYSTEM_ERROR).appendMsg("测试异常").errThrow();
56      }
57
58  }
59
```

Step3. 用于演示的请求协议。

请求方法：POST

请求参数格式如下。

```
1  {
2      "id":111
3  }
```

成功响应格式如下。

```
1  {
2      "success":true,
3      "errorCode":0,
4      "result":{
5          "age":1,
6          "id":1,
7          "name":"oinone eip test"
8      }
9  }
```

异常响应格式如下。

```
1  {
2      "success":false,
3      "errorCode":"20140000",
4      "errorMsg":"系统异常，测试异常"
5  }
```

Step4. 初始化 EIP，代码如下。

构建 SecondModuleBizInit 实现 InstallDataInit、UpgradeDataInit、ReloadDataInit 接口，调用 EIP 注解模式的初始化方法：EipResolver.resolver (SecondModule.MODULE_MODULE,null)。

```java
14      @Component
15      public class SecondModuleBizInit implements InstallDataInit, UpgradeDataInit, Rel
        oadDataInit {
16
17          @Override
18          public boolean init(AppLifecycleCommand command, String version) {
19              initEip();
20              return Boolean.TRUE;
21          }
22
23          @Override
24          public boolean reload(AppLifecycleCommand command, String version) {
25              initEip();
26              return Boolean.TRUE;
27          }
28
29          @Override
30          public boolean upgrade(AppLifecycleCommand command, String version, String ex
        istVersion) {
31              initEip();
32              return Boolean.TRUE;
33          }
34
35          @Override
36          public List<String> modules() {
37              return Collections.singletonList(SecondModule.MODULE_MODULE);
38          }
39
40          @Override
41          public int priority() {
42              return 0;
43          }
44
45          private void initEip() {
46              EipResolver.resolver(SecondModule.MODULE_MODULE,null);
47          }
48      }
```

Step5. 重启看效果。

Second 模块重新 mvn install。

重启 pamris-second-boot 工程。

请求地址一：http://localhost:8094/openapi/pamirs/queryById4Open?tenant=pamirs。

地址一重启 pamris-second-boot 工程，代码如下。

```
1  curl --location --request POST 'http://localhost:8094/openapi/pamirs/queryById4Ope
   n?tenant=pamirs' \
2  --header 'Content-Type: application/json' \
3  --data-raw '{
4      "id": "111",
5  }'
```

返回结果如下。

```
1  {"errorCode":0,"result":{"age":1,"id":1,"name":"oinone eip test"},"success":true}
```

请求地址二：http://localhost:8094/openapi/pamirs/error?tenant=pamirs。

地址二重启 pamris-second-boot 工程，代码如下。

```
1  curl --location --request POST 'http://localhost:8094/openapi/pamirs/error?tenant=
   pamirs' \
2  --header 'Content-Type: application/json' \
3  --data-raw '{
4      "id": "111",
5  }'
```

返回结果如下。

```
1  {"success":false,"errorCode":"20140000","errorMsg":"系统异常，测试异常"}
```

### 4. 集成接口（举例）

在下面的例子中，我们将用集成接口调用开放接口，集成接口配置在 Demo 模块。

Step1. 用于演示的集成配置模型定义包含以下几点：

（1）配置模型必须继承 IEipAnnotationSingletonConfig 接口。

（2）自定义 construct 函数，实现修改配置时可以自动刷新对应的集成接口和开放接口。

（3）host 为服务端："域名 + 端口"，如 www.oinone.top:80。

（4）shcema 为请求协议：Http 或 Https。

（5）类上可以增加 @Open.Advanced 和 @Integrate.Advanced 配置达到通用配置的作用，优先级低于方法上的注解。

```java
11  @Model.model(TestEipConfig.MODEL_MODEL)
12  @Model(displayName = "演示集成接口配置模型")
13  public class TestEipConfig extends IdModel implements IEipAnnotationSingletonConf
    ig<TestEipConfig> {
14
15      public static final String MODEL_MODEL = "demo.TestEipConfig";
16
17      @Field(displayName = "服务端域名")
18      private String host;
19      @Field(displayName = "请求协议Http或Https")
20      private String schema;
21
22      @Function(openLevel = FunctionOpenEnum.API,summary = "演示集成接口配置模型")
23      @Function.Advanced(type = FunctionTypeEnum.QUERY)
24      public TestEipConfig construct(TestEipConfig config){
25          TestEipConfig config1 = config.singletonModel();
26          if(config1!=null){
27              return config1;
28          }
29          return config.construct();
30      }
31  }
```

Step2. 用于演示的接口定义包含以下几点：

（1）TestIntegrateService 接口定义了内部访问函数 callQueryById、callQueryByData、callQueryByIdError。

（2）TestIntegrateServiceImpl 实现 TestIntegrateService 接口，并把方法定义为集成接口：

① config = TestEipConfig.class 设置通用配置类，可以在 TestEipConfig 增加 @Integrate.Advanced。优先级低于方法上的注解。它在集成接口定义中时是必须的。

② @Integrate.Advanced (path=" ")，集成接口最终请求地址：TestEipConfig.schema+ "://" + TestEipConfig.host+path。

③ 方法体返回空值就可以，实际执行时该方法被拦截并不会执行。

④ @Integrate.ConvertParam：

● inParam 表示入参，outParam 表示出参。比如 inParam 为 data.id，outParam 为 Id，即把 data.id 的值，赋值给 id。

● url 和 head 的参数转化参见"paramConverter 的特殊转换"。

⑤ finalResultKey 是指最终的请求参数 key，会从上下文中找对应的 key 值作为请求参数。

⑥ 更多注解配置见 6.2 节中的"EIP 注解说明"。

TestIntegrateService 接口定义内部访问函数代码如下。

```
 9  @Fun(TestIntegrateService.FUN_NAMESPACE)
10  public interface TestIntegrateService {
11      String FUN_NAMESPACE ="demo.TestIntegrateService";
12      @Function
13      EipResult<SuperMap> callQueryById(String id);
14      @Function
15      EipResult<SuperMap> callQueryByData(TestOpenApiModel data);
16      @Function
17      EipResult<SuperMap> callQueryByIdError(TestOpenApiModel data);
18
19  }
20
```

TestIntegrateServiceImpl 实现 TestIntegrateService 接口，代码如下。

```
13  @Fun(TestIntegrateService.FUN_NAMESPACE)
14  @Component
15  public class TestIntegrateServiceImpl implements TestIntegrateService {
16
17      @Override
18      @Function
19      @Integrate(config = TestEipConfig.class)
20      @Integrate.Advanced(path = "/openapi/pamirs/queryById4Open?tenant=pamirs")
21      public EipResult<SuperMap> callQueryById(String id) {
22          return null;
23      }
24
25      @Override
26      @Function
27      @Integrate(config = TestEipConfig.class)
28      @Integrate.Advanced(path = "/openapi/pamirs/queryById4Open?tenant=pamirs")
29      @Integrate.RequestProcessor(
30          convertParams={
31              @Integrate.ConvertParam(inParam="data.id",outParam="id")
32          }
33      )
34      public EipResult<SuperMap> callQueryByData(TestOpenApiModel data) {
35          return null;
36      }
37
38      @Function
39      @Integrate(config = TestEipConfig.class)
40      @Integrate.Advanced(path = "/openapi/pamirs/error?tenant=pamirs")
41      @Integrate.RequestProcessor(
42              finalResultKey = "out",
43          convertParams={
44              @Integrate.ConvertParam(inParam="data.id",outParam="out.id")
45          }
46      )
47      public EipResult<SuperMap> callQueryByIdError(TestOpenApiModel data) {
48          return null;
49      }
50
51  }
```

Step3. 新建 TestEipAction，代码如下。

```
17    @Model.model(TestEipConfig.MODEL_MODEL)
18    @Component
19    @Slf4j
20    public class TestEipAction {
21
22        @Autowired
23        private TestIntegrateService testIntegrateService;
24
25        @Action(displayName = "调用集成接口callQueryById",contextType = ActionContextTypeEnum.CONTEXT_FREE)
26        public TestEipConfig callQueryById(TestEipConfig data){
27            EipResult<SuperMap> eipResult = testIntegrateService.callQueryById("111");
28            PamirsSession.getMessageHub().error(JsonUtils.toJSONString(eipResult));
29            return data;
30        }
31
32        @Action(displayName = "调用集成接口callQueryByIdError",contextType = ActionContextTypeEnum.CONTEXT_FREE)
33        public TestEipConfig callQueryByIdError(TestEipConfig data){
34            EipResult<SuperMap> eipResult = testIntegrateService.callQueryByIdError((TestOpenApiModel)new TestOpenApiModel().setId(111L));
35            PamirsSession.getMessageHub().error(JsonUtils.toJSONString(eipResult));
36            return data;
37        }
38
39        @Action(displayName = "调用集成接口callQueryByData",contextType = ActionContextTypeEnum.CONTEXT_FREE)
40        public TestEipConfig callQueryByIdSuccess(TestEipConfig data){
41            EipResult<SuperMap> eipResult = testIntegrateService.callQueryByData((TestOpenApiModel)new TestOpenApiModel().setId(111L));
42            PamirsSession.getMessageHub().error(JsonUtils.toJSONString(eipResult));
43            return data;
44        }
45    }
```

Step4. 构建测试入口，代码如下。

```
1    @UxMenu("集成测试")@UxRoute(TestEipConfig.MODEL_MODEL) class TestEipConfigMenu{}
```

Step5. 初始化 EIP，代码如下。

在 DemoModuleBizInit 类中增加私有方法 initEip，并在 init、reload、upgrade 中调用该私有化。

```
1    private void initEip() {
2        EipResolver.resolver(DemoModule.MODULE_MODULE,null);
3    }
```

Step6. 重启看效果。

（1）选择"集成测试"菜单，单击"新增"按钮，配置集成接口信息如图 6-13 所示。

图 6-13　配置集成接口信息

(2)选择"集成测试"菜单,依次单击三个方法,如图 6-14 所示。

图 6-14　单击集成测试菜单

(3)模块切换进入集成接口,选择"日志"菜单。对比请求的不同点。

① 设置 finalResultKey 的表达,从接口上下文中获取最终对应的 value 作为请求参数,不然会传全部接口上下文过去。

② callQueryByIdError,对应的开放接口和集成接口对结果认知差别。

● 开放接口返回异常信息。

● 集成接口没有对返回信息做判定,直接认定为成功。集成接口如何判定对方返还的内容是否错误,请见下面的"自定义异常判定"。

**5. 自定义异常判定(举例)**

Step1. 新建异常判定函数,代码如下。

新建异常判定处理类 TestExceptionPredictFunction,实现 IEipExceptionPredict 接口。从 context.getInterfaceContextValue(DEFAULT_ERROR_CODE_KEY) 中获取成功与否的值,并进行判断。

```
12  @Fun(TestExceptionPredictFunction.FUN_NAMESPACE)
13  public class TestExceptionPredictFunction implements IEipExceptionPredict<SuperMa
    p> {
14
15      public static final String FUN_NAMESPACE ="demo.TestExceptionPredictFunctio
    n";
16      public static final String FUN ="testFunction";
17
18      @Override
19      public boolean test(IEipContext<SuperMap> context) {
20          return !Boolean.TRUE.toString().equals(StringHelper.valueOf(context.getEx
    ecutorContextValue(DEFAULT_ERROR_CODE_KEY)));
21      }
22
23      @Function
24      @Function.fun(FUN)
25      public Boolean testFunction(IEipContext<SuperMap> context) {
26          return test(context);
27      }
28  }
```

Step2. 修改 TestIntegrateServiceImpl 类,代码如下。

修改 TestIntegrateServiceImpl 的 callQueryByIdError 函数的注解,增加 @Integrate.ExceptionProcessor 注解。

(1)errorCode ="success",从返回结果中获取键值为"success"的值,作为接口上下文对应键值为 IEipContext.DEFAULT_ERROR_CODE_KEY 的值。

(2)exceptionPredictFun 指定函数名。

(3)exceptionPredictNamespace 指定函数命名空间。

```
1    …… 依赖类引用
2
3    @Fun(TestIntegrateService.FUN_NAMESPACE)
4    @Component
5    public class TestIntegrateServiceImpl implements TestIntegrateService {
6
7        ……其他代码
8
9        @Function
10       @Integrate(config = TestEipConfig.class)
11       @Integrate.Advanced(path = "/openapi/pamirs/error?tenant=pamirs")
12       @Integrate.RequestProcessor(
13               finalResultKey = "out",
14               convertParams={
15                       @Integrate.ConvertParam(inParam="data.id",outParam="out.id")
16               }
17       )
18       @Integrate.ExceptionProcessor(
19               errorCode ="success",
20               exceptionPredictFun = TestExceptionPredictFunction.FUN,
21               exceptionPredictNamespace = TestExceptionPredictFunction.FUN_NAMESPAC
22       E )
23       public EipResult<SuperMap> callQueryByIdError(TestOpenApiModel data) {
24           return null;
25       }
26
27   }
28
```

Step3. 重启看效果，如图 6-15 所示。

（1）选择"集成测试"菜单，单击"调用集成接口 callQueryByIdError"按键。

（2）切换模块进入集成接口，选择"日志"菜单。本例中对集成接口是否成功调用判定为 false。

图 6-15　重启看效果

**6. 自定义安全策略（AccessToken）**

开发接口提供默认的三种实现方式，例子采用 DEFAULT_NO_ENCRYPT_AUTHENTICATION_PROCESSOR_FUN。

（1）DEFAULT_AUTHENTICATION_PROCESSOR_FUN：根据 EipApplication 配置的加密类型进行加、解密处理的认证。

（2）DEFAULT_NO_ENCRYPT_AUTHENTICATION_PROCESSOR_FUN：不进行加、解密处理的认证。

（3）DEFAULT_MD5_SIGNATURE_AUTHENTICATION_PROCESSOR_FUN：使用 MD5 进行简单验签的认证。

Step1. 给开放接口加上认证处理配置，代码如下。

```java
1    ……依赖类引用
2    
3    @Fun(TestOpenApiModelService.FUN_NAMESPACE)
4    @Component
5    public class TestOpenApiModelServiceImpl implements TestOpenApiModelService {
6    
7        @Override
8        @Function
9        public TestOpenApiModel queryById(Long id) {
10           return new TestOpenApiModel().queryById(id);
11       }
12   
13       @Function
14       @Open
15       @Open.Advanced(
16           authenticationProcessorFun= EipFunctionConstant.DEFAULT_NO_ENCRYPT_AUTHENTICATION_PROCESSOR_FUN,
17           authenticationProcessorNamespace = EipFunctionConstant.FUNCTION_NAMESPACE
18       )
19       public OpenEipResult<TestOpenApiResponse> queryById4Open(IEipContext<SuperMap> context , ExtendedExchange exchange) {
20           String id = Optional.ofNullable(String.valueOf(context.getInterfaceContext().getIteration("id"))).orElse("");
21           TestOpenApiModel temp = queryById(Long.valueOf(id));
22           TestOpenApiResponse response = new TestOpenApiResponse();
23           if(temp != null ) {
24               response.setAge(temp.getAge());
25               response.setId(temp.getId());
26               response.setName(temp.getName());
27           }else{
28               response.setAge(1);
29               response.setId(1L);
30               response.setName("oinone eip test");
31           }
32           OpenEipResult<TestOpenApiResponse> result = new OpenEipResult<TestOpenApiResponse>(response);
33           return result;
34       }
35   
36       @Function
37       @Open(path = "error")
38       @Open.Advanced(
39           httpMethod = "post",
40           authenticationProcessorFun= EipFunctionConstant.DEFAULT_NO_ENCRYPT_AUTHENTICATION_PROCESSOR_FUN,
41           authenticationProcessorNamespace = EipFunctionConstant.FUNCTION_NAMESPACE
42       )
43       public OpenEipResult<TestOpenApiResponse> queryById4OpenError(IEipContext<SuperMap> context , ExtendedExchange exchange) {
44           throw PamirsException.construct(EipExpEnumerate.SYSTEM_ERROR).appendMsg("测试异常").errThrow();
45       }
46   
47   
48   
49   }
```

Step2. 修改 TestEipAction，增加返回结果判断，代码如下。

```java
1   @Model.model(TestEipConfig.MODEL_MODEL)
2   @Component
3   @Slf4j
4   public class TestEipAction {
5
6       @Autowired
7       private TestIntegrateService testIntegrateService;
8
9       @Action(displayName = "调用集成接口callQueryById",contextType = ActionContextTypeEnum.CONTEXT_FREE)
10      public TestEipConfig callQueryById(TestEipConfig data){
11          EipResult<SuperMap> eipResult = testIntegrateService.callQueryById("111");
12          if(!eipResult.getSuccess()){
13              throw PamirsException.construct(EipExpEnumerate.SYSTEM_ERROR).appendMsg("调用集成接口callQueryById 失败").errThrow();
14          }
15          PamirsSession.getMessageHub().error(JsonUtils.toJSONString(eipResult));
16          return data;
17      }
18
19      @Action(displayName = "调用集成接口callQueryByIdError",contextType = ActionContextTypeEnum.CONTEXT_FREE)
20      public TestEipConfig callQueryByIdError(TestEipConfig data){
21          EipResult<SuperMap> eipResult = testIntegrateService.callQueryByIdError((TestOpenApiModel)new TestOpenApiModel().setId(111L));
22          if(!eipResult.getSuccess()){
23              throw PamirsException.construct(EipExpEnumerate.SYSTEM_ERROR).appendMsg("调用集成接口callQueryByIdError 失败").errThrow();
24          }
25          PamirsSession.getMessageHub().error(JsonUtils.toJSONString(eipResult));
26          return data;
27      }
28
29      @Action(displayName = "调用集成接口callQueryByData",contextType = ActionContextTypeEnum.CONTEXT_FREE)
30      public TestEipConfig callQueryByIdSuccess(TestEipConfig data){
31          EipResult<SuperMap> eipResult = testIntegrateService.callQueryByData((TestOpenApiModel)new TestOpenApiModel().setId(111L));
32          if(!eipResult.getSuccess()){
33              throw PamirsException.construct(EipExpEnumerate.SYSTEM_ERROR).appendMsg("调用集成接口callQueryByData 失败").errThrow();
34          }
35          PamirsSession.getMessageHub().error(JsonUtils.toJSONString(eipResult));
36          return data;
37      }
38  }
```

**Step3. 重启看效果。**

因为开放接口增加了安全相关校验，集成接口并没有调整，所以结果都会抛出错误提示，如图6-16所示。

图6-16 重启看效果

Step4. 为集成接口定义 auth 处理类。

Step4.1. 修改 TestEipConfig，代码如下。

增加了 appkey 和 appSecret 两个字段，用于存储开放接口的 appkey 和 appSecret。注意我们是用集成接口去访问开放接口的，将集成接口想象成一个应用，将开放接口想象为一个应用，用集成接口来访问开放接口，模拟两个应用通过集成平台相互调用。

站在开放接口角度，但凡需要访问开放接口，就需要先申请 appkey 和 appSecret，并通过 appkey 和 appSecret 来换取 accessToken，再用 accessToken 来请求访问。

站在集成接口角度，我们需要访问对方的接口，必须遵循对方接口的安全规范。

```
12  @Model.model(TestEipConfig.MODEL_MODEL)
13  @Model(displayName = "演示集成接口配置模型")
14  @Integrate.ExceptionProcessor(
15          errorCode ="success",
16          exceptionPredictFun = "testFunction",
17          exceptionPredictNamespace = "demo.TestExceptionPredictFunction"
18  )
19  public class TestEipConfig extends IdModel implements IEipAnnotationSingletonConfi
20
21      public static final String MODEL_MODEL = "demo.TestEipConfig";
22
23      @Field(displayName = "服务端域名")
24      private String host;
25      @Field(displayName = "请求协议Http或Https")
26      private String schema;
27
28      @Field(displayName = "appkey")
29      private String appKey;
30
31      @Field(displayName = "appSecret")
32      @Field.String(size = 4096)
33      private String appSecret;
34
35      @Function(openLevel = FunctionOpenEnum.API,summary = "演示集成接口配置模型")
36      @Function.Advanced(type = FunctionTypeEnum.QUERY)
37      public TestEipConfig construct(TestEipConfig config){
38          TestEipConfig config1 = config.singletonModel();
39          if(config1!=null){
40              return config1;
41          }
42          return config.construct();
43      }
44  }
45
```

Step4.2. 修改 TestIntegrateService 和 TestIntegrateServiceImpl。

（1）定义和实现 fetchOinoneAccessToken，代码如下。

```
1  @Function
2  EipResult<SuperMap> fetchOinoneAccessToken(String appkey,String appSecret) ;
3
```

（2）修改集成注解增加 auth 处理器，代码如下。

```java
2    @Override
3    @Function
4    @Integrate(config = TestEipConfig.class)
5    @Integrate.Advanced(path = "/openapi/pamirs/queryById4Open?tenant=pamirs")
6    @Integrate.RequestProcessor(
7            finalResultKey = "out",
8            convertParams={
9                    @Integrate.ConvertParam(inParam="id",outParam="out.id")
10           },
11           authenticationProcessorNamespace = TestAuthFunction.FUN_NAMESPACE,
12           authenticationProcessorFun =  TestAuthFunction.FUN
13   )
14   public EipResult<SuperMap> callQueryById(String id) {
15       return null;
16   }
17
18   @Override
19   @Function
20   @Integrate(config = TestEipConfig.class)
21   @Integrate.Advanced(path = "/openapi/pamirs/queryById4Open?tenant=pamirs")
22   @Integrate.RequestProcessor(
23           finalResultKey = "out",
24           convertParams={
25                   @Integrate.ConvertParam(inParam="data.id",outParam="out.id")
26           },
27           authenticationProcessorNamespace = TestAuthFunction.FUN_NAMESPACE,
28           authenticationProcessorFun =  TestAuthFunction.FUN
29   )
30   public EipResult<SuperMap> callQueryByData(TestOpenApiModel data) {
31       return null;
32   }
33
34   @Function
35   @Integrate(config = TestEipConfig.class)
36   @Integrate.Advanced(path = "/openapi/pamirs/error?tenant=pamirs")
37   @Integrate.RequestProcessor(
38           finalResultKey = "out",
39           convertParams={
40                   @Integrate.ConvertParam(inParam="data.id",outParam="out.id")
41           },
42           authenticationProcessorNamespace = TestAuthFunction.FUN_NAMESPACE,
43           authenticationProcessorFun =  TestAuthFunction.FUN
44
45   )
46   public EipResult<SuperMap> callQueryByIdError(TestOpenApiModel data) {
47       return null;
48   }
49
50   @Function
51   @Integrate(config = TestEipConfig.class)
52   @Integrate.Advanced(path = "/openapi/get/access-token?tenant=pamirs")
53   @Integrate.RequestProcessor(
54           convertParams={
55                   @Integrate.ConvertParam(inParam="appkey",outParam= IEipContext.HEADER_PARAMS_KEY+".appkey"),
56                   @Integrate.ConvertParam(inParam="appSecret",outParam=IEipContext.HEADER_PARAMS_KEY+".appSecret")
57           }
58   )
59   @Integrate.ResponseProcessor(
60           convertParams={
61                   @Integrate.ConvertParam(inParam="access_token",outParam="accessToken"),
62           }
63   )
64   public EipResult<SuperMap> fetchOinoneAccessToken(String appkey,String appSecret)
   {
65       return null;
66   }
```

Step4.3. 新建 TestAuthFunction，代码如下。

（1）获取 TestEipConfig 的配置信息。

（2）先从缓存中拿到 accessToken。

① 拿到 accessToken 直接放到请求头中。

② 如拿不到，调用对方接口获取 accessToken（这里对方接口也就是我们自己的开放平台的接口）获取 accessToken。

```
20  @Fun(TestAuthFunction.FUN_NAMESPACE)
21  @Component
22  public class TestAuthFunction implements IEipAuthenticationProcessor<SuperMap> {
23
24      public static final String FUN_NAMESPACE ="demo.TestAuthFunction";
25      public static final String FUN ="testAuthentication";
26      @Function
27      @Function.fun(FUN)
28      public Boolean testAuthentication(IEipContext<SuperMap> context,ExtendedExchange exchange) {
29          return authentication(context,exchange);
30      }
31
32      @Override
33      public boolean authentication(IEipContext<SuperMap> context, ExtendedExchange exchange) {
34          //获取第三方配置
35          TestEipConfig testEipConfig = new TestEipConfig().singletonModel();
36          TestIntegrateService testIntegrateService = BeanDefinitionUtils.getBean(TestIntegrateService.class);
37          RedisTemplate redisTemplate = (RedisTemplate)BeanDefinitionUtils.getBean("redisTemplate");
38          String accessToken = (String)redisTemplate.opsForValue().get(testEipConfig.getAppKey());
39          if(StringUtils.isNotBlank(accessToken)){
40              context.putInterfaceContextValue(IEipContext.HEADER_PARAMS_KEY+".accessToken",accessToken);
41          }else {
42              //调用获取accessToken接口
43              EipResult<SuperMap> eipResult = testIntegrateService.fetchOinoneAccessToken(testEipConfig.getAppKey(), testEipConfig.getAppSecret());
44              if (eipResult.getSuccess()) {
45                  accessToken = String.valueOf(eipResult.getContext().getInterfaceContextValue("accessToken"));
46                  if (StringUtils.isBlank(accessToken)) {
47                      return Boolean.FALSE;
48                  }
49                  context.putInterfaceContextValue(IEipContext.HEADER_PARAMS_KEY + ".accessToken", accessToken);
50                  //Oinone的accessToken默认失效时间为15分钟
51                  redisTemplate.opsForValue().set(testEipConfig.getAppKey(), accessToken,13*60, TimeUnit.SECONDS);
52              } else {
53                  return Boolean.FALSE;
54              }
55          }
56          return Boolean.TRUE;
57      }
58  }
```

Step4.4. 修改 TestEipAction，代码如下。

借用 TestEipConfig 模型的列表页提供一个新增 EipApplication 的入口。

```
1    ……引用依赖类
2    @Model.model(TestEipConfig.MODEL_MODEL)
3    @Component
4    @Slf4j
5    public class TestEipAction {
6        ……其他代码
7
8        @Action(displayName = "创建一个集成应用",contextType = ActionContextTypeEnum.CONTEXT_FREE)
9        public TestEipConfig createEipApplication(TestEipConfig data){
10
11           EipApplication eipApplication = new EipApplication().setName("测试的集成应用").queryOne();
12           if(eipApplication ==null){
13               eipApplication = new EipApplication().setName("测试的集成应用");
14               eipApplication.create();
15           }else {
16               eipApplication.updateById();
17           }
18           return data;
19       }
20   }
```

Step5. 重启看效果。

（1）选择"集成测试"菜单，单击"创建一个集成应用"按钮，如图 6-17 所示。在数据库中查看创建出来的 EipApplication 记录和 EipAuthentication 记录，如图 6-18 所示。

图 6-17　单击"创建一个集成应用"按钮

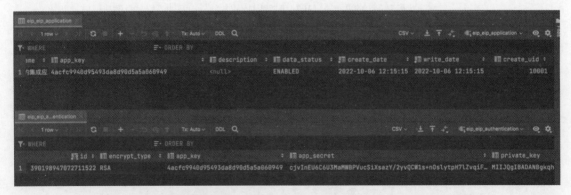

图 6-18　数据库中查看创建出来的 EipApplication 记录和 EipAuthentication 记录

（2）选择"集成测试"菜单，编辑记录，填写 appkey 和 appSecret，如图 6-19、图 6-20 所示。

图 6-19　编辑记录

图 6-20　填写 appkey 和 appSecret

（3）选择"集成测试"菜单，依次单击恢复原有的效果，只有调用集成接口 callQueryByIdError 报错，其他调用不再报错了，才可以进入集成接口模块的日志菜单查看请求日志，如图 6-21 所示。

图 6-21　选择"集成测试"菜单

### 7. 自定义安全策略（AccessToken+ 加密）

例子采用 DEFAULT_AUTHENTICATION_PROCESSOR_FUN 安全策略，根据 EipApplication 配置的加密类型进行加解密处理的认证。安全模式："AccessToken+ 加密"。

当开放接口采用 DEFAULT_AUTHENTICATION_PROCESSOR_FUN 的策略，需要外部调用方请求参数为 {result:" 加密后的请求体内容 "} 的结构，result 的值为加密后的请求体内容。

Step1. 更改开放接口认证方式，代码如下。

认证方式换成 DEFAULT_AUTHENTICATION_PROCESSOR_FUN，@Open.Advanced (authenticationProcessorFun= EipFunctionConstant.DEFAULT_AUTHENTICATION_PROCESSOR_FUN)。

读者们可以把 @Open.Advanced (authenticationProcessorFun) 的配置挪到 TestEipConfig 中去，这样就不用为每个接口配置了，自己动手试试吧。

```
1   //…… 包引用
2   @Fun(TestOpenApiModelService.FUN_NAMESPACE)
3   @Component
4   public class TestOpenApiModelServiceImpl implements TestOpenApiModelService {
5
6       //…… 其他代码
7
8       @Function
9       @Open
10      @Open.Advanced(
11              authenticationProcessorFun= EipFunctionConstant.DEFAULT_AUTHENTICATION_PROCESSOR_FUN,
12              authenticationProcessorNamespace = EipFunctionConstant.FUNCTION_NAMESPACE
13      )
14      public OpenEipResult<TestOpenApiResponse> queryById4Open(IEipContext<SuperMap> context , ExtendedExchange exchange) {
15          //…… 其他代码
16      }
17
18      @Function
19      @Open(path = "error")
20      @Open.Advanced(
21              httpMethod = "post",
22              authenticationProcessorFun= EipFunctionConstant.DEFAULT_AUTHENTICATION_PROCESSOR_FUN,
23              authenticationProcessorNamespace = EipFunctionConstant.FUNCTION_NAMESPACE
24      )
25      public OpenEipResult<TestOpenApiResponse> queryById4OpenError(IEipContext<SuperMap> context , ExtendedExchange exchange) {
26          throw PamirsException.construct(EipExpEnumerate.SYSTEM_ERROR).appendMsg("测试异常").errThrow();
27      }
28
29  }
30
```

Step2. 重启看效果。

期待效果，因为开放接口修改了安全相关校验，集成接口这边并没有调整，所以都会抛出错误提示，如图 6-22 所示。

图 6-22　集成接口未调整所以都会抛错

Step3. 为集成接口定义新的 auth 处理类。

Step3.1. 修改 TestEipConfig。

增加两个字段 encryptType 和 publicKey，代码如下。这里加密类型为 RSA，我们为集成接口申请的 EipApplication 没有指定加密类型，默认为 RSA。Oinone 的开放接口总共支持两种加密类型：RSA 和 AES。

```
1
2        @Field(displayName = "encryptType")
3        private EncryptTypeEnum encryptType;
4
5        @Field(displayName = "publicKey")
6        @Field.String(size = 4096)
7        private String publicKey;
8
```

Step3.2. 新建 TestInOutConvertFunction。

新建 TestInOutConvertFunction 实现 IEipInOutConverter 接口，代码如下。

```
22   @Fun(TestInOutConvertFunction.FUN_NAMESPACE)
23   @Component
24   public class TestInOutConvertFunction implements IEipInOutConverter {
25
26       public static final String FUN_NAMESPACE ="demo.third.TestInOutConvertFunction";
27       public static final String FUN ="exchangeObject";
28
29       @Function
30       @Function.fun(FUN)
31       @Override
32       public Object exchangeObject(ExtendedExchange exchange, Object inObject) throws Exception {
33           //获取第三方配置
34           TestEipConfig testEipConfig = new TestEipConfig().singletonModel();
35           //数据加密
36           String data;
37           if (inObject instanceof InputStream) {
38               data = (String) JSON.parseObject((InputStream)inObject, String.class, new Feature[0]);
39           } else if (inObject instanceof String) {
40               data = (String)inObject;
41           } else {
42               data = JsonUtils.toJSONString(inObject);
43           }
44           switch (testEipConfig.getEncryptType()) {
45               case RSA:
46                   try {
47                       data = EncryptHelper.encryptByKey(EncryptHelper.getPublicKey(testEipConfig.getEncryptType().value(), testEipConfig.getPublicKey()), data);
48                   } catch (NoSuchPaddingException | NoSuchAlgorithmException | InvalidKeyException | BadPaddingException | IllegalBlockSizeException | InvalidKeySpecException e) {
49                       e.printStackTrace();
50                       data = null;
51                   }
52                   break;
53               case AES:
54                   try {
55                       data = EncryptHelper.encryptByKey(EncryptHelper.getSecretKeySpec(testEipConfig.getEncryptType().value(), testEipConfig.getPublicKey()), data);
56                   } catch (NoSuchPaddingException | NoSuchAlgorithmException | InvalidKeyException | BadPaddingException | IllegalBlockSizeException e) {
57                       e.printStackTrace();
58                       data = null;
59                   }
60                   break;
61               default:
```

```
62                    throw new IllegalStateException("Unexpected value: " + testEipCon
     fig.getEncryptType());
63            }
64            //放到指定的路径中，因为集成接口的finalResultKey为"out"，这样刚好构建{result:""}的
     请求结构，不同公司不一样
65            return "{result:\""+data+"\"}";
66       }
67   }
```

Step3.3. 修改 TestIntegrateServiceImpl。

为集成方法增加 inOutConverter 的注解，代码如下。

```
1    @Override
2    @Function
3    @Integrate(config = TestEipConfig.class)
4    @Integrate.Advanced(path = "/openapi/pamirs/queryById4Open?tenant=pamirs")
5    @Integrate.RequestProcessor(
6            finalResultKey = "out",
7            convertParams={
8                    @Integrate.ConvertParam(inParam="id",outParam="out.id")
9            },
10           authenticationProcessorNamespace = TestAuthFunction.FUN_NAMESPACE,
11           authenticationProcessorFun =  TestAuthFunction.FUN,
12           inOutConverterFun = TestInOutConvertFunction.FUN,
13           inOutConverterNamespace = TestInOutConvertFunction.FUN_NAMESPACE
14   )
15   public EipResult<SuperMap> callQueryById(String id) {
16       return null;
17   }
18
19   @Override
20   @Function
21   @Integrate(config = TestEipConfig.class)
22   @Integrate.Advanced(path = "/openapi/pamirs/queryById4Open?tenant=pamirs")
23   @Integrate.RequestProcessor(
24           finalResultKey = "out",
25           convertParams={
26                   @Integrate.ConvertParam(inParam="data.id",outParam="out.id")
27           },
28           authenticationProcessorNamespace = TestAuthFunction.FUN_NAMESPACE,
29           authenticationProcessorFun =  TestAuthFunction.FUN,
30           inOutConverterFun = TestInOutConvertFunction.FUN,
31           inOutConverterNamespace = TestInOutConvertFunction.FUN_NAMESPACE
32   )
33   public EipResult<SuperMap> callQueryByData(TestOpenApiModel data) {
34       return null;
35   }
36
37   @Function
38   @Integrate(config = TestEipConfig.class)
39   @Integrate.Advanced(path = "/openapi/pamirs/error?tenant=pamirs")
40   @Integrate.RequestProcessor(
41           finalResultKey = "out",
42           convertParams={
43                   @Integrate.ConvertParam(inParam="data.id",outParam="out.id")
44           },
45           authenticationProcessorNamespace = TestAuthFunction.FUN_NAMESPACE,
46           authenticationProcessorFun =  TestAuthFunction.FUN,
47           inOutConverterFun = TestInOutConvertFunction.FUN,
48           inOutConverterNamespace = TestInOutConvertFunction.FUN_NAMESPACE
49   )
```

```
50    public EipResult<SuperMap> callQueryByIdError(TestOpenApiModel data) {
51        return null;
52    }
```

Step4. 重启看效果。

（1）选择"集成测试"菜单，单击"创建一个集成应用"按钮，在数据库中查看创建出来的 EipApplication 记录和 EipAuthentication 记录，为记录填写 publicKey，如图 6-23、图 6-24 所示。

图 6-23　选择"集成测试"菜单

图 6-24　为记录填写 publicKey

（2）再次选择"集成测试"菜单，依次单击恢复原有的效果，只有调用集成接口 callQueryByIdError 报错，其他调用不再报错了，才可以进入集成接口模块的日志菜单查看请求日志，如图 6-25 所示。

图 6-25　只有调用集成接口 callQueryByIdError 报错，其他调用不再报错

**8. 自定义安全策略（AccessToken+ 签名）**

例子采用 DEFAULT_MD5_SIGNATURE_AUTHENTICATION_PROCESSOR_FUN。

签名支持：md5 和 hmac。

签名方式和签名摘要：请求方指定，在请求体中设置 signatureMethod、signature。

Step1. 更改开放接口认证方式，代码如下。

认证方式换成 DEFAULT_MD5_SIGNATURE_AUTHENTICATION_PROCESSOR_FUN，@Open.Advanced(authenticationProcessorFun= EipFunctionConstant.DEFAULT_MD5_SIGNATURE_AUTHENTICATION_PROCESSOR_FUN)。

```java
1   //…… 包引用
2
3   @Fun(TestOpenApiModelService.FUN_NAMESPACE)
4   @Component
5   public class TestOpenApiModelServiceImpl implements TestOpenApiModelService {
6       //…… 其他代码
7
8       @Function
9       @Open
10      @Open.Advanced(
11          authenticationProcessorFun= EipFunctionConstant.DEFAULT_MD5_SIGNATURE_AUTHENTICATION_PROCESSOR_FUN,
12          authenticationProcessorNamespace = EipFunctionConstant.FUNCTION_NAMESPACE
13      )
14      public OpenEipResult<TestOpenApiResponse> queryById4Open(IEipContext<SuperMap> context, ExtendedExchange exchange) {
15          //…… 其他代码
16      }
17
18      @Function
19      @Open(config = TestEipConfig.class, path = "error")
20      @Open.Advanced(
21          httpMethod = "post",
22          authenticationProcessorFun= EipFunctionConstant.DEFAULT_MD5_SIGNATURE_AUTHENTICATION_PROCESSOR_FUN,
23          authenticationProcessorNamespace = EipFunctionConstant.FUNCTION_NAMESPACE
24      )
25      public OpenEipResult<TestOpenApiResponse> queryById4OpenError(IEipContext<SuperMap> context, ExtendedExchange exchange) {
26          throw PamirsException.construct(EipExpEnumerate.SYSTEM_ERROR).appendMsg("测试异常").errThrow();
27      }
28
29  }
```

Step2. 重启看效果。

期待效果，因为开放接口修改了安全相关校验，集成接口这边并没有调整，所以都会抛出错误提示，如图 6-26 所示。

图 6-26 重启看效果

Step3. 为集成接口定义新的 auth 处理类。

Step3.1. 新建 SignUtil。

新建 SignUtil 签名工具类，代码如下。

```java
16  public class SignUtil {
17
18      public static String signTopRequest(Map<String, String> params, String secret, String signMethod) throws IOException {
19          // 第一步：检查参数是否已经排序
20          String[] keys = params.keySet().toArray(new String[0]);
21          Arrays.sort(keys);
22
23          // 第二步：把所有参数名和参数值串在一起
24          StringBuilder query = new StringBuilder();
25          if ("md5".equals(signMethod)) { //签名的摘要算法，可选值为：hmac, md5, hmac-sha256
26              query.append(secret);
27          }
28          for (String key : keys) {
29              String value = params.get(key);
30              if (StringUtils.isAllBlank(new CharSequence[]{key, value})) {
31                  query.append(key).append(value);
32              }
33          }
34
35          // 第三步：使用MD5/HMAC加密
36          byte[] bytes;
37          if ("hmac".equals(signMethod)) {
38              bytes = encryptHMAC(query.toString(), secret);
39          } else {
40              query.append(secret);
41              bytes = encryptMD5(query.toString());
42          }
43
44          // 第四步：把二进制转化为大写的十六进制(正确签名应该为32位大写字符串，此方法需要时使用)
45          return byte2hex(bytes);
46      }
47
48      public static byte[] encryptHMAC(String data, String secret) throws IOException {
49          byte[] bytes = null;
50          try {
51              SecretKey secretKey = new SecretKeySpec(secret.getBytes(StandardCharsets.UTF_8), "HmacMD5");
52              Mac mac = Mac.getInstance(secretKey.getAlgorithm());
53              mac.init(secretKey);
54              bytes = mac.doFinal(data.getBytes(StandardCharsets.UTF_8));
55          } catch (GeneralSecurityException gse) {
56              throw new IOException(gse.toString());
57          }
58          return bytes;
59      }
60
61      private static byte[] encryptMD5(String data) throws IOException {
62          return encryptMD5(data.getBytes(StandardCharsets.UTF_8));
63      }
64
65      private static byte[] encryptMD5(byte[] data) throws IOException {
66          try {
67              MessageDigest md = MessageDigest.getInstance("MD5");
68              return md.digest(data);
69          } catch (NoSuchAlgorithmException var3) {
70              throw new IOException(var3.toString(), var3);
71          }
```

```
72        }
73
74      public static String byte2hex(byte[] bytes) {
75          StringBuilder sign = new StringBuilder();
76          for (int i = 0; i < bytes.length; i++) {
77              String hex = Integer.toHexString(bytes[i] & 0xFF);
78              if (hex.length() == 1) {
79                  sign.append("0");
80              }
81              sign.append(hex.toUpperCase());
82          }
83          return sign.toString();
84      }
85  }
```

Step3.2. 新建 TestMd5InOutConvertFunction，代码如下。

（1）跟加密一样，必须在所有内容都处理完成后进行处理，要不然参数个数是对不上的。

（2）例子中新增一种设置 Header 的方法——exchange.getMessage ().setHeader，在前面我们学习 AccessToken 的时候用的方法是：context.putInterfaceContextValue (IEipContext.HEADER_PARAMS_KEY+".accessToken")。

```
18  @Fun(TestMd5InOutConvertFunction.FUN_NAMESPACE)
19  @Component
20  public class TestMd5InOutConvertFunction extends DefaultInOutConverter implements
     IEipInOutConverter {
21
22      public static final String FUN_NAMESPACE ="demo.third.TestMd5InOutConvertFunction";
23      public static final String FUN ="exchangeObject";
24
25      @Function
26      @Function.fun(FUN)
27      public Object exchangeObject(ExtendedExchange exchange, Object inObject) throws Exception {
28          if (inObject instanceof Map) {
29              Map<String, Object> map = (Map)inObject;
30              if (MapUtils.isNotEmpty(map)) {
31
32                  TestEipConfig testEipConfig = new TestEipConfig().singletonModel();
33                  String signatureMethod ="md5";
34                  Map<String, String> params = new HashMap(map.size());
35                  for(Map.Entry<String ,Object> entry: map.entrySet()) {
36                      String key = entry.getKey();
37                      Object value = entry.getValue();
38                      params.put(key, String.valueOf(value));
39                  }
40                  try {
41                      String signature = SignUtil.signTopRequest(params, testEipConfig.getAppKey(), signatureMethod);
42                      exchange.getMessage().setHeader("signatureMethod",signatureMethod);
43                      exchange.getMessage().setHeader("signature",signature);
44                  } catch (IOException e) {
45                      e.printStackTrace();
46                  }
47              }
48          }
49          return super.exchangeObject(exchange, inObject);
50      }
51
52  }
```

Step3.3. 修改 TestIntegrateServiceImpl。

更换 inOutConverter 的注解为 TestMd5InOutConvertFunction，代码如下。

```
@Override
@Function
@Integrate(config = TestEipConfig.class)
@Integrate.Advanced(path = "/openapi/pamirs/queryById4Open?tenant=pamirs")
@Integrate.RequestProcessor(
        finalResultKey = "out",
        convertParams={
                @Integrate.ConvertParam(inParam="id",outParam="out.id")
        },
        authenticationProcessorNamespace = TestAuthFunction.FUN_NAMESPACE,
        authenticationProcessorFun =  TestAuthFunction.FUN,
        inOutConverterFun = TestMd5InOutConvertFunction.FUN,
        inOutConverterNamespace = TestMd5InOutConvertFunction.FUN_NAMESPACE
)
public EipResult<SuperMap> callQueryById(String id) {
    return null;
}

@Override
@Function
@Integrate(config = TestEipConfig.class)
@Integrate.Advanced(path = "/openapi/pamirs/queryById4Open?tenant=pamirs")
@Integrate.RequestProcessor(
        finalResultKey = "out",
        convertParams={
                @Integrate.ConvertParam(inParam="data.id",outParam="out.id")
        },
        authenticationProcessorNamespace = TestAuthFunction.FUN_NAMESPACE,
        authenticationProcessorFun =  TestAuthFunction.FUN,
        inOutConverterFun = TestMd5InOutConvertFunction.FUN,
        inOutConverterNamespace = TestMd5InOutConvertFunction.FUN_NAMESPACE
)
public EipResult<SuperMap> callQueryByData(TestOpenApiModel data) {
    return null;
}

@Function
@Integrate(config = TestEipConfig.class)
@Integrate.Advanced(path = "/openapi/pamirs/error?tenant=pamirs")
@Integrate.RequestProcessor(
        finalResultKey = "out",
        convertParams={
                @Integrate.ConvertParam(inParam="data.id",outParam="out.id")
        },
        authenticationProcessorNamespace = TestAuthFunction.FUN_NAMESPACE,
        authenticationProcessorFun =  TestAuthFunction.FUN,
        inOutConverterFun = TestMd5InOutConvertFunction.FUN,
        inOutConverterNamespace = TestMd5InOutConvertFunction.FUN_NAMESPACE

)
public EipResult<SuperMap> callQueryByIdError(TestOpenApiModel data) {
    return null;
}
```

Step4. 重启看效果。

选择"集成测试"菜单，依次单击恢复原有的效果，只有调用集成接口 callQueryByIdError 报错，其他调用不再报错了，才可以进入集成接口模块的日志菜单查看请求日志，如图 6-27 所示。

图 6-27 重启看效果

### 9. 自定义序列化方式

Step1. 新增一个 xml 的开放接口。

为 TestOpenApiModelServiceImpl 新增一个方法 queryById4ReturnXml，代码如下，使用 @Open 注解，路径配置为 xml。同时记得重启 second 应用，发布 openApi。

```
@Function
@Open(path = "xml")
@Open.Advanced(
        httpMethod = "post",
        authenticationProcessorFun= EipFunctionConstant.DEFAULT_MD5_SIGNATURE_AUTHENTICATION_PROCESSOR_FUN,
        authenticationProcessorNamespace = EipFunctionConstant.FUNCTION_NAMESPACE
)
public String queryById4ReturnXml(IEipContext<SuperMap> context , ExtendedExchange exchange) {
    return "<result><id>12</id><success>true</success></result>";
}
```

Step2. Demo 工程引入 dom4j 的依赖包。

在总工程中修改 pom.xml 增加对 dom4j 的依赖管理，代码如下。

```
<dependency>
    <groupId>dom4j</groupId>
    <artifactId>dom4j</artifactId>
    <version>1.1</version>
</dependency>
```

在 pamirs-demo-core 工程中修改 pom.xml 增加对 dom4j 的依赖，代码如下。

```
<dependency>
    <groupId>dom4j</groupId>
    <artifactId>dom4j</artifactId>
</dependency>
```

Step3. 自定义序列化函数 TestSerializableFunction，代码如下。

新建 TestSerializableFunction 实现 IEipSerializable<SuperMap> 接口。

```java
22      @Fun(TestSerializableFunction.FUN_NAMESPACE)
23      @Component
24      public class TestSerializableFunction implements IEipSerializable<SuperMap> {
25
26          public static final String FUN_NAMESPACE ="demo.third.TestSerializableFuncti
    on";
27          public static final String FUN ="serializable";
28
29          @Override
30          @Function.Advanced(displayName = "自定义xml序列化方式")
31          @Function.fun(FUN)
32          public SuperMap serializable(Object inObject) {
33              if (inObject == null) {
34                  return new SuperMap();
35              } else {
36                  SuperMap result;
37                  if (inObject instanceof String) {
38                      String inObjectString = (String)inObject;
39                      if (StringUtils.isNotBlank(inObjectString)) {
40                          result = this.stringToSuperMap(inObjectString);
41                      } else {
42                          result = new SuperMap();
43                      }
44                  } else if (inObject instanceof InputStream) {
45                      result = this.inputStreamToString((InputStream)inObject);
46                  } else if (inObject instanceof SuperMap) {
47                      result = (SuperMap) inObject;
48                  }
49                  else{
50                      result = new SuperMap();
51                  }
52                  return result;
53              }
54          }
55
56          protected SuperMap inputStreamToString(InputStream inputStream) {
57              StringBuilder sb = new StringBuilder();
58              String line;
59              try (BufferedReader bufferedReader = new BufferedReader(new InputStreamR
    eader(inputStream))) {
60                  while ((line = bufferedReader.readLine()) != null) {
61                      sb.append(line);
62                  }
63                  return serializable(sb.toString());
64              } catch (IOException e) {
65                  return new SuperMap();
66              }
67          }
68
69
70          protected SuperMap stringToSuperMap(String s) {
71              SuperMap result = new SuperMap();
72              try {
73                  Document document = DocumentHelper.parseText(s);
74                  Element root = document.getRootElement();
75                  iterateNodes(root, result);
76                  return result;
77              } catch (Exception e) {
78                  e.printStackTrace();
```

```java
79          }
80          return result;
81      }
82
83      public static void iterateNodes(Element node, SuperMap superMap){
84          //获取当前元素的名称
85          String nodeName = node.getName();
86          if(superMap.containsKey(nodeName)){
87              //该元素在同级下有多个
88              Object object = superMap.getIteration(nodeName);
89              List<Object> list = Lists.newArrayList();
90              if(object instanceof JSONArray){
91                  list = (List) object;
92              }else {
93                  list = Lists.newArrayList();
94                  list.add(object);
95              }
96              //获取该元素下所有子元素
97              List<Element> listElement = node.elements();
98              if(listElement.isEmpty()){
99                  //该元素无子元素，获取元素的值
100                 String nodeValue = node.getTextTrim();
101                 list.add(nodeValue);
102                 superMap.putIteration(nodeName, list);
103                 return;
104             }
105             //有子元素
106             SuperMap subMap = new SuperMap();
107             //遍历所有子元素
108             for(Element e:listElement){
109                 //递归
110                 iterateNodes(e, subMap);
111             }
112             list.add(subMap);
113             subMap.putIteration(nodeName, list);
114             return;
115         }
116         List<Element> listElement = node.elements();
117         if(listElement.isEmpty()){
118             //该元素无子元素，获取元素的值
119             String nodeValue = node.getTextTrim();
120             superMap.putIteration(nodeName, nodeValue);
121             return;
122         }
123         //有子节点，新建一个JSONObject来存储该节点下子节点的值
124         SuperMap subMap = new SuperMap();
125         //遍历所有一级子节点
126         for(Element e:listElement){
127             //递归
128             iterateNodes(e, subMap);
129         }
130         superMap.putIteration(nodeName, subMap);
131     }
132
133 }
```

Step4. 新增一个异常判定类，代码如下。

新增 TestXmlExceptionPredictFunction 异常判定类，因为原来的异常判定类配置在了 TestEipConfig，所以所有方法都会默认使用 TestEipConfig 类上定义的异常判定类。

```java
10    @Fun(TestXmlExceptionPredictFunction.FUN_NAMESPACE)
11    public class TestXmlExceptionPredictFunction implements IEipExceptionPredict<SuperMap> {
12
13        public static final String FUN_NAMESPACE ="demo.TestXmlExceptionPredictFunction";
14        public static final String FUN ="testFunction";
15
16        @Override
17        public boolean test(IEipContext<SuperMap> context) {
18            return Boolean.FALSE;
19        }
20
21        @Function
22        @Function.fun(FUN)
23        public Boolean testFunction(IEipContext<SuperMap> context) {
24            return test(context);
25        }
26    }
27
```

Step5. 修改 TestIntegrateService 和 TestIntegrateServiceImpl。

TestIntegrateService 增加方法定义，代码如下。

```java
1    @Function
2    EipResult<SuperMap> callQueryReturnXml(TestOpenApiModel data) ;
```

TestIntegrateServiceImpl 实现 callQueryReturnXml 并定义为集成接口，同时配置自定义序列化函数，代码如下。

```java
1    ……引用依赖类
2
3    @Fun(TestIntegrateService.FUN_NAMESPACE)
4    @Component
5    public class TestIntegrateServiceImpl implements TestIntegrateService {
6            ……其他代码
7
8        @Function
9        @Integrate(config = TestEipConfig.class)
10       @Integrate.Advanced(
11               path = "/openapi/pamirs/xml?tenant=pamirs",
12               httpMethod = "post"
13       )
14       @Integrate.RequestProcessor(
15               finalResultKey = "out",
16               convertParams={
17                       @Integrate.ConvertParam(inParam="data.id",outParam="out.id")
18               },
19               authenticationProcessorNamespace = TestAuthFunction.FUN_NAMESPACE,
20               authenticationProcessorFun =  TestAuthFunction.FUN,
21               inOutConverterFun = TestMd5InOutConvertFunction.FUN,
```

```
22              inOutConverterNamespace = TestMd5InOutConvertFunction.FUN_NAMESPACE
23          )
24          @Integrate.ResponseProcessor(
25              finalResultKey = "result",
26              serializableFun = TestSerializableFunction.FUN,
27              serializableNamespace = TestSerializableFunction.FUN_NAMESPACE
28          )
29          @Integrate.ExceptionProcessor(
30              errorCode ="success",
31              exceptionPredictFun = TestXmlExceptionPredictFunction.FUN,
32              exceptionPredictNamespace = TestXmlExceptionPredictFunction.FUN_NAMES
33   PACE
34          )
35      public EipResult<SuperMap> callQueryReturnXml(TestOpenApiModel data) {
36          return null;
37      }
38
39  }
40
```

Step6. 增加测试入口。

修改 TestEipAction 类增加一个 callQueryByXml 的 Action 定义，调用 testIntegrateService 集成接口的 callQueryReturnXml 方法，代码如下。

```
1   @Action(displayName = "调用集成接口callQueryReturnXml",contextType = ActionContextTypeEnum.CONTEXT_FREE)
2   public TestEipConfig callQueryByXml(TestEipConfig data){
3       EipResult<SuperMap> eipResult = testIntegrateService.callQueryReturnXml((TestOpenApiModel)new TestOpenApiModel().setName("cpc").setId(111L));
4       if(!eipResult.getSuccess()){
5           throw PamirsException.construct(EipExpEnumerate.SYSTEM_ERROR).appendMsg("调用集成接口callQueryReturnXml 失败").errThrow();
6       }
7       PamirsSession.getMessageHub().error(JsonUtils.toJSONString(eipResult));
8       return data;
9   }
```

Step7. 重启看效果，如图 6-28 所示。

图 6-28　重启看效果

### 10. Eip 注解说明

（1）开放接口注解。

Open
├── name 显示名称
├── config 配置类
├── path 路径
├── Advanced 更多配置
│   ├── httpMethod 请求方法，默认：post
│   ├── inOutConverterFun 输入、输出转换器函数名称

```
|       └── inOutConverterNamespace 输入、输出转换器函数命名空间
|       ├── authenticationProcessorFun 认证处理器函数名称
|       ├── authenticationProcessorNamespace 认证处理器函数命名空间
|       ├── serializableFun 序列化函数名称
|       ├── serializableNamespace 序列化函数命名空间
|       ├── deserializationFun 反序列化函数名称
|       └── deserializationNamespace 反序列化函数命名空间
```

（2）集成接口注解。

```
Integrate
├── name 显示名称
├── config 配置类
├── Advanced 更多配置
|       ├── host 请求"域名+端口"
|       ├── path 请求路径 以"/"开头
|       ├── schema 请求协议，Http 或者 Https
|       └── httpMethod 请求方法，默认 post
├── ExceptionProcessor 异常配置
|       ├── exceptionPredictFun 异常判定函数名
|       ├── exceptionPredictNamespace 异常判定函数命名空间
|       ├── errorMsg 异常判定 Msg 的键值
|       └── errorCode 异常判定 errorCode 的键值
├── RequestProcessor 请求处理配置
|       ├── finalResultKey 请求的最终结果键值
|       ├── inOutConverterFun 输入、输出转换器函数名称
|       ├── inOutConverterNamespace 输入、输出转换器函数命名空间
|       ├── paramConverterCallbackFun 参数转换回调函数名称
|       ├── paramConverterCallbackNamespace 参数转换回调函数命名空间
|       ├── authenticationProcessorFun 认证处理器函数名称
|       ├── authenticationProcessorNamespace 认证处理器函数命名空间
|       ├── serializableFun 序列化函数名称
|       ├── serializableNamespace 序列化函数命名空间
|       ├── deserializationFun 反序列化函数名称
|       ├── deserializationNamespace 反序列化函数命名空间
|       ├── convertParams 参数转化集合
|       └── ConvertParam 参数转化
|              ├── inParam 输入参数的键值
|              └── outParam 输出参数的键值
├── ResponseProcessor 请求处理配置
|       ├── finalResultKey 响应的最终结果键值
|       ├── inOutConverterFun 输入、输出转换器函数名称
|       ├── inOutConverterNamespace 输入、输出转换器函数命名空间
```

```
|       ├── paramConverterCallbackFun 参数转换回调函数名称
|       ├── paramConverterCallbackNamespace 参数转换回调函数命名空间
|       ├── authenticationProcessorFun 认证处理器函数名称
|       ├── authenticationProcessorNamespace 认证处理器函数命名空间
|       ├── serializableFun 序列化函数名称
|       ├── serializableNamespace 序列化函数命名空间
|       ├── deserializationFun 反序列化函数名称
|       ├── deserializationNamespace 反序列化函数命名空间
|       └── convertParams 参数转化集合
|           └── ConvertParam 参数转化
|               ├── inParam 输入参数的键值
|               └── outParam 输出参数的键值
```

## 11. Eip 处理函数说明

Eip 处理函数说明见表 6-10。

表 6-10　Eip 处理函数说明

| Eip 处理函数 | 作用 | 需实现接口 | 说明 |
| --- | --- | --- | --- |
| inOutConverter | 输入、输出转换器函数 | IEipInOutConverter | inOutConverter 核心处理请求体的构造或响应体的构造 |
| paramConverter | 参数转换回调函数 | IEipParamConverter | 如果自定义 paramConverter，会让默认 ConvertParam 配置的 (inParam, outParam) 转化失效了，需要自行处理 |
| authenticationProcessor | 认证处理器函数 | IEipAuthenticationProcessor<SuperMap> | |
| serializable | 序列化函数 | IEipSerializable<SuperMap> | 序列化方法将任意对象转换为上下文承载对象 |
| deserialization | 反序列化函数 | IEipDeserialization<SuperMap> | 目前这个函数意义不大，因为如果设置了 finalResultKey，就会不进行反序列化。现在都是通过 inoutconverter 处理的 |
| exceptionProcessor | 异常判定函数 | IEipExceptionPredict<SuperMap> | |

（1）paramConverter 的特殊转换。

paramConverter 的特殊转换见表 6-11。

表 6-11　paramConverter 的特殊转换

| paramConverter 的特殊转换 | | |
| --- | --- | --- |
| URL_QUERY_PARAMS_KEY | IEipContext.URL_QUERY_PARAMS_KEY.xxx 是请求参数，会拼接在 url 的问号后面 | convertParams={<br>@Integrate.ConvertParam (inParam="id",outParam=IEipContext.URL_QUERY_PARAMS_KEY+".id")<br>} |
| HEADER_PARAMS_KEY | IEipContext.HEADER_PARAMS_KEY.xxx 是 http 请求头参数 | convertParams={<br>@Integrate.ConvertParam (inParam="id",outParam=IEipContext.HEADER_PARAMS_KEY+".id")<br>} |

(2)调用流程。

请求调用流程如图 6-29 所示。

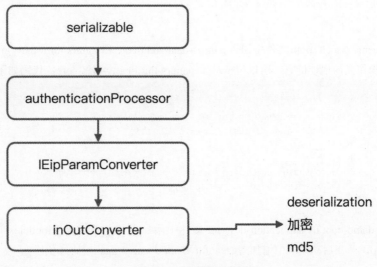

图 6-29 请求调用流程

响应调用流程如图 6-30 所示。

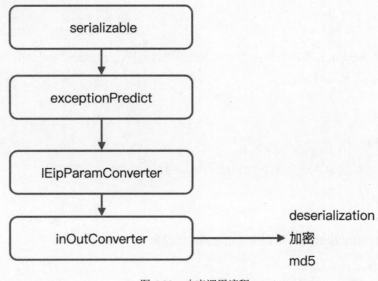

图 6-30 响应调用流程

## 6.3 数据审计

在业务应用中我们经常需要为一些核心数据的变更做审计追踪，记录字段的前后变化、操作 IP、操作人、操作地址等。数据审计模块为此提供了支撑和统一管理。它在成熟的企业的核心业务系统中，需求是比较旺盛的。接下来我们开始学习数据审计模块。

**1. 准备工作**

（1）在 pamirs-demo-core 的 pom 文件中引入 pamirs-data-audit-api 包依赖，代码如下。

```
1  <dependency>
2      <groupId>pro.shushi.pamirs.core</groupId>
3      <artifactId>pamirs-data-audit-api</artifactId>
4  </dependency>
```

（2）pamirs-demo-boot 的 pom 文件中引入 pamirs-data-audit-core 和 pamirs-third-party-map-core 包依赖，数据审计会记录操作人的地址信息，所以也依赖了 pamirs-third-party-map-core，代码如下。

```
1  <dependency>
2      <groupId>pro.shushi.pamirs.core</groupId>
3      <artifactId>pamirs-data-audit-core</artifactId>
4  </dependency>
5  <dependency>
6      <groupId>pro.shushi.pamirs.core.map</groupId>
7      <artifactId>pamirs-third-party-map-core</artifactId>
8  </dependency>
```

（3）pamirs-demo-boot 的 application-dev.yml 文件中增加配置 pamirs.boot.modules，增加 data_audit 和 third_party_map，即在启动模块中增加 data_audit 和 third_party_map 模块，代码如下。

```
1  pamirs:
2    boot:
3      modules:
4        - data_audit
5        - tp_map
```

（4）为 third_party_map 模块增加高德接口 API，将代码中的 e439dda234467b07709f28b57f0a9bd5 部分换成自己的 key，代码如下。

```
1  pamirs:
2    eip:
3      map:
4        gd:
5          key: e439dda234467b07709f28b57f0a9bd5
```

**2. 进行数据审计**

注解式（举例）。

Step1. 新增 PetTalentDataAudit 数据审计定义类，代码如下。

```
 7  @DataAudit(
 8      model = PetTalent.MODEL_MODEL,//需要审计的模型
 9      modelName = "宠物达人" ,//模型名称,默认模型对应的displayName
10      //操作名称
11      optTypes = {PetTalentDataAudit.PETTALENT_CREATE,PetTalentDataAudit.PETTAL
    ENT_UPDATE},
12      fields={"nick","picList.id","picList.url","petShops.id","petShops.shopNam
    e"}//需要审计的字段,关系字段用"."连接
13  )
14  public class PetTalentDataAudit {
15      public static final String PETTALENT_CREATE ="宠物达人创建";
16      public static final String PETTALENT_UDPATE ="宠物达人修改";
17
18
```

Step2. 修改 PetTalentAction 的 update 方法，代码如下。

做审计日志埋点：手工调用 OperationLogBuilder.newInstance ().record () 方法。需要注意的是这里需要把原有记录的数据值先查出来做对比。

```
1   @Function.Advanced(type= FunctionTypeEnum.UPDATE)
2   @Function.fun(FunctionConstants.update)
3   @Function(openLevel = {FunctionOpenEnum.API})
4   public PetTalent update(PetTalent data){
5       //记录日志
6       OperationLogBuilder.newInstance(PetTalent.MODEL_MODEL, PetTalentDataAudi
    t.PETTALENT_UDPATE).record(data.queryById().fieldQuery(PetTalent::getPicList).fie
    ldQuery(PetTalent::getPetShops),data);
7
8       PetTalent existPetTalent = new PetTalent().queryById(data.getId());
9       if(existPetTalent !=null){
10          existPetTalent.fieldQuery(PetTalent::getPicList);
11          existPetTalent.fieldQuery(PetTalent::getPetShops);
12          existPetTalent.relationDelete(PetTalent::getPicList);
13          existPetTalent.relationDelete(PetTalent::getPetShops);
14      }
15      data.updateById();
16      data.fieldSave(PetTalent::getPicList);
17      data.fieldSave(PetTalent::getPetShops);
18      return data;
19  }
```

Step3. 重启看效果。

修改宠物达人记录对应的字段，然后进入审计模块查看日志，如图 6-31、图 6-32 所示。

图 6-31　修改宠物达人记录对应的字段

图 6-32　进入审计模块查看日志

## 6.4 国际化之多语言

多语言是国际化中大家最常面对的问题,我们需要对应用的页面结构元素进行翻译,也需要对系统内容进行翻译,比如菜单、数据字典等,甚至还会对业务数据进行翻译。但不管翻译需求是什么,在实现上基本可以归类为前端翻译和后端翻译。前端翻译顾名思义是在前端根据用户选择语言对内容进行翻译,反之就是后端翻译。本文会带着大家了解 Oinone 的前端翻译与后端翻译。

**1. 准备工作**

(1) 在 pamirs-demo-boot 的 pom 文件中引入 pamirs-translate 包依赖,代码如下。

```
1  <dependency>
2      <groupId>pro.shushi.pamirs.core</groupId>
3      <artifactId>pamirs-translate</artifactId>
4  </dependency>
```

(2) pamirs-demo-boot 的 application-dev.yml 文件中增加配置 pamirs.boot.modules 增加 translate,代码如下。

```
1  pamirs:
2    boot:
3      modules:
4        - translate
```

**2. 后端翻译(使用)**

这里通过对菜单的翻译来带大家了解翻译模块。

Step1. 新增翻译记录,如图 6-33 所示。

图 6-33　新增翻译记录

切换应用到 translate 模块,单击"新增翻译"按钮。

(1) 选择新增翻译生效模块。

(2) 选择翻译的模型为菜单模型。

(3) 源语言选择中文,目标选择 English。

(4) 添加翻译项目:

① 源术语为"商店"。

② 翻译值为"shop"。

③ 状态为"激活"。

Step2. 查看效果。

应用切换到 Demo 模块，在右上角切换语言为英语，如图 6-34 所示。

图 6-34 右上角切换语言为英语

### 3. 后端翻译（自定义模型的翻译）

在前面菜单的翻译中，似乎我们什么都没做就可以正常通过翻译模块完成多语言的切换了。是不是所有翻译任务真如我们想象的一样，当然不是。菜单翻译的简单是因为 Menu 模型的 displayName 字段加上了 @Field (translate = true) 注解。自定义模型的翻译步骤如下：

Step1. 为 PetType 模型的 name 字段增加翻译注解，代码如下。

```
 7    @Model.MultiTable(typeField = "kind")
 8    @Model.model(PetType.MODEL_MODEL)
 9    @Model(displayName="品种",labelFields = {"name"})
10  ▾ public class PetType extends IdModel {
11
12        public static final String MODEL_MODEL="demo.PetType";
13
14        @Field(displayName = "品种名" , translate = true)
15        private String name;
16
17        @Field(displayName = "宠物分类")
18        private String kind;
19    }
```

Step2. 重启应用查看效果。

（1）切换应用到 translate 模块，单击"新增翻译"按钮，如图 6-35 所示。

图 6-35 单击"新增翻译"按钮

（2）切换应用到 Demo 模块，切换中英文，查看效果，如图 6-36 所示。

图 6-36　切换中英文，查看效果

**4. 前端翻译**

还记得本书前端第一个自定义动作吗？会弹出"Oinone 第一个自定义 Action，啥也没干"，我们要对它进行翻译。

Step1. 修改前端 DoNothingActionWidget.ts，代码如下。

（1）增加 import translateValueByKey。

（2）提示语用 translateValueByKey 加上翻译 const confirmRs = executeConfirm (translateValueByKey ('Oinone 第一个自定义 Action，啥也没干 ')||'Oinone 第一个自定义 Action，啥也没干 ')。

前端更多翻译工具请见 4.2.7 节。

```
import {
  Action,
  ActionContextType,
  ActionWidget,
  executeConfirm,
  IClientAction,
  SPI,
  ViewType,
  Widget,
  translateValueByKey
} from '@kunlun/dependencies';

@SPI.ClassFactory(ActionWidget.Token({ name: 'demo.doNothing' }))
export class DoNothingActionWidget extends ActionWidget {
  @Widget.Method()
  public async clickAction() {
    const confirmRs = executeConfirm(translateValueByKey('oinone第一个自定义Action, 啥也没干')||'oinone第一个自定义Action, 啥也没干');
  }
}

//定义动作元数据
Action.registerAction('*', {
  displayName: '啥也没干',
  name: 'demo.doNothing',
  id: 'demo.doNothing',
  contextType: ActionContextType.ContextFree,
  bindingType: [ViewType.Table]
} as IClientAction);
```

Step2. 新增翻译记录，如图 6-37 所示。

图 6-37　新增翻译记录

前端翻译的翻译记录对应的模型可以随意放至任意模型中。但要注意以下几点：

（1）不要找有字段配置 translate = true 的模型，因为会影响后端翻译性能。

（2）最好统一到一个模型中，便于后续管理。

这里大家可以自定义一个没有业务访问且本身不需要翻译的模型来挂载翻译记录，避免性能损失。

Step3. 刷新远程资源生成前端语言文件，如图 6-38 所示。

图 6-38　刷新远程资源生成前端语言文件

Step4. 新增或修改 .env。

前端在项目根目录下新增或修改 .env，可以参考 .env.example 文件。通过 .env 文件为前端配置 oss 文件路径，针对 I18N_OSS_URL 配置项，代码如下。真实前端访问翻译语言文件的路径规则为：http://bucket.downloadUrl/mainDir/ 租户 /translate/ 模块 / 语言文件。

（1）yaml 文件中 oss 配置的文件路径：http://pamirs.oss-cn-hangzhou.aliyuncs.com/upload/demo/。

（2）对于租户 /translate/ 模块 / 语言文件，前端会自动根据上下文组织。

```
1    # 后端API配置
2    # API_BASE_URL=http://127.0.0.1:8090
3    # 下面是国际化语言的cdn配置，默认用当前请求链接下的路径：/pamirs/translate/${module}/i18n_
     ${lang}.js
4    I18N_OSS_URL=http://pamirs.oss-cn-hangzhou.aliyuncs.com/upload/demo
```

Step5. 重启前端应用看效果。

对语言进行中英文切换，进入宠狗达人页面，单击"第一个自定义 Action"按钮，查看前端翻译效果，如图 6-39、图 6-40 所示。

图 6-39　查看翻译效果（1）

图 6-40　查看翻译效果（2）

## 6.5　权限体系

做好企业级软件，首先得过权限这一关。

在企业的 IT 部门沟通中，权限是避免不了的。自嘲一下，在我们刚出来创业时，为了收获客户对我们技术能力的信任，每当与跟客户沟通时都会说我们是阿里出来的，但在权限设计这个环节，这种沟通方式不那么灵验，反而被打上了不懂 B 端权限设计的标签，会被问很多问题。笔者就很奇怪难道大厂就没有内部管理系统了？大厂只有 C 端交易，没有 B 端交易？但从侧面说明权限特别重要。做好企业级软件，首先得过权限这一关。

**1. 整体介绍**

（1）对于平台运行来说，权限是必须的，但 auth 模块不是必须的，auth 模块只是我们提供的一种默认实现，客户可以根据平台的 spi 机制进行替换。auth 模块利用了平台的 hook 特性做到与业务无关，在我们开发上层应用的时候，是不用感知它的存在的。

（2）auth 模块涉及数据权限、功能权限。

① 数据权限：行权限和列权限。备注：数据权限的控制只能用于存储模型。

● 表级权限：表达的语义是，是否对该表可读 / 写（修改和新增）。

● 列权限：表达的语义是，是否对该列可读 / 写（修改和新增）。

● 行权限：表达的语义是，是否对该行可读 / 写（修改）/ 删除。

② 功能权限：表达的语义是 ServerAction/Function 是否可执行 / 展示，viewAction 是否可展示，菜单是否可以显示。

③ 范围说明：

● 配置多个权限项的时候，取并集。

● 配置多个角色的时候，取并集。

（3）模型设计如图 6-41 所示。

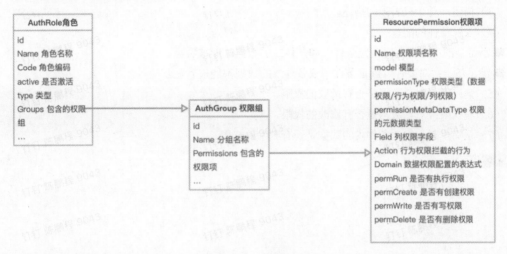

图 6-41　模型设计

**2. 产品体验**

Step1. 创建角色。

通过 App Finder 切换至权限应用，单击"新增"按钮创建一个名为 Oinone 的角色，如图 6-42、图 6-43 所示。

图 6-42　创建一个名为 Oinone 的角色

图 6-43　角色创建完成

Step2. 创建数据权限项，如图 6-44 所示。

在权限项列表菜单，单击"创建"按钮，新增一个名为宠物达人数据权限项，同时宠物达人的数据权限设置为只能查看性别为男的记录。

配置说明：

（1）名称：代表该项配置的权限的名字（必填，必须是全系统唯一）。

(2)权限模型：选择需要拦截的数据所在的表，即为模型，可以搜索使用。

(3)描述：对该权限项的描述。

(4)权限条件的配置：

● 满足全部：对要同时满足条件一和条件二的数据才能被看见。

● 满足任一：对要任意满足条件一或条件二的数据都能被看见。

(5)读权限：对该数据是否有读取的权限。

(6)写权限：对该数据是否有修改的权限。

(7)删除权限：对该条件内的数据是否有删除的权限。

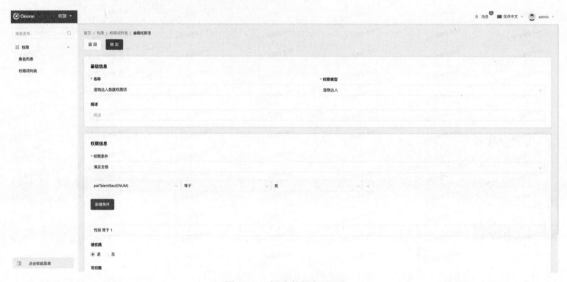

图 6-44　创建数据权限项

Step3. 为角色配置权限。

（1）编辑 Oinone 角色，只开通 OinoneDemo 工程应用，如图 6-45 所示。

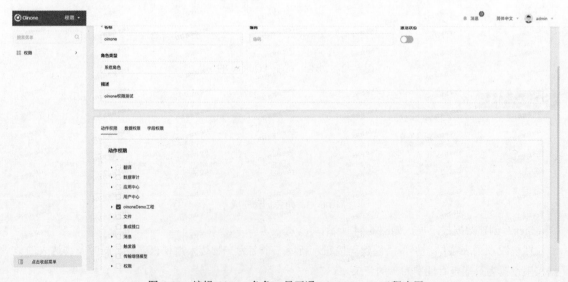

图 6-45　编辑 Oinone 角色，只开通 OinoneDemo 工程应用

（2）选中数据权限选项卡，单击"添加"按钮，勾选宠物达人数据权限项，单击"确定"按钮，如图 6-46 所示。

图 6-46　勾选宠物达人数据权限项，单击"确定"按钮

（3）整体单击"保存"按钮，回到列表页记得单击"权限生效"按钮，如图 6-47 所示。

图 6-47　回到列表页记得单击"权限生效"按钮

**Step4.** 新建用户绑定角色。

切换到用户中心模块，单击"创建"按钮填写必要信息，并在角色项选中 Oinone 权限组，如图 6-48 所示。

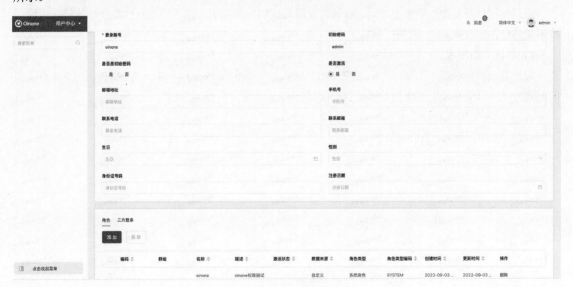

图 6-48　新建用户绑定角色

退出 admin 用户，用 Oinone 登录，如图 6-49 所示，权限效果：

图 6-49　用 Oinone 登录

（1）只能看见 demo 模块。

（2）Oinone 登录只能看到性别为男的宠物达人记录，如图 6-50 所示。

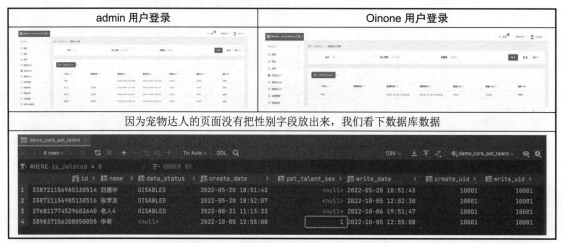

图 6-50　admin 用户登录与 Oinone 用户登录对比

**3. auth 模块扩展**

在日常项目开发中，难免会碰到一些针对权限管理的特殊需求，或是为提升性能设计的特殊逻辑。接下来将给大家介绍 auth 模块扩展性。

1）权限全局配置

对所有权限角色都做限制，而且不想让用户感知，可以实现 PermissionFunApi 接口，API 接口实现的配置方式，只能用于支持全局的数据权限配置。

（1）实现接口 PermissionFunApi，代码如下。

（2）将实现托管给 SpringAOP。

```
7  @Component
8  public class PetTalentPermissionFunApi implements PermissionFunApi {
9      @Override
10     public String permissionDomain(Object... args) {
11         //获取当前组织中
12         return "name == '张学友'";
13     }
14
15     @Override
16     public String nameSpace() {
17         return PetTalent.MODEL_MODEL;
18     }
19 }
```

第一个自定义 Action 如图 6-51 所示。

图 6-51　第一个自定义 Action

添加达人数据如图 6-52 所示。

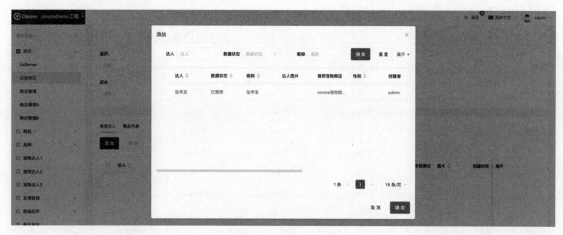

图 6-52　添加达人数据

2）不参与权限控制

如果某一接口不想做权限控制，则可以在启动工程的 application-dev.yml 文件中配置不需要权限过滤的接口，代码如下。

```
pamirs:
  auth:
    fun-filter:
      - namespace: demo.PetTalent
        fun: queryPage
```

换一个没有配置宠物达人权限的用户（除管理员以外）进入系统，也可以看到数据。注意，权限全局控制还是生效的。

3）API

获取当前用户对该模型的行权限，代码如下。

```
Result<String> result = CommonApiFactory.getApi(AuthApi.class).canReadAccessData("Model");
```

返回值如下所示。

```
{
    'data':"name=in=('hahaha')"
    'success': true
    ...
}
```

用法如下：
- 场景：前端发起的请求都会经过权限拦截，后端代码逻辑发起的数据请求都是不经过任何权限的过滤，但是某些特殊情况下需要在后端代码逻辑发起的数据请求中也带上权限过滤。
- 入参：请求的模型。
- 出参：Result 数据结构中 data 会存储一段字符串，该字符串为 Rsql。

将该 Rsql 追加到 wrapper 中，代码如下。

```
Result<String> result = CommonApiFactory.getApi(AuthApi.class).canReadAccessData("base.UeModule");
String data=result.getData();
QueryWrapper<UeModule> wrapper = Pops.query();
wrapper.setEntity(ueModule);
if (!StringUtils.isBlank(data)) {
    wrapper.apply(data);
}
```

## 6.6 消息

在我们系统研发过程中经常需要发送短信、邮件、站内信等，笔者在本节将给大家介绍下如何使用 Oinone 的消息模块。

**1. 准备工作**

如果通过我们工程脚手架工具生成消息模块，则已经引入了相关设置，无须做更多的配置，如果不是则需要按以下步骤先配置依赖和增加启动模块。

（1）pamirs-demo-boot 的 pom 文件中引入 pamirs-message-core 包依赖，代码如下。

```xml
<dependency>
    <groupId>pro.shushi.pamirs.core</groupId>
    <artifactId>pamirs-message-core</artifactId>
</dependency>
```

（2）pamirs-demo-boot 的 application-dev.yml 文件中增加配置 pamirs.boot.modules 并增加 message，即在启动应用中增加 message 模块，代码如下。

```yaml
pamirs:
  boot:
    modules:
      - message
```

**2. 消息参数设置**

发送邮件和短信需要设置对应的发送邮箱服务器和短信云，短信目前默认使用阿里云短信。我们通过代码示例来完成对应邮箱和短信的参数设置。

Step1. 增加 pamirs-message-api 依赖。

在 pamirs-demo-core 的 pom 文件中引入 pamirs-message-api 包依赖，代码如下。

```xml
<dependency>
    <groupId>pro.shushi.pamirs.core</groupId>
    <artifactId>pamirs-message-api</artifactId>
</dependency>
```

Step2. 消息参数设置，代码如下。

请自行替换邮箱服务器和短信通道的账号信息。

```java
@Component
public class DemoMessageInit implements InstallDataInit, UpgradeDataInit {

    private void initEmail(){
        EmailSenderSource emailSenderSource = new EmailSenderSource();
        emailSenderSource.setName("邮件发送服务");
        emailSenderSource.setType(MessageEngineTypeEnum.EMAIL_SEND);
        //优先级
        emailSenderSource.setSequence(10);
        //发送账号 FIXME 自行替换
        emailSenderSource.setSmtpUser("");
        //发送密码 FIXME 自行替换
        emailSenderSource.setSmtpPassword("");
        //发送服务器地址和端口
        emailSenderSource.setSmtpHost("smtp.exmail.qq.com");
        emailSenderSource.setSmtpPort(465);
        //" None: SMTP 对话用明文完成." +
        //" TLS (STARTTLS): SMTP对话的开始时要求TLS 加密（建议)" +
        //" SSL/TLS: SMTP对话通过专用端口用 SSL/TLS 加密（默认是：465)")
        emailSenderSource.setSmtpSecurity(EmailSendSecurityEnum.SSL);
        emailSenderSource.createOrUpdate();
    }

    private void initSms(){
        SmsChannelConfig smsChannelConfig = new SmsChannelConfig();
        smsChannelConfig.setType(MessageEngineTypeEnum.SMS_SEND);
        smsChannelConfig.setChannel(SMSChannelEnum.ALIYUN);
        //短信签名名称
        smsChannelConfig.setSignName("oinone");
        //阿里云账号信息 FIXME 自行替换
        smsChannelConfig.setAccessKeyId("");
        smsChannelConfig.setAccessKeySecret("");
        smsChannelConfig.setEndpoint("https://dysmsapi.aliyuncs.com");
        smsChannelConfig.setRegionId("cn-hangzhou");
        smsChannelConfig.setTimeZone("GMT");
        smsChannelConfig.setSignatureMethod("HMAC-SHA1");
        smsChannelConfig.setSignatureVersion("1.0");
        smsChannelConfig.setVersion("2017-05-25");
        smsChannelConfig.createOrUpdate();
```

```
57          //初始化短信模板
58          //目前支持阿里云短信通道：获取短信模板，如没有短信模板，需要先创建模板，并审核通过
59          SmsTemplate smsTemplate = new SmsTemplate();
60          smsTemplate.setName("通知短信");
61          smsTemplate.setTemplateType(SMSTemplateTypeEnum.NOTIFY);
62          smsTemplate.setTemplateCode("SMS_244595482");//从阿里云获取，自行提供 FIXME
63          smsTemplate.setTemplateContent("尊敬的&{name},你的&{itemName}库存为&{quantit
    y}");
64          smsTemplate.setChannel(SMSChannelEnum.ALIYUN);
65          smsTemplate.setStatus(SMSTemplateStatusEnum.SUCCESS);
66          smsTemplate.createOrUpdate();
67      }
68
69      @Override
70      public boolean init(AppLifecycleCommand command, String version) {
71          initEmail();
72          initSms();
73          return Boolean.TRUE;
74      }
75
76      @Override
77      public boolean upgrade(AppLifecycleCommand command, String version, String ex
    istVersion) {
78          initEmail();
79          initSms();
80          return Boolean.TRUE;
81      }
82
83      @Override
84      public List<String> modules() {
85          return Collections.singletonList(DemoModule.MODULE_MODULE);
86      }
87
88      @Override
89      public int priority() {
90          return 0;
91      }
92  }
93
```

### 3. 消息发送举例

以各种站内信、邮件、短信方式发送宠物商品库存记录信息为例，构建宠物商品库存信息发送服务，并在宠物商品库存模型的表格页面创建发送入口。

Step1. 创建宠物商品库存信息发送服务接口。

在 Api 工程新建 PetItemInventoryMessageService 接口，并定义发送站内信、邮件、短信的三个 Function，代码如下。

```
 7  @Fun(PetItemInventoryMessageService.FUN_NAMESPACE)
 8  public interface PetItemInventoryMessageService {
 9
10      String FUN_NAMESPACE = "demo.PetItemInventoryMessageService";
11      @Function
12      void sendMail(PetItemInventroy data);
13      @Function
14      void sendEmail(PetItemInventroy data);
15      @Function
16      void sendSms(PetItemInventroy data);
17
18  }
```

Step2. 发送站内信。

在 Core 工程新建 PetItemInventoryMessageService 接口的实现类，代码如下。

```java
36  @Fun(PetItemInventoryMessageService.FUN_NAMESPACE)
37  @Slf4j
38  @Component
39  public class PetItemInventoryMessageServiceImpl implements PetItemInventoryMessa
    geService {
40
41      public static final String INVALID_ADDRESSES_CODE = "Invalid Addresses";
42      public static final String INVALID_ADDRESSES_MSG = "非法的邮箱地址";
43      public static final String CAN_NO_CONNECT_TO_SMTP_CODE = "Could not connect
    to SMTP";
44      public static final String CAN_NO_CONNECT_TO_SMTP_MSG = "连接邮件服务失败";
45
46      @Override
47      @Function
48      public void sendMail(PetItemInventroy data) {
49          //接收对象列表
50          PamirsUser user = DemoSession.getUser();
51
52          List<Long> userIds = new ArrayList<>();
53          userIds.add(user.getId());
54
55          String subject = "商品库存信息";
56          String body = data.getItemName() + "_" + data.getQuantity();
57          MessageSender mailSender = (MessageSender) MessageEngine.get(MessageEngi
    neTypeEnum.MAIL_SEND).get(null);
58          PamirsMessage message = new PamirsMessage()
59                  .setName(subject)
60                  .setSubject(subject)
61                  .setBody(body)
62                  .setMessageType(MessageTypeEnum.NOTIFICATION);
63
64          List<PamirsMessage> messages = new ArrayList<>();
65          messages.add(message);
66          SystemMessage systemMessage = new SystemMessage();
67          systemMessage.setPartners(userIds.stream().map(i -> (PamirsUser) new Pam
    irsUser().setId(i)).collect(Collectors.toList()))
68                  .setType(MessageGroupTypeEnum.SYSTEM_MAIL)
69                  .setMessages(messages);
70          try {
71              mailSender.sendSystemMail(systemMessage);
72          } catch (Exception e) {
73              // TODO：可增加失败业务处理
74              throw PamirsException.construct(DemoExpEnumerate.SYSTEM_ERROR, e).ap
    pendMsg("站内信发送失败").errThrow();
75          }
76      }
77
78      @Override
79      @Function
80      public void sendEmail(PetItemInventroy data) {
81          //发送一个邮件
82          List<String> receiveEmails = new ArrayList<>();
83          receiveEmails.add("testhaha@shushi.pro");//收件人邮箱,自行替换 FIXME
84          EmailSender emailSender = null;
85          EmailPoster emailPoster = null;
86
87          emailSender = (EmailSender) MessageEngine.get(MessageEngineTypeEnum.EMAI
    L_SEND).get(null);
```

```
 88
 89             String title = data.getItemName();//"邮件标题";
 90             String body = data.getItemName() + "_" + data.getQuantity();//"邮件内容";
 91             String sender = DemoSession.getUserName();//"发件人";
 92             String replyEmail = "回复邮箱";
 93
 94             emailPoster = new EmailPoster().setSender(sender).setTitle(title).setBody(body).setReplyTo(replyEmail);
 95
 96             EmailPoster finalEmailPoster = emailPoster;
 97             EmailSender finalEmailSender = emailSender;
 98
 99             StringBuilder errorMessages = new StringBuilder();
100             receiveEmails.stream().distinct().collect(Collectors.toList()).forEach(
101                     email -> {
102                         try {
103                             if (!finalEmailSender.send(finalEmailPoster.setSendTo(email))) {
104                                 log.error("发送邮件失败:emailPoster:{}", JsonUtils.toJSONString(finalEmailPoster));
105                                 String message = "发送邮件失败,错误信息: " + "系统异常" + ", 邮箱: " + email + ";";
106                                 errorMessages.append(message);
107                             }
108                         } catch (Exception e) {
109                             log.error("发送邮件失败:emailPoster:{},异常:{}", JsonUtils.toJSONString(finalEmailPoster), e);
110                             String errorMsg = transferEmailThrowMessage(e);
111                             String message = "发送邮件失败,错误信息: " + errorMsg + ", 邮箱: " + email + ";";
112                             errorMessages.append(message);
113                         }
114                     }
115             );
116
117             if (StringUtils.isNotBlank(errorMessages.toString())) {
118                 // TODO: 可增加处理邮件发送失败的业务逻辑
119             }
120         }
121
122     private String transferEmailThrowMessage(Exception e) {
123         String errorMessage = e.getMessage();
124         Throwable cause = e.getCause();
125         if (cause != null) {
126             if (cause instanceof SendFailedException) {
127                 String message = cause.getMessage();
128                 if (INVALID_ADDRESSES_CODE.equals(message)) {
129                     errorMessage = INVALID_ADDRESSES_MSG;
130                 }
131             } else if (cause instanceof MessagingException) {
132                 String message = cause.getMessage();
133                 if (StringUtils.isNotBlank(message) && message.contains(CAN_NO_CONNECT_TO_SMTP_CODE)) {
134                     errorMessage = CAN_NO_CONNECT_TO_SMTP_MSG;
135                 }
136             }
137         }
138         return errorMessage;
139     }
140
141     @Override
142     @Function
```

```java
143 ▼  public void sendSms(PetItemInventroy data) {
144         StringBuilder errorMessages = new StringBuilder();
145
146         List<String> receivePhoneNumbers = new ArrayList<>();
147         receivePhoneNumbers.add("13777899044");//接收手机号,自行提供 FIXME
148
149         //目前支持阿里云短信通道:获取短信模板,如没有短信模板,需要先创建模板,并审核通过
150         //如下示例
151         SmsTemplate smsTemplate = new SmsTemplate().setTemplateType(SMSTemplateT
    ypeEnum.NOTIFY).setTemplateCode("SMS_246455054").queryOne();//从阿里云获取
152
153         // 占位符处理
154         Map<String, String> vars = new HashMap<>();
155         vars.put("name", DemoSession.getUserName());
156         vars.put("itemName", data.getItemName());
157         vars.put("quantity", data.getQuantity().toString());
158
159         SMSSender smsSender = (SMSSender) MessageEngine.get(MessageEngineTypeEnu
    m.SMS_SEND).get(null);
160 ▼     receivePhoneNumbers.stream().distinct().forEach(it -> {
161 ▼         try {
162 ▼             if (!smsSender.smsSend(smsTemplate, it, vars)) {
163                 String message = "发送短信失败,错误信息:" + "系统异常" + ",手机
    号:" + it + ";";
164                 errorMessages.append(message);
165             }
166 ▼         } catch (Exception e) {
167             String message = "发送短信失败,错误信息:" + e.getMessage() + ",手
    机号:" + it + ";";
168             errorMessages.append(message);
169         }
170     });
171 ▼     if (StringUtils.isNotBlank(errorMessages.toString())) {
172         // TODO: 可增加失败业务处理
173     }
174  }
175 }
```

Step3. 为宠物商品库存模型的表格页面新增发送入口。

把以下代码复制到 PetItemInventroyAction 类中。

```java
1  @Autowired
2  private PetItemInventoryMessageService petItemInventoryMessageService;
3  @Action(displayName = "发送站内信",bindingType = ViewTypeEnum.TABLE,contextType = A
   ctionContextTypeEnum.SINGLE)
4  public PetItemInventroy sendMail(PetItemInventroy data) {
5      petItemInventoryMessageService.sendMail(data);
6      return data;
7  }
8
9
10
11 /**
12  * 发送邮件
13  */
14 @Action(displayName = "发送邮件",bindingType = ViewTypeEnum.TABLE,contextType = Ac
   tionContextTypeEnum.SINGLE)
15 public PetItemInventroy sendEmail(PetItemInventroy data){
16     petItemInventoryMessageService.sendEmail(data);
17     return data;
18 }
19
```

```
20
21  /**
22   * 发送短息
23   */
24  @Action(displayName = "发送短信",bindingType = ViewTypeEnum.TABLE,contextType = Ac
    tionContextTypeEnum.SINGLE)
25  public PetItemInventroy sendSms(PetItemInventroy data){
26      petItemInventoryMessageService.sendSms(data);
27      return data;
28  }
```

Step4. 重启看效果。

请分别单击"发送邮件""发送短信""发送站内信"按钮测试效果，如图 6-53 所示。

图 6-53　自行测试效果

# 第 7 章　Oinone 的设计器

设计器专为非专业研发人员设计，在 Oinone 3.0 版本中已经完成元数据完整在线化，真正做到低无一体。对于设计器的定位我们开篇就介绍过，它是 LCDP 的产品化呈现，相当于露在外面的冰山，核心还是在 LCDP 本身。本章先带大家目睹设计器的一些产品页面，如果您想体验，可以在 Oinone 官网（https://www.oinone.top）注册，不过需要先联系工作人员或合作伙伴获取邀请码。

1. 设计器总览
2. 实战训练（积分发放）
3. 实战训练（全员营销为例）
4. Oinone 的低无一体

## 7.1　设计器总览

**1. 模型设计器**

Oinone 以模型为驱动，有了模型、数据字典、数据编码等设计功能，我们就可以完整地定义产品数据模型，模型设计器整体呈现区别于普通 ER 图，以当前模型为核心视角展开，可以单击关联模型切换主视角。这样的好处在于突出当前设计，聚焦设计本身。同时模型上预留了几个核心入口，如分类管理、继承拓扑图、页面设计、逻辑设计等。另外我们在体验上区分了专家模式和经典模式，顾名思义，专家模式的功能会更加丰富，对专业知识的要求也会更高。专家模式下一般会增加一些与业务无关的配置，如索引设置等调优行为，如图 7-1 ～图 7-6 所示。

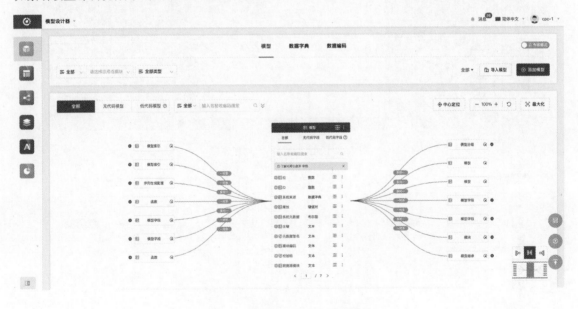

图 7-1　模型设计器——系统模型

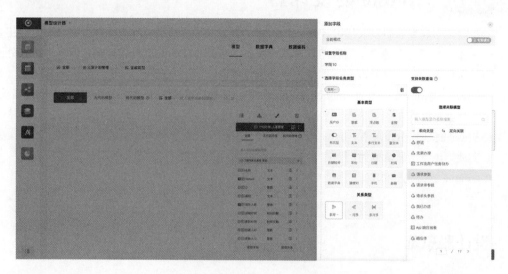

图 7-2　模型设计器——添加字段

图 7-3　模型设计器——创建模型

图 7-4　模型设计器——系统字典

图 7-5　模型设计器——自定义编码

图 7-6　模型设计器——编辑数据编码

**2. 逻辑设计器**

从图灵完备的角度上说，设计器要支持的功能越完备，使用就越复杂。我们优先从图灵完备的角度出发，所以我们第一版逻辑设计器相对比较复杂，第二版规划中会如模型设计器推出专家版和经典版，如图 7-7、图 7-8 所示。

图 7-7　逻辑设计器——逻辑设计

图 7-8 逻辑设计器——逻辑运行

### 3. 界面设计器

界面设计器第一版会先支撑后端页面在线自定义，后面将陆续推出前端页面、多端能力。为了支持多端和 2C 页面的设计，我们对前后端协议做了比较大的改造。目前设计器已经支持完全基于 V3 的前后端协议，如图 7-9、图 7-10 所示。

图 7-9 界面设计器——页面视图

图 7-10 界面设计器——组件库

### 4. 数据可视化

数据可视化支持从内部系统模型获取数据内容，根据业务需求自定义图表，目的是为企业提供更高效的数据分析工具。

与市场同类产品相比，Oinone 的数据可视化产品不需要前置维护数据源、进行数据转换；可以智能获取业务系统模型，系统自动解析选择的模型、接口、表格中的字段后进行数据分析；降低对数据分析人员研发能力要求的同时，也提升了数据分析的效率，如图 7-11、图 7-12 所示。

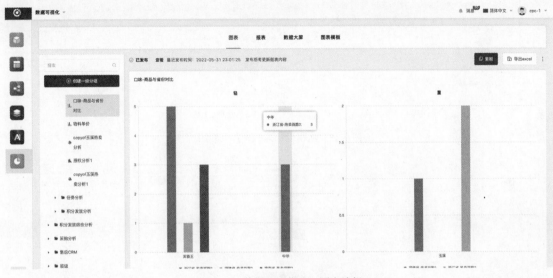

图 7-11　数据可视化——图表编辑

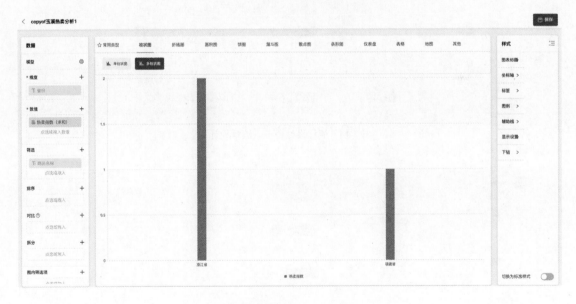

图 7-12　数据可视化——图表自定义

### 5. 流程设计器

Oinone 流程设计器为业务流程和审批流程提供了可自动执行的流程模型，通过定义流转过程中的各个动作、规则，以此实现流程自动化。在 Oinone 流程设计器中，流程可以跨应用设计，不同应用的模型之间可以通过同一流程执行，如图 7-13～图 7-15 所示。

图 7-13　流程设计器——首页视图

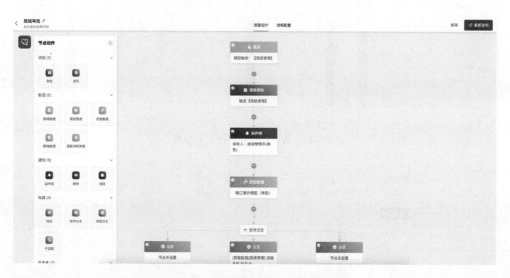

图 7-14　流程设计器——流程设计

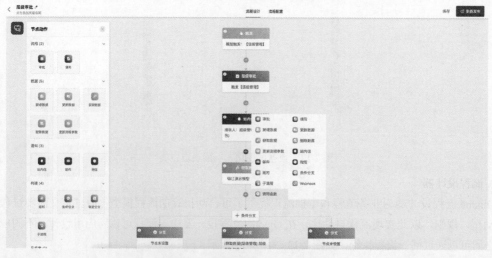

图 7-15　流程设计器——节点设置

## 7.2 实战训练（积分发放）

**1. 背景介绍**

当我们碰到一个全新的场景，除了写代码以外也可以通过设计器来完成，大致步骤如下：

（1）分析业务场景，规划对应的模型并通过模型设计器进行配置。

（2）通过界面设计器，设计出必要管理页面。

（3）通过流程设计器，设计对应业务流程。

（4）通过数据可视化，设计相应的数据看板。

**2. 场景说明**

本节通过例子完成一个积分成本分摊的业务场景。

积分支出："谁受益谁支出 + 导购手动确认"原则，通过门店应用的积分规则，来实现"自动化 + 手动积分"形式。

案例背景：某家具企业经营多种家具类型，不同系列、不同品类的家具在不同的事业部经营，如图 7-16 所示。

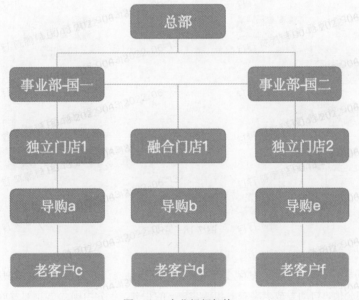

图 7-16　企业组织架构

注：

独立门店：只能售卖本事业部下的商品。

融合门店：可以售卖多事业部下的商品。

需求：需要建立一套积分规则，遵循"谁受益谁支出 + 导购手动确认"原则，通过门店应用的积分规则，来实现"自动化 + 手动"积分。

（1）场景一。

导购 a 邀请老客户 c 通过裂变分享新客户 x 在独立门店 1 下单。系统根据独立门店 1 的积分规则，自动发放积分给老客户 c 和新客户 x，积分由国一事业部承担，如图 7-17 所示。

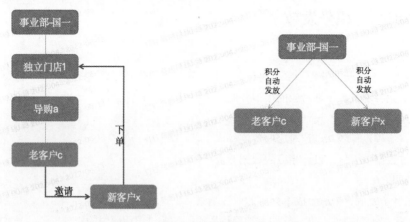

图 7-17 场景一

（2）场景二。

融合门店 1 导购 b 邀请老客户 d 裂变分享新客户 y 在融合门店下单。该订单可能涉及多个事业部，由导购 b 手动选择最大量值的积分规则进行发放，如图 7-18 所示。

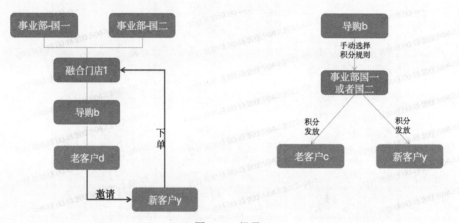

图 7-18 场景二

（3）场景三。

独立门店 1 的导购 a 邀请老客户 c 裂变分享新客户 z 在独立门店 2 下单。系统根据独立门店 2 的积分规则，来计算积分值，自动发放老客户 c 和新客户 z 的积分，积分由国二事业部发放，如图 7-19 所示。

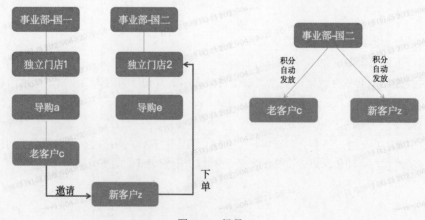

图 7-19 场景三

**3. 实战训练**

Step1. 分析业务场景规划对应模型。

本场景涉及基础对象模型包括：事业部、门店、导购、会员等。

（1）事业部：它是积分成本的载体。

（2）门店：类型分为独立门店和融合门店，独立门店必须隶属于一个事业部，同时配置默认积分发放规则，融合门店可能属于多个事业部当发生积分发放时，需要店员手工选择成本事业部和积分发放规则。

（3）导购员：导购员必须隶属于一个门店等。

（4）会员：消费者需要记录它隶属导购员，以及是由哪个会员推荐过来的等。

涉及业务对象模型包括：积分发放规则、积分发放记录、邀请下单记录。

（1）积分发放规则：会员积分发放规则，对应邀请老客户的积分发放规则等。

（2）积分发放记录：本次发放积分、本次发放积分规则、发放对象（会员）、成本事业部、关联门店、关联导购、关联老客户（可空）、关联下单记录编码等。

（3）邀请下单记录：导购、下单会员、下单门店、商品信息、下单金额等。

对应模型如图 7-20 所示。

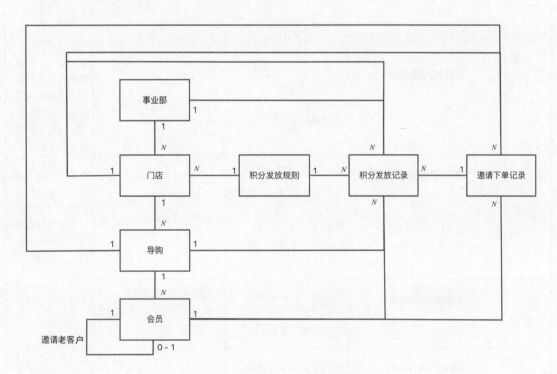

图 7-20　分析业务场景规划对应模型

Step2. 利用模型设计器设计模型，见表 7-1。

我们建模型的时候应用选择全员营销，本次实战逻辑放到全员营销应用下，目前 Oinone 的 apps 还未开放新建应用或模块的入口，后续开放可以新建一个新的应用并把逻辑放到独立的应用下。

表 7-1　利用模型设计器设计模型

| 模型 | 设计器呈现 | 自定义字段列表 | 关系字段说明 | 说明 |
|---|---|---|---|---|
| 事业部 | | 事业部负责人（文本）<br>门店列表（o2m） | 与门店建立 o2m 绑定关系，绑定时选择双向绑定，双向绑定意思是在事业部这边建立 o2m 到门店的关系字段，在门店那边建立 m2o 的关系字段 | 关系字段需要在有对方模型的情况下再建，比如事业部中的门店列表是在后续追加新增的 |
| 门店 | | 导购列表（o2m）<br>事业部（m2o）<br>默认积分发放规则（m2o）<br>门店类型（枚举） | 分别与事业部、导购、积分发放规则建立 m2o、o2m、m2o 关系 | 门店类型：需要先建对应的数据字典 |
| 导购 | | 绑定用户（m2o）<br>门店（m2o）<br>是否离职（布尔型） | 与门店建立 m2o 关系 | 绑定用户，用于后续业务流程设计中的填写规则，打马赛克的信息可以忽略，在其他场景测试用 |
| 会员 | | 会员累计积分（浮点数）<br>推荐客户（m2o）<br>所属导购员（m2o）<br>是否为新客（布尔型） | 与导购、会员建立 m2o 关系 | 会员 m2o 的字段是自关联，用于存储推荐会员，打马赛克的信息可以忽略，在其他场景测试用 |
| 积分发放规则 | | 推荐客户发放比例（浮点数）<br>发放倍数（整数） | | |
| 积分发放记录 | | 最终发放积分（浮点数）<br>关联积分规则（m2o）<br>事件编码（文本）<br>推荐导购员（m2o）<br>推荐会员（m2o）<br>关系门店（m2o）<br>成本事业部（m2o）<br>会员（m2o）<br>成本事业部名称（文本）<br>会员名称（文本） | 与积分发放规则、导购员、会员、门店、事业部建立 m2o 关系 | 会员有两个 m2o，分别是用户记录发放会员和发放会员的推荐会员也就是老客，事件编码是用户维护触发本次积分发放记录产生的源头单据编码如：邀请下单记录的编码 |

续表

| 模型 | 设计器呈现 | 自定义字段列表 | 关系字段说明 | 说明 |
|---|---|---|---|---|
| 邀请下单记录 | （券购邀请下单，自定义字段：成本事业部、选择积分发放规则、下单门店、购买商品、下单金额、会员、导购） | 成本事业部（m2o）<br>选择积分发放规则（m2o）<br>下单门店（m2o）<br>购买商品（文本）<br>下单金额（整数）<br>会员（m2o）<br>导购（m2o） | 与成本事业部、积分发放规则、下单门店、会员、导购等建立 m2o 的关系 | 会员、下单门店、导购属于必要信息<br>成本事业部、积分发放规则是业务流程中自动计算回填的数据 |

**Step3.** 利用界面设计器，设计出必要的管理页面。

以事业部为例，构建管理页面。其他模型依次按步骤建立管理页面。

（1）进入界面设计器，应用选择全员营销，模型选择事业部，选择"添加页面"下拉框中的"直接创建"选项，如图 7-21 所示。

图 7-21　选择"直接创建"选项

（2）设置页面标题、模型（自动带上可切换）、业务类型（运营管理后续会扩展其他类型）、视图类型（表单）后单击"确认"按钮进入事业部表单设计页面，如图 7-22 所示。

图 7-22　页面设置

（3）进入页面设计器，对事业部表单页面进行设计，如图 7-23 所示。

① 左侧为物料区：分为组件、模型。

● "组件"选项卡下为通用物料区，我们可以为页面增加对应布局、字段（如同在模型设计器增加字段）、动作、数据、多媒体等等。

● "模型"选项卡下为页面对应模型的自定义字段、系统字段以及模型已有动作。

② 中间是设计区域。

③ 右侧为属性面板,在设计区域选择中组件会显示对应组件的可配置参数。

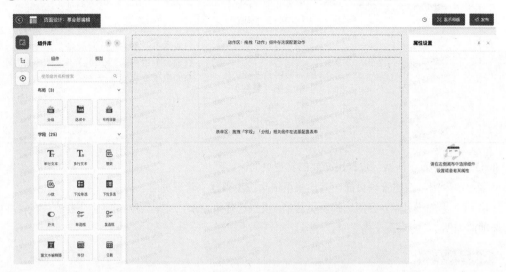

图 7-23　对事业部表单页面进行设计

(4) 在左侧"组件"选项卡下,拖入布局组件"分组",并设置组件标题属性为基础信息,如图 7-24 所示。

图 7-24　设置组件标题属性为基础信息

(5) 在左侧"组件"选项卡下,再次拖入布局组件"分组",并设置组件标题属性为门店列表,如图 7-25 所示。

图 7-25　设置组件标题属性为门店列表

（6）在左侧"模型"选项卡下，分别把自定义字段中的"事业部负责人"和系统字段中的"名称"拖入"基础信息"分组，把自定义字段中的"门店列表"字段拖入门店列表分组，如图 7-26 所示。

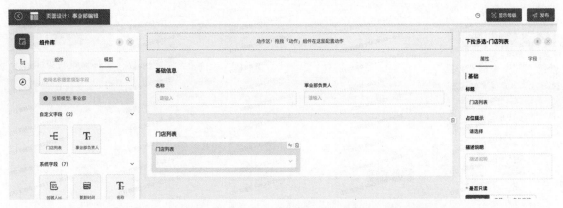

图 7-26　把自定义字段中的"门店列表"字段拖入门店列表分组

（7）在设计区域切换"门店列表"展示组件为"表格"，如图 7-27 所示。

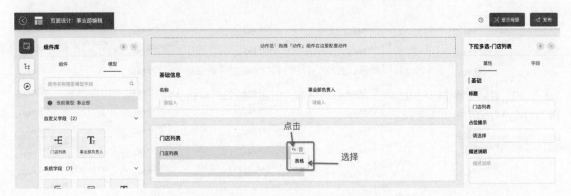

图 7-27　在设计区域切换"门店列表"展示组件为"表格"

（8）此时"门店列表"展示形式变成了表格形式，选中"门店列表"组件，会发现左侧"模型"选项卡下的当前模型切换成了"门店"，同时我们在右属性面板区置空其"标题属性"，如图 7-28 所示。

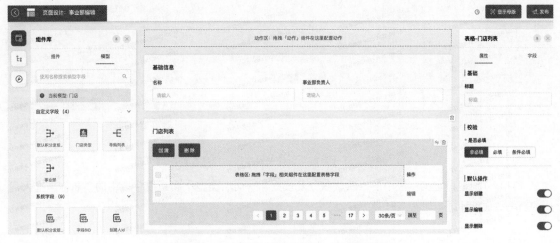

图 7-28　在右属性面板区置空其"标题属性"

（9）设计区选中"门店列表"的表格组件，分别把自定义字段中的"默认积分发规则""门店类型""导购列表"和系统字段中的"名称"拖入"门店列表"表格组件的表格字段设计区，如图 7-29 所示。

图 7-29　将"名称"拖入"门店列表"表格组件的表格字段设计区

（10）设计区选中"门店列表"的表格组件的"创建"按钮，单击"打开弹窗"设计关系字段"门店"的新增页面，如图 7-30 所示。

图 7-30　单击"打开弹窗"设计关系字段"门店"的新增页面

（11）分别把自定义字段中的"默认积分发规则""门店类型"和系统字段中的"名称"拖入门店的新增页面设计区，如图 7-31 所示。

图 7-31 新页面设计区

（12）选中弹出框中"取消"取消按钮，设置其"按钮样式"属性为"次要按钮"，如图 7-32 所示。

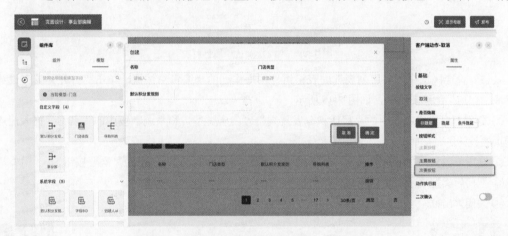

图 7-32 设置其"按钮样式"属性为"次要按钮"

（13）关闭弹出框，回到主设计区，如图 7-33 所示。

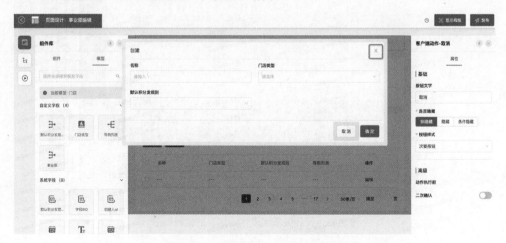

图 7-33 关闭弹出框，回到主设计区

（14）设计区选中"门店列表"的表格组件的"删除"按钮，设置其"按钮样式"属性为"次要按钮"，"二次确认"属性打开，如图 7-34 所示。

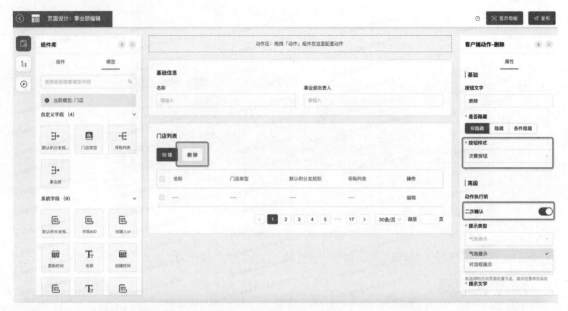

图 7-34　"二次确认"属性打开

（15）设计区选中"门店列表"的表格组件中操作列的"编辑"按钮，单击"打开弹窗"设计关系字段"门店"的编辑页面，如图 7-35 所示。

① 分别把自定义字段中的"默认积分发规则""门店类型"和系统字段中的"名称"拖入门店的新增页面设计区。

② 选中弹出框中"取消"选项，取消按钮，设置其"按钮样式"属性为"次要按钮"。

③ 把门店类型展示组件切换为"单选框"。

④ 关闭弹出框。

图 7-35　设计关系字段"门店"的编辑页面

（16）设计区选择非"门店列表"组件，如基础信息，模型切换为主模型"事业部"，在左侧"模型"选项卡下，把动作分类下的提交类型"创建"动作拖入中间设计区的动作区，如图 7-36 所示。

图 7-36　将"创建"动作拖入中间设计区的动作区

（17）选择展示名为"确定"的创建动作按钮，在右侧属性面板中设置：是否隐藏属性为"条件隐藏"，隐藏条件为"ID 非空"，如图 7-37 所示。

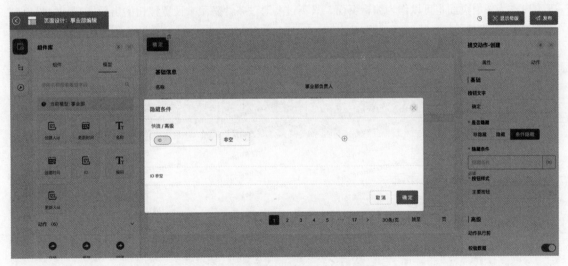

图 7-37　将隐藏条件设置为"ID 非空"

（18）在左侧"模型"选项卡下，把动作分类下的提交类型"更新"动作拖入中间设计区的动作区，如图7-38所示。

图7-38　将"更新"动作拖入中间设计区的动作区

（19）选择展示名为"确定"的更新动作按钮，在右侧属性面板中设置：是否隐藏属性为"条件隐藏"，隐藏条件为"ID 为空"。之所以同时拖入"创建"和"更新"动作并都命名为确认，是期望这个页面既可以作为新增页面也可以作为编辑页面，只不过要通过条件隐藏来设置按钮的出现规则，如图7-39所示。

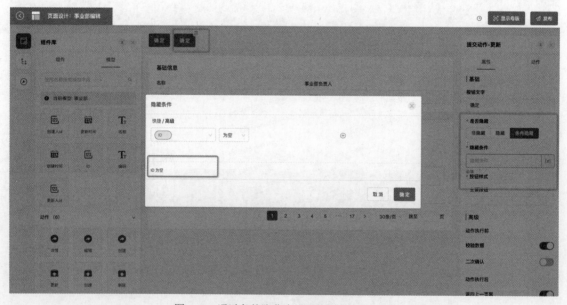

图7-39　通过条件隐藏来设置按钮的出现规则

（20）在左侧"组件"选项卡下，把动作分类下的"客户端动作"拖入中间设计区的动作区，如图 7-40 所示。

图 7-40　将"客户端动作"拖入中间设计区的动作区

（21）选择设计区"客户端动作"，在右侧属性面板中设置：动作名称为"返回"，客户端行为为"返回上一个页面"，并单击"保存"按钮，如图 7-41 所示。

图 7-41　设计区"客户端动作"

（22）选择设计区"返回"动作，在右侧属性面板中设置：按钮样式为"返回"，将"二次确认"属性打开并设置提示文字"返回页面操作将不被保存"，可以单击"预览"按钮，二次确认看效果，如图 7-42 所示。

图 7-42　可单击"预览"按钮，二次确认看效果

（23）单击"发布"按钮，页面成功发布，每发布一次会有一个历史版本，可以通过历史版本进行恢复，如图 7-43 所示。

图 7-43　单击"发布"按钮，页面发布成功

（24）单击右上角历史版本图标，进入历史版本查看页面，如图 7-44 所示。

（25）在历史版本页面可以选择对应历史版本记录，并单击"恢复此版本"按钮来完成页面的历史版本切换，如图 7-45 所示。

第 7 章 Oinone 的设计器 ·:· 531

图 7-44 进入历史版本查看页面

图 7-45 可选择对应历史版本记录并切换

（26）接下来我们为事业部模型创建表格管理页面，入口同编辑页面。设置页面标题、模型（自动带上可切换）、业务类型（运营管理后续会扩展其他类型）、视图类型（表格）后单击"确认"按钮，进入事业部表格设计页面，如图 7-46 所示。

图 7-46 为事业部模型创建表格管理页面

（27）进入页面设计器，对事业部表格页面进行设计，如图7-47所示。

图7-47 对事业部表格页面进行设计

（28）在左侧"模型"选项卡下，分别把自定义字段中的"事业部负责人"和系统字段中的"名称"拖入表格组件的表格字段设计区，如图7-48所示。

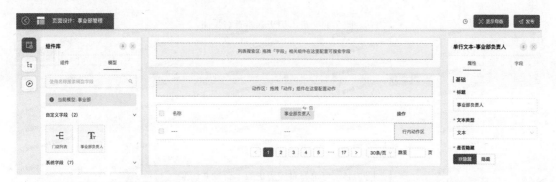

图7-48 分别把"事业部负责人"和"名称"拖入表格组件的字段设计区

（29）在左侧"组件"选项卡下，把动作分类下的"跳转动作"拖入中间设计区的动作区，并在右侧属性面板中设置动作名称为"新增"，数据控制类型为"不进行数据处理"，打开方式为"当前窗口打开"，动作跳转页面为"事业部编辑"页面，并单击"保存"，如图7-49所示。

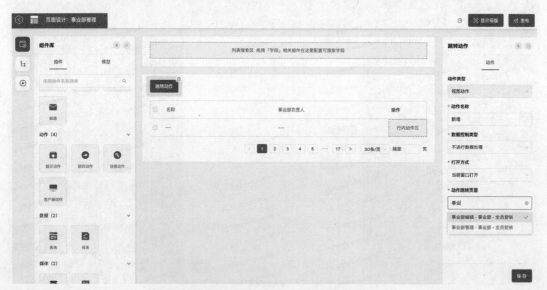

图7-49 数据控制类型为"不进行数据处理"

（30）在左侧"组件"选项卡下，把动作分类下的"跳转动作"拖入中间设计区的行内动作区，并在右侧属性面板中设置动作名称为"编辑"，数据控制类型为"处理单条数据"，打开方式为"当前窗口打开"，动作跳转页面为"事业部编辑"页面，并单击"保存"按钮，如图 7-50 所示。

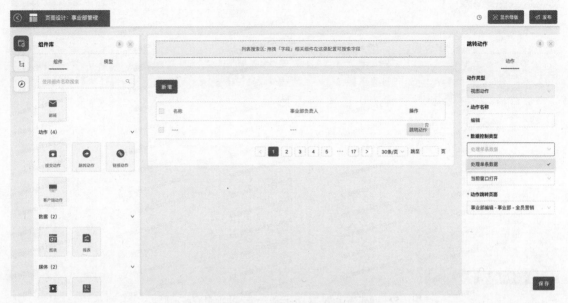

图 7-50　数据控制类型为"处理单条数据"

（31）在左侧"模型"选项卡下，把动作分类下的"删除"拖入中间设计区的动作区，并在右侧属性面板中设置动作名称为"删除"，按钮样式为"次要按钮"，"二次确认"属性打开，如图 7-51 所示。

图 7-51　把动作分类下的"删除"拖入中间设计区的动作区

（32）单击右上角"显示母版"进入页面最终展示形式，单击"添加"菜单项，并在输入框中输入"事业部管理"，如图 7-52 所示。

图 7-52　单击右上角"显示母版"进入页面最终展示形式

（33）单击菜单右侧设置图标，选择"绑定已有页面"，进行菜单与页面的绑定操作，如图 7-53 所示。

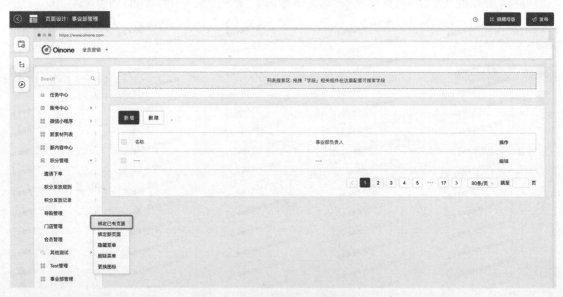

图 7-53　进行菜单与页面的绑定操作

（34）在绑定页面中，模型选择"事业部"，视图选择"事业部管理"，单击"确认"按钮提交，如图 7-54 所示。

图 7-54　绑定页面设置

（35）拖动"事业部管理"菜单到"积分管理"父菜单下，如图 7-55 所示。

图 7-55　拖动"事业部管理"菜单到"积分管理"父菜单下

（36）最后别忘了单击右上角"发布"按钮，对"事业部管理"表格页面进行发布，回到界面设计器首页查看刚刚建好的两个页面，如图 7-56 所示。

图 7-56　回到界面设计器首页查看刚刚建好的两个页面

（37）以事业部为例分别对门店、导购、会员、积分发放规则、积分发放记录、邀请下单记录等模型进行页面设计，这里不再赘述，请按照自身学习需要，尝试进行界面设计。

**Step4.** 通过流程设计器，设计对应业务流程。

我们先来整理下核心流程，即邀请下单流程，如图 7-57 所示。

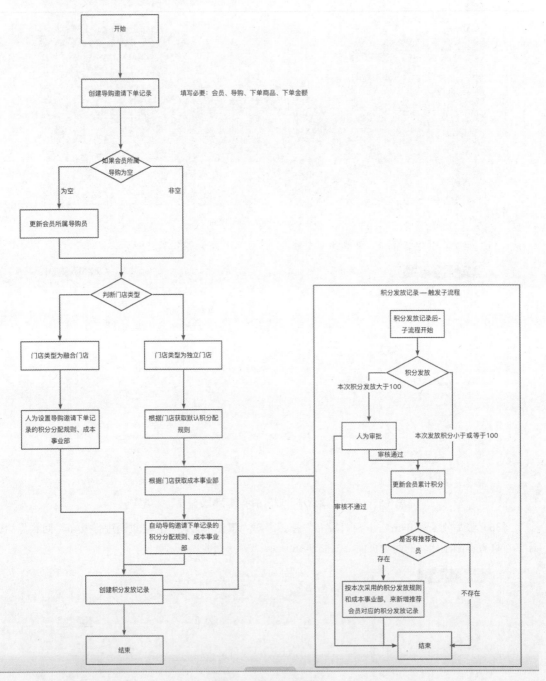

图 7-57　邀请下单流程

Step4.1. 创建导购邀请下单记录触发流程。

（1）进入流程设计器，单击"创建"按钮，如图 7-58 所示。

图 7-58　进入流程设计器，单击"创建"按钮

注意：流程中需要获取关系类型的字段（如：O2O、O2M、M2O 及 M2M），这种类型都是复杂的对象字段，除关联字段（一般为 ID）以外的字段，需要通过"数据获取"节点单独获取"关系字段"的对象数据。所以在流程设计中经常会用到"数据获取"节点。

（2）左上角编辑流程名称为"导购邀请下单触发流程"，单击第一个"触发"节点，触发方式选择"模型触发"选项，模型选择"导购邀请下单"选项，触发场景选择"新增数据时"选项，单击该节点的"保存"按钮，如图 7-59 所示。

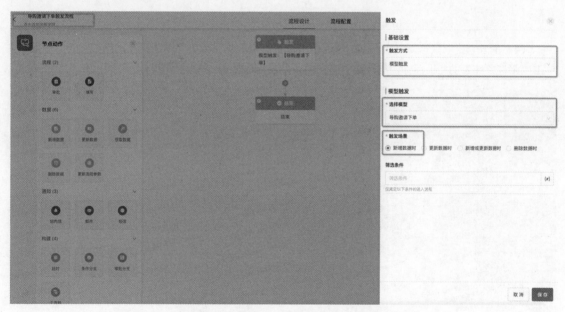

图 7-59　左上角编辑流程名称为"导购邀请下单触发流程"及其他设置

（3）单击流程图节点间的"+"图标选择增加"获取数据"节点，或者拖动左侧物料区"获取数据"到特定的"+"图标，如图 7-60、图 7-61 所示。

图 7-60　单击流程图节点间的 + 图标选择增加"获取数据"节点

图 7-61　拖动左侧物料区"获取数据"到特定的"+"图标

（4）单击"获取数据"，在右侧属性面板中设置获取数据条数为"单条"，选择模型为"导购"，单击"筛选条件"的 {x} 图标，进行数据获取的条件设置，如图 7-62 所示。

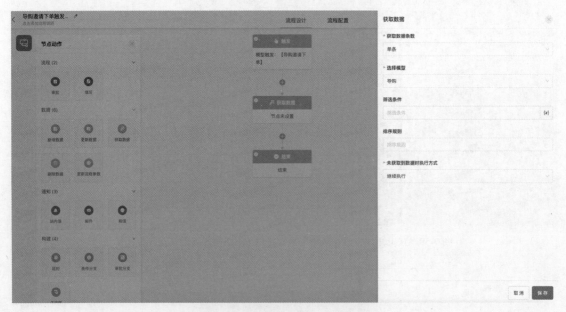

图 7-62　进行数据获取的条件设置

（5）选择条件字段为"ID"，条件操作符为"等于"，条件为变量的导购字段的 ID。当上下文只有一个变量时默认不需要选择，这里默认的是"触发 [ 导购邀请下单记录 ]"，设置好以后单击"确认"

按钮，回到属性面板设置未获取到数据时执行方式为"终止流程"，并单击节点"保持"按钮，如图7-63所示。

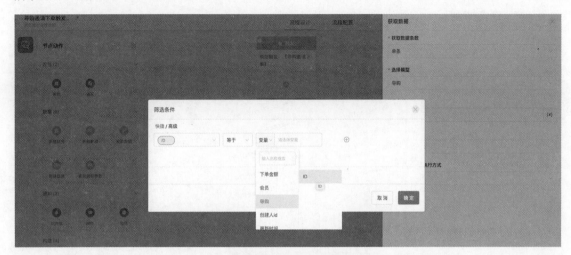

图7-63 筛选条件设置

（6）再增加一个获取数据节点，在右侧属性面板中设置。

① 设置获取数据条数为"单条"，选择模型为"会员"。

② 单击"筛选条件"的 {x} 图标，进行数据获取的条件设置，选择条件字段为"ID"，条件操作符为"等于"，条件变量为"导购邀请下单记录"的"会员"关系字段的"ID"字段。因为上下文中存在多个变量时需要选择对应变量，跟获取导购数据有点区别，在获取导购数据时，上下文中只有此次导购邀请下单记录所以不需要选择对应变量，设置好以后单击"确认"按钮，如图7-64所示。

③ 回到属性面板设置未获取到数据时执行方式为"终止流程"。

④ 最后单击节点"保持"按钮。

图7-64 增加一个"获取数据"节点，在右侧属性面板中设置

（7）新增"条件分支"，分别对分支 1 和分支 2 进行条件设置，条件为"获取数据 [ 会员 ]"的关系字段"所属导购员"的"ID"字段，两个条件的"ID"字段一个设置"为空"，一个设置"非空"，如图 7-65、图 7-66 所示。

图 7-65　新增"条件分支"并进行相应设置

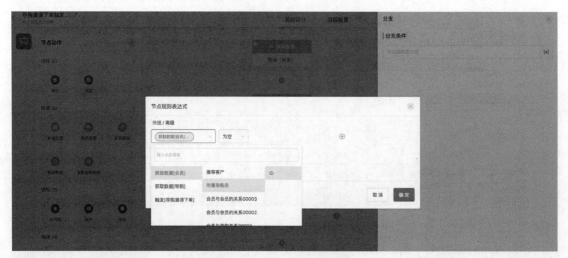

图 7-66　节点规则表达式设置

（8）在"条件分支"的"获取数据 [ 会员 ]"的关系字段"所属导购员"的"ID"字段"为空"的分支流程中，增加"更新数据"节点，在右侧属性面板中：

① "更新模型"选择"获取数据 [ 会员 ]"。

② "字段列表"单击"创建"按钮。

③ "字段选择"更新"所属导购员"的"ID"。

● 表达式设置为："获取数据 [ 导购 ]"的"ID"，单击"确认"按钮。

● 最终完成对"获取数据 [ 会员 ]"的关系字段"所属导购员"的"ID"字段等于"获取数据 [ 导购 ]"的"ID"字段的设置，最后单击"保持"按钮。如图 7-67 所示。

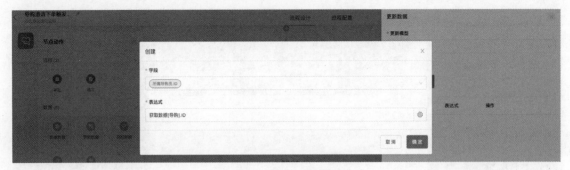

图 7-67　创建字段和表达式

（9）再增加一个"获取数据"节点，在右侧属性面板中设置：

① "获取数据条数"为"单条"，选择模型为"门店"。

② 单击"筛选条件"的 {x} 图标，进行数据获取的条件设置，选择条件字段为"ID"，条件操作符为"等于"，条件变量为"导购邀请下单记录"的关系字段"下单门店"的"ID"字段。设置好以后单击"确认"按钮。

③ 回到属性面板设置"未获取到数据时执行方式"为"终止流程"。

④ 最后单击"保持"按钮。如图 7-68 所示。

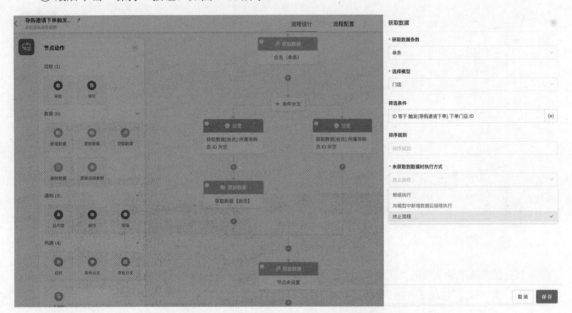

图 7-68　增加一个"获取数据"节点，在右侧属性面板中设置

（10）新增"条件分支"，分别对分支 1 和分支 2 进行条件设置，条件为："获取数据 [ 门店 ]"的"门店类型"字段，两个条件的"门店类型"字段一个设置为"独立门店"，一个设置为"融合门店"。这里条件值没有选择上下文变量，而是选择了选项。选项即字段对应的枚举值，在这个例子中"门店类型"字段类型为数据字典，如图 7-69、图 7-70 所示。

图 7-69　设置节点规则表达式

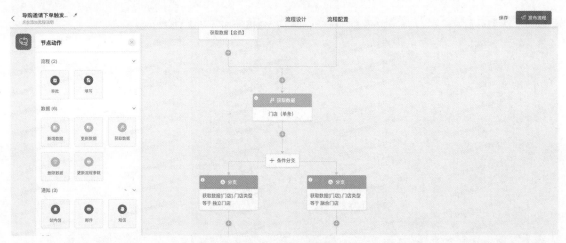

图 7-70　流程设计

（11）在"条件分支"的"获取数据 [ 门店 ]"的"门店类型"字段等于"独立门店"的分支流程中，增加"更新数据"节点，在右侧属性面板中设置：

① "更新模型"选择"触发 [ 导购邀请下单 ]"。

② "字段列表"单击"创建"按钮。

● 字段选择 更新"成本事业部"的"ID"。

● 表达式设置为"获取数据 [ 门店 ]"的关系字段"事业部"的"ID"字段，单击"确认"按钮。

③ "字段列表"单击"创建"按钮。

● 字段选择 更新"选择积分发放规则"的"ID"。

● 表达式设置为："获取数据 [ 门店 ]"的关系字段"默认积分发放规则"的"ID"字段，单击"确认"按钮。

（12）最终完成的"触发 [ 导购邀请下单 ]"更新设置：

① "触发 [ 导购邀请下单 ]"的关系字段"成本事业部"的"ID"字段等于"获取数据 [ 门店 ]"的

关系字段"事业部"的"ID"字段，如图 7-71 所示。

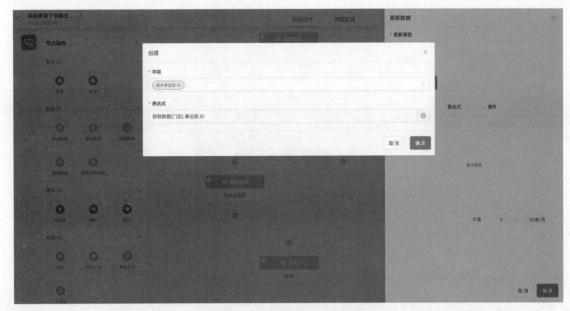

图 7-71　成本事业部 ID 字段设置

②"触发 [ 导购邀请下单 ]"的关系字段"选择积分发放规则"的"ID"字段等于"获取数据 [ 门店 ]"的关系字段"积分发放规则"的"ID"字段，如图 7-72 所示。

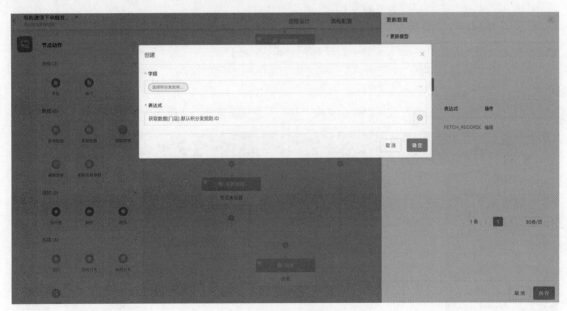

图 7-72　积分发放规则字段设置

③最后单击"保持"按钮，如图7-73所示。

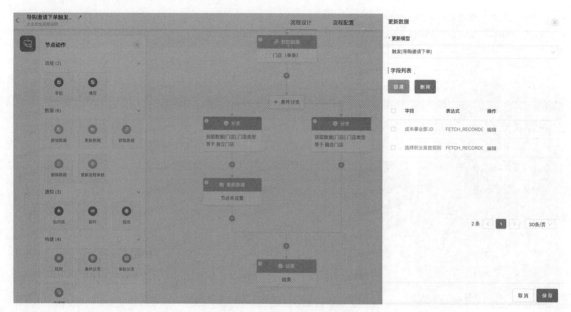

图 7-73　单击"保持"按钮

（13）在"条件分支"的"获取数据[门店]"的"门店类型"字段等于"融合门店"的分支流程中，增加"填写"节点，在右侧属性面板中设置：

①"填写模型"选择模型为"触发[导购邀请下单]"。

②"选择视图"选择"流程中邀请下单"选项，这里需要专门为流程增加一个表单页面，因为正常的"导购邀请下单"的表单不需要填写积分发放规则和成本部门。

③填写人，选择导购的"绑定用户"，同时加上角色为"超级管理员"方便测试如图7-74、图7-75所示。

④数据权限除了"积分发放规则"和"成本部门"需要编辑外，其他都为查看，如图7-76所示。

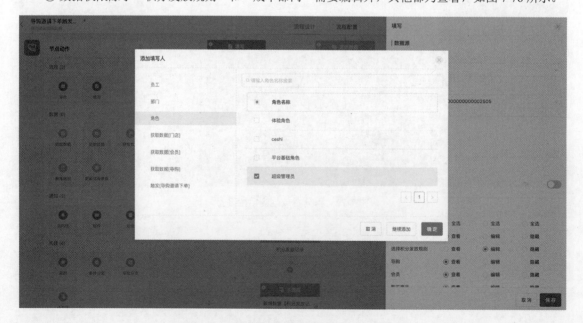

图 7-74　加上角色为"超级管理员"方便测试

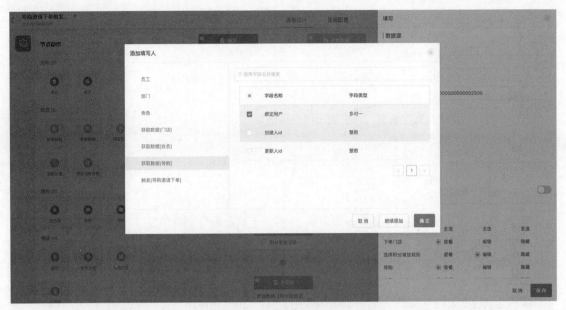

图 7-75 绑定用户多对一

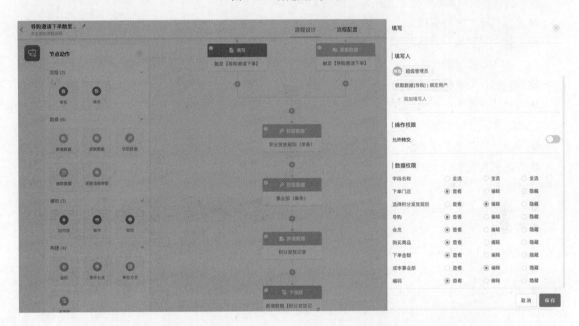

图 7-76 数据权限除了"积分发放规则"和"成本部门"需要编辑外，其他都为查看

（14）再增加一个"获取数据"节点，在右侧属性面板中设置：

① 设置"获取数据条数"为"单条"，选择模型为"积分发放规则"。

② 单击"筛选条件"的 {x} 图标，进行数据获取的条件设置，选择条件字段为"ID"，条件操作符为"等于"，条件变量为"触发 [ 导购邀请下单 ]"的关系字段"选择积分发放规则"的"ID"字段。设置好以后单击"确认"按钮。

③ 回到属性面板设置"未获取到数据时执行方式"为"终止流程"。

④ 最后单击"保持"按钮。

（15）再增加一个"获取数据"节点，在右侧属性面板中设置：

① 设置"获取数据条数"为单条，选择模型为"事业部"。

② 单击"筛选条件"的 {x} 图标，进行数据获取的条件设置，选择条件字段为"ID"，条件操作符为"等于"，条件为变量"触发 [ 导购邀请下单 ]"的关系字段"成本事业部"的"ID"字段。设置好以后单击"确认"按钮。

③ 回到属性面板设置"未获取到数据时执行方式"为"终止流程"。

④ 最后单击"保持"按钮。

（16）增加"新增数据"节点，在右侧属性面板中设置。

① "新增模型"选择模型为"积分发放记录"。

② "新增数据的节点"设置为"节点执行完即时增加业务数据"。

③ "新增数据"单击"创建"按钮。

● 字段选择"最终发放积分"。

● 表达式设置为："触发 [ 导购邀请下单 ]"的"下单金额"字段"乘以"，"获取数据 [ 积分发放规则 ]"的"发放倍数"字段，单击"确认"按钮。

④ "新增数据"单击"创建"按钮。

● 字段选择关系字段"积分发放规则"的"ID"。

● 表达式设置为："获取数据 [ 积分发放规则 ]"的"ID"字段，单击"确认"按钮。

⑤ "新增数据"单击"创建"按钮。

● 字段选择"事件编码"。

● 表达式设置为："触发 [ 导购邀请下单 ]"的"编码"字段，单击"确认"按钮。

⑥ "新增数据"单击"创建"按钮。

● 字段选择关系字段"推荐导购员"的"ID"

● 表达式设置为："获取数据 [ 导购 ]"的"ID"字段，单击"确认"按钮。

⑦ "新增数据"单击"创建"按钮。

● 字段选择关系字段"推荐会员"的"ID"

● 表达式设置为："获取数据 [ 会员 ]"的关系字段"推荐客户"的"ID"字段，单击"确认"按钮。

⑧ "新增数据"单击"创建"按钮。

● 字段选择关系字段"关系门店"的"ID"。

● 表达式设置为："获取数据 [ 门店 ]"的"ID"字段，单击"确认"按钮。

⑨ "新增数据"单击"创建"按钮。

● 字段选择关系字段"成本事业部"的"ID"。

● 表达式设置为："触发 [ 导购邀请下单 ]"的关系字段"成本事业部"的"ID"字段，单击"确认"按钮。

⑩ "新增数据"单击"创建"按钮。

● 字段选择关系字段"会员"的"ID"。

● 表达式设置为："获取数据 [ 会员 ]"的"ID"字段，单击"确认"按钮。

⑪ "新增数据"单击"创建"按钮 -- 为数据分析预留字段。

● 字段选择"成本事业部名称"。

● 表达式设置为："获取数据 [ 事业部 ]"的"名称"字段，单击"确认"按钮。

⑫ "新增数据"单击"创建"按钮，为数据分析预留字段。

● 字段选择"会员名称"。

● 表达式设置为："获取数据 [ 会员 ]"的"名称"字段，单击"确认"按钮。

⑬ 最终完成的"新增数据 [ 积分发放记录 ]"新增设置，最后单击"保持"按钮。

具体操作如图 7-77 ~ 图 7-87 所示。

图 7-77　字段表达式设置

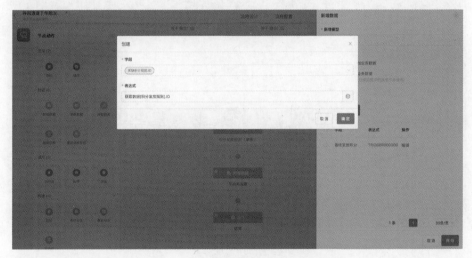

图 7-78　关联积分规则 ID 字段及表达式设置

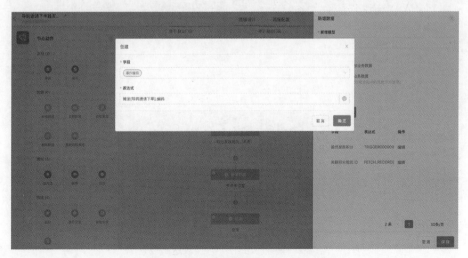

图 7-79　事件编码字段及表达式设置

图 7-80　推荐导购员 ID 字段及表达式设置

图 7-81　推荐会员 ID 字段及表达式设置

图 7-82　关系门店 ID 字段及表达式设置

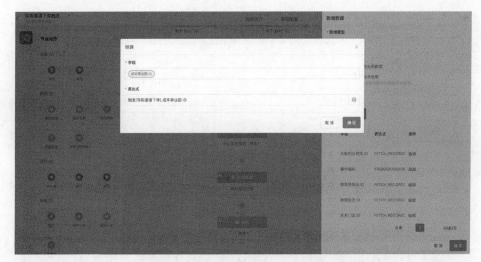

图 7-83 成本事业部 ID 字段及表达式设置

图 7-84 会员 ID 字段及表达式设置

图 7-85 名称事业部名称字段及表达式设置

图 7-86 会员名称字段及表达式设置

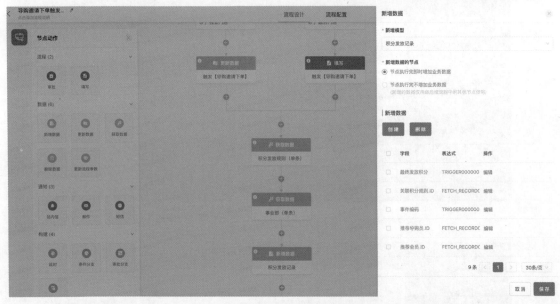

图 7-87 流程新增数据设置

（17）新增"子流程"节点，这里之所以设置为子流程，就跟上面我们业务流程图中画的一样，"导购邀请下单触发流程"已经完成，会员积分累加从业务上来说是独立的，不影响"导购邀请下单触发流程"，如图 7-88 所示。我们在右侧属性面板中设置：

① "选择模型"设置为"新增数据 [ 积分发放记录 ]"。

② "选择子流程"设置为"创建新的子流程"，并设置子流程名称为"会员积分累加子流程"。

③ "子流程执行方式"设置为"子流程和后续节点同时进行"。

④ 最后单击"保持"按钮。

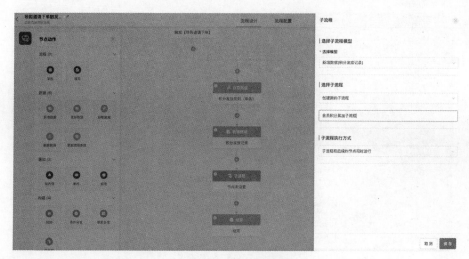

图 7-88　子流程模型设置

**Step4.2.** 创建积分发放记录子流程：积分累积子流程。

（1）单击子流程的跳转图标，跳转到子流程的设计页面。子流程的触发模型由主流程指定，无法更改，如图 7-89～图 7-91 所示。

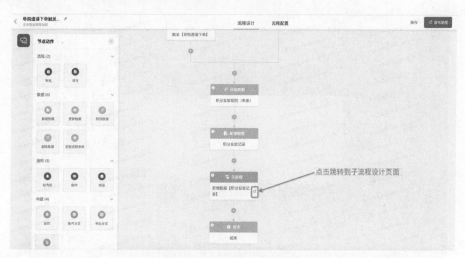

图 7-89　单击子流程的跳转图标

图 7-90　子流程设计

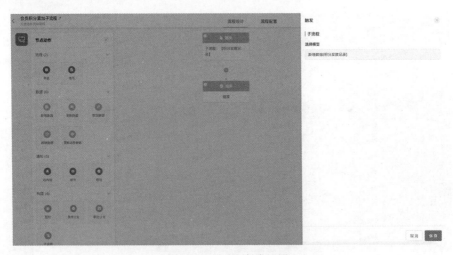

图 7-91　子流程触发设置

（2）新增"条件分支"，分别对分支 1 和分支 2 进行条件设置，条件为"触发 [ 积分发放记录 ]"的"最终发放积分"字段一个设置为小于或等于 100，一个设置大于 100。这里条件值没有选择上下文变量，而是选择了数值，因为"最终发放积分"字段为数字的普通类型，所以可以直接填数字，如图 7-92、图 7-93 所示。

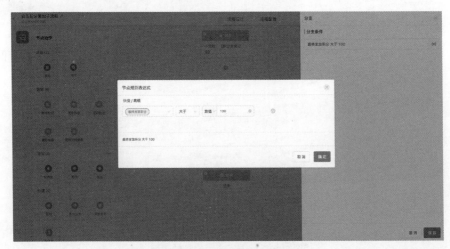

图 7-92　节点规则表达式设置

图 7-93　流程设计

(3)在"条件分支"的"触发[积分发放记录]"的"最终发放积分"字段一个设置为大于100的分支流程中,增加"审批"节点,在右侧属性面板中设置(如图7-94所示):

①"审批模型"选择模型为"触发[积分发放记录]"。

②"审批人"选择角色为"超级管理员"。

③"数据"权限全部设置为"查看"。

④其他配置项默认,需要了解更多请移至Oinone官网或联系官方工作人员。

⑤最后单击"保持"按钮。

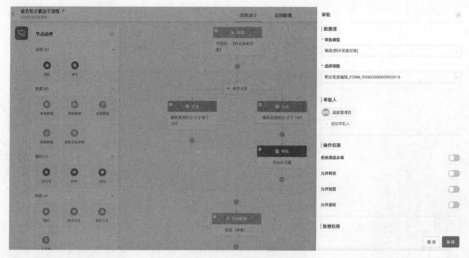

图7-94 审批模型属性设置

(4)再增加一个"获取数据"节点,在右侧属性面板中设置(如图7-95所示):

①设置"获取数据条数"为单条,选择模型为"会员"。

②单击"筛选条件"的"{x}"图标,进行数据获取的条件设置,选择条件字段为"ID",条件操作符为"等于",条件变量为"会员"的"ID"字段。设置好以后单击"确认"按钮。当上下文只有一个变量时默认不需要选择,这里默认的是"触发[积分发放记录]"。

③回到属性面板设置"未获取到数据时执行方式"为"终止流程"。

④最后单击"保持"按钮。

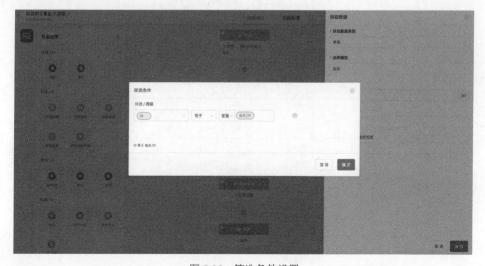

图7-95 筛选条件设置

（5）增加"更新数据"节点，在右侧属性面板中设置（如图 7-96、图 7-97 所示）：

① "更新模型"选择"获取数据 [ 会员 ]"。

② "字段列表"单击"创建"按钮。

● 字段选择 更新"会员累计积分"字段

● 表达式设置为"获取数据 [ 会员 ]"的"会员累计积分"字段加上"触发 [ 积分发放记录 ]"的"最终发放积分"字段，单击"确认"按钮。

最终完成的"触发 [ 导购邀请下单 ]"更新设置。

③ "获取数据 [ 会员 ]"的"会员累计积分"字段等于自身加上"触发 [ 积分发放记录 ]"的"最终发放积分"字段。

④ 最后单击"保持"按钮。

图 7-96 字段表达式设置

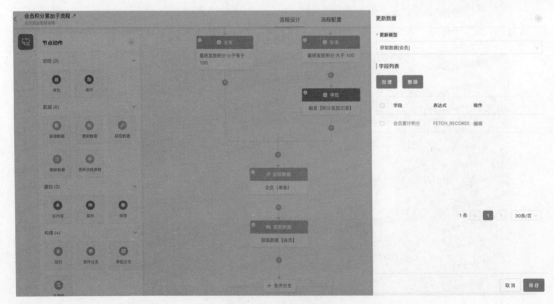

图 7-97 自定义设置

(6)新增"条件分支",分别对分支1和分支2进行条件设置,条件为"获取数据[会员]"的关系字段"推荐客户"的"ID"字段,两个条件的"ID"字段一个设置为"为空",一个设置为"非空",如图7-98所示。

图7-98　分支条件设置

(7)在"条件分支"的"获取数据[会员]"的关系字段"推荐客户"的"ID"字段非空,增加"获取数据"节点,在右侧属性面板中设置:

① 设置"获取数据条数"为"单条",选择模型为"积分发放规则"。

② 单击"筛选条件"的 {x} 图标,进行数据获取的条件设置,选择条件字段为"ID",条件操作符为"等于",条件变量为"触发[积分发放记录]"的关系字段"关联积分规则"的"ID"字段。设置好以后单击"确认"按钮,如图7-99所示。

③ 回到属性面板设置"未获取到数据时执行方式"为"终止流程"。

④ 最后单击"保持"按钮。

图7-99　筛选条件设置

(8)在"条件分支"的"获取数据[会员]"的关系字段"推荐客户"的"ID"字段非空,增加"获取数据"节点,在右侧属性面板中设置:

①设置"获取数据条数"为单条,选择模型为"会员"。

②单击"筛选条件"的{x}图标,进行数据获取的条件设置,选择条件字段为"ID",条件操作符为"等于",条件变量为"获取数据[会员]"的关系字段"推荐客户"的"ID"字段。设置好以后单击"确认"按钮。

③回到属性面板,设置"未获取到数据时执行方式"为"终止流程"。

④最后单击"保持"按钮。

⑤编辑节点名字为"推荐客户",如图7-100所示。

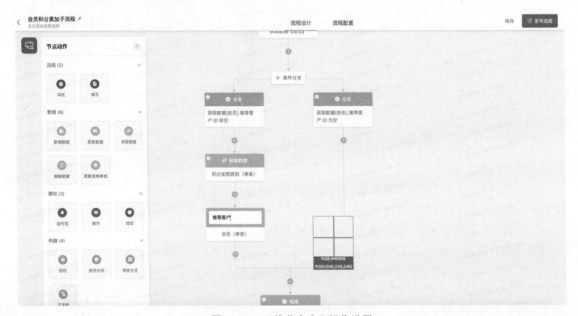

图7-100 "推荐客户"操作设置

(9)在"条件分支"的"获取数据[会员]"的关系字段"推荐客户"的"ID"字段非空,增加"新增数据"节点,在右侧属性面板中设置(如图7-101~图7-109所示):

①"新增模型"选择模型为"积分发放记录"。

②"新增数据的节点"设置为"节点执行完即时增加业务数据"。

③"新增数据"单击"创建"按钮:

● 字段选择"最终发放积分"。

● 表达式设置为"触发[积分发放记录]"的"最终发放积分"字段"乘以""获取数据[积分发放规则]"的"推荐客户发放比例"字段,单击"确认"按钮。

④"新增数据"单击"创建"按钮:

● 字段选择关系字段"积分发放规则"的"ID"。

● 表达式设置为"获取数据[积分发放规则]"的"ID"字段,单击"确认"按钮。

⑤"新增数据"单击"创建"按钮:

● 字段选择"事件编码"。

● 表达式设置为"触发[积分发放记录]"的"事件编码"字段,单击"确认"按钮。

⑥"新增数据"单击"创建"按钮:
● 字段选择关系字段"推荐导购员"的"ID"。
● 表达式设置为"触发[积分发放记录]"关系字段"推荐导购员"的"ID",单击"确认"按钮。
⑦"新增数据"单击"创建"按钮:
● 字段选择关系字段"关系门店"的"ID"。
● 表达式设置为"触发[积分发放记录]"的关系字段"关系门店"的"ID",单击"确认"按钮。
⑧"新增数据"单击"创建"按钮:
● 字段选择关系字段"成本事业部"的"ID"。
● 表达式设置为:"触发[积分发放记录]"的关系字段"成本事业部"的"ID"字段,单击"确认"按钮。
⑨"新增数据"单击"创建"按钮:
● 字段选择关系字段"会员"的"ID"。
● 表达式设置为"推荐客户[会员]"的"ID"字段,单击"确认"按钮。
⑩"新增数据"单击"创建"按钮——为数据分析预留字段:
● 字段选择"成本事业部名称"。
● 表达式设置为"触发[积分发放记录]"的"成本事业部名称"字段,单击"确认"按钮。
⑪"新增数据"单击"创建"按钮——为数据分析预留字段:
● 字段选择"会员名称"。
● 表达式设置为:"推荐客户[会员]"的"名称"字段,单击"确认"按钮。
⑫最终完成的"新增数据[积分发放记录]"新增设置,最后单击"保持"按钮。

图 7-101　字段表达式设置

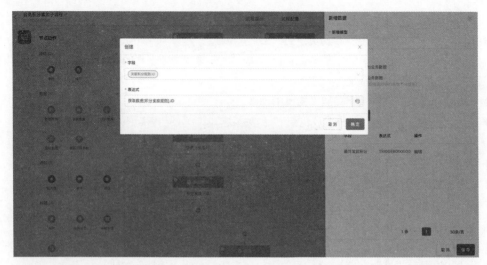

图 7-102　关联积分规则 ID 字段及表达式设置

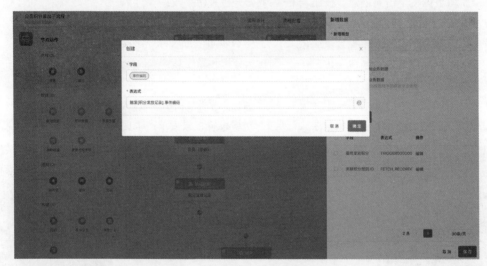

图 7-103　事件编码字段及表达式设置

图 7-104　推荐导购员 ID 字段及表达式设置

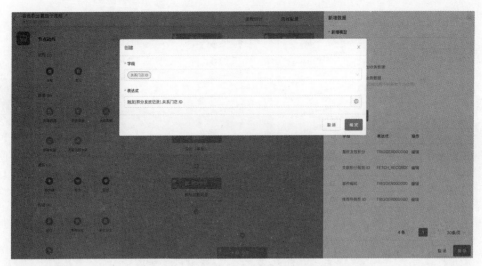

图 7-105　关系门店 ID 字段及表达式设置

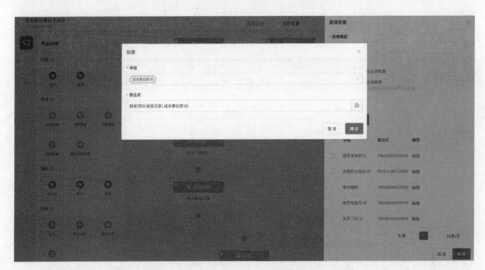

图 7-106　成本事业部 ID 字段及表达式设置

图 7-107　会员 ID 字段及表达式设置

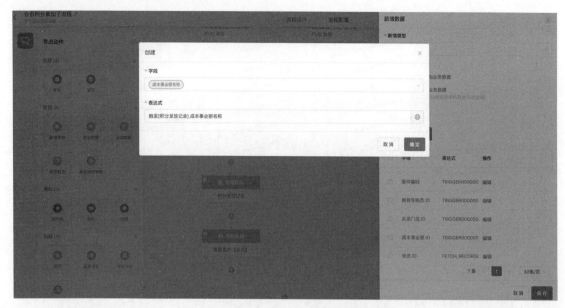

图 7-108　成本事业部名称字段及表达式设置

图 7-109　会员名称字段及表达式设置

（10）增加"更新数据"节点，在右侧属性面板中设置（如图 7-110、图 7-111 所示）：
①"更新模型"选择"推荐客户 [ 会员 ]"。
②"字段列表"单击"创建"按钮：
● 字段选择更新"会员累计积分"字段。
● 表达式设置为"推荐客户 [ 会员 ]"的"会员累计积分"字段加上"新增 [ 积分发放记录 ]"的"最终发放积分"字段，单击"确认"按钮。
最终完成的"触发 [ 导购邀请下单 ]"更新设置：
③"推荐客户 [ 会员 ]"的"会员累计积分"字段等于自身加上"新增 [ 积分发放记录 ]"的"最终发放积分"字段。

④ 最后单击"保持"按钮。

图 7-110　字段表达式设置

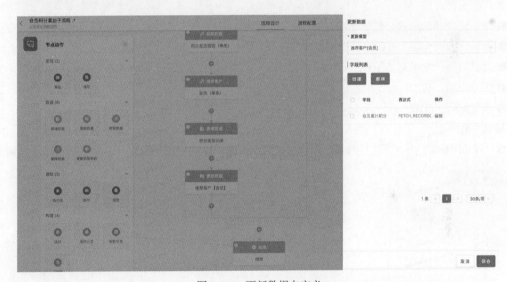

图 7-111　更新数据自定义

Step4.3. 两个流程确保都保持并发布过。

流程列表页查看流程状态图标，如图 7-112 所示。

图 7-112　流程列表页查看流程状态图标

Step5. 基础数据管理。

流程前提：基础数据通过管理页面已经建好，随后查看实战效果的基础数据。

（1）事业部：国一事业部、国二事业部。

（2）门店：国一独立门店 01（绑定：国一事业部；积分发放规则：正常发放）、国二独立门店 01（绑定：国二事业部；积分发放规则：双倍发放）、融合门店 01。

（3）导购员：国一独立门店导购——cpc（绑定：国一独立门店 01）、国二独立门店导购——温振（绑定：国二独立门店 01）、融合门店导购——梦瑶（绑定：融合门店 01）。

（4）会员：会员 1001（导购绑定：融合门店导购——梦瑶）、会员 1002（导购绑定：国二独立门店导购——温振；推荐会员绑定：会员 1001）、会员 1003（导购绑定：未绑定；推荐会员绑定：会员 1001）。

（5）积分发放规则：双倍发放（发放倍数：2；推荐客户发放比例：0.5）、正常发放（发放倍数：1；推荐客户发放比例：0.2）。

Step6. 完成业务流程，查看效果。

Step6.1. 用例设计 1。

新建"导购邀请下单记录"如下：

● 门店选择：国一独立门店 01。
● 导购选择：国一独立门店导购——cpc。
● 会员选择：会员 1003。
● 商品填写：测试商品 1001。
● 下单金额：12。
● 事件编码：O1001。

期望结果：

● "会员 1003"的"所属导购员"字段更新为："国一独立门店导购——cpc"，如图 7-113 所示。

图 7-113  "会员 1003"的"所属导购员"字段更新为"国一独立门店导购——cpc"

● "会员 1003"的"会员累计积分"字段在原有值上增加"12"。
● "会员 1001"的"会员累计积分"字段在原有值上增加"2.4"。
● 新增两条"积分发放记录"，如图 7-114 所示。

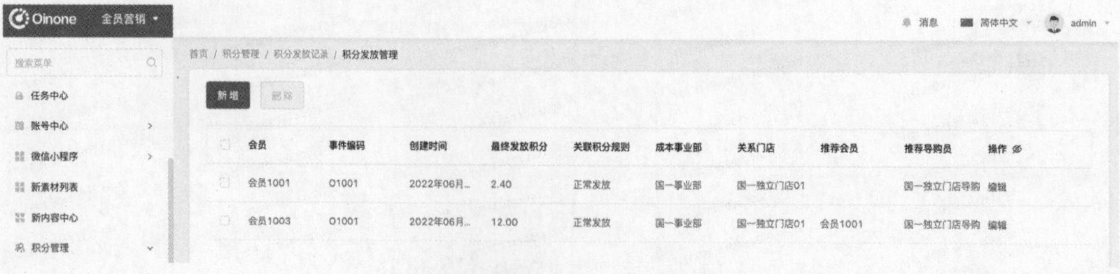

图 7-114  新增两条"积分发放记录"

国一独立门店 01 会员记录表如表 7-2 所示。

表 7-2 国一独立门店 01 会员记录表

| 会员 | 关系门店 | 最终发放积分 | 关联积分规则 | 事件编码 | 推荐导购员 | 成本事业部 | 推荐会员 |
| --- | --- | --- | --- | --- | --- | --- | --- |
| 会员 1003 | 国一独立门店 01 | 12 | 正常发放 | O1001 | 国一独立门店导购——cpc | 国一事业部 | 会员 1001 |
| 会员 1001 | 国一独立门店 01 | 2.4 | 正常发放 | O1001 | 国一独立门店导购——cpc | 国一事业部 | |

Step6.2. 用例设计 2。

新建"导购邀请下单记录"如下：

- 门店选择：融合门店 01。
- 导购选择：融合门店导购——梦瑶。
- 会员选择：会员 1001。
- 商品填写：测试商品 1002。
- 下单金额：12。
- 事件编码：O1002。

期望结果：

- 出现填写节点，"选择积分发放规则"：双倍发放；"成本事业部"：国二事业部。
- "会员 1001"的"会员累计积分"字段在原有值上增加"24"，如图 7-115 所示。

图 7-115 新增一条会员 1001 的"积分发放记录"

- 新增一条"积分发放记录"。

会员 1001 信息表如表 7-3 所示。

表 7-3 会员 1001 信息表

| 会员 | 关系门店 | 最终发放积分 | 关联积分规则 | 事件编码 | 推荐导购员 | 成本事业部 | 推荐会员 |
| --- | --- | --- | --- | --- | --- | --- | --- |
| 会员 1001 | 融合门店 01 | 24 | 双倍发放 | O1002 | 融合门店导购——梦瑶 | 国二事业部 | |

Step6.3. 用例设计 3。

新建"导购邀请下单记录"如下：

- 门店选择：国二独立门店 01。
- 导购选择：国二独立门店导购——温振。
- 会员选择：会员 1002。
- 商品填写：测试商品 1003。

- 下单金额：60。
- 事件编码：O1003。

期望结果：

- 出现审批，审批通过
- "会员1002"的"会员累计积分"字段在原有值上增加"120"。
- "会员1001"的"会员累计积分"字段在原有值上增加"60"。
- 新增两条"积分发放记录"，如图7-116所示。

图7-116　新增两条"积分发放记录"

国二独立门店01会员记录表如表7-4所示。

表7-4　国二独立门店01会员记录表

| 会员 | 关系门店 | 最终发放积分 | 关联积分规则 | 事件编码 | 推荐导购员 | 成本事业部 | 推荐会员 |
|---|---|---|---|---|---|---|---|
| 会员1002 | 国二独立门店01 | 120 | 双倍发放 | O1003 | 国二独立门店导购——温振 | 国二事业部 | 会员1001 |
| 会员1001 | 国二独立门店01 | 60 | 双倍发放 | O1003 | 国二独立门店导购——温振 | 国二事业部 | |

Step6.4. 用例设计4。

新建"导购邀请下单记录"如下：

- 门店选择：国一独立门店01。
- 导购选择：国二独立门店导购——温振。
- 会员选择：会员1002。
- 商品填写：测试商品1003。
- 下单金额：60。
- 事件编码：O1004。

期望结果：

- "会员1002"的"会员累计积分"字段在原有值上增加"60"。
- "会员1001"的"会员累计积分"字段在原有值上增加"12"。
- 新增两条"积分发放记录"，如图7-117所示。

国一独立门店01会员记录表如表7-5所示。

表7-5　国一独立门店01会员记录表

| 会员 | 关系门店 | 最终发放积分 | 关联积分规则 | 事件编码 | 推荐导购员 | 成本事业部 | 推荐会员 |
|---|---|---|---|---|---|---|---|
| 会员1002 | 国一独立门店01 | 60 | 正常发放 | O1004 | 国二独立门店导购——温振 | 国一事业部 | 会员1001 |
| 会员1001 | 国一独立门店01 | 12 | 正常发放 | O1004 | 国二独立门店导购——温振 | 国一事业部 | |

图 7-117　新增两条"积分发放记录"

Step7. 通过数据可视化，设计相应的数据看板。

Step7.1. 事业部积分发放分析。

（1）进入数据可视化"图表"设计页，单击"创建一级分组"按钮，在分组栏中会多出一行，并填入积分发放综合分析，如图 7-118 所示。

图 7-118　数据可视化"图表"设计页

（2）鼠标放到一级分组上会出现增加二级分组和删除入口，单击"+"图标，创建两个二级分组分别为：积分发放分析和订单下单分析，如图 7-119 所示。

图 7-119　创建两个二级分组：积分发放分析和订单下单分析

（3）选中二级分组"积分发放分析"出现"添加图标"操作入口，单击"添加图表"按钮，如图7-120所示。

图7-120　选中二级分组"积分发放分析"出现"添加图标"操作入口

（4）在创建图表弹出框中设置（如图7-121所示）：

① "图表标题"填入"事业部积分成本分析"。
② "模型"选择"积分发放记录"选项。
③ "方法"选择"根据条件分页查询记录列表和总数"选项。
④ 最后单击"确定"按钮。

图7-121　事业部积分成本分析图表弹出框

（5）上一步单击"确定"按钮后自动进入"事业部积分成本分析"图表设计页面，如图7-122、图7-123所示。

① "维度"选择"成本事业部名称"选项。
② "数值"选择"最终发放积分"选项。
③ 单击右上角"保存"按钮，并单击左上角"< 事业部积分成本分析"按钮，返回图表管理入口。

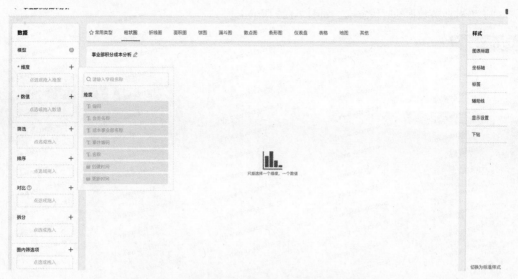

图 7-122 "事业部积分成本分析"图表设计页面（1）

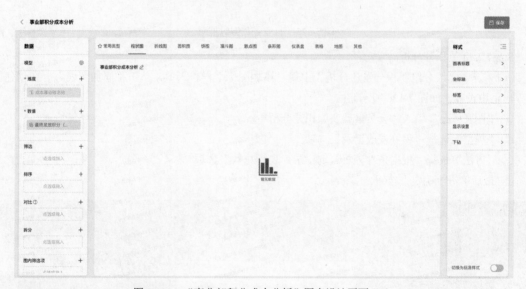

图 7-123 "事业部积分成本分析"图表设计页面（2）

（6）单击右上角"发布"按钮，即完成一个图表的开发，如图 7-124 所示。

图 7-124 待发布图表

(7)当有业务数据的时候展示,如图 7-125 所示。

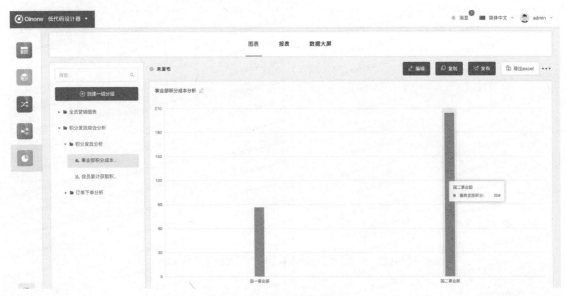

图 7-125 有业务数据时展示

Step7.2. 会员积分分析。

(1)选中二级分组"积分发放分析"出现"添加图标"操作入口,单击"添加图表"。在创建图表弹出框中设置(如图 7-126 所示):

① "图表标题"填入"会员累计获取积分分析"。
② "模型"选择"积分发放记录"选项。
③ "方法"选择"根据条件分页查询记录列表和总数"选项。
④ 最后单击"确定"按钮。

图 7-126 会员累计获取积分分析图表弹出框

(2)进入"会员累计获取积分分析"图表设计页面中设置(如图 7-127 所示):

① "维度"选择"会员名称"选项。
② "数值"选择"最终发放积分"选项。
③ "对比"选择"成本事业部名称"选项。

④ 单击右上角"保存"按钮,并单击左上角"< 会员累计获取积分分析"按钮返回图表管理入口。

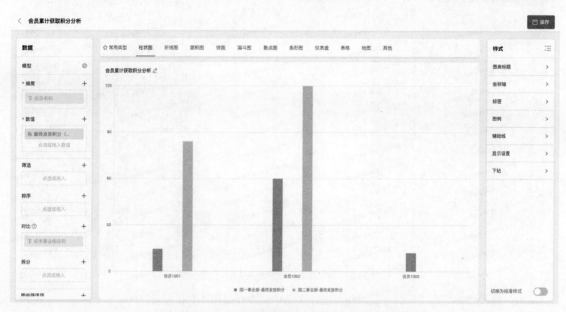

图 7-127 "会员累计获取积分分析"图表设计页面

(3) 单击右上角"发布"按钮,即完成该图表的开发,如图 7-128 所示。

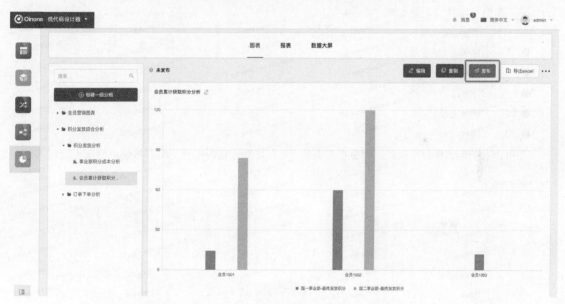

图 7-128 单击右上角"发布"按钮,即完成该图表的开发

Step7.3. 商品下单金额分析

(1) 选中二级分组"订单下单分析"出现"添加图标"操作入口,单击"添加图表"。在创建图表弹出框中设置(如图 7-129 所示):

① "图表标题"填入"商品购买金额分析"。

② "模型"选择"导购邀请下单"选项。

③ "方法"选择"根据条件分页查询记录列表和总数"选项。

④ 最后单击"确定"按钮。

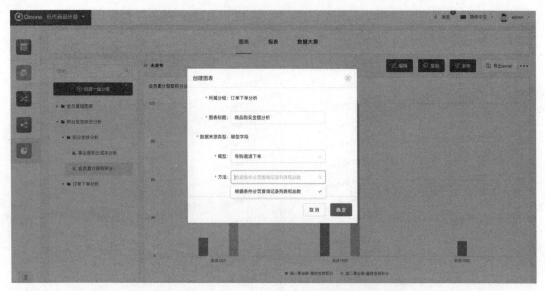

图 7-129　商品购买金额分析分析图表弹出框

（2）进入"商品购买金额分析"图表设计页面中设置（如图 7-130 所示）：

① 图表形式切换为"饼图"。

② "维度"选择"购买商品"选项。

③ "数值"选择"下单金额"选项。

④ 右侧属性面板设置：

● "展示标签"为"开启"。

● 勾选"购买商品"。

● "展示图例"为"开启"。

● 勾选"购买商品"。

● 勾选"数值"。

⑤ 单击右上角"保存"按钮，并单击左上角"< 会员累计获取积分分析"按钮返回图表管理入口。

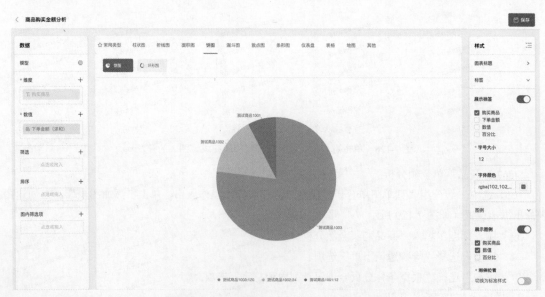

图 7-130　"商品购买金额分析"图表设计页面

(3) 单击右上角"发布"按钮，即完成该图表的开发，如图7-131所示。

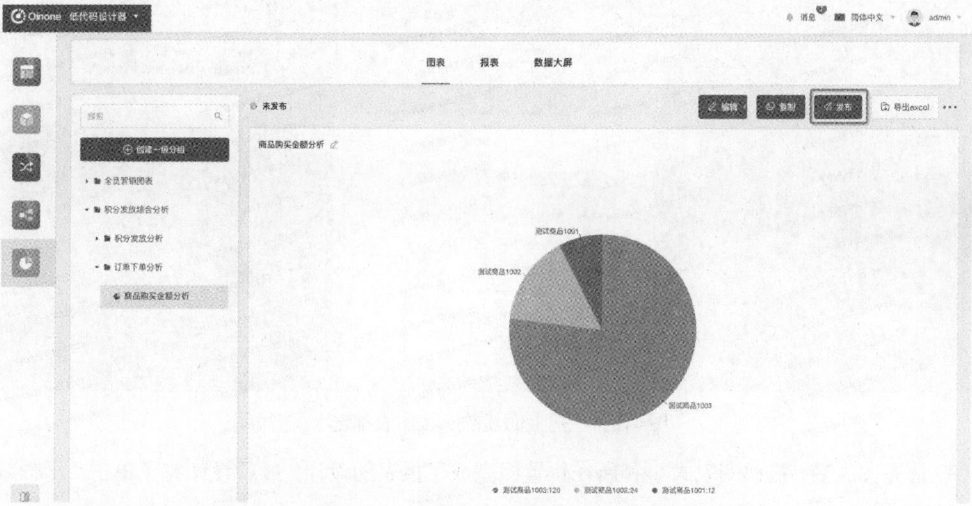

图7-131　单击右上角"发布"按钮，即完成该图表的开发

Step7.4. 设计综合分析报表。

（1）进入数据可视化"报表"设计页，单击创建一、二级分组，分别为"积分分析报表""积分综合报表"，选中二级分组"积分综合报表"出现"添加图标"操作入口，在分组栏中会多出一行，并填入积分发放综合分析，如图7-132所示。

图7-132　进入数据可视化"报表"设计页

（2）选中二级分组"积分发放综合分析"，单击右上角"选择图表"按钮，把上面建的三张图表都选上，如图7-133、图7-134所示。

图7-133　单击右上角"选择图表"

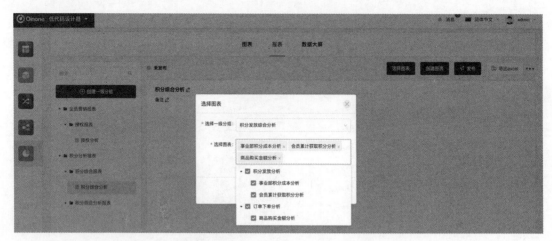

图 7-134　把上面建的三张图表都选上

（3）选完图表后，拖动图表大小和所在位置即完成了报表的设计，最后设计完了单击"发布"按钮，如图 7-135、图 7-136 所示。

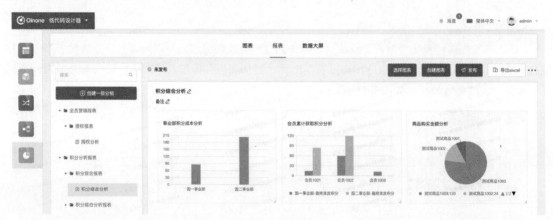

图 7-135　报表设计

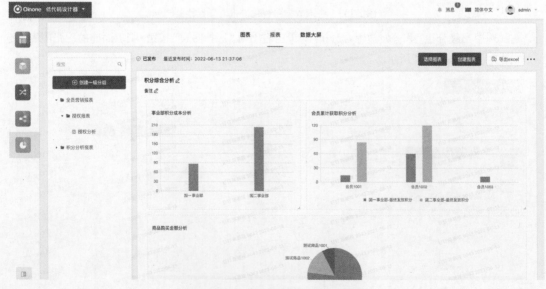

图 7-136　可拖动图表大小和所在位置

Step7.5. 把报表挂载到页面上并绑上菜单。

简单描述：新建名为"积分综合分析页"的页面，选择模型"积分发放记录"，页面类型设置为"详情"，并在左侧"组件"选项卡下找到"报表"组件，拖入到设计器区并在其属性面板中选择展示报表。新建菜单"积分数据分析"并绑定已有页面"积分综合分析页"，最后单击"发布"按钮，如图7-137所示。

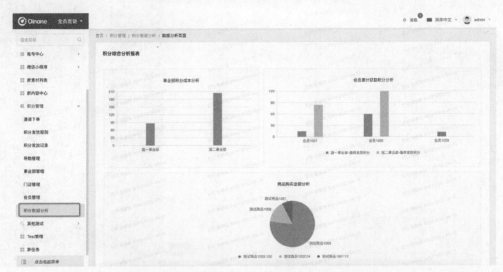

图 7-137　新建菜单"积分数据分析"并绑定已有页面

## 7.3　实战训练（全员营销为例）

### 7.3.1　去除资源上传大小限制

**1. 场景说明**

全员营销标准软件产品对于视频上传限制为35MB，该大小适合管理短视频类的素材。该软件产品的使用方为营销部门，营销部门不仅管理短视频素材，也负责管理电商商品的视频素材、公司团建的素材等，且不少素材的内存都远大于35MB。

业务需求：将全员营销中的内容中心扩展成营销部门的内容统一管理中心，管理营销部门工作范围内的所有内容素材，且不受35MB的限制。

**2. 实战训练**

新增一个资源管理页面，替换原来的资源管理，并构建新的上传行为。

Step1. 通过界面设计器，设计出必要管理页面。

（1）进入界面设计器，应用选择全员营销，模型通过搜索"素材"选择"Gemini 素材"，单击"添加页面"下的"直接创建"按钮，如图7-138所示。

图 7-138　选择"Gemini 素材"

（2）设置页面标题、模型（自动带上可切换）、业务类型（运营管理后续会扩展其他类型）、视图类型（表格）后单击"确定"按钮进入"内容中心 - 新素材管理"设计页面，如图 7-139 所示。

图 7-139　页面信息设置

（3）进入页面设计器，对"内容中心 - 新素材管理"表格页面进行设计（如图 7-140 所示）：

① 左侧为物料区：分为组件、模型。

● "组件"选项卡下为通用物料区，我们可以为页面增加对应布局、字段（如同在模型设计器增加字段）、动作、数据、多媒体等。

● "模型"选项卡下为页面对应模型的自定义字段、系统字段以及模型已有动作。

② 中间是设计区域。

③ 右侧为属性面板，在设计区域选择中组件会显示对应组件的可配置参数。

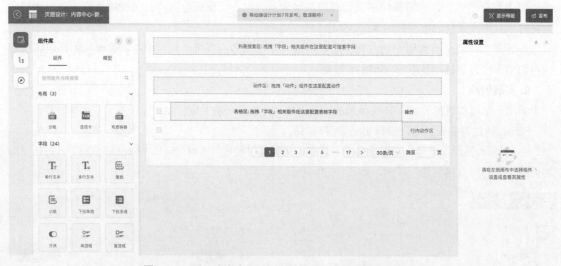

图 7-140　对"内容中心 - 新素材管理"表格页面进行设计

（4）在左侧"模型"选项卡下，分别把系统字段中的"素材名称""素材链接""素材来源""素材类型""更新时间""创建时间"等字段拖入设计区域的表格区，如图 7-141 所示。

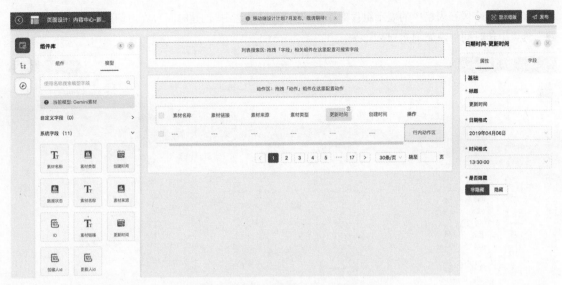

图 7-141　将相应字段拖入设计区域的表格区

（5）设置字段在表格中的展示组件，在设计区域切换"素材链接"展示组件为"超链接"，如图 7-142 所示。

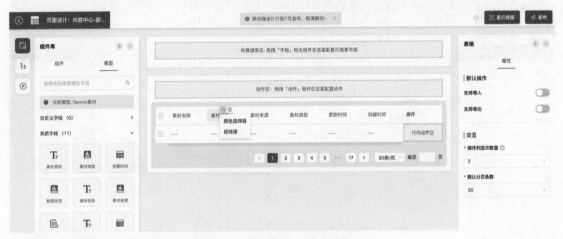

图 7-142　设置字段在表格中的展示组件

（6）设置字段在表格中的展示组件的属性，在设计区域选中"素材名称"，在右侧属性面板中设置"标题"为"内容名称"，如图 7-143 所示。

图 7-143　设置字段在表格中展示组件的属性

（7）设置字段在表格中的展示组件的属性，在设计区域选中"创建时间"，在右侧属性面板中设置"标题"为"上传时间"，如图 7-144 所示。

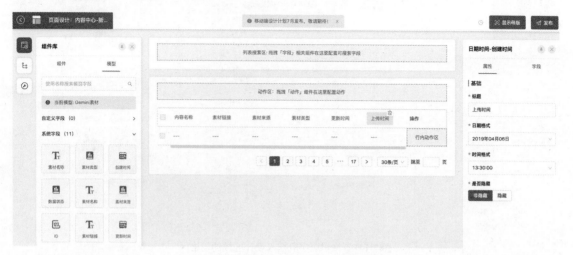

图 7-144　设置"标题"为"上传时间"

（8）在左侧"模型"选项卡下，把动作分类下的提交类型"下载"和"删除"动作拖入中间设计区的动作区，并单击"删除"按钮，在右侧属性面板中设置"按钮样式"为"次要按钮"，如图 7-145 所示。

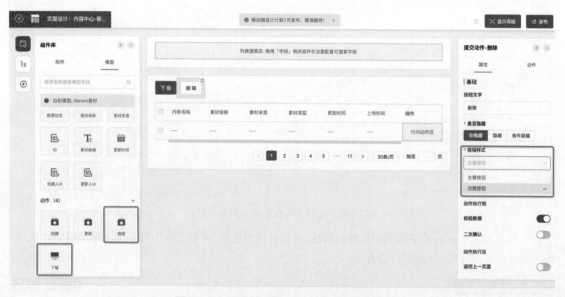

图 7-145　设置"按钮样式"为"次要按钮"

（9）在左侧"组件"选项卡下，把动作分类下的"跳转动作"拖入中间设计区的动作区，并在右侧属性面板中设置动作名称为"上传素材"，数据控制类型为"不进行数据处理"，打开方式为"弹窗打开"，弹出内容为"创建新页面"，"弹窗模型"通过搜索"素材"选择"Gemini 素材代理 - 上传"，并单击"保存"按钮，如图 7-146、图 7-147 所示。

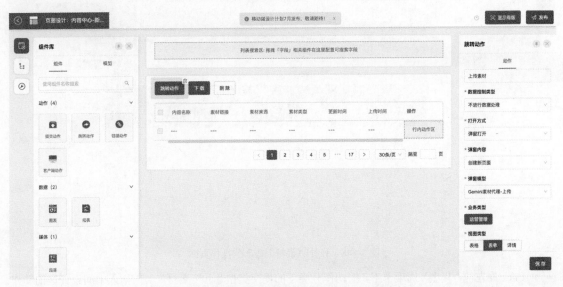

图 7-146  右侧属性面板中设置动作名称为"上传素材"

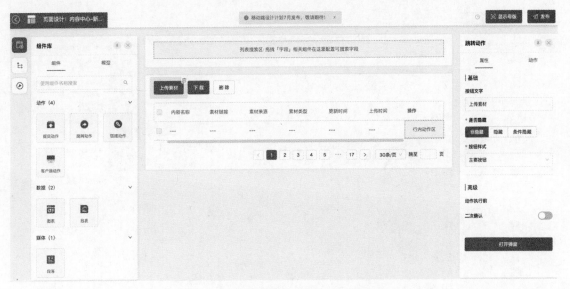

图 7-147  设置"按钮样式"为"次要按钮"

（10）设计区选中"上传素材"按钮，单击"打开弹窗"设计"素材上传"的操作页面，此时会发现左侧"模型"选项卡下的当前模型切换成了"Gemini 素材代理 - 上传"，如图 7-148 所示。

① 分别把系统字段中的"上传素材链接列表"拖入"Gemini 素材代理 - 上传"的弹窗页面设计区。

② 选中"上传素材链接列表"切换展示组件为"文件上传"。

③ 选中"上传素材链接列表"并在右侧属性面板中：

● 设置"校验"分组下，设置"最大上传文件体积"为空，即不设置。

- 设置"校验"分组下,设置"限制上传文件类型"为打开,并勾选"图片"和"视频"。
- 设置"交互"分组下的宽度属性为"1"。

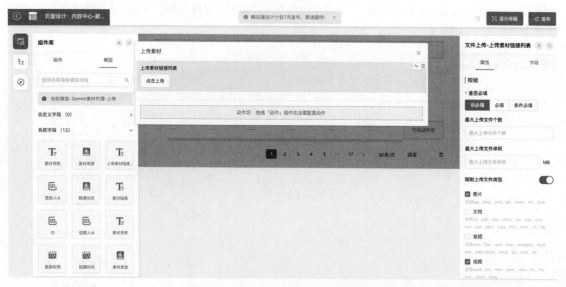

图 7-148　设计"素材上传"的操作页面

（11）在左侧"模型"选项卡下,把动作分类下的提交类型"上传素材"动作拖入中间设计区的动作区,如图 7-149 所示。

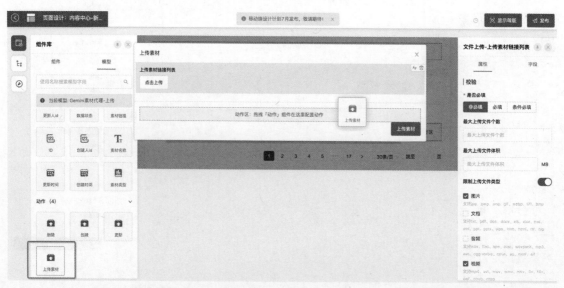

图 7-149　把动作分类下的提交类型"上传素材"动作拖入中间设计区的动作区

（12）在左侧"组件"选项卡下，把动作分类下的"客户端动作"类型拖入中间设计区的动作区，选中并在右侧属性面板中设置"动作名称"为"返回"，设置"客户端行为"为"关闭弹窗"，单击"保存"按钮来完成动作的基础设置，如图 7-150 所示。

图 7-150　将"客户端动作"类型拖入中间设计区的动作区

（13）单击"返回"按钮，并在右侧属性面板中设置"按钮样式"为"次要按钮"，如图 7-151 所示。

图 7-151　设置"按钮样式"为"次要按钮"

（14）关闭弹窗返回主模型设计区，如图 7-152 所示。

图 7-152　关闭弹窗返回主模型设计区

（15）单击右上角"显示母版"进入页面最终展示形式，单击"添加"菜单项，并在输入框中输入"新内容中心"，如图 7-153、图 7-154 所示。

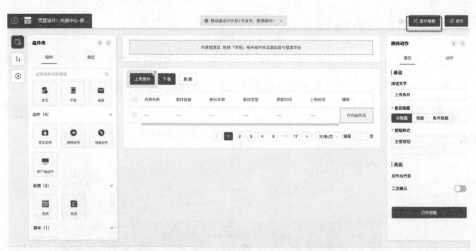

图 7-153　单击右上角"显示母版"可进入页面最终展示形式

图 7-154　单击添加菜单项并在输入框中输入"新内容中心"

（16）单击菜单右侧设置图标，选择"绑定已有页面"，进行菜单与页面的绑定操作，如图7-155所示。

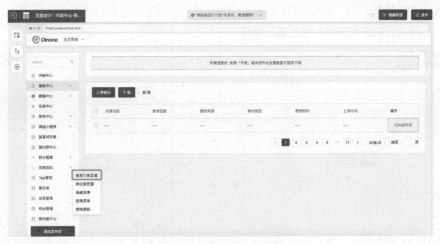

图 7-155　选择"绑定已有页面"

（17）在绑定页面中，模型选择"Gemini素材"，视图选择"内容中心-新素材管理"，单击"确认"按钮提交，如图7-156所示。

图 7-156　绑定页面选择

（18）最后别忘了单击右上角"发布"按钮对"内容中心-新素材管理"表格页面进行发布，回到界面设计器首页查看刚刚建好的表格页面。

Step2. 测试完成以后隐藏原"内容中心"菜单。

（1）进入"界面设计器"管理页面，通过单击"设计图标"进入任意页面的设计页面，如图7-157所示。

图 7-157　单击"设计图标"可进入任意页面的设计页面

（2）单击右上角"显示母版"按钮进入页面最终展示形式，找菜单"内容中心"单击菜单右侧设置图标，选择"隐藏菜单"，如图7-158所示，因为"内容中心"菜单是标准产品自带菜单，只能进行隐藏操作，无法进行如绑定页面和调整菜单顺序。

图7-158　选择"隐藏菜单"

Step3. 回到全员营销应用，刷新页面体验效果。

整个实战训练到此结束，若想了解更多细节可移至 Oinone 官网：https://oinone.top/ 或联系 Oinone 官方工作人员。

## 7.3.2　原业务加审批流程

**1. 场景说明**

场景描述：全员营销标准产品的功能并未有任务发放的审批流，在实际执行中，当营销专员配置好任务后，需部门领导对整个活动如该任务内容、形式、参与人员进行审批。

业务需求：在发布任务这个流程中增加审批节点。

**2. 实战训练**

Step1. 原业务分析。

（1）单击菜单"任务中心"，通过 URL 上的 Model 参数找到对应模型编码为"gemini.biz.GeminiTaskProxy"，如图 7-159 所示。

图 7-159　找到对应模型编码为"gemini.biz.GeminiTaskProxy"

（2）进入模型设计器主页面，应用选择"全员营销"、选择"系统模型"，通过搜索关键字"任务"选择"Gemini 任务代理"，展示方式从图模式切换到表单模式，对比"模型编码"如图 7-160、图 7-161 所示。

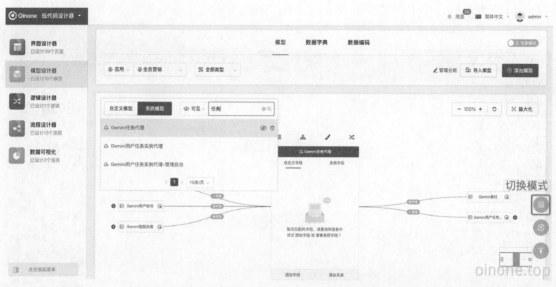

图 7-160　通过搜索关键字"任务"选择"Gemini 任务代理"

图 7-161　模型信息编辑

（3）但目前模型为代理模型，代理模型是用于代理存储模型的数据管理器能力，同时又可以扩展出非存储数据信息的交互功能的模型。因为在代理模型中新增的字段都是非存储字段，所以如果要增加"审核状态"的字段一定要在存储模型增加。其父模型的查看有以下两种方式。

① 表单模式下可以直接看父模型。

② 在图模式和表单模式下单击继承关系，如图 7-162 所示。

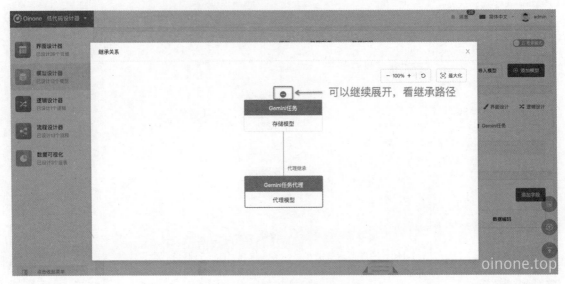

图 7-162　可继续展开看继承路径

（4）单击"Gemini 任务"，进入"Gemini 任务"的模型设计界面，可以看出该模型所在模块为"全员营销核心业务"，从"系统字段"中找到"任务状态"字段，单击查看字段详情，我们可以看到"业务类型"为"数据字典"，"字典类型"为"任务状态"，如图 7-163 所示。

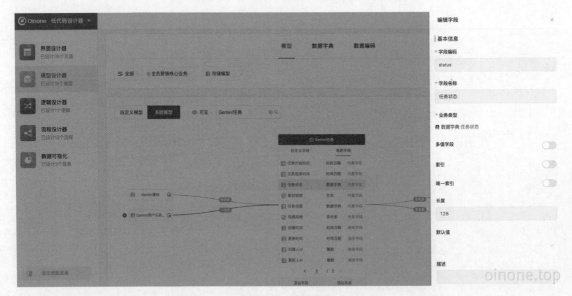

图 7-163　编辑字段

（5）在模型设计器的管理页面上方单击"数字字典"选项卡，模块选择"全员营销核心业务"，选择"系统字典"就可以查看到"任务状态"数字字典，如图 7-164、图 7-165 所示。

图 7-164　选择"系统字典"可查看"任务状态"数字字典

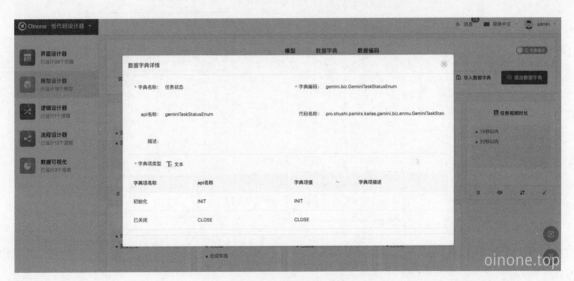

图 7-165　数据字典详情

(6) 总结如下:

① 给"Gemini 任务"模型增加一个"任务审批状态",记录审批状态。

② 在任务创建的时候,修改"任务状态"为"关闭",确保任务未审批通过时,用户无法操作该任务。

③ 审批通过后,恢复"任务状态"为"初始化"。

④ 我们先来整理一下核心流程,即任务审批流程,如图 7-166 所示。

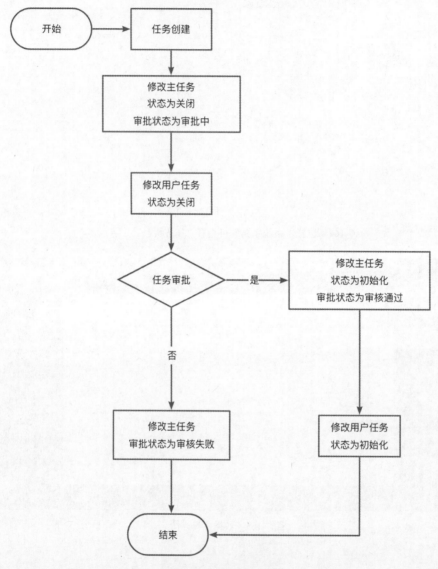

图 7-166　任务审批流程

**Step2.** 利用模型设计器设计模型。

(1) 在模型设计器的管理页面上方单击"数字字典"选项卡,模块选择为"全员营销核心业务",单击添加"数据字典"按钮,设置对应数据项,如图 7-167 所示:

① 设置"字典名称"为"审批状态"。

② 设置"字典项类型"为"文本"。

③ 通过"添加数据字典项"按钮增加对应数据字典项,如审核中、审核失败、审核成功。

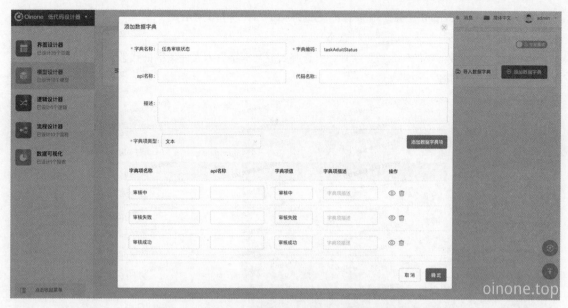

图 7-167 单击添加"数据字典"按钮,设置对应数据项

(2)在模型设计器的管理页面上方单击"模型"选项卡,模块选择"全员营销核心业务",选择"系统模型",搜索任务选择"Gemini 任务",单击添加字段,如图 7-168 所示。

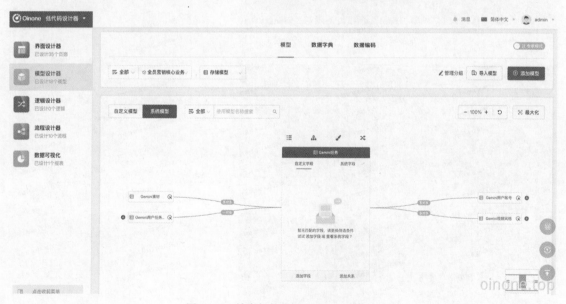

图 7-168 搜索任务选择"Gemini 任务"

（3）为模型"Gemini 任务"添加字段，如图 7-169、图 7-170 所示。
① 设置"字段名称"为"任务审批状态"。
② 设置"字段业务类型"为数据字典，并选择关联数据字典为"任务审批状态"。
③ 最后单击"创建"按钮完成操作。

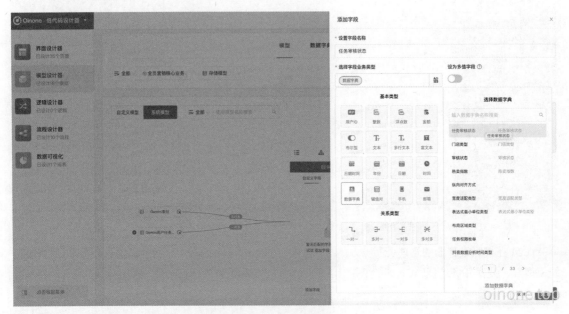

图 7-169　字段设置

图 7-170　设置"字段业务类型"为数据字典

（4）回到"Gemini 任务"设计区，我们可以看到在模型的"自定义字段"选项卡下方多了一个"任务审批状态"字段，如图 7-171 所示。

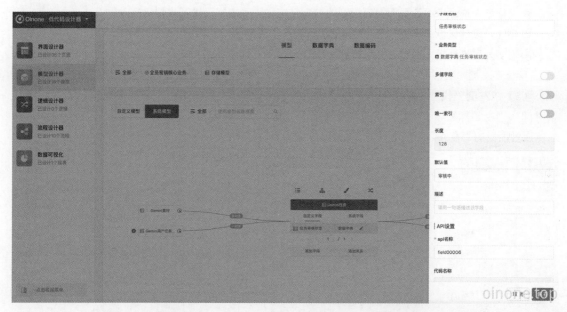

图 7-171　模型的"自定义字段"选项卡下方多了一个"任务审批状态"字段

Step3. 利用界面设计器，设计出必要的审核页面。

（1）进入界面设计器，应用选择"全员营销"，模型选择"Gemini 任务"，单击"添加页面"下的"直接创建"。

（2）设置页面标题、模型（自动带上可切换）、业务类型（运营管理后续会扩展其他类型）、视图类型（表单）后单击"确认"按钮进入"Gemini 任务"表单设计页面，如图 7-172 所示。

图 7-172　在界面设计器中进行页面创建

（3）进入页面设计器，对"Gemini 任务"表单页面进行设计（如图 7-173 所示）：

① 左侧为物料区：分为组件、模型；

- "组件"选项卡下为通用物料区，我们可以为页面增加对应布局、字段（如同在模型设计器增加字段）、动作、数据、多媒体等。
- "模型"选项卡下为页面对应模型的自定义字段、系统字段以及模型已有动作。

② 中间是设计区域。

③ 右侧为属性面板，在设计区域选择中组件会显示对应组件的可配置参数。

（4）在左侧"组件"选项卡下，拖入布局组件"分组"，并设置组件"标题属性"为基础信息。

（5）在左侧"模型"选项卡下，分别将系统字段中的"任务标题""任务开始时间""任务结束时间""视频标题""视频风格""任务描述"拖入"基础信息"分组，并单击"任务描述"，在右侧属性面板的"交互"分组中设置宽度为1。最后别忘了单击"发布"按钮完成页面的发布。

图 7-173　对"Gemini 任务"表单页面进行设计

Step4. 通过流程设计器，设计对应业务流程。

（1）进入流程设计器，单击"创建"按钮。

注意：流程中需要获取"关系字段"的除关联字段（一般为ID）以外的字段需要通过"数据获取"节点单独获取"关系字段"的对象数据。所以在流程设计中经常会用到"数据获取"节点，如图7-174所示。

图 7-174　进入流程设计器，创建对应业务流程

（2）左上角编辑流程名称为"任务审批流程"，单击第一个"触发"节点，"触发方式"选择"模型触发"，"模型"选择"Gemini 任务"，"触发场景"选择"新增或更新数据时"，"筛选条件"设置"任务审批状态"为为空或"任务审批状态"等于"审核中"，单击该节点的"保存"按钮，如图 7-175、图 7-176 所示。

图 7-175　左上角编辑流程名称为"任务审批流程"

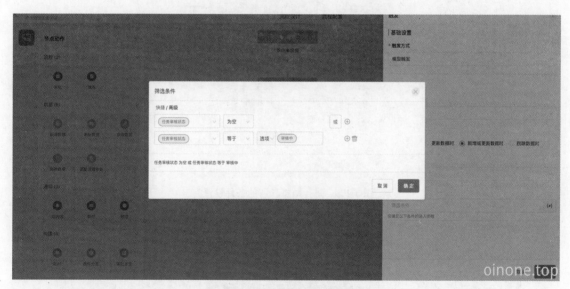

图 7-176　流程筛选条件设置

（3）单击流程图节点间的"+"图标选择增加"获取数据"节点，或者拖动左侧物料区"获取数据"到特定的"+"图标，如图 7-177、图 7-178 所示。

图 7-177　单击流程图节点间的 + 图标选择增加"获取数据"节点

图 7-178　拖动左侧物料区"获取数据"到特定的 + 图标

（4）单击"获取数据"，在右侧属性面板中设置"获取数据条数"为多条，选择"模型"为"Gemini用户任务实例"，单击"筛选条件"的 {x} 图标，进行数据获取的条件设置，如图 7-179 所示。

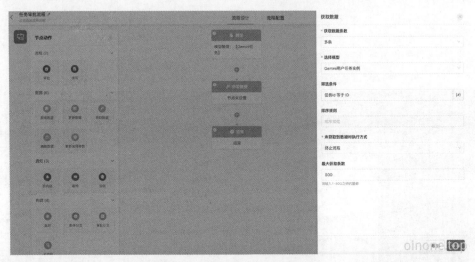

图 7-179　进行数据获取的条件设置

（5）选择条件字段为"任务 ID"，条件操作符为"等于"，条件为变量的导购字段的 ID。当上下文只有一个变量时默认不需要选择，这里默认的是"模型触发：[Gemini 任务 ]"，设置好以后单击"确认"，回到属性面板设置"未获取到数据时执行方式"为"终止流程"，并单击"保持"按钮，如图 7-180 所示。

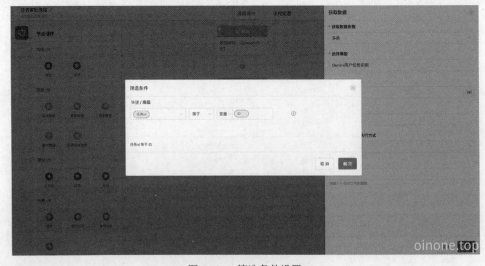

图 7-180　筛选条件设置

（6）增加"更新数据"节点，在右侧属性面板中设置（如图7-181、图7-182所示）：
① "更新模型"选择"模型触发：[Gemini 任务 ]"。
② "字段列表"单击"创建"按钮：
● 字段选择 更新"任务状态"字段。
● 表达式设置为："已关闭"。
③ "字段列表"单击"创建"按钮：
● 字段选择更新"任务审核状态"字段。
● 表达式设置为"审核中"。
最终完成的"模型触发：[Gemini 任务 ]"更新设置：
① "模型触发：[Gemini 任务 ]"的"任务状态"字段等于数字字典的"已关闭"，任务审核状态为"审核中"。
② 最后单击"保持"按钮。

图 7-181　字段表达式设置

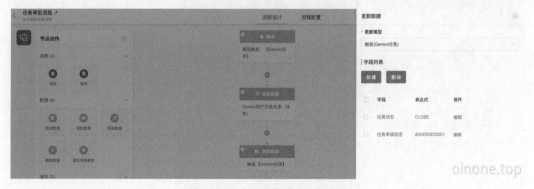

图 7-182　"更新数据"节点设置

（7）再增加"更新数据"节点，在右侧属性面板中设置（如图7-183所示）：
① "更新模型"选择"获取数据 [Gemini 用户任务实例 ]"。
② "字段列表"单击"创建"按钮：
● 字段选择更新"任务状态"字段。

● 表达式设置为"已关闭"。

最终完成的"获取数据 [Gemini 用户任务实例 ]"更新设置：

① "获取数据 [Gemini 用户任务实例 ]"的"任务状态"字段等于数字字典的"已关闭"。

② 最后单击"保持"按钮。

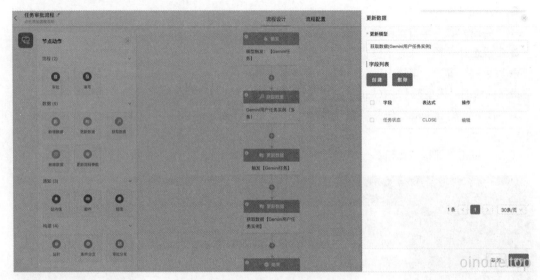

图 7-183　再增加"更新数据"节点

（8）增加"审批"节点，在右侧属性面板中设置（如图 7-184 所示）：

① "审批模型"选择模型为"模型触发：[Gemini 任务 ]"。

② "选择视图"选择前面新建的"流程中的任务编辑页"。

③ "审批人"选择角色为"超级管理员"。

④ "数据"权限全部设置为"查看"。

⑤ 其他配置项默认，需要了解更多请查看产品使用手册。

⑥ 最后单击"保持"按钮。

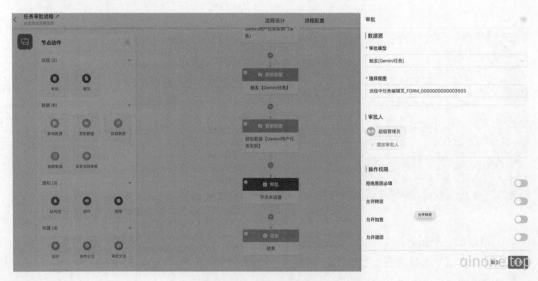

图 7-184　增加"审批"节点

(9)新增"审核分支",在"通过"分支中增加两个数据更新节点,跟审核前的两个数据更新节点对应。

① "模型触发:[Gemini 任务]"的"任务状态"字段等于数字字典的"初始化",任务审核状态为"审核通过",如图 7-185 所示。

② "获取数据[Gemini 用户任务实例]"的"任务状态"字段等于数字字典的"初始化",如图 7-186 所示。

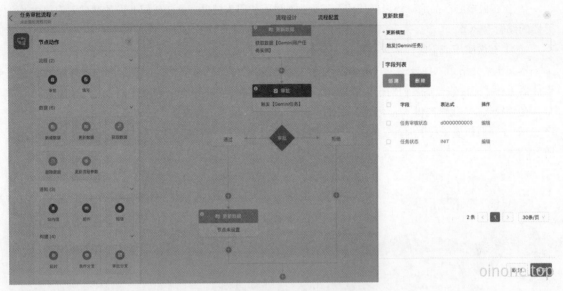

图 7-185 新增"审核分支"节点

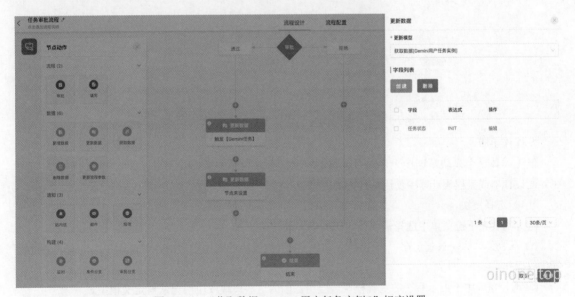

图 7-186 "获取数据[Gemini 用户任务实例]"相应设置

(10)流程确保保持并发布过,单击右上角"发布流程"完成流程的保存与发布。

Step5. 检验效果。

(1)创建任务后,任务状态为"关闭"状态,任务列表中的任务状态为多个状态的计算值。

(2)审核通过后,任务状态为"进行中"状态,任务列表中的任务状态为多个状态的计算值。

## 7.4 Oinone 的低无一体

### 1. 基础介绍

前面我们学习了基于低代码开发平台进行快速开发，以及通过 Oinone 的设计器进行零代码开发两种模式。当然低无一体不是简单地说两种模式，还指低无两种模式可以融合。

（1）在做核心产品的时候以低代码开发为主，以无代码为辅助。见低代码开发的基础入门篇中"设计器的结合"一节。

（2）在做实施或临时性需求则是以无代码为主，以低代码为辅助。

本文主要介绍第二种模式，它是 Oinone 官网在 SaaS 模式下的专有特性。满足客户安装标品后通过设计器进行适应性修改后，但对于一些特殊场景还是需要通过代码进行完善或开发。

在该模式下，我们提供了 jar 和代码托管两种模式，客户只要选择需要进行代码开发的模块，单击"生产 SDK"，下载扩展工程模板，按 Oinone 低代码开发平台规范进行研发，后上传扩展工程即可，如图 7-187 所示。

图 7-187  按照 Oinone 低代码开发平台规范进行研发

### 2. 操作手册

低无一体这个模块是连接无代码设计器的桥梁，可以为一个模块或应用设计低代码的逻辑，可以在界面设计器或流程设计器中使用低代码的逻辑。

（1）选择模块。

首先需要在下拉菜单中选择需要低代码的模块或应用。

① 下拉菜单选中只展示在"应用中心"中已安装的模块或应用，可前往"应用中心"安装后继续低代码操作。

② 选择模块中不展示系统的基础模块或应用，因为这些模块或应用无法自定义模型。

（2）模块信息。

模块信息展示的是选择模块的基础信息：模块名称、模块编码、模块作者、模块版本、包的前缀、工程模板下载地址，下载地址仅在上传 jar 包模式时用到。

（3）低无一体操作。

低无一体支持两种使用模式：上传 jar 包模式、源码托管模式。

上传控制工程或创建研发分支动作完成会生成一条数据，可以对单条数据进行部署、卸载、修改、删除。

## 上传 jar 包模式

在这个模式下,需要做四步动作(如图 7-188 所示):

● 生成 SDK,单击按钮之后,会把模块的当前模型状态打包成一个 SDK 包,SDK 最新生成时间更新。当模型变更但未生成 SDK 时,使用低无一体就会出错,请重新生成 SDK 并修改扩展工程。生成 SDK 通常需要 1 分钟左右,若第一次使用低无一体模块,可能需要更长时间,请耐心等待。

● 下载扩展工程模板,单击按钮之后,会将 SDK 包和工程模板生成一个下载链接,复制模块信息中的卸载地址打开即可下载。

● 技术人员在工程模板的基础上写低代码逻辑。

● 上传扩展工程,单击按钮展开弹窗,在弹窗中设置标签、备注,并将最终的 jar 包上传,完成上传之后表格中就会新增一条数据。

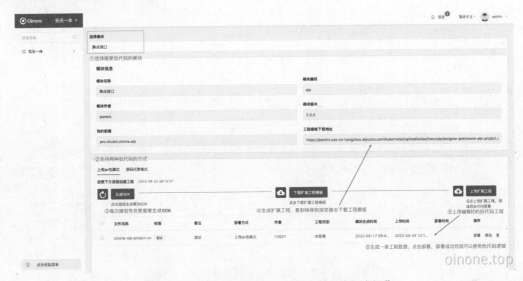

图 7-188　上传 jar 包模式下,按照步骤完成相应操作

● 上传 jar 包模式下,模板工程中代码需要注意的点参考图 7-189。

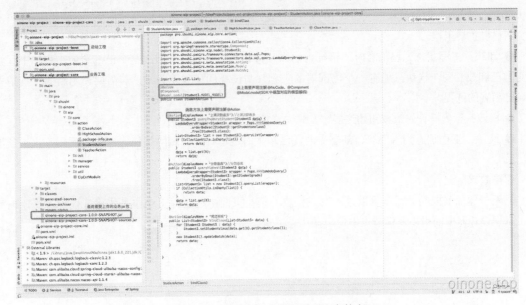

图 7-189　模板工程中代码需要注意的点

**源码托管模式**

在这个模式下，需要做三步动作（如图 7-190 所示）：

● 生成 SDK，单击按钮之后，会把模块的当前模型状态打包成一个 SDK 包，SDK 最新生成时间更新。当模型变更但未生成 SDK 时，使用低无一体就会报错，请重新生成 SDK 并修改扩展工程。生成 SDK 通常需要 1 分钟左右，若第一次使用低无一体模块，可能需要更长时间，请耐心等待。

● 创建研发分支，单击按钮展开弹窗。首次创建时需要设置 git 账号名称、git 账号邮箱来创建一个账号，另外在弹窗中设置分支名称、标签、备注，完成创建后表格中就会新增一条数据。

● 通过表格中的 Gitlab 地址，技术人员写低代码逻辑。

图 7-190　源码托管模式下，按照步骤完成响应操作

**行内操作**

● 部署：工程状态为未部署、部署失败、已卸载时展示行内的部署按钮，单击之后进行部署，工程状态变为部署中。部署过程需要 5～10 分钟，请耐心等待。部署完成之后，会生成一个新的模块："原模块名称"扩展工程。

● 卸载：工程状态为已部署时展示行内的卸载按钮，单击之后会卸载这个已部署的工程，工程状态变为已卸载。同一模块只能有一个已部署的工程（与选择的模式无关），若需要使用新的工程请先卸载已部署的工程。

● 修改：行内操作修改按钮始终展示，只允许修改标签、备注。

● 删除：工程状态为未部署、部署失败、已卸载时展示行内的删除按钮，单击之后删除这一条工程记录。

**部署效果**

低无一体部署成功之后，可以进入对应模块的模型页面中使用提交动作来使用低代码逻辑，也可以在流程设计器中的引用逻辑节点中使用低代码逻辑。

# 附录 A 下载说明

| 章节说明 | 下载内容 | 下载地址 | 备注 |
|---|---|---|---|
| 3.1.1 环境准备 > 环境准备（Mac 版） | 安装 JDK 1.8 | https://www.oracle.com/Java/technologies/downloads/#Java8l | |
| | 安装 MySQL 8.0.26 | https://dev.MySQL.com/downloads/MySQL/ | |
| | 安装 Idea 社区版 2020.2.4 | https://www.jetbrains.com/idea/download/other.html | |
| | 安装 Idea 插件 | 请移至 Oinone 官网（https://www.oinone.top/）对应页面下载或联系 Oinone 官方客服 | 根据各自 Idea 版本下载对应插件，下载文件后去除 .txt 后缀 |
| | 安装 git 2.2.0 | https://sourceforge.net/projects/git-osx-installer/files/git-2.15.0-intel-universal-mavericks.dmg/download?use_mirror=nchc | |
| | 安装 GraphQL 的客户端工具 Insomnia | 请移至 Oinone 官网（https://www.oinone.top/）对应页面下载或联系 Oinone 官方客服 | 下载文件后修改文件名去除 .txt 后缀 |
| | 安装 maven | https://archive.apache.org/dist/maven/maven-3/3.8.1/binaries/ | |
| | 安装脚本 zk | https://archive.apache.org/dist/zookeeper/zookeeper-3.5.8/apache-zookeeper-3.5.8-bin.tar.gz | |
| | 安装脚本 rocketmq | https://archive.apache.org/dist/rocketmq/4.7.1/rocketmq-all-4.7.1-bin-release.zip | |
| | 安装脚本 redis | https://download.redis.io/releases/redis-5.0.2.tar.gz | |
| | 安装 nvm | https://github.com/nvm-sh/nvm/blob/master/README.md | |
| 3.1.1 环境搭建 > 环境准备（Windows 版） | 安装 JDK 1.8 | https://www.oracle.com/Java/technologies/downloads/#Java8 | |
| | 安装 Apache Maven 3.8+ | https://maven.apache.org/download.cgi | |
| | 下载 settings-develop.xml | 请移至 Oinone 官网 https://www.oinone.top/ 对应页面下载或联系 Oinone 官方客服 | 下载到 C:\Users\你的用户名\.m2 目录中并重命名为 settings.xml |
| | 安装 Jetbrains IDEA 2020.2.4 | https://www.jetbrains.com/idea/download/other.html | |
| | 安装 Jetbrains IDEA 2020.2.4 需下载插件 | https://pan.baidu.com/share/init?surl=HNzSxxH0KncvglkfITUrsA | 提取密码：mdji |
| | 安装 Idea 插件 | 请移至 Oinone 官网（https://www.oinone.top/）对应页面下载或联系 Oinone 官方客服 | 根据各自 Idea 版本下载对应插件，下载文件后去除 .txt 后缀 |
| | 安装 MySQL 8 | https://dev.MySQL.com/downloads/MySQL/ | |
| | 安装 Git | https://git-scm.com/download/win | |
| | 安装 GraphQL 测试工具 Insomnia | https://github.com/Kong/insomnia/releases | |
| | 安装 RocketMQ | https://rocketmq.apache.org/download/ | |
| | 安装 ElasticSearch 版本 8.4.1 | https://www.elastic.co/cn/downloads/past-releases/elasticsearch-7-6-1 | |

续表

| 章节说明 | 下载内容 | 下载地址 | 备注 |
|---|---|---|---|
| 3.1.1 环境搭建＞环境准备（Windows版） | 安装 Redis | https://download.redis.io/releases/ | |
| | ZooKeeper 安装 | https://dlcdn.apache.org/zookeeper/zookeeper-3.8.0/apache-zookeeper-3.8.0-bin.tar.gz | |
| | 安装 nodejs 版本 12.12.0 | https://nodejs.org/dist/v12.12.0/node-v12.12.0-win-x64.zip | |
| | 安装 cnpm | https://www.npmjs.com/package/cnpm | |
| 3.2.1 Oinone 一模块为组织＞构建第一个 Module | 安装 archetype-project-generate.sh 脚本 | 请移至 Oinone 官网（https://www.oinone.top/）对应页面下载或联系 Oinone 官方客服 | |
| 3.5.5 Oinone 以交互为外在＞设计器的结合 | 安装 Docker | https://www.docker.com/get-started/ | |
| | 下载结构包：Oinone-op-ds.zip | 请移至 Oinone 官网（https://www.oinone.top/）对应页面下载或联系 Oinone 官方客服 | |
| 4.1.10 后端高级特性＞函数之触发与定时 | 下载 canal 中间件：pamirs-middleware-canal-deployer-3.0.1.zi | 请移至 Oinone 官网（https://www.oinone.top/）对应页面下载或联系 Oinone 官方客服 | |
| 4.1.11 后端高级特性＞函数之异步执行 | 下载 tbSchedule 的控制台 jar 包：pamirs-middleware-schedule-console-3.0.1.jar.txt | 请移至 Oinone 官网（https://www.oinone.top/）对应页面下载或联系 Oinone 官方客服 | |
| | 下载 schedule.json | 请移至 Oinone 官网（https://www.oinone.top/）对应页面下载或联系 Oinone 官方客服 | 下载以下文件放在 pamirs-demo-boot 的 src/main/resources/init 目录下 |
| 4.1.25 后端高级特性＞框架之搜索引擎 | ES 安装 | 方式一：官方下载安装包：https://www.elastic.co/cn/downloads/past-releases/elasticsearch-8-4-1 | 下载后去除后缀 .txt，然后解压文件 |
| | | 方式二：请移至 Oinone 官网（https://www.oinone.top/）对应页面下载或联系 Oinone 官方客服 | |
| "其他说明"本书第 3～6 章部分源代码过长未展示完整版，如您需要了解完整代码可参见云盘对应说明或联系 Oinone 官方客服 | | | |